D0077065

Mechanics of Machines

Mechanics of Machines

Samuel Doughty
Professor of Mechanical Engineering
University of Wisconsin—Platteville

John Wiley & Sons, Inc.
New York • Chichester • Brisbane • Toronto • Singapore

Library of Congress Cataloging-in-Publication Data

Doughty, Samuel.
Mechanics of machines.

Bibliography: p.
1. Mechanical engineering. I. Title.
TJ170.D68 1987 621.8 87-23042
ISBN 0-471-84276-1

Printed and bound in the United States of America by Braun-Brumfield, Inc.

10 9 8 7 6 5 4 3

To Ann Elizabeth and Susan Beth

Preface

A course in kinematics of mechanisms, with or without additional work in statics and dynamics of machines, has long been a staple item in mechanical engineering curricula. This course is often considered the first of the professional courses in the mechanics stem, and traditionally deals with mechanisms significantly more kinematically complex than those considered in the introductory courses in statics and dynamics. This book is intended to serve as a modern, computer-oriented text for this critical course. The book provides more than enough material for a three-credit, one-semester, junior-level course, allowing the instructor some freedom to select from among its topics. Because of the emphasis on analytical formulations and computer solutions, this text will be of interest as well to practicing engineers in the area of machine design.

The book begins with a chapter describing the computer-oriented viewpoint and introduces the idea of degrees of freedom. Chapters 2 and 3 deal with kinematics of single and multidegree of freedom mechanisms. The first part of Chapter 4 is concerned with the analytical design of cams, and the second part of the chapter considers the analysis of specified cam systems, including the special case of circular cams. The kinematics of involute gear teeth is considered in the beginning of Chapter 5, followed by a look at gear train analysis and design. The statics of mechanisms is discussed in Chapter 6, with emphasis on the principle of virtual work. Chapter 7 considers the dynamics of single degree of freedom machines using Eksergian's equation, and describes the numerical simulation of such systems. The dynamics of multidegree of freedom systems is addressed by way of the Lagrange equation in Chapter 8, with a further presentation on numerical simulation. The analysis of reactions and internal forces is taken up in Chapter 9, comparing static, kinetostatic, and dynamic approaches to the problem. The appendices provide supplementary and review material on matrices, numerical methods, and other topics in mechanics and mathematics.

From the very beginning, the book is thoroughly computer-oriented. The student is encouraged to look at all problems—whether examples in the text or homework problems—from the standpoint of preparing them for a computer solution. In some cases, the student will actually carry the solution all the way through to a computer solution; in other cases, the student will only formulate the problem in a manner appropriate for computer solution. (There are also many problems to be solved completely in closed form for practice without the need to have a computer available.) Although computers have been around for many years, the last decade has seen a real revolution in computing with the development of microcomputers. It is reasonable to think that every engineer has or shortly will have free access to a computer of some sort, and that to be computer is most likely a microcomputer. Most microcomputers today have the capacity to solve the kinds of problems encountered in the mechanics of machines. It goes without saying that any modern mainframe computer is sufficient as well, so the choice of whether to use a mainframe or a microcomputer is mainly dependent on what is available. In terms of programming languages, virtually all microcomputers support some form of BASIC, and some support other languages such as FORTRAN and Pascal as well. The universality of BASIC makes it a useful language for the examples in this text, but there is nothing about the methods of analysis that is tied to BASIC. Any of the methods presented here can be performed satisfactorily using any high-level programming language. Most of the programs presented in BASIC in the body of the text are also given in FORTRAN in Appendix A10.

This textbook takes the position vector loop approach to position analysis, in which the configuration of the mechanism is described in terms of closed position vector loops. Scalar components of these vector equations are used because these are solved more easily for the required information using numerical techniques. The derivatives of these same position vector loops provide the velocity and acceleration descriptions for the mechanism. The numerical techniques required to solve these systems of equations are discussed where required, as well as in detailed presentations in Appendices 1 and 2. Position-dependent velocity coefficient functions are defined, relating the secondary velocities to the primary velocity. Both the velocity coefficients and the velocity coefficient derivatives are required to express the accelerations. The velocity coefficients and the velocity coefficient derivatives are used to connect the kinematics to the later work in kinetics.

In the area of kinetics, a major objective of this book has been to provide an energy-based approach, appropriate to the course level. In Chapter 8, multidegree of freedom dynamics is presented using the well-known Lagrange equation as the basis. The presentation is directed toward application of the Lagrange equation to machine systems, so the derivation is deferred to Appendix A9. For single degree of freedom mechanisms (Chapter 7), the less well-known Eksergian's equation is presented. It provides a powerful tool that offers useful insight into the way various parameters affect the motion. Chapter 6 presents statics from the viewpoint of the Principle of Virtual Work. This method is valuable in its own right and also as preparation for the dynamics presentations of Chapters 7 and 8.

The discussion on cams is approached first from the standpoint of design, by looking at some length into the problem of analytical design of a cam profile to generate a prescribed follower response. The required design decisions are identified, and their effects on size, state of stress, and other matters are considered. The later part of Chapter 4 deals with the analysis of prescribed cam and follower systems.

The typical machine design textbook contains quite a bit of discussion on the design of gears, often including material on involute kinematics, gear tooth strength, and train ratio calculations. An important kinematic design question that is missing from most textbooks on kinematics or machine design is the matter of how to determine workable combinations of gear tooth numbers to achieve a specified train ratio. Design approaches to this problem are included in the discussion on gears and gear trains in Chapter 5.

There are several ways to organize a course using this book. There is certainly more material included than can be covered in a one-semester course. At the University of Wisconsin—Platteville the course is usually taught using Chapters 1 through 3 for kinematics, part of Chapter 4 on cams, Chapter 5 on gears, Chapter 6 for virtual work, and dynamics using Chapter 7. If cams and gears are covered elsewhere, a very satisfactory course can be constructed using Chapters 1 through 3 for kinematics, and Chapters 6 through 9 for kinetics and analysis of forces. Another variation is to focus on single degree of freedom systems using Chapters 1 and 2 for kinematics, Chapter 4 dealing with cams, gears in Chapter 5 (excluding planetary systems), and Chapters 6 and 7 for kinetics. The main considerations in structuring a course are to be sure to get the necessary material regarding position analysis, velocity coefficients, and velocity coefficient derivatives—presented in Chapters 2 and 3—that are required for all of the work in kinetics, Chapters 6 through 9.

Most of the problems at the end of each chapter ask the student to prepare the description for an eventual computer solution. One of the early steps on such preparation is the choice of appropriate variables to describe the problem. In most cases, there are several possible choices. For some problems all correct choices are equally useful, whereas for other problems one choice is definitely better than the others. The ability to choose a workable set of variables, preferably the best set, is a skill that is only learned by practice. Consequently, the need to avoid guiding the choice of variables makes it virtually impossible to tabulate answers to the problems. Complete solutions are, however, available in the Instructor's Manual. It is also important for students to develop an ability to validate their own mathematical descriptions; this is rarely ever done when a student is working toward an answer tabulated in the back of a textbook.

The mathematical level of the book is appropriate to junior-level engineering students who are well grounded in calculus. A firm grasp of differential calculus is absolutely essential to understanding and applying the methods developed in the text, and integral calculus is essential for understanding many of the derivations. Matrix notation is used extensively, but only a very modest level of knowledge is required in this area. The review of matrix methods provided in Appendix A1 covers all of the topics required for use here.

Many of the examples and homework problems are drawn from my own industrial experience and consulting assignments. From a teaching viewpoint, my experience has been that students are attracted to problem descriptions that resemble real machines. I believe that anyone teaching this material and wishing to draw on personal industrial experience will find this book complementary to that effort. Comments and suggestions from those using the book will be welcome so that it may be improved in the future.

Samuel Doughty

Contents

APPENDICES

CHAPTER 1

Introduction

1.1 GENERAL INTRODUCTORY COMMENTS

The term "mechanics of machines" is understood to include the kinematics, statics, and dynamics of mechanical systems of the type commonly found in machinery. Introductory courses in statics and dynamics usually include relatively little information about this area primarily because of kinematic complexity and, with regard to dynamics, the complications associated with varying effective inertia for mechanisms. In this text, these problems are addressed directly.

This subject has a long history extending back to pre-Christian times, and has been of increasing importance since the beginning of the Industrial Revolution. Although electronics have now replaced some complicated mechanisms, as in the case of computing mechanisms, there will always be a need for mechanisms and machines as long as physical products are produced.

Until recently, the analysis of mechanisms has been almost completely dependent on graphical techniques. The advent of high-speed digital computation has changed most aspects of engineering analysis and design, and has opened many new options in the analysis of machines. Consequently, the methods presented here rely heavily on digital computation instead of on graphical constructions. This text will discuss numerical methods of solving various classes of equations along with the physics of the situations that give rise to these equations.

Four major physical concepts underlie all of the topics discussed in this book. Each of these concepts has been known for many years, but their individual utility has been limited. The application of these concepts together, and implementation of the resulting descriptions via the digital computer, is a relatively recent synthesis. In using this book, it is well to be prepared to recognize these four concepts whenever they appear. For this purpose, they are introduced here.

Kinematics is the study of motion without regard for the forces involved in the motion. In the area of kinematics, the first of the four major concepts is the use of *Position Vector Loop Equations*. In typical machinery systems, the components form closed loops that change shape as the components move, but remain closed; each loop can be described as a vector sum identically zero at all times. If enough information has been specified to determine the mechanism configuration uniquely, these vector equations, or their scalar equivalents, can be solved for all the remaining position variables. The details of this process will be discussed at length later, but for the present, it is important to remember that the position vector loop equations provide a means for determining all required position information.

The use of *Velocity Coefficients* and *Velocity Coefficient Derivatives* is the second major concept, again originating in the general subject area of kinematics. Many readers will be familiar with the fact that for two pulleys connected with a taut belt, the rotation rate of the drive pulley and that of the driven pulley have a fixed ratio; this is a consequence of the invariant geometry of the system. The actual value of the speed ratio is fixed by the particular geometry of the system, in this case the pulley radii. In a system where the geometry varies as the system moves, such as a slider–crank assembly, the ratio of the output to input speeds is variable, *depending on the instantaneous position of the system*. The key point here is that the speed or velocity ratio is dependent on the position but is independent of the actual speed. Thus, the speeds of all points can be expressed as appropriate position-dependent, velocity coefficient functions multiplying a common reference speed. The situation regarding accelerations is somewhat more complicated, but it will be shown that for many systems the acceleration is the sum of two terms. One term involves the acceleration of the reference point and the position-dependent velocity coefficient, and another term involves the derivative of the velocity coefficient and the square of the reference speed. Therefore, all speeds and accelerations are rather simply related to the speed and acceleration of a reference point through the use of the *position-dependent* velocity coefficients and velocity coefficient derivatives.

The third concept, usually associated with the area of statics, is called the *Principle of Virtual Work*. It is one of the oldest energy principles of mechanics, and describes the conditions for equilibrium in a manner fully equivalent to the more familiar statements regarding the vanishing of the force and moment sums on a body. When force and moment sums are computed for a body such as a dam or a tower, there is no difficulty in that application *because the position of the body is specified*. In many machinery problems, the positions of the components under load are not specified in the beginning, but must be determined as part of the solution. This is just the situation for which the Principle of Virtual Work is best suited. Its application is heavily dependent on the kinematics of the system under consideration and the ideas of functional dependence and independence. These needs are well accommodated through the use of velocity coefficients.

The dynamics of a machine system can be described by applying Newton's Second law, or by using equations of motion derived from energy considerations. Thus, the fourth major concept is the use of *Energy-Based Equations of Motion*. These equations apply equally well to conservative and nonconservative systems. The well-known Lagrange equations of motion apply to systems with any number of independent coordinates. For problems involving only one independent coordinate, there is an alternate energy formulation available, which is associated with the name Eksergian. This method is often easier to use and gives more insight than the Lagrange equation. In using either of these approaches, the expressions for the kinetic and potential energies as well as those for the virtual work of the external forces are central. The kinetic energy expression is greatly facilitated through the use of the velocity coefficients. The velocity coefficients and the potential energy are, in turn, dependent on the system position that is determined from the position vector loop equations.

Each of the four major concepts has a long history, and they can certainly be applied together in manual calculations. The greatest benefit comes, however, through joint computer implementation of these concepts. The application of the position vector loop equations often leads to a very complicated system of nonlinear transcendental equations. In such cases, closed-form solutions are virtually impossible, but numerical solutions by iterative techniques are usually readily available. The calculation of the velocity coefficients involves the simultaneous solution of a system of linear equations. Although in concept this is straightforward, it becomes excessively laborious for more than two equations. When applying the principle of virtual work, it is often necessary to solve a system of nonlinear equations, and numerical solution by an iterative technique using a computer is the only easy way to achieve a solution. It is important to note that the equations of motion for most mechanisms are highly nonlinear differential equations, irrespective of the manner in which they are obtained. Consequently, solving them through analytic methods is usually not possible; numerical integration is required and this is practical only when performed by a computer.

1.2 DEGREES OF FREEDOM

The term "degrees of freedom" has long been used in classical dynamics to refer to the number of *independent* variables that are required to describe a mechanical system. The concept is introduced at this point, and understanding of the term will grow with progress through the discussions in the following chapters.

Consider a particle that is free to move relative to a rectangular Cartesian coordinate system in two dimensions. If two numbers are specified that represent the X and Y coordinates of the particle, then the position of the particle relative to the coordinate system is fully described. Alternatively, polar coordinates could be specified for the particle, again two values. In either case, two items of information are required, and the particle is said to have two degrees of freedom, often abbreviated 2DOF. For a rigid body in the same two-dimensional space, there are three degrees of freedom. One way to consider these degrees of freedom is to associate two of them with the coordinates of one point on the body, and to associate the third degree of freedom with an angle describing the orientation of the body. Be sure to notice that the number of degrees of freedom speaks of the number of coordinates required, not about any particular choices for the coordinates. The angle used is only a coordinate in the sense that it partially describes the position of the body, although not in the sense of a rectangular Cartesian coordinate.

There are many other ways to associate the degrees of freedom with coordinates describing the position of the body, but there are two important requirements to be satisfied by any such choice. The first is that the entire set of coordinates is sufficient to describe the position completely. The second is that each coordinate is independent. Using the association described above for the rigid body, the coordinate X can be varied without any change required in Y or the angle; the same statement can be made about each of the other coordinates, so they are independent. A set of coordinates that is (1) *complete* in that they are sufficient to specify the system position fully and (2) each *independent* of the others is called a set of *generalized coordinates.* Generalized coordinates may include rectangular Cartesian coordinates, angles, arc lengths, or any other measures that contribute to the description of the system configuration, subject to the requirement that they form a complete, independent set. For the present, the number of generalized coordinates should be considered equal to the number of degrees of freedom. Exceptions to this statement will be discussed later under the topic of "Constraints."

For machinery kinematics problems, there will usually be more kinematic variables than degrees of freedom. After choosing generalized coordinates to be associated with the degrees of freedom and considered as *primary variables,* the remaining kinematic variables will be called *secondary variables.* For purely kinematic problems, the primary variables are considered as assigned values ("inputs"), whereas the secondary variables are among the problem unknowns. The position vector loop equations provide the means to determine the secondary variables based on the assigned values for the primary variables (generalized coordinates). In the case of a dynamics problem, the governing differential equations describe the manner in which the generalized coordinates (primary variables) vary with time and, again, the position vector loop equations determine the secondary coordinates.

1.3 USE OF MATRIX NOTATION

In solving mechanics of machines problems, there are many physical quantities of a vector character to consider. It is often convenient to use vector notation to analyse such a problem, but when it comes time to compute, such notation is not suitable. The digital computer can not deal with vectors as such. It can, however, accommodate one-dimensional arrays, and these can represent a vector quite well. If the vector of interest is a position vector in a two-dimensional space, then a (2×1) column matrix, represented by a (2×1) array in the computer, can store each component of the vector. There are other times when many variables are described simply as $X_1, X_2, \ldots, X_n$ where n is a positive integer; this case can also be represented by a column matrix that translates to an $(n \times 1)$ array in the computer. In fact, systems of linear algebraic equations of any size can be neatly represented and manipulated in the matrix form.[1]

Matrix notation is only a *notation* but it has much to recommend it. By itself, it does nothing that cannot be done in other ways, but it provides an orderly, systematic means to deal with the computational bookkeeping problem. It adapts so readily to computer implementation that if a problem is formulated using matrix notation, the computer code required for the solution is usually quite easy to write.

1.4 COMPUTER-AIDED PROBLEM SOLVING

Engineering analysis always revolves around problem solving of one type or another, and the computer can be a great aid in many cases. However, the computer can never do the entire job, and the mistaken attitude that it can often leads to much lost time! This misconception becomes rather obvious when the various steps to problem solving are considered, so a quick review of the process may be beneficial.

The typical steps in solving a mechanics of machines problem are as follows:

1. *Problem Definition.* This is the task of determining what problem is to be solved and what information is available to begin that solution. Many errors are made at this point, and these errors usually lead to major time loss.

[1] Throughout this book, matrix notation is used extensively. The reader who is not familiar with matrices should review Appendix A1. This appendix provides a summary of the necessary aspects of matrix notation and manipulation. For further details, a text on matrix algebra should be consulted.

2. *Conceptual Solution.* Based on the information gathered in step 1, an overall plan of attack must be established. This plan is formulated with the assumption that no unexpected difficulties will arise; for this reason, it may require modification. Nevertheless, there is no point to beginning calculation until a tentative route to the solution is determined. If a computer solution is indicated, a rough flow chart should be drafted at this point.
3. *Detailed Analysis.* The conceptual solution process often presumes that the governing kinematic equations can be written; this is the point at which they are actually written. All of the governing equations for the problem are assembled and manipulated to set the stage for the computational work that follows. Note that this step often involves the substitution of one equation into another, the use of calculus, and the recasting of equations into matrix form. These are all operations, however, that cannot be done readily in the computer. At the end of this step, detailed equations for use in a computer program should be available.
4. *Computer Implementation.* This is the time for writing and executing computer programs. If adequate preparation has been made, this process will move quite rapidly. Note that it is often a good idea to test any program after the first major calculation has been programmed. As an example, in a problem involving a complete kinematic analysis for a linkage, it is usually worthwhile to check the position solution and debug as necessary before programming the velocity and acceleration solutions.
5. *Interpretation of the Results.* This is the most essential part of any analysis. Engineering analysis is always done for a purpose—to answer some type of question. Typical acceptable answers are (a) "The maximum velocity is 187 in./s," (b) "The minimum bearing force occurs when the crank position is 1.277 radians," or (c) "The arm will fail due to excessive dynamic loading." It is **never** sufficient to submit a stack of computer output as an engineering report. It is, however, always necessary to draw a conclusion from the analysis, and to provide the analysis and the computer results that support the conclusion.

The foregoing steps are typical and, as such, one may need to deviate from them during the course of solving some problems. It should be noted that the computer is actually used only in step 4. The availability of the computer is assumed in steps 2 and 3 and this availability influences the overall planning (step 2) and the final form sought in the analysis (step 3). In days before the general availability of high-speed digital computers, the computational step was one of the longest; now it should not require much more time than any other step.

1.5 COMPUTER LANGUAGES

There are a variety of computer languages used for engineering computations. For many years, FORTRAN was the standard engineering language, but more recently others such as BASIC and PASCAL have come into wide use.

In developing an engineering textbook, it is often useful to include sample programs that detail the logic required to implement various solution techniques. The BASIC language has been used for all programs in the body of this text. Similar programs in FORTRAN are provided in Appendix A10 of the text. If the user prefers to work in yet another language, it is a fairly simple matter to transfer the logic of the program from one high-level language to another. There is nothing about the computer techniques that is tied to any particular computer language.

As mentioned previously, matrix notation and the associated matrix operations will play a major role in the solution of mechanics of machines problems. Some versions of BASIC have included matrix operations such as "MAT A = B*C", which will form the matrix product of the arrays B and C and store the result in the array A. If available, these operations should certainly be used; they are usually written in machine language and will execute very quickly. Unfortunately, most versions of BASIC for microcomputers do not include these matrix operations. To facilitate their use, however, we provide a set of matrix operations subroutines in BASIC in Appendix A1.6; corresponding subroutines in FORTRAN are given in Appendix A10. These subroutines may be keyed in and debugged a single time with the final results stored on diskette. Development of computer programs that require matrix operations begins with copying these subroutines. Note that the BASIC routines begin at line 5000, so there is adequate room for most programs to be stored ahead of the matrix operations routines. The FORTRAN matrix operations subroutines should be compiled and called exactly as any other FORTRAN subroutines.

1.6 CONCLUSION

The discussions about the use of matrix notation, the steps in computer-aided problem solving, and the matter of computer languages indicate the general flavor of the material in the remainder of this book. The material will be presented at a moderately sophisticated mathematical level, with a clear orientation toward computer implementation of the analytical results. The four major concepts underlying much of the presentation have been discussed; it will help to be alert to these as they appear throughout the book, and to look for the interplay between these ideas. The discussion on degrees of freedom and generalized variables has laid the foundation of an understanding that will be built upon throughout the remainder of the book.

PROBLEM SET

Determine the number of degrees of freedom for each of the following systems, and fully document the reasoning in each case. In making this determination, make the following assumptions:

a. All bodies are rigid, so there are no deformations;
b. Belts do not slip or creep;
c. All single-pin joints are pivots, allowing rotation.

If there appear to be ambiguities in any of the problems, make a reasonable assumption to remove the ambiguity and clearly document that assumption.

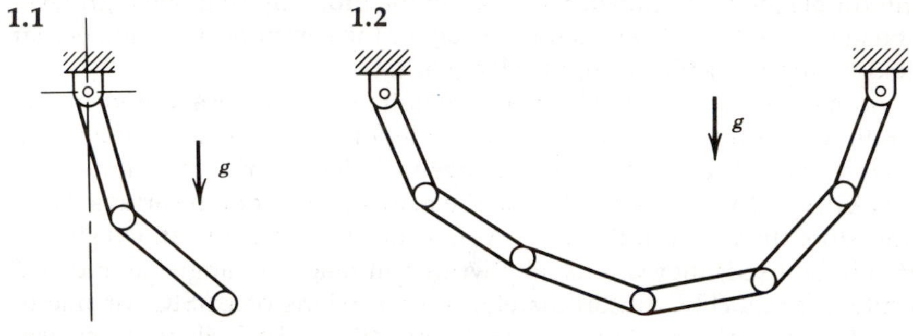

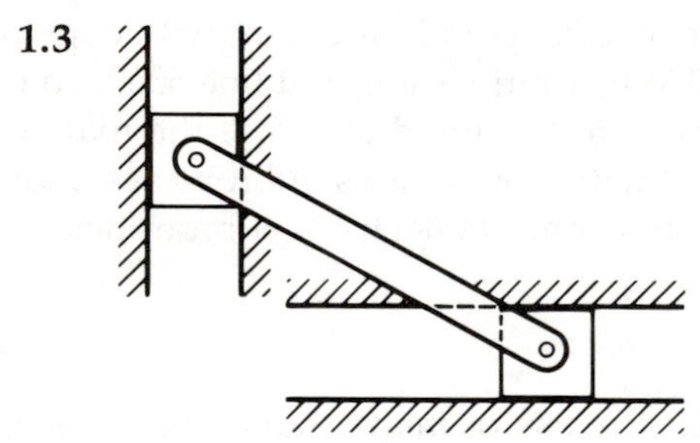

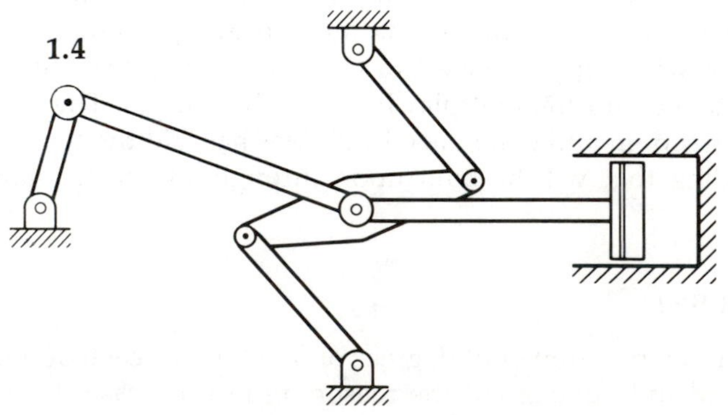

1.5

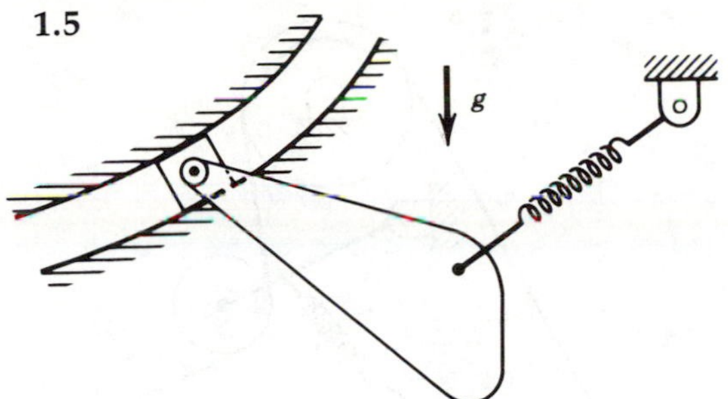

1.6

1.7

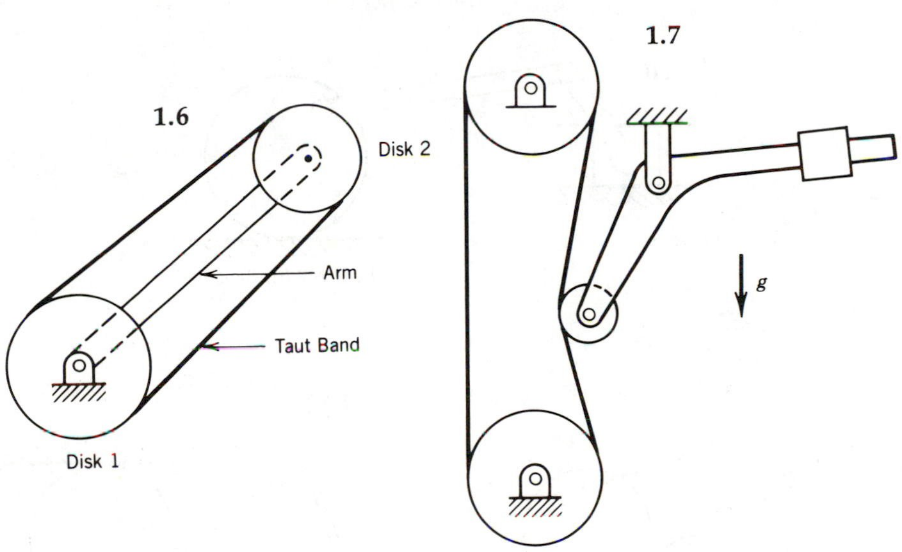

1.8

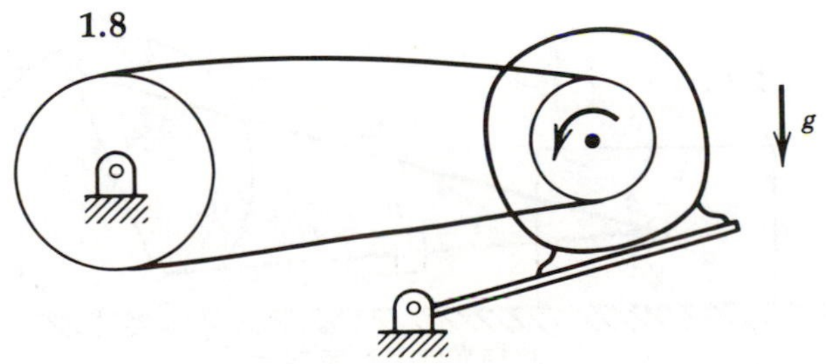

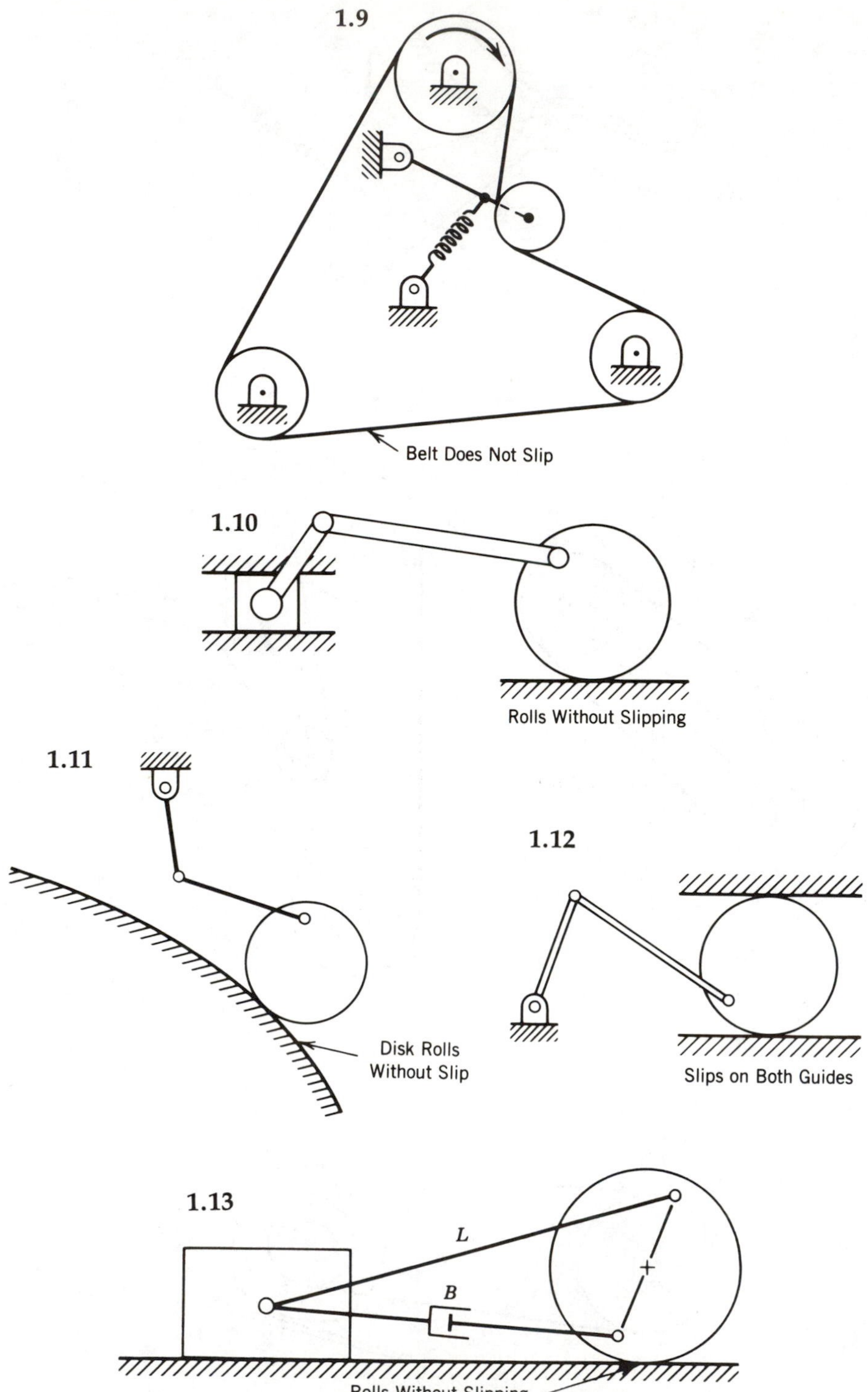
1.9
Belt Does Not Slip
1.10
Rolls Without Slipping
1.11
Disk Rolls
Without Slip
1.12
Slips on Both Guides
1.13
L
B
Rolls Without Slipping

1.14

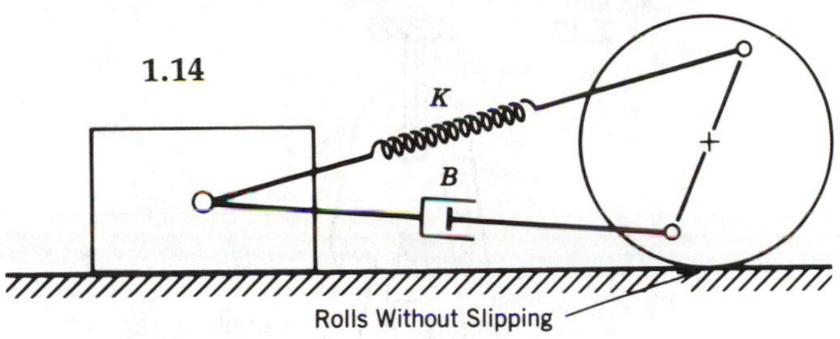
K
B
Rolls Without Slipping

1.15

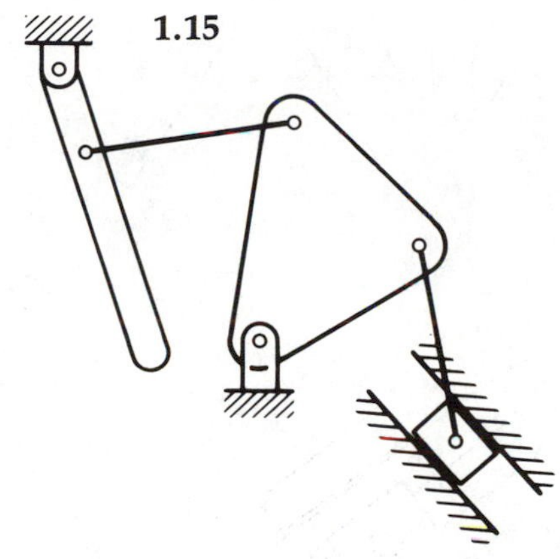

1.16

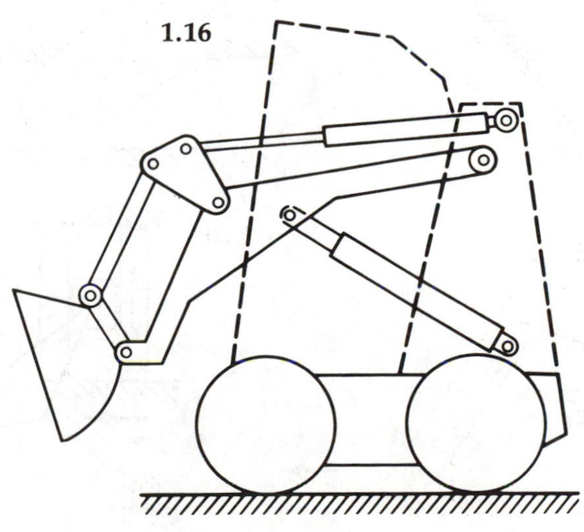

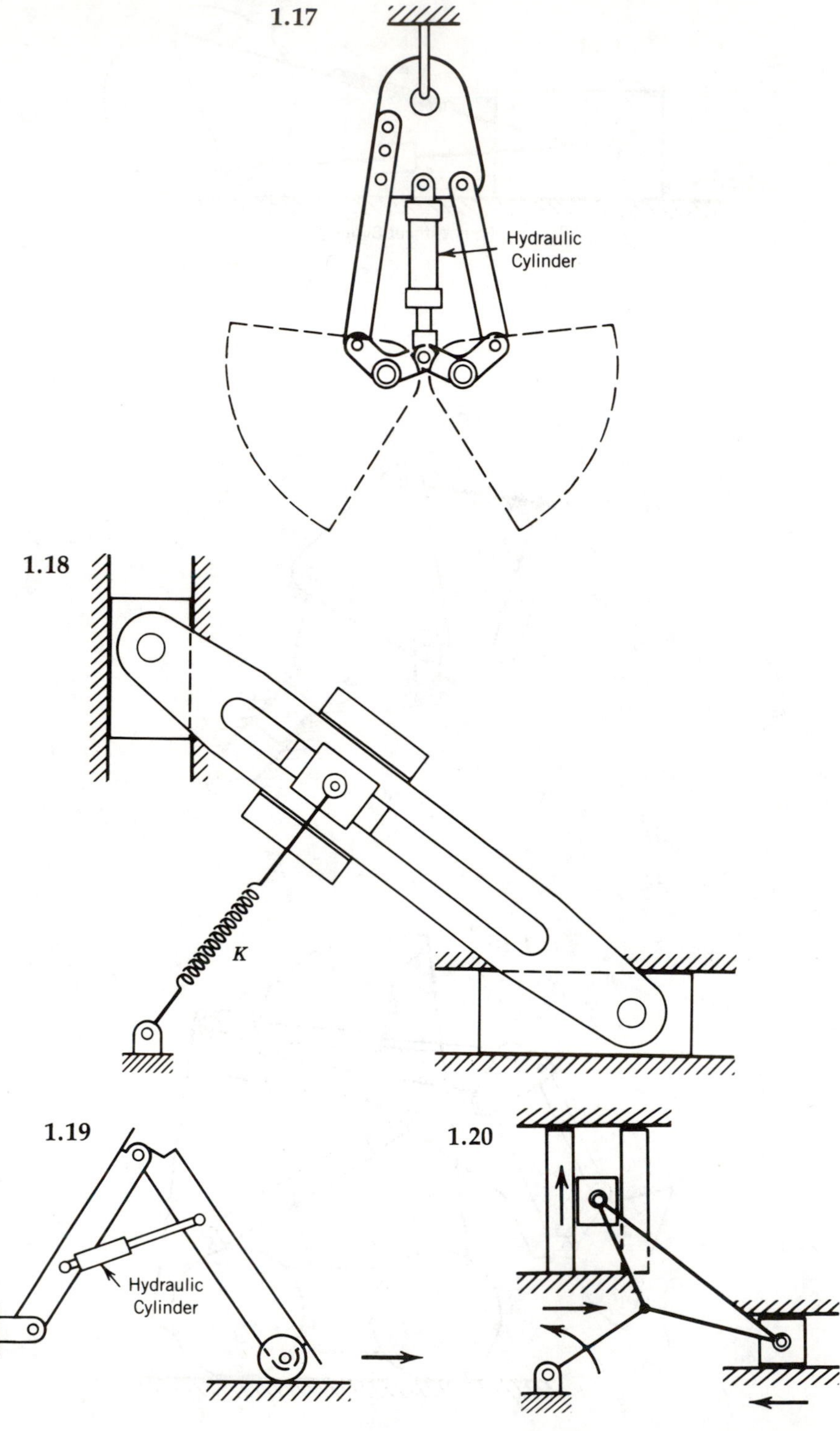
1.17
Hydraulic
Cylinder
1.18
K
1.19
Hydraulic
Cylinder
1.20

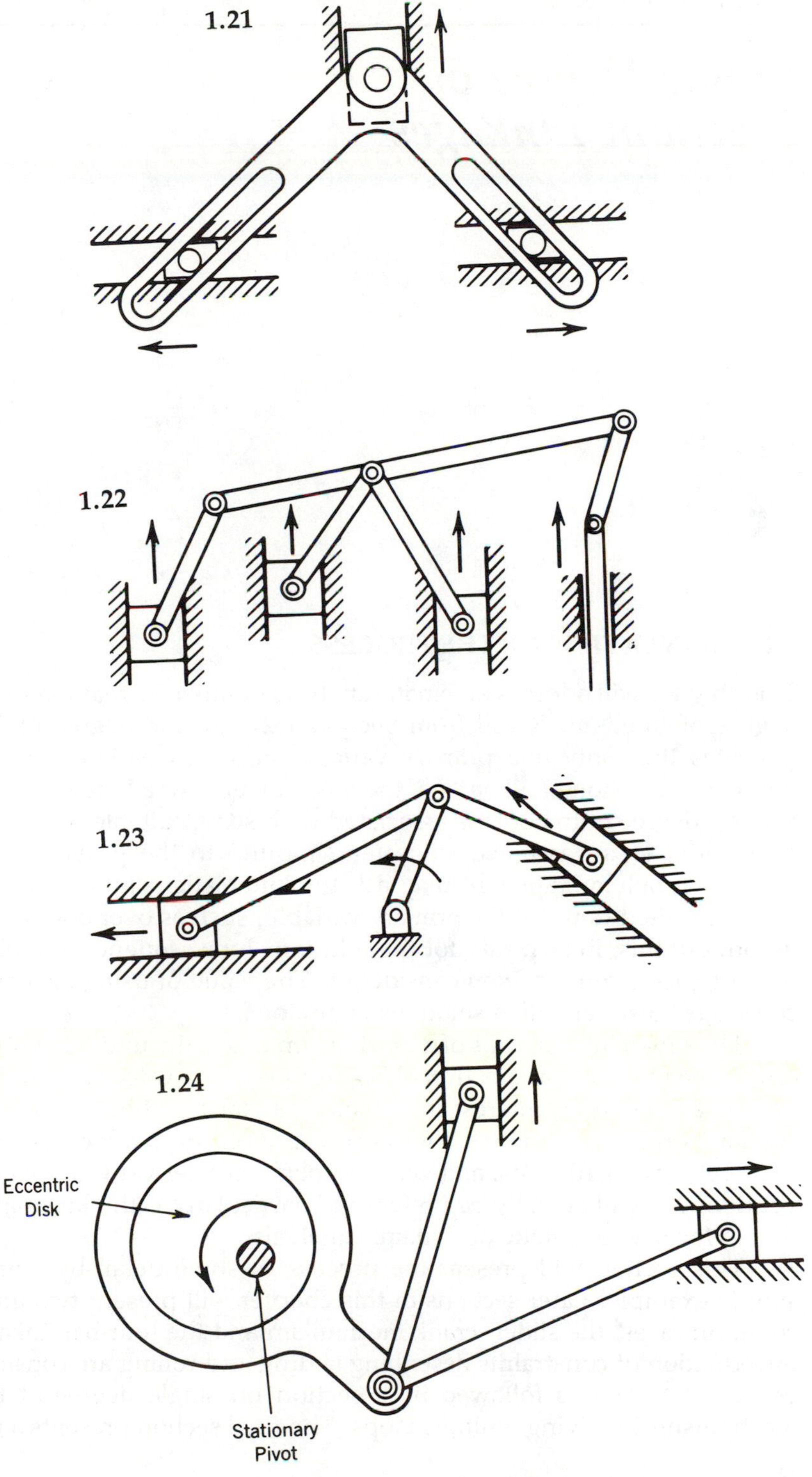
1.21
1.22
1.23
1.24
Eccentric
Disk
Stationary
Pivot

CHAPTER 2

Single Degree of Freedom Linkages

2.1 AN OVERVIEW OF THE PROCESS

This chapter will address kinematic analysis of linkages that have only one degree of freedom. Recall from Section 1.2 that one degree of freedom indicates that only one primary variable needs to be specified to fully define the position of all parts of the mechanism. For a kinematic analysis, the one degree of freedom is associated with some suitable coordinate and considered as an assigned value (an "input") to the problem. In some cases, a problem statement may ask for kinematic solutions throughout some specified range of the primary variable, such as over one crank revolution, but this, in fact, asks for the solutions for a sequence of problems—one for each crank position considered. The value of using a computer to carry out these repetitive solutions is obvious.

The kinematic analysis of a mechanism is usually understood to mean development of equations describing the position, velocity, and acceleration at all points of interest in the mechanism for chosen values of the primary variable, its speed, and its acceleration. In many cases, the position, velocity coefficients, and velocity coefficient derivatives (concepts not yet defined) will actually be preferable, particularly if the kinematic solution is for use in a static or dynamic analysis.

This section will present the process in some detail by means of a simple example. Later sections in this chapter will present two important common cases, the slider–crank mechanism and the four-bar linkage. The introduction of constraints describing sliding and rolling are considered in Section 2.4. This is followed by a section on single degree of freedom mechanisms involving multiple loops. The final section presents a generalized development of the whole kinematic analysis process for single degree of freedom systems.

For the first example, consider the Crank–Lever Reciprocating Drive shown pictorially in Figure 2.1 and in "kinematic skeleton" form in Figure 2.2. The crank pivoted at P_2 rotates through a full circle causing the arm pivoted at P_1 to have a reciprocating rotational motion. Figure 2.1 shows the crank tip sliding in a slot in the arm for this case, whereas the kinematic skeleton, Figure 2.2, shows the arm as a rod passing through a block at the crank tip. The device could actually be built either way and the two constructions are kinematically completely equivalent. This simple mechanism is found in many feeder devices and other machines. The center to center distance C and the crank radius R, labeled in Figure 2.2, are known dimensions. With C and R known, if a value is given for the crank position angle q, the mechanism configuration is fully specified, so this is evidently a single degree of freedom (SDOF) system. The position variables that remain unknown are the arm angle, A, and the location of the crank–arm contact, length B. The first step in the analysis will be to establish the position vector loop equations.

Position Analysis. Consider the three position vectors **B**, **C**, and **R** shown in Figure 2.3. From the closure of the triangle, it is evident that

$$\mathbf{B} - \mathbf{R} - \mathbf{C} = 0$$

or, in scalar form,

$$B \cos A - R \cos q - C = 0 \qquad \text{horizontal component}$$

$$B \sin A - R \sin q - 0 = 0 \qquad \text{vertical component}$$

These are the (nonlinear) position vector loop equations for this mechanism. The input q and the two dimensions C and R are known, whereas the variables A and B are unknown.

The next step is to solve the equations for A and B. Often this step will require numerical solution, but not always. The possibility of an analytic solution should always be considered, and in this case this can be done.

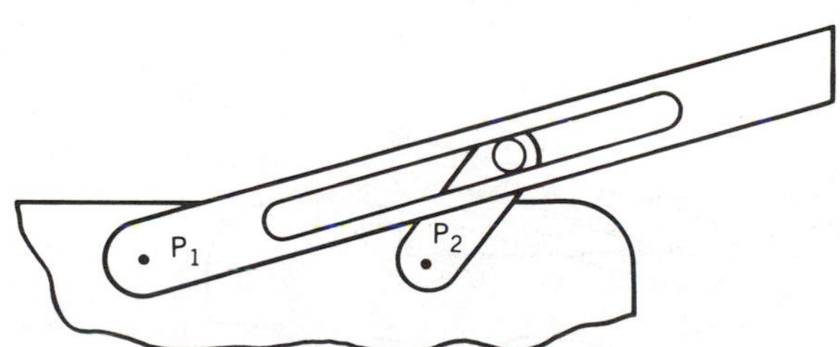

FIGURE 2.1 Pictorial Representation of Crank-Slot Reciprocating Drive

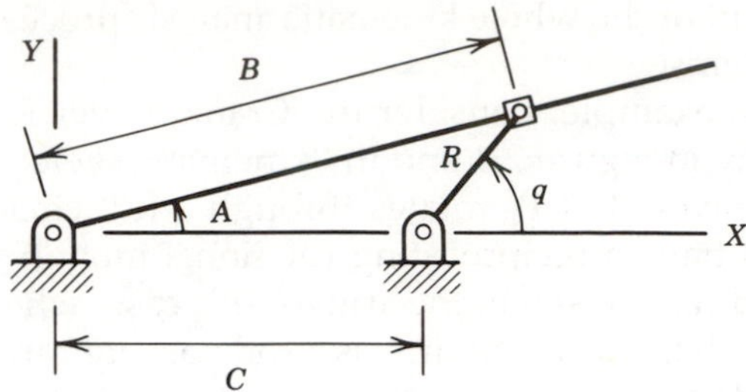

FIGURE 2.2 Kinematic Skeleton Representation of Crank-Slot Reciprocating Drive

Eliminating B between the two equations gives

$$\tan A = \frac{R \sin q}{C + R \cos q}$$

from which A can be determined using the principal value of the arctangent function. With A known, the distance B can be readily computed by solving either equation for B:

$$B = \frac{C + R \cos q}{\cos A} \qquad \text{or} \qquad B = \frac{R \sin q}{\sin A}$$

Note that there will be times when sin A is zero and, for some proportions of the linkage, cos A can also go to zero. Even if both of these events can occur, they will not happen simultaneously, so that B can always be evaluated using one of the above expressions. For computer implementation, an appropriate test (`IF ABS (COS(A))<0.1 THEN ......`) should be included to choose the second expression for B in the event that the first is approaching an indeterminant condition. If the calculations are to be done numerically, this is quite far enough. If further analysis is to be done with these results, then it may be of value to eliminate A algebraically.

Velocity Analysis. In performing the velocity analysis, it is assumed that the results of the position analysis are available as well as the original

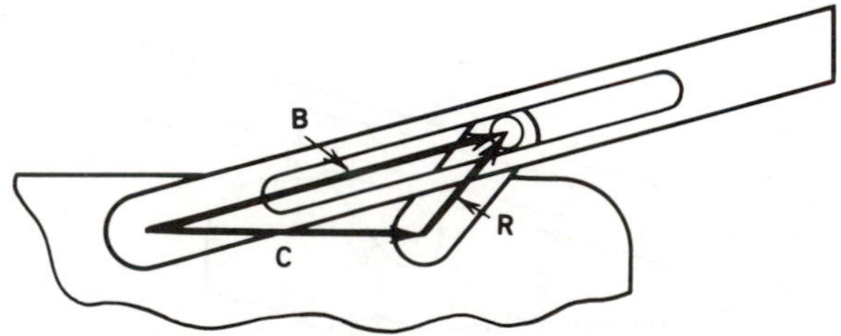

FIGURE 2.3 Position Vector Loop for Crank-Slot Reciprocating Drive

known data. For this problem, this means that the known values now include q, C, R, A, and B, while the new unknowns to be addressed are $\dot{A}$ and $\dot{B}$. The velocity loop equations are obtained by differentiation of the position loop equations, thus:

$$\dot{B}\cos A - B\dot{A}\sin A + R\dot{q}\sin q = 0$$
$$\dot{B}\sin A + B\dot{A}\cos A - R\dot{q}\cos q = 0$$

In view of the quantities already known, these equations are actually a pair of simultaneous linear algebraic equations in the two unknowns, $\dot{A}$ and $\dot{B}$. This is more clearly evident if the equations are cast in matrix form

$$\begin{bmatrix} \cos A & -B\sin A \\ \sin A & B\cos A \end{bmatrix} \begin{Bmatrix} \dot{B} \\ \dot{A} \end{Bmatrix} = R\dot{q} \begin{Bmatrix} -\sin q \\ \cos q \end{Bmatrix}$$

To solve these equations analytically, it is necessary to premultiply by the inverse of the coefficient matrix on the left.[1] Note first that the determinant of the coefficient matrix is simply B. Making the premultiplication by the inverse of the coefficient matrix gives

$$\begin{Bmatrix} \dot{B} \\ \dot{A} \end{Bmatrix} = \frac{R\dot{q}}{B} \begin{bmatrix} B\cos A & B\sin A \\ -\sin A & \cos A \end{bmatrix} \begin{Bmatrix} -\sin q \\ \cos q \end{Bmatrix}$$
$$= \dot{q} \begin{Bmatrix} R\sin(A-q) \\ (R/B)\cos(A-q) \end{Bmatrix}$$

Note that this shows that each of the unknown velocities, $\dot{A}$ and $\dot{B}$, is given by the product of $\dot{q}$ with a factor dependent on the position. The velocity coefficients for this case are

$$\dot{B}/\dot{q} = K_b(q) = R\sin(A-q)$$
$$\dot{A}/\dot{q} = K_a(q) = (R/B)\cos(A-q)$$

For a specific value of $\dot{q}$, the corresponding values of $\dot{A}$ and $\dot{B}$ are readily calculated from the solution above; without specifying the value of $\dot{q}$, the velocity analysis may be made in generalized form by evaluating the position dependent velocity coefficients, $K_b(q)$ and $K_a(q)$.

Acceleration Analysis. To develop the acceleration analysis, the results of both the position and velocity analyses are presumed to be known. Thus, the list of known information now includes q, C, R, A, B, $\dot{q}$, $\dot{A}$, and $\dot{B}$. The unknowns to be determined in this step are $\ddot{A}$ and $\ddot{B}$.

There are two approaches to the acceleration analysis; both will be described here because each gives a different insight. First, differentiation

[1] For the case of two equations and the associated (2 × 2) coefficient matrix, the Appendix A1.5 provides the system solution in terms of a closed-form inverse.

of the scalar velocity loop equations and rearrangement gives the acceleration loop equations as follows:

$$\ddot{B}\cos A - 2\dot{B}\dot{A}\sin A - B\ddot{A}\sin A - B\dot{A}^2\cos A + R\ddot{q}\sin q + R\dot{q}^2\cos q = 0$$

$$\ddot{B}\sin A + 2\dot{B}\dot{A}\cos A + B\ddot{A}\cos A - B\dot{A}^2\sin A - R\ddot{q}\cos q + R\dot{q}^2\sin q = 0$$

Once again, the result is a set of linear, simultaneous equations in the unknowns due to the fact that all of the nonlinear terms involve only known quantities. The linear relation is more apparent when the equations are put into matrix form:

$$\begin{bmatrix} \cos A & -B\sin A \\ \sin A & B\cos A \end{bmatrix} \begin{Bmatrix} \ddot{B} \\ \ddot{A} \end{Bmatrix} = \begin{Bmatrix} 2\dot{B}\dot{A}\sin A + B\dot{A}^2\cos A - R\ddot{q}\sin q - R\dot{q}^2\cos q \\ -2\dot{B}\dot{A}\cos A + B\dot{A}^2\sin A + R\ddot{q}\cos q - R\dot{q}^2\sin q \end{Bmatrix}$$

When this system of linear equations is solved and the $\dot{A}$ and $\dot{B}$ factors are replaced in terms of appropriate velocity coefficients multiplied by $\dot{q}$, the acceleration expressions are

$$\ddot{B} = \ddot{q}R\sin(A - q) + \dot{q}^2[K_a{}^2B - R\cos(A - q)]$$

$$\ddot{A} = \frac{\ddot{q}R}{B}\cos(A - q) + \dot{q}^2\left[-\frac{2K_aK_b}{B} + \frac{R}{B}\sin(A - q)\right]$$

This result shows that the acceleration $\ddot{A}$ and $\ddot{B}$ each consist of two terms, one proportional to $\ddot{q}$ and another proportional to $\dot{q}^2$. The fact that this is true, and the significance of the two coefficient factors, is more evident when the second approach to the acceleration analysis is employed. For that purpose, consider the velocity relations written as

$$\dot{B} = \dot{q}K_b(q)$$

$$\dot{A} = \dot{q}K_a(q)$$

If these equations are differentiated with respect to time, employing the chain rule because K_b and K_a are functions of q and q is a function of time, the accelerations are

$$\ddot{B} = \frac{d(\dot{q})}{dt}K_b + \dot{q}\frac{d(K_b)}{dq}\frac{dq}{dt} = \ddot{q}K_b + \dot{q}^2\frac{d(K_b)}{dq}$$

$$\ddot{A} = \frac{d(\dot{q})}{dt}K_a + \dot{q}\frac{d(K_a)}{dq}\frac{dq}{dt} = \ddot{q}K_a + \dot{q}^2\frac{d(K_a)}{dq}$$

With this approach to the analysis, the two-term form is clearly expected, and the coefficients are seen to be (1) the velocity coefficient multiplying $\ddot{q}$ and (2) the derivative of the velocity coefficient with respect to q multiplying $\dot{q}^2$. Looking back to the results of the first approach, it can be seen that the coefficients there are just as described.

Numerical Values. For numerical evaluation, consider a particular crank–lever reciprocating drive defined by

$C = 2.5$ in.
$R = 0.75$ in.

For this particular system, the previous analysis has been implemented numerically and the results plotted for one full crank revolution at constant crank speed (shown in Figure 2.4). Figure 2.4(a) shows the plot of normalized curves for A, $\dot{A}$, and $\ddot{A}$ (actually $A/|A|_{\max}$, $\dot{A}/|\dot{A}|_{\max}$, and $\ddot{A}/|\ddot{A}|_{\max}$), and Figure 2.4(b) presents similar curves for B and its derivatives. The crank angle q is expressed in radians for both plots.

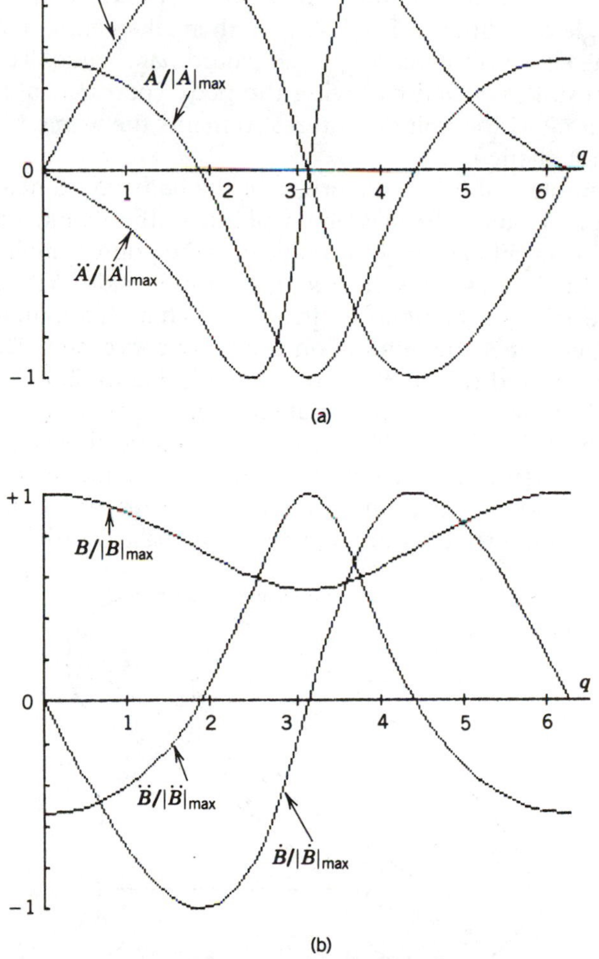

FIGURE 2.4 Normalized Kinematic Response for Crank-Slot Reciprocating Drive at Constant Crank Speed
(a) A and Derivatives
(b) B and Derivatives

Position, Velocity, and Acceleration for Points of Interest. The results determined so far give a full description of the angle A, the length coordinate B, and their derivatives, but nothing has been said regarding the position, velocity, or acceleration of any points on either of the components. Because there are only two components and each turns about a fixed center, the following analysis is specific to rotation about a fixed center. However, the analysis for a more general case follows along these same lines.

The first matter of concern is the description of the particular point of interest, P. Begin by defining a *body coordinate system,* denoted (U, V), *fixed on the body* so that it moves with the body. Although any choice can be made for the body coordinate axes, it is usually convenient to choose the U–axis such that its orientation is described by one of the primary or secondary angle coordinates. The V–axis is then taken at right angles to the U–axis, in the sense of increasing angle coordinate value. The particular point of interest is specified by giving the body coordinates, (U_p, V_p) for the point; as long as the point of interest remains the same, the values U_p and V_p will not change.

For the problem at hand, the origin of the body coordinates is chosen at the fixed pivot point. To describe a point on the crank, the U–axis is taken along the terminal side of the angle q. If the point of interest is on the slotted lever, the U–axis is taken along the terminal side of the angle A. To generalize the analysis that follows, the angle is denoted E; note that E can represent either q or A, depending on the body concerned. This choice of body coordinates and the angle E are shown in Figure 2.5.

Next define a stationary coordinate system, termed the *base coordinate system* and denoted by (X, Y). It is also convenient to choose the origin for the base coordinates at the pivot point of the rotating member, as is shown in Figure 2.5. To locate the point of interest with respect to the stationary coordinate system, it is necessary to express the base coordinates, (X_p, Y_p)

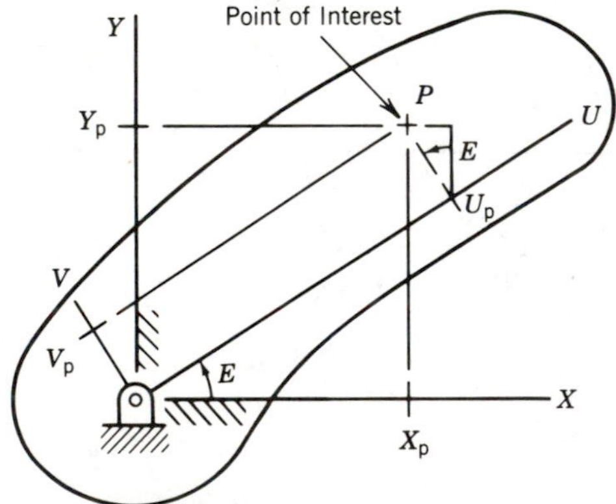

FIGURE 2.5 Body Coordinates and Base Coordinates for a Particular Point of Interest on a Body Rotating About a Fixed Point

for the point. The abscissa of the point is expressed using two terms, $U_p \cos E$ and $V_p \sin E$, which are in opposite senses. The ordinate of the point also involves two terms, $U_p \sin E$ and $V_p \cos E$, which are in the same sense. The complete relations are

$$X_p = U_p \cos E - V_p \sin E$$
$$Y_p = U_p \sin E + V_p \cos E$$

These two equations provide the means to express the stationary, base coordinates for the point of interest in terms of the body coordinates U_p and V_p, and the angle E.

The velocity components for the point of interest are $\dot{X}_p$ and $\dot{Y}_p$, determined by differentiation of the expressions for X_p and Y_p. Because U_p and V_p do not change with time but the angle E does change, the velocity components are

$$V_{px} = \dot{X}_p = \dot{E}(-U_p \sin E - V_p \cos E)$$
$$V_{py} = \dot{Y}_p = \dot{E}(U_p \cos E - V_p \sin E)$$

These relations give the two velocity components for the point, one for each direction of motion. The velocity coefficients for the point are

$$K_{px} = -U_p \sin E - V_p \cos E$$
$$K_{py} = U_p \cos E - V_p \sin E$$

In terms of the velocity coefficients, the velocity components are

$$V_{px} = \dot{X}_p = \dot{E}K_{px}$$
$$V_{py} = \dot{Y}_p = \dot{E}K_{py}$$

For the acceleration components, the velocity components are differentiated with respect to time

$$A_{px} = \ddot{X}_p = \ddot{E}(-U_p \sin E - V_p \cos E) + \dot{E}^2(-U_p \cos E + V_p \sin E)$$
$$A_{py} = \ddot{Y}_p = \ddot{E}(U_p \cos E - V_p \sin E) + \dot{E}^2(-U_p \sin E - V_p \cos E)$$

By inspection, it is evident that these may be written as

$$A_{px} = \ddot{X}_p = \ddot{E}K_{px} + \dot{E}^2 L_{px}$$
$$A_{py} = \ddot{Y}_p = \ddot{E}K_{py} + \dot{E}^2 L_{py}$$

where L_{px} and L_{py} are derivatives of the velocity coefficients K_{px} and K_{py} with respect to E. This last observation can be readily verified by direct calculation. Note again that for a particular point in the body there are two velocity coefficients, K_{px} and K_{py}, and, consequently, two velocity coefficient derivatives, L_{px} and L_{py}.

This analysis applies for any point on a body rotating about a fixed center. Although both bodies in the example problem are of that type, many machinery components do not move about fixed centers. A method applicable to those situations will be demonstrated in Sections 2.2 and 2.3.

Summary of Observations from the Example Problem

The kinematic analysis of this SDOF system involved the following steps:

1. Developing the position equations and solving those equations for the secondary variables;
2. Differentiating the position equations to obtain the velocity equations that are solved for the secondary velocities, or for the velocity coefficients;
3. Determining secondary accelerations by solving the equations resulting from differentiation of the velocity equations. This may be done by the straightforward differentiation indicated, or by using the velocity coefficients and velocity coefficient derivatives, obtained by differentiation from step 2;
4. Defining body coordinates to specify the location of a particular point of interest;
5. Determining base coordinates for the point of interest in terms of the body coordinates, secondary variables, and the primary variable;
6. Determining velocity components, acceleration components, velocity coefficients, and velocity coefficient derivatives as required, beginning with the base coordinate expressions.

The position loop equations were written in a scalar form that could be expressed as

$$f_1(q, A, B) = 0 \qquad \text{horizontal component}$$

$$f_2(q, A, B) = 0 \qquad \text{vertical component}$$

When the velocity equations were cast in matrix form, the coefficient matrix for the unknown velocities was

$$\begin{bmatrix} \dfrac{\partial f_1}{\partial B} & \dfrac{\partial f_1}{\partial A} \\ \\ \dfrac{\partial f_2}{\partial B} & \dfrac{\partial f_2}{\partial A} \end{bmatrix} = \begin{bmatrix} \cos A & -B \sin A \\ \sin A & B \cos A \end{bmatrix}$$

This is called the *Jacobian matrix* for this system. This same Jacobian matrix is seen again as the coefficient of the unknown accelerations for the acceleration equations written in matrix form. If a numerical solution had been required for the position equations, the Jacobian matrix would have been used in the numerical solution process (see Section 2.3 and Appendix A3). The occurrence of the Jacobian matrix in these three different relations is not accidental; it should be expected, and its failure to appear at these points is an indication of an error in the analysis. Whenever the Jacobian is to be evaluated numerically the same code, in the form of a subroutine, should be used for each evaluation. In each case, the system of equations (position loop equations, velocity loop equations, or acceleration loop equations) is solvable only so long as the determinant of the Jacobian matrix is non-zero. Positions for which the Jacobian determinant is zero are known as *singular points;* these positions require special treatment.

2.2 KINEMATICS OF THE SLIDER–CRANK MECHANISM

One of the most important common mechanisms is the slider–crank mechanism. It is found in pumps, compressors, steam engines, feeders, crushers, punches, and injectors. The slider–crank mechanism is central to the diesel and gasoline internal combustion engines that are so much a part of modern life. In most cases the crank rotates continuously in the same direction, although in some cases the crank motion may be oscillatory. The analysis presented here is in a general form so that the results apply to any slider–crank device.

The kinematic skeleton for a typical slider–crank mechanism is shown in Figure 2.6. The crank turns about a fixed pivot at the origin of coordinates, and the slider has reciprocating motion along a line parallel to the X–axis. As shown, the path of the slider pivot is displaced a distance C above the X–axis. For many applications the offset C will be zero; it may also be negative. The link joining the crank and the slider is called the *connecting rod,* and its length is denoted as L.

Number of Degrees of Freedom. Before beginning the kinematic analysis, it is important to determine the number of degrees of freedom associated with the mechanism. For the slider–crank, if a value is specified for the crank angle q, then the configuration of the entire assembly is determined and the mechanism has one degree of freedom. To see this, consider a graphical construction process shown in Figure 2.7. All dimensional data is

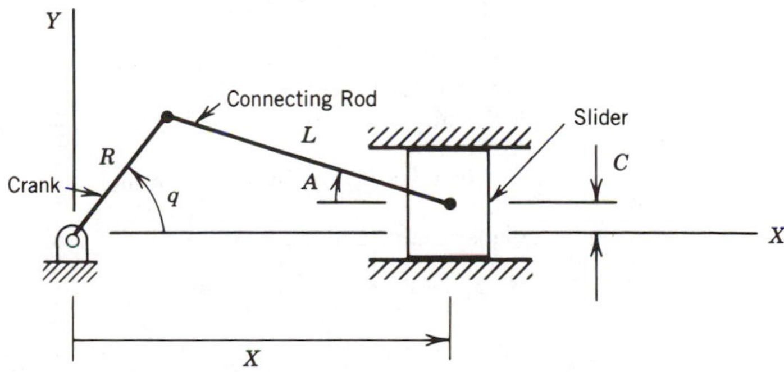

FIGURE 2.6 Typical Slider-Crank Mechanism

(*a*) Locating Crank Throw

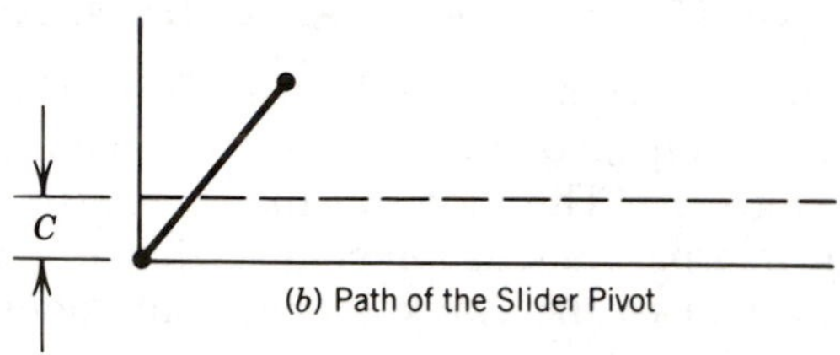

(*b*) Path of the Slider Pivot

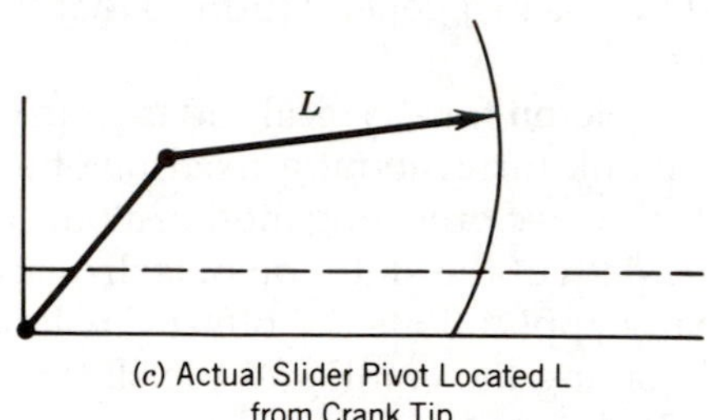

(*c*) Actual Slider Pivot Located L from Crank Tip

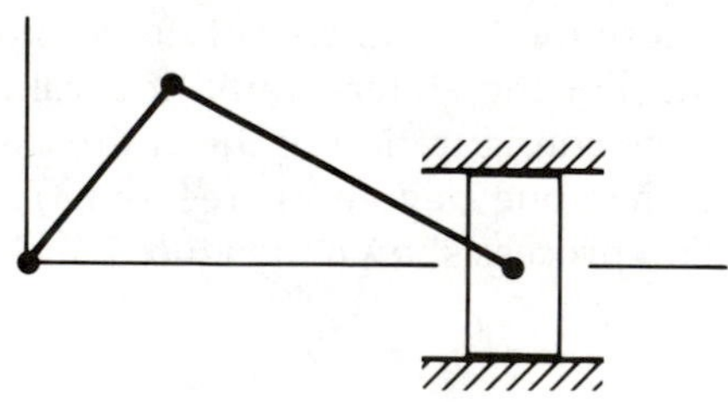

(*d*) Completed Slider Crank Configuration

FIGURE 2.7 Construction Demonstrating that the Slider-Crank is a SDOF Mechanism
(a) Locating Crank Throw
(b) Path of the Slider Pivot
(c) Actual Slider Pivot Located a Distance L from Crank Tip
(d) Completed Slider Crank Configuration

assumed known before the construction begins. The important question is "What is the minimum number of additional values that must be specified to determine the configuration of the mechanism?" The steps for the construction are as follows:

a. With q given, the location of the crank throw is determined by the polar coordinates (R, q). (The crank throw is the outer end of the crank to which one end of the connecting rod is attached.)
b. The path of the slider pivot is a line parallel to the X–axis and is displaced from it by a distance C.
c. The present location of the slider pivot is the point along the slider pivot locus that is a distance L from the crank throw.
d. The figure is now complete.

In this problem, it is evident immediately that the crank rotation will be a useful primary variable (generalized coordinate), so the construction begins with a value assigned for that variable. Because we could complete the construction by assigning only one variable, the answer to the initial question is, "One additional value is sufficient." Therefore the slider–crank mechanism has a single degree of freedom.

The graphical thought process described has broad applicability; it can be adapted to determine the number of degrees of freedom for a variety of mechanisms. In each case, it is necessary to answer the question, "How many variables must be specified (in addition to the dimensional data) to graphically construct the mechanism?" The word *construct* means that known lengths or angles can be measured and laid out accordingly. Other points must be located and lines defined strictly in the sense of Euclidean geometric construction, performed with a compass and a straightedge. When this approach is applied to systems with multiple degrees of freedom, the need to specify values for other variables becomes apparent as the construction proceeds. Unless the system is very complicated, it is not necessary to actually draw the mechanism; instead, a mental construction is usually sufficient.

Position Analysis. As just discussed, the slider–crank mechanism position is fully determined by specifying a single variable, here assumed to be the crank angle. The secondary kinematic variables of interest are the connecting rod obliquity angle, A, and the slider position, X (see Figure 2.6). There is a single position vector loop that runs from the crank pivot to the crank throw, along the connecting rod to the slider pivot, and back to the crank pivot. The scalar position loop equations are

$$R \cos q + L \cos A - X = 0$$
$$R \sin q - L \sin A - C = 0$$

These equations must be solved for A and X, with the following results:

$$A = \text{Arcsin}[(R \sin q - C)/L]$$

$$X = R \cos q + L \cos A$$

Because A will always lie in the first or fourth quadrants, the principal value of the inverse sine function will be correct. The solutions may be checked by substitution in the position loop equations.

Velocity Analysis. The object of the analysis in this step is to obtain the position-dependent velocity coefficients, K_a and K_x. Differentiation of the position loop equations gives the velocity loop equations

$$-R\dot{q} \sin q - L\dot{A} \sin A - \dot{X} = 0$$

$$R\dot{q} \cos q - L\dot{A} \cos A = 0$$

Prior to solving for the unknown velocities, these equations are recast in the matrix form

$$\begin{bmatrix} -L \sin A & -1 \\ -L \cos A & 0 \end{bmatrix} \begin{Bmatrix} \dot{A} \\ \dot{X} \end{Bmatrix} = \dot{q} \begin{Bmatrix} R \sin q \\ -R \cos q \end{Bmatrix}$$

The (2×2) coefficient matrix on the left side of the equation is the Jacobian matrix for the slider–crank mechanism. This is a system of linear, simultaneous algebraic equations that must be solved. (The solution for the case of two equations in two unknowns is given in detail in Appendix A1.5.) For the present system of equations, the solution is

$$\begin{Bmatrix} \dot{A} \\ \dot{X} \end{Bmatrix} = \frac{-\dot{q}}{L \cos A} \begin{bmatrix} 0 & 1 \\ L \cos A & -L \sin A \end{bmatrix} \begin{Bmatrix} R \sin q \\ -R \cos q \end{Bmatrix}$$

$$= \dot{q} \begin{Bmatrix} \dfrac{R \cos q}{L \cos A} \\ -R \sin q - R \cos q \tan A \end{Bmatrix}$$

The expression for $\dot{X}$ may be simplified further by using the second of the position equations to replace $(R \sin q)$ with $(L \sin A + C)$. After some algebra, the final result is

$$\dot{X} = -\dot{q}(C + X \tan A)$$

To express the velocity coefficients, the ratios $\dot{A}/\dot{q}$ and $\dot{X}/\dot{q}$ are formed:

$$\begin{Bmatrix} \dot{A}/\dot{q} \\ \dot{X}/\dot{q} \end{Bmatrix} = \begin{Bmatrix} K_a \\ K_x \end{Bmatrix} = \begin{Bmatrix} \dfrac{R \cos q}{L \cos A} \\ -(C + X \tan A) \end{Bmatrix}$$

Acceleration Analysis. As mentioned in Section 2.1, there are two ways to approach the secondary accelerations. The following discussion will present both approaches, beginning with the direct differentiation of the velocity loop equations with respect to time. Performing the indicated differentiations gives

$$-R\ddot{q}\sin q - R\dot{q}^2\cos q - L\ddot{A}\sin A - L\dot{A}^2\cos A - \ddot{X} = 0$$

$$R\ddot{q}\cos q - R\dot{q}^2\sin q - L\ddot{A}\cos A + L\dot{A}^2\sin A = 0$$

When cast in matrix form, the Jacobian matrix is evident again in the (2 × 2) coefficient matrix on the left side of the equation:

$$\begin{bmatrix} -L\sin A & -1 \\ -L\cos A & 0 \end{bmatrix}\begin{Bmatrix} \ddot{A} \\ \ddot{X} \end{Bmatrix} = \ddot{q}\begin{Bmatrix} R\sin q \\ -R\cos q \end{Bmatrix} + \dot{q}^2\begin{Bmatrix} R\cos q \\ R\sin q \end{Bmatrix} + \dot{A}^2\begin{Bmatrix} L\cos A \\ -L\sin A \end{Bmatrix}$$

This system of equations is solved by the established procedure to express the accelerations, $\ddot{A}$ and $\ddot{X}$. From the second approach to the acceleration calculations, the accelerations will take the form

$$\begin{Bmatrix} \ddot{A} \\ \ddot{X} \end{Bmatrix} = \ddot{q}\begin{Bmatrix} K_a \\ K_x \end{Bmatrix} + \dot{q}^2\begin{Bmatrix} L_a \\ L_x \end{Bmatrix}$$

where K_a and K_x are the velocity coefficients, and L_a and L_x are the derivatives of K_a and K_x with respect to q. By direct differentiation of K_a and K_x with respect to q, the velocity coefficient derivatives are

$$L_a = \frac{-R\sin q}{L\cos A} + K_a{}^2\tan A$$

$$L_x = -R\cos q - K_a{}^2 L\cos A - L_a L\sin A$$

This result can be confirmed by completing the solution to the system of simultaneous equations in the first approach to the acceleration calculation. As stated before, the complete accelerations are given by a sum involving these velocity coefficient derivatives and the velocity coefficients, multiplied by $\dot{q}^2$ and $\ddot{q}$, respectively.

Numerical Values. Consider a slider–crank mechanism defined by the following dimensions:

$R = 3.5$ in.

$L = 10.3$ in.

$C = 0.35$ in.

The position values A and X, velocity coefficients, and velocity coefficient derivatives are to be evaluated for the crank angle $q = \pi/3$ radians.

The preceding analysis has been implemented in a short computer program, which follows. Notice that the problem data is embedded in the program from lines 1130 to 1160. The program asks for the crank angle as input at line 1240. The input value used is 1.0472 radians, which is the decimal equivalent to $\pi/3$.

```
1000 REM  SLIDER-CRANK KINEMATIC ANALYSIS
1010 REM  USING CLOSED FORM KINEMATIC RELATIONS
1020 REM
1030 REM
1040 REM  R       = CRANK RADIUS
1050 REM  L       = CONNECTING ROD LENGTH
1060 REM  C       = SLIDER PIVOT OFFSET
1070 REM  A       = CONNECTING ROD ANGLE
1080 REM  Q       = CRANK ROTATION ANGLE
1090 REM            ASSOCIATED WITH THE SDOF
1100 REM  X       = SLIDER POSITION
1110 REM
1120 REM
1130 REM  SYSTEM DATA
1140 R=3.5
1150 L=10.3
1160 C=0.35
1170 REM
1200 REM  POSITION ANALYSIS
1210 HOME
1220 PRINT"SLIDER-CRANK KINEMATIC ANALYSIS"
1230 PRINT"ENTER CRANK ANGLE IN RADIANS"
1240 INPUT Q
1250 PRINT
1260 N=R*SIN(Q)-C
1270 D=SQR(L^2-N^2)
1290 A=ATN(N/D)
1300 X=R*COS(Q)+L*COS(A)
1310 REM
1400 REM  VELOCITY ANALYSIS
1410 KA=(R*COS(Q))/(L*COS(A))
1420 KX=-(C+X*TAN(A))
1430 REM
1500 REM  ACCELERATION ANALYSIS
1510 LA=KA^2*TAN(A)-R*SIN(Q)/(L*COS(A))
1520 LX=-R*COS(Q)-KA^2*L*COS(A)-LA*L*SIN(A)
1530 REM
1600 REM  OUTPUT
1610 PRINT"FOR Q=";Q;" RAD"
1620 PRINT"        A =";A
1630 PRINT"        X =";X
1640 PRINT"        KA=";KA
1650 PRINT"        KX=";KX
1660 PRINT"        LA=";LA
1670 PRINT"        LX=";LX
1680 END
```

When the program is executed, with the crank angle entered as 1.0472 radians, the results are:

```
SLIDER-CRANK KINEMATIC ANALYSIS
ENTER CRANK ANGLE IN RADIANS
? 1.0472

FOR Q=1.0472 RAD
        A =.263333222
        X =11.6949279
        KA=.17596822
        KX=-3.50288039
        LA=-.296439699
        LX=-1.2631532
```

This means that at the instant when $q = 1.0472$ radians, the obliquity is 0.263333 radians and the distance from the crank pivot to the slider pivot is 11.6949 inches. At that same moment, the obliquity is increasing at the rate 0.175968 radians per crank radian. The slider is moving at the rate 3.50288 inches per crank radian; the movement is toward the crank as indicated by the negative sign. The obliquity rate is decreasing 0.296440 radians per crank radian per crank radian. The slider position rate is decreasing 1.26315 inches per crank radian per crank radian. The phrase "per crank radian" means per one radian of crank rotation. Because LA and LX are not zero, the non-constant nature of the velocity coefficients is clearly evident.

Position, Velocity, and Acceleration for Points of Interest

To this point, the analysis has established values for the secondary kinematic variables and their derivatives. However, the motion of particular points on the various components remains to be determined. All points on the crank are on a rigid body rotating about a fixed center; for these points the analysis given under this same heading in Section 2.1 applies (because this is, again, motion on a circular path). All points in the slider differ in position only by a constant, so the velocity and acceleration for these points are $\dot{X}$ and $\ddot{X}$, respectively. Points on the connecting rod require further consideration. The usual practice throughout this text is to work in terms of the scalar components of position, velocity, and acceleration. However, for the discussion here and at the end of Section 2.3, vector expressions will be employed so as to emphasize the relation between the vector components and the actual vector quantities.

Consider the slider–crank mechanism shown in Figure 2.8 where the outline of the connecting rod has been included. Notice that a body coordinate system, (U, V), has also been added to the figure. The U–axis of this system is along the line of centers from the crank throw to the slider pivot. The V–axis is perpendicular to the U–axis as shown; this coordinate system moves with the connecting rod. A particular point on the connecting

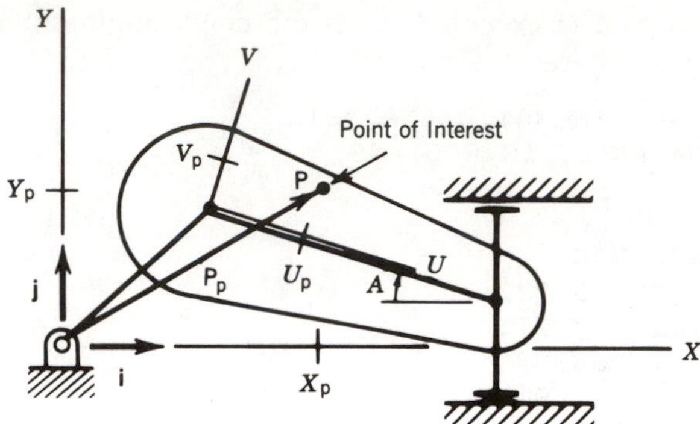

FIGURE 2.8 Position Vector, Body Coordinates, and Base Coordinates for a Point of Interest on the Slider—Crank Connecting Rod

rod is specified by giving its body coordinates (U_p, V_p). These coordinates do not change with time as the system moves; they change only when a different point is to be considered.

Position. The position vector of the point of interest is $\mathbf{P}_p$, where the subscript p is a reminder of the point this vector describes;

$$\begin{aligned}\mathbf{P}_p &= \mathbf{i}X_p + \mathbf{j}Y_p \\ &= \mathbf{i}(R\cos q + U_p\cos A + V_p\sin A) \\ &\quad + \mathbf{j}(R\sin q - U_p\sin A + V_p\cos A)\end{aligned}$$

The variables (X_p, Y_p) are the *base coordinates* of the point. After the secondary kinematic variables A and X have been evaluated, this position vector can be readily evaluated for any specified pair of body coordinates (U_p, V_p).

Velocity. The velocity vector for the designated point, relative to stationary coordinates, is $\mathbf{V}_p$. This is obtained by differentiation of the position vector:

$$\begin{aligned}\mathbf{V}_p = \dot{\mathbf{P}}_p &= \mathbf{i}(-R\dot{q}\sin q - U_p\dot{A}\sin A + V_p\dot{A}\cos A) \\ &\quad + \mathbf{j}(R\dot{q}\cos q - U_p\dot{A}\cos A - V_p\dot{A}\sin A) \\ &= \dot{q}(\mathbf{i}K_{px} + \mathbf{j}K_{py})\end{aligned}$$

where

$$K_{px} = -R\sin q - K_a(U_p\sin A - V_p\cos A)$$
$$K_{py} = R\cos q - K_a(U_p\cos A + V_p\sin A)$$

The velocity coefficients K_{px} and K_{py} depend on the designated point through (U_p, V_p) and on the secondary variable, A, and its velocity coefficient, K_a.

Acceleration. The acceleration of the designated point is obtained by differentiating the last velocity expression noted above with respect to time. Because K_{px} and K_{py} are functions of K_a, which is itself a function of q, the chain rule must be used for differentiation, as follows:

$$\begin{aligned}\mathbf{A}_p = \dot{\mathbf{V}}_p = \ddot{\mathbf{P}}_p &= \ddot{q}(\mathbf{i}K_{px} + \mathbf{j}K_{py}) + \dot{q}^2(\mathbf{i}L_{px} + \mathbf{j}L_{py}) \\ &= \mathbf{i}(\ddot{q}K_{px} + \dot{q}^2L_{px}) + \mathbf{j}(\ddot{q}K_{py} + \dot{q}^2L_{py})\end{aligned}$$

where

$$\begin{aligned}L_{px} &= \frac{d(K_{px})}{dq} \\ &= -R\cos q - L_a(U_p \sin A - V_p \cos A) - K_a^2(U_p \cos A + V_p \sin A) \\ L_{py} &= \frac{d(K_{py})}{dq} \\ &= -R\sin q - L_a(U_p \cos A + V_p \sin A) - K_a^2(-U_p \sin A + V_p \cos A)\end{aligned}$$

The velocity coefficient derivatives are dependent on the coordinates of the point of interest, (U_p, V_p), and on the secondary variable quantities A, K_a, and L_a.

Vector notation has been used in the preceding development to emphasize the relation between the position, velocity, and acceleration vectors on the one hand, and the scalar position variables, velocity coefficients, and velocity coefficient derivatives on the other. The scalar variables are the quantities that can be handled efficiently in the computer, but the vector quantities can be expressed whenever needed.

For design studies, it may be of interest to study the path (locus of positions), velocity, and acceleration of several points on the connecting rod as the crank sweeps out a full revolution. For each crank position, the secondary kinematic variables, the velocity coefficients, and the velocity coefficient derivatives must first be evaluated. Then the position, velocity, and acceleration of any number of points on the connecting rod may be computed by implementing the preceding calculations for each point. This process can then be repeated for a sequence of crank positions spaced around a full revolution.

2.3 KINEMATICS OF THE FOUR-BAR LINKAGE

One of the most versatile common mechanisms is the four-bar linkage. It is found in a wide variety of machines, including oil field pump jacks, electric shavers, and centerless grinders. The variety of motions that can be gener-

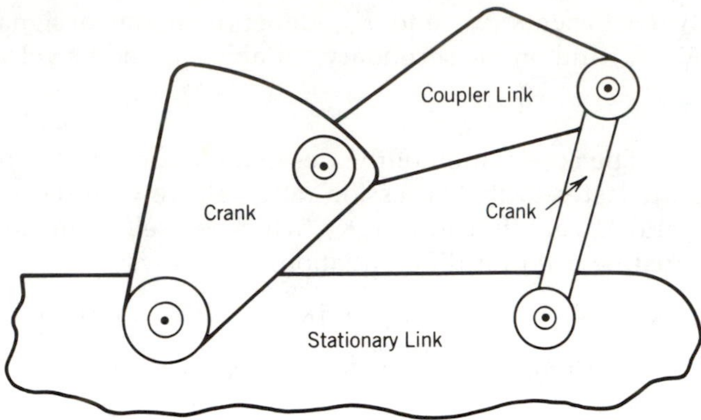

FIGURE 2.9 Typical Four-Bar Linkage

ated with the four-bar linkage include approximate straight lines (Watt and Scott Russel linkages), closed curves, and even circles (Galloway mechanism). This range of adaptability has attracted the interest of designers over the years through the present day.

The four-bar linkage consists of four bars or links of constant length, as the name implies, and can be visualized as a closed assembly of four, pin-jointed links. A typical four-bar linkage is shown in Figure 2.9. One of the four links shown is completely stationary, and physically does not look like a link at all. Nevertheless, it functions as a link, maintaining the separation of the two pivot points where the remainder of the linkage is supported. The two links joined to the stationary link are referred to as *cranks* because their motions are purely circular about their stationary pivots. The final link joining the two cranks is called the *coupler link.* For most purposes, the four-bar linkage is used in one of two ways: (1) the mechanism may be used to transfer power from one crank (the *input crank*) to a second crank (the *output crank*) with a particular motion or (2) a point on the coupler link may be driven in a desired motion by the input crank rotation. It is also possible, but relatively uncommon, for the input link to be the coupler.

The four-bar linkage is a single degree of freedom mechanism. This is readily seen by mentally applying the "graphical construction" test described in Section 2.2. It will be convenient to associate the one degree of freedom with the angular position of the input crank. The following analysis will first develop the position, velocity, and acceleration relations for the secondary kinematic variables in terms of the motion of the input crank. Closed-form position solutions are again possible, but in this case they are unwieldy. Instead, this opportunity is taken to introduce a general numerical technique for solving the position equations (the Newton–Raphson method).

Position Analysis. The link lengths, C_1, C_2, C_3, and C_4, are understood to be given for the four-bar linkage shown in Figure 2.10. The configuration of

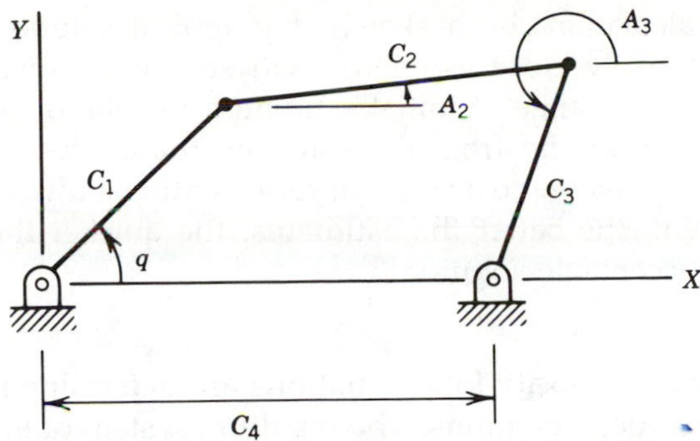

FIGURE 2.10 Kinematic Skeleton for Typical Four-Bar Linkage

the four-bar linkage is then fully determined when the input crank angle q is specified. The remaining secondary kinematic variables are the angular positions for the coupler and the second crank, A_2 and A_3. These will be determined from the following position loop equations:

$$f_1(q, A_2, A_3) = C_1 \cos q + C_2 \cos A_2 + C_3 \cos A_3 - C_4 = 0$$

$$f_2(q, A_2, A_3) = C_1 \sin q + C_2 \sin A_2 + C_3 \sin A_3 = 0$$

As previously mentioned, a numerical solution procedure called the Newton–Raphson method, will be used to solve the position loop equations. The details of the Newton–Raphson solution are given in Appendix A2 and should be reviewed at this point if not familiar. In the notation of the appendix, the vector of unknowns, $\{S\}$, is here called $\{A\}$, where

$$\{S\} = \{A\} = \text{Col}\,(A_2, A_3)$$

and the residual vector is $\{F\}$, where

$$\{F\} = \text{Col}(f_1(q, A_2 A_3),\ f_2(q, A_2, A_3))$$

The Jacobian matrix for this system is determined by partial differentiation, with the result

$$\begin{bmatrix} \dfrac{\partial f_1}{\partial A_2} & \dfrac{\partial f_1}{\partial A_3} \\[2ex] \dfrac{\partial f_2}{\partial A_2} & \dfrac{\partial f_2}{\partial A_3} \end{bmatrix} = \begin{bmatrix} -C_2 \sin A_2 & -C_3 \sin A_3 \\ C_2 \cos A_2 & C_3 \cos A_3 \end{bmatrix}$$

Using the residual vector $\{F\}$ and the Jacobian matrix just defined, the position loop solution may be computed iteratively to any required degree of precision. This solution method is demonstrated in an example problem following the velocity and acceleration analyses.

One requirement for using the Newton–Raphson solution is an initial estimate for each of the unknowns. These estimates can be determined by a very rough calculation, by a sketched graphical solution, or by pure guess; any reasonably close estimates will suffice because the solution converges for a wide range of initial estimates. On the other hand, if the initial estimates are too far from the solution, the process will either not converge, or will converge to a solution representing a different configuration. In any event, the better the estimates, the quicker the process will converge to an acceptable solution.

Velocity Analysis. Velocity loop equations are determined by differentiating the position loop equations. The resulting system of linear equations can be cast in matrix form as

$$\begin{bmatrix} -C_2 \sin A_2 & -C_3 \sin A_3 \\ C_2 \cos A_2 & C_3 \cos A_3 \end{bmatrix} \begin{Bmatrix} \dot{A}_2 \\ \dot{A}_3 \end{Bmatrix} = \dot{q} \begin{Bmatrix} C_1 \sin q \\ -C_1 \cos q \end{Bmatrix}$$

Notice that the coefficient matrix on the left side is again the Jacobian matrix. For given values of q and $\dot{q}$, this system of linear equations is readily solved (see Appendix A1.4). Alternatively, the velocity coefficients, K_2 and K_3, may be determined by dividing both sides of the equation by $\dot{q}$:

$$\begin{Bmatrix} K_2 \\ K_3 \end{Bmatrix} = \begin{Bmatrix} \dot{A}_2/\dot{q} \\ \dot{A}_3/\dot{q} \end{Bmatrix} = \begin{bmatrix} -C_2 \sin A_2 & -C_3 \sin A_3 \\ C_2 \cos A_2 & C_3 \cos A_3 \end{bmatrix}^{-1} \begin{Bmatrix} C_1 \sin q \\ -C_1 \cos q \end{Bmatrix}$$

The velocity coefficient solution is implemented in the example program that follows the acceleration analysis.

Acceleration Analysis. Time differentiation of the velocity loop equations provides the acceleration loop equations. With the results cast in matrix form and after some rearrangement, the system of linear equations to be solved for the secondary accelerations is

$$\begin{bmatrix} -C_2 \sin A_2 & -C_3 \sin A_3 \\ C_2 \cos A_2 & C_3 \cos A_3 \end{bmatrix} \begin{Bmatrix} \ddot{A}_2 \\ \ddot{A}_3 \end{Bmatrix} = \ddot{q} \begin{Bmatrix} C_1 \sin q \\ -C_1 \cos q \end{Bmatrix} + \dot{q}^2 \begin{Bmatrix} C_1 \cos q \\ C_1 \sin q \end{Bmatrix} + \dot{A}_2^2 \begin{Bmatrix} C_2 \cos A_2 \\ C_2 \sin A_2 \end{Bmatrix} + \dot{A}_3^2 \begin{Bmatrix} C_3 \cos A_3 \\ C_3 \sin A_3 \end{Bmatrix}$$

Notice that the coefficient matrix on the left side is again the Jacobian matrix. The right side involves the squares of the secondary velocities, as well as $\ddot{q}$ and $\dot{q}^2$. These may be expressed in terms of $\dot{q}^2$ by using the velocity coefficients found previously. With these substitutions, the secondary accelerations are

$$\begin{Bmatrix} \ddot{A}_2 \\ \ddot{A}_3 \end{Bmatrix} = \begin{bmatrix} -C_2 \sin A_2 & -C_3 \sin A_3 \\ C_2 \cos A_2 & C_3 \cos A_3 \end{bmatrix}^{-1} \left(\ddot{q} \begin{Bmatrix} C_1 \sin q \\ -C_1 \cos q \end{Bmatrix} \right.$$

$$+ \dot{q}^2 \begin{Bmatrix} C_1 \cos q + K_2^2 C_2 \cos A_2 + K_3^2 C_3 \cos A_3 \\ C_1 \sin q + K_2^2 C_2 \sin A_2 + K_3^2 C_3 \sin A_3 \end{Bmatrix} \Bigg)$$

$$= \ddot{q} \begin{Bmatrix} K_2 \\ K_3 \end{Bmatrix} + \dot{q}^2 \begin{Bmatrix} L_2 \\ L_3 \end{Bmatrix}$$

Note that the inverse of the Jacobian matrix multiplied into the coefficient of $\ddot{q}$ does, in fact, give the velocity coefficients, K_2 and K_3. The inverse of the Jacobian matrix multiplied into the coefficient of $\dot{q}^2$ gives the velocity coefficient derivatives, L_2 and L_3. This last step can be verified by (1) completing the solution for the velocity coefficients (see Appendix A1.5), (2) differentiating the expressions for K_2 and K_3 with respect to q to produce L_2 and L_3, and (3) completing the solution for the coefficient of $\dot{q}^2$, above. When L_2 and L_3 are compared to the coefficients of $\dot{q}^2$, they are found to be the same.

Numerical Values. For a four-bar linkage defined by the link lengths

$C_1 = 5$ in.
$C_2 = 9$ in.
$C_3 = 7$ in.
$C_4 = 10$ in.

as shown in Figure 2.11, the secondary positions, velocity coefficients, and velocity coefficient derivatives are to be evaluated when the assigned crank angle is $q = \pi/3$ radians.

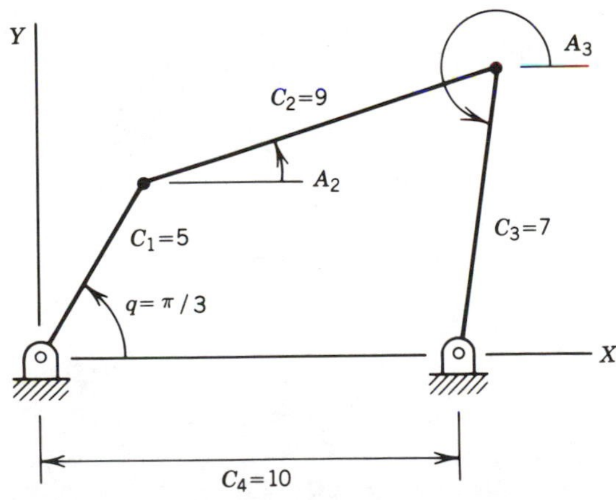

FIGURE 2.11 Four-Bar Linkage for Numerical Example of Section 2.3

The previous analysis applies directly to this problem, and it has been implemented in the computer program that follows shortly. Notice the organization of the program:

1. lines 1000 to 1490 perform all initializations
2. lines 1500 to 1990 perform the position solution
3. lines 2000 to 2150 are the velocity coefficient evaluations
4. lines 2500 to 2640 perform the velocity coefficient derivative evaluations
5. lines 2800 to 2880 are the output
6. lines 3000 to 3110 specify the problem

All the problem data and governing equations are embedded in the final segment of the program, which is called as a subroutine from lines 1520 and 2030. The Matrix Operations Package is also required (see Appendix A1.6).

```
1000 REM  FOUR BAR LINKAGE ANALYSIS
1010 REM  USING NEWTON-RAPHSON SOLUTION
1020 REM  FOR POSITION ANALYSIS
1030 REM
1040 REM  F      = VECTOR OF FUNCTION VALUES
1050 REM  A      = VECTOR OF ANGLE VALUES
1060 REM  D      = VECTOR OF ANGLE CHANGES
1070 REM  FA     = SQUARE MATRIX OF PARTIAL
1080 REM           DERIVATIVES OF F WRT A
1090 REM
1200 DIM F(2),A(2),D(2),FA(2,2)
1210 DIM O1(2),O2(2,3),O3(2,2),O4(2,2)
1220 REM FUNCTION ERROR LIMIT, EF
1230 EF=0.0001
1240 REM MIN CORRECTION, EA
1250 EA=0.0001
1260 REM MAX NUMBER OF ITERATIONS, I9
1270 I9=20
1280 REM
1400 REM  INITIALIZATION
1410 PRINT"ENTER INPUT CRANK ANGLE, Q"
1420 INPUT Q
1430 PRINT "ENTER INITIAL ESTIMATES FOR A2 & A3"
1440 PRINT"A2=?"
1450 INPUT A2
1460 PRINT"A3=?"
1470 INPUT A3
1480 A(1)=A2
1490 A(2)=A3
1500 REM  BEGIN POSITION SOLUTION
1510 FOR I8=1 TO I9
1520 GOSUB 3000
1530 E=0
```

```
1540 FOR I=1 TO 2
1550 IF E<F(I)^2 THEN E=F(I)^2
1560 NEXT I
1570 IF E<EF  THEN GOTO 1980
1580 REM IF E<EF THEN THE SOLUTION HAS CONVERGED
1600 REM  NEWTON RAPHSON DETAILS
1640 FOR I=1 TO 2
1650 FOR J=1 TO 2
1660 O2(I,J)=FA(I,J)
1670 NEXT J
1680 NEXT I
1690 FOR I=1 TO 2
1700 O2(I)=-F(I)
1710 NEXT I
1720 Z1=2
1722 Z2=1
1723 Z3=1
1730 GOSUB 5400
1740 FOR I=1 TO 2
1750 D(I)=O2(I,3)
1760 NEXT I
1780 REM  SEARCH OUT & TEST LARGEST CHANGE
1790 E=0
1800 FOR I=1 TO 2
1810 IF E<D(I)^2 THEN E=D(I)^2
1820 NEXT I
1830 IF E<EA  THEN GOTO 1980
1840 REM  IF E<EA THEN THE MIN STEP
1850 REM  CRITERION HAS BEEN SATISFIED
1860 REM
1870 REM  MAKE THE CHANGE
1880 FOR I=1 TO 2
1890 A(I)=A(I)+D(I)
1900 NEXT I
1910 NEXT I8
1920 REM  IF THE PROCESS GETS TO THIS
1930 REM  POINT, IT HAS FAILED TO CONVERGE
1940 PRINT"POSITION SOLUTION FAILED"
1950 END
1980 A2=A(1)
1990 A3=A(2)
2000 REM  VELOCITY SOLUTION
2010 REM
2020 REM  RE-EVALUATE THE JACOBIAN
2030 GOSUB 3000
2040 REM  SET UP LINEAR EQNS FOR VEL COEF
2050 O2(I,3)=C1*SIN(Q)
2060 O2(2,3)=-C1*COS(Q)
2070 FOR I=1 TO 2
2080 FOR J=1 TO 2
2090 O2(I,J)=FA(I,J)
```

```
2100 NEXT J
2110 NEXT I
2120 REM  Z1=2 PREVIOUSLY SET
2130 GOSUB 5400
2140 K2=O2(I,3)
2150 K3=O2(2,3)
2500 REM  ACCELERATION SOLUTION
2510 REM  RE-EVALUATE THE JACOBIAN
2520 GOSUB 3000
2540 FOR I=1 TO 2
2550 FOR J=1 TO 2
2560 O2(I,J)=FA(I,J)
2570 NEXT J
2580 NEXT I
2590 REM  NOW SET UP COEF OF Q DOT SQUARED
2600 O2(1,3)=C1*COS(Q)+C2*K2^2*COS(A2)+C3*K3^2
     *COS(A3)
2610 O2(2,3)=C1*SIN(Q)+C2*K2^2*SIN(A2)+C3*K3^2
     *SIN(A3)
2620 GOSUB 5400
2630 L2=O2(1,3)
2640 L3=O2(2,3)
2650 REM
2660 REM  RECALL THAT COEF OF Q DOT DOT
2670 REM  ARE K2, K3, THE VEL COEF
2790 REM
2800 REM  OUTPUT
2810 PRINT"         RESULTS"
2820 PRINT"FOR Q=";Q
2830 PRINT"          A2=";A2
2840 PRINT"          A3=";A3
2850 PRINT"          K2=";K2
2860 PRINT"          K3=";K3
2870 PRINT"          L2=";L2
2880 PRINT"          L3=";L3
2890 END
3000 REM  SYSTEM GEOMETRY
3010 C1=5
3020 C2=9
3030 C3=7
3040 C4=10
3050 F(1)=C1*COS(Q)+C2*COS(A(1))+C3*COS(A(2))-C4
3060 F(2)=C1*SIN(Q)+C2*SIN(A(1))+C3*SIN(A(2))
3070 FA(1,1)=-C2*SIN(A(1))
3080 FA(1,2)=-C3*SIN(A(2))
3090 FA(2,1)=C2*COS(A(1))
3100 FA(2,2)=C3*COS(A(2))
3110 RETURN
```

The crank angle is requested as an input in line 1420. This entry will be expected in radians, so the value entered will be 1.0472, the decimal equivalent for $\pi/3$. Initial estimates for A_2 and A_3 are also requested at lines 1450 and 1470. Based on a quick sketch, the coupler link is nearly horizontal while the second crank is near the vertical. Consequently, the initial estimates for A_2 and A_3 are 0.0 and 4.0 radians, respectively. Using only radian measure for angles aids in keeping units correct and otherwise obtaining correct computer results.

When the program is executed, the results are

```
ENTER INPUT CRANK ANGLE, Q
?1.0472
ENTER INITIAL ESTIMATES FOR A2 and A3
A2=?
?0.0
A3=?
?4.0
   RESULTS
FOR Q=1.0472
   A2=.287574457
   A3=4.54223869
   K2=-.214362456
   K3=.548338675
   L2=.342966741
   L3=.495098891
```

From the results, it is evident that the initial estimates were reasonably good. The first velocity coefficient, K_2, is negative, indicating that A_2 decreases as q increases, which is evident by intuition. The second velocity coefficient, K_3, is positive, indicating that both cranks are turning in the same sense.

Coupler Point Position, Velocity, and Acceleration Analysis

The variety of motions possible with the four-bar linkage was one of the reasons for interest in this mechanism. The stationary link does not move at all, and all points on the cranks have purely circular motion. The great variety of possible motions is associated with points on the coupler link, known as *coupler points*. The analysis to be developed here will provide the position, velocity, and acceleration for any point on the coupler link after the secondary kinematic variables have been evaluated by the methods presented previously. The development will be similar to that given in Section 2.2 for the slider–crank connecting rod.

A typical four-bar linkage with the outline of the coupler link added to the kinematic skeleton is shown in Figure 2.12. A body coordinate system, fixed on the coupler link and denoted (U, V), is also shown in the figure. Any particular coupler point P is readily described by giving its body coordinates, U_p and V_p. As the motion progresses, the base coordinates, X_p and Y_p, for the point change but the body coordinates, U_p and V_p, do not change. The position vector for the designated coupler point is denoted

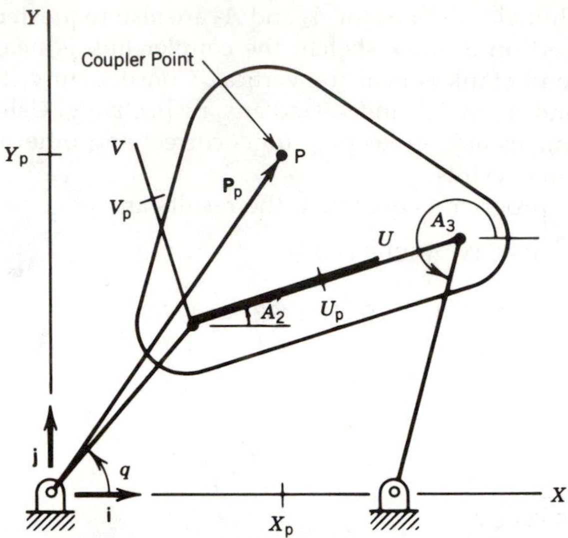

FIGURE 2.12 Coupler Point Position Vector and Body Coordinate System on the Coupler Link

$\mathbf{P}_p$, where the subscript p is again a reminder that the quantity is associated with point P.

$$\begin{aligned}\mathbf{P}_p &= \mathbf{i}X_p + \mathbf{j}Y_p \\ &= \mathbf{i}(C_1 \cos q + U_p \cos A_2 - V_p \sin A_2) \\ &\quad + \mathbf{j}(C_1 \sin q + U_p \sin A_2 + V_p \cos A_2)\end{aligned}$$

For an assigned value of the input crank angle q, and following the associated solution for the secondary kinematic variables, this position vector may be evaluated readily.

The velocity of the coupler point is obtained by differentiating the position vector with respect to time. This differentiation will involve the time derivative $\dot{A}_2$, which can be expressed as $\dot{q}K_2$, with the result

$$\begin{aligned}\mathbf{V}_p &= \mathbf{i}(-\dot{q}C_1 \sin q - \dot{A}_2 U_p \sin A_2 - \dot{A}_2 V_p \cos A_2) \\ &\quad + \mathbf{j}(\dot{q}C_1 \cos q + \dot{A}_2 U_p \cos A_2 - \dot{A}_2 V_p \sin A_2) \\ &= \dot{q}(\mathbf{i}K_{px} + \mathbf{j}K_{py})\end{aligned}$$

where

$$\begin{aligned}K_{px} &= -C_1 \sin q - U_p K_2 \sin A_2 - V_p K_2 \cos A_2 \\ K_{py} &= C_1 \cos q + U_p K_2 \cos A_2 - V_p K_2 \sin A_2\end{aligned}$$

The variables K_{px} and K_{py} are the horizontal and vertical velocity coefficients for the point P. Note that they depend on the secondary kinematic quantities A_2 and K_2. They also depend on the particular coupler point through the body coordinates U_p and V_p.

For the acceleration of the coupler point another time differentiation is required. From the final velocity expression shown previously, the acceleration is

$$\begin{aligned}\mathbf{A}_p &= \ddot{q}(\mathbf{i}K_{px} + \mathbf{j}K_{py}) + \dot{q}^2(\mathbf{i}L_{px} + \mathbf{j}L_{py}) \\ &= \mathbf{i}(\ddot{q}K_{px} + \dot{q}^2L_{px}) + \mathbf{j}(\ddot{q}K_{py} + \dot{q}^2L_{py})\end{aligned}$$

where the velocity coefficient derivatives L_{px} and L_{py} are

$$\begin{aligned}L_{px} &= \frac{d(K_{px})}{dq} \\ &= -C_1 \cos q - U_pL_2 \sin A_2 - U_pK_2^2 \cos A_2 \\ &\quad - V_pL_2 \cos A_2 + V_pK_2^2 \sin A_2 \\ L_{py} &= \frac{d(K_{py})}{dq} \\ &= -C_1 \sin q + U_pL_2 \cos A_2 - U_pK_2^2 \sin A_2 \\ &\quad - V_pL_2 \sin A_2 - V_pK_2^2 \cos A_2\end{aligned}$$

The variables L_{px} and L_{py} are the velocity coefficient derivatives for the horizontal and vertical directions, respectively.

As noted in Section 2.2, vector notation has been used to emphasize the relation between the vector quantities position, velocity, and acceleration, and the scalar position variables, velocity coefficients, and velocity coefficient derivatives. With computer evaluation, the scalar variables are handled easily and there is often no need to evaluate the vector variables. This development again shows how the vector components are expressed in terms of the scalar variables, so they can be evaluated readily if needed.

For design studies, it is often important to determine the path of a coupler point as well as the velocity and acceleration at each point along the path. The path of the point is generated readily simply by plotting the locus of its positions as the input crank angle varies. The previous analysis provides a means to evaluate the velocity (or velocity coefficients) and acceleration (or velocity coefficient derivatives) for each point used to plot the path.

2.4 CONSTRAINTS

When the word *constraint* is used in describing the motion of a mechanical system, it is an admission that there are forces acting in the system that cannot be known before the motion is fully known. These forces act to *constrain* or limit the motion in some way. Consequently, the resulting limitation, instead of those forces that produce that limitation, must be described. If the system motion is to be described using Newton's Second law and the appropriate kinematical relations, discussion of constraints implies that there are force terms required in the sum of forces that cannot be determined until after the equations of motion have been solved! What can be done? The idea of a constraint relation is that these unknown and unknowable forces can be included in the motion description in terms of the *effects* they produce. Often this takes the form of one or more additional equations that describe the result of the unknown forces. An example will help establish this concept.

Consider a two-dimensional problem involving a wheel moving in contact with a horizontal surface. Initially, it appears that the system has two degrees of freedom, and it is convenient to associate one with the position of the center of the wheel, X, and the second with an angular coordinate, A, describing the rotation of the wheel. Without friction between the wheel and the supporting surface, the wheel motion may be any combination of rolling and sliding, so this is completely correct. If, on the other hand, the surface is perfectly rough, there will be no sliding. The phrase "perfectly rough" implies the existence of whatever force is needed to assure that no sliding occurs, but the exact magnitude of that force cannot be determined until the motion is known. For the perfectly rough surface, the *effect* of the friction force is to cause a direct relation between the horizontal displacement, X, and the rotation, A. This relation is

$$dX - RdA = 0$$

or

$$X - RA + Constant = 0$$

Either of these may be called the *equation of constraint,* which has the effect in this example of reducing the number of degrees of freedom from two to one. This reduction is the usual case, but not the only case; read on.

Constraint equations are classified in two types: *holonomic* constraint or *nonholonomic* constraint. These are sometimes considered roughly equivalent to "less difficult" and "more difficult," although these are not necessarily correct interpretations. Constraint relations take one of three forms: (1) finite equalities (non-differential equations), (2) finite inequalities, and (3) differential relations (equations). The first form, the finite equality, can always be used to reduce the number of degrees of freedom of the system. This form and its effect were illustrated earlier. This is the constraint type of greatest interest in the mechanics of machines; it is called a *holonomic constraint*. It will be discussed in detail after the other two types have been described. The second constraint type, the inequality relation, is always a *nonholonomic constraint,* but it usually causes little difficulty. The third type, the differential relation, *may be either a holonomic or nonholonomic constraint,* depending on the exact nature of the relation. The nonholonomic differential constraint is the source of the "less difficult/more difficult" perception.

As an example of the inequality constraint, consider the motion of a projectile toward a spherical target. The equations of motion formulated for this motion will apply only while the projectile is outside the target. After the projectile strikes the target and begins to penetrate, different equations of motion will apply. In terms of a rectangular coordinate system with the target sphere of radius R centered on the origin, if the projectile position is given by (X,Y,Z), the appropriate constraint is simply

$$X^2 + Y^2 + Z^2 > R^2$$

For the motion problem, the differential equations describing the flight of the projectile toward the target are solved by ignoring the effect of the constraint. Rather, the constraint limits the range of applicability of the solution; the solution is valid only for positions satisfying the constraint. If the solution is to be done numerically, after each time step the validity of the constraint is tested. Whenever the constraint is no longer satisfied, the equations of motion must be changed for the penetration problem. This is the typical handling of inequality constraints—a relatively simple type of nonholonomic constraint.

In the case of the differential relation, the constraint condition that it describes may or may not be holonomic. Suppose that the differential relation is in the form

$$C_x\, dX + C_y\, dY + C_z\, dZ = 0$$

If this is the exact differential of a function $F(X,Y,Z)$, then the differential constraint relation is equivalent to

$$\begin{aligned} dF &= \frac{\partial F}{\partial X}\, dX + \frac{\partial F}{\partial Y}\, dY + \frac{\partial F}{\partial Z}\, dZ \\ &= C_x\, dX + C_y\, dY + C_z\, dZ \\ &= 0 \end{aligned}$$

which can be integrated to obtain

$$F(X,Y,Z) = 0$$

Because this last relation is a finite equality, a holonomic constraint condition is described by either the integrated finite form or the original differential form. The conditions under which a differential form can be integrated are described in texts on the theory of differential equations.[2] In the event that no function F exists whose differential is the given differential constraint equation, then the constraint is nonholonomic.

In mechanical systems, nonholonomic constraints are most often associated with rolling contacts. There are three classical problems commonly used to illustrate nonholonomic constraints: (1) a sphere rolling on a rough plane, (2) a coin rolling on a rough plane, and (3) two wheels that have a common axle rolling on a rough plane. A variation on the third problem is

[2] See E. L. Ince, *Ordinary Differential Equations*, pp. 52–55, Dover, New York, 1944.

the motion on ice of two skate blades rigidly joined by a rod that is perpendicular to them. Only the sphere rolling on a rough plane will be discussed here.

2.4.1 Sphere Rolling on a Rough Plane

Consider a solid sphere of radius R, rolling without slipping on a rough plane. For any rigid body moving in three dimensions, there are six degrees of freedom requiring six generalized coordinates. It is common to associate three of these generalized coordinates with the location of the center of mass, while the other three are associated with angular coordinates that describe the orientation of the body. For the sphere under discussion, the six generalized coordinates are reduced to five because the sphere is to remain in contact with the rough plane; one of the translational coordinates for the center of mass is fixed by contact with the plane. How many degrees of freedom does this system have?

Consider a stationary rectangular cartesian coordinate system with X– and Y–axes in the plane and the Z–axis normal to the plane; the associated unit vectors are **i**, **j**, and **k**. The angular velocity of the sphere will be

$$\boldsymbol{\omega} = \omega_x \mathbf{i} + \omega_y \mathbf{j} + \omega_z \mathbf{k}$$

where ω_x, ω_y, and ω_z are each functions of $\dot{A}_1$, $\dot{A}_2$, and $\dot{A}_3$, the time derivatives of the angular coordinates. If X and Y denote the center of mass coordinates and R is the radius of the sphere, the velocity of the point on the surface of the sphere in contact with the plane is $\mathbf{V}_p$:

$$\mathbf{V}_p = \dot{X}\mathbf{i} + \dot{Y}\mathbf{j} + \boldsymbol{\omega}\mathrm{x}(-R\mathbf{k})$$

The sphere remains in contact with the plane, so that Z–component of this velocity equation must be identically zero. The constraint expressing rolling without slipping requires that the other two components also be zero, resulting in two differential equations of constraint:

$$f_1(\dot{X}, \dot{A}_1, \dot{A}_2, \dot{A}_3) = 0$$

$$f_2(\dot{Y}, \dot{A}_1, \dot{A}_2, \dot{A}_3) = 0$$

These two equations of constraint are, in fact, nonholonomic. Thus, there are five generalized coordinates required to describe the system configuration (X, Y, A_1, A_2, A_3), but the system has only three degrees of freedom. In general, for systems involving nonholonomic constraints, the number of coordinates required are more than the number of degrees of freedom.

The presentation of the rolling sphere problem has, of necessity, been somewhat sketchy so as to avoid the need for Euler angles and other advanced concepts. For those who wish to pursue the matter further, these three classical examples are discussed in a number of other texts.[3–7]

[3] H. Goldstein, *Classical Mechanics*, p. 13, Addison–Wesley, Reading, MA, 1959.

[4] D. T. Greenwood, *Principles of Dynamics*, pp. 234–235, Prentice–Hall, Englewood Cliffs, NJ, 1965.

Two-Dimensional Rolling Constraint

One of the common constraints encountered in the mechanics of machines is the two-dimensional *rolling constraint*. This refers to the two-dimensional problem of one body rolling over another, such as a wheel rolling along the ground or one gear rolling on another. The problem of the wheel rolling along the ground was briefly discussed at the beginning of this section. In that case, the equation of constraint was first written in differential form, and then in the equivalent finite (non-differential) form. For two-dimensional rolling problems such as this, the equation of constraint can always be written in finite form, and thus, the two-dimensional rolling constraint is always holonomic. This means that the constraint equation can be used to eliminate one of the coordinates as well as a degree of freedom, keeping the number of degrees of freedom and the number of generalized coordinates equal.

In the earlier discussion, to assure rolling motion, certain surfaces were described as *perfectly rough.* The term perfectly rough implies sufficient friction to assure no sliding under all conditions. In reality, there are no perfectly rough plane surfaces. Problems in elementary mechanics are often solved by first assuming that there is no sliding and, after the motion has been determined on the basis of this assumption, calculating the magnitude of the required friction force. If this required friction force exceeds that available, then the system must slip, and the problem must be reworked with slipping included.

One device often used in machinery to assure that slip does not occur, and thus to approximate the perfectly rough surface, is to cut mating teeth on the two bodies where they are in contact. If the two bodies are wheels, the result is a pair of gears. If one body is a straight surface and the other is a wheel, with the addition of teeth they become a rack and pinion. For properly formed teeth, the teeth do not affect the motion but only serve to assure that there is no sliding. The matter of proper tooth forms will be considered in Chapter 5. For the present, it will be sufficient to consider a gear as a perfectly rough wheel. The effective radius of this wheel is known in gear terminology as the *pitch radius.*

The introduction of a rolling constraint modifies the established procedure for kinematic analysis. When constraints are involved, the combined set of position loop equations and constraint equations must be solved simultaneously for the secondary variables. For velocity and acceleration analyses, the constraint equation(s) must be differentiated along with the

---

[5] C. W. Kilmister and J. E. Reeve, *Rational Mechanics,* p. 200, Elsevier, Amsterdam/New York, 1966.

[6] B. J. Torby, *Advanced Dynamics for Engineers,* pp. 261–262, Holt, Rinehart and Winston, New York, 1984.

[7] J. L. Synge, "Classical Dynamics," pp. 40–41, in *Encyclopedia of Physics,* S. Flugge, ed., Vol. III/1, Springer–Verlag, New York/Berlin, 1960.

position loop equations and again included in the solutions. The kinematic analysis for a system involving a rolling constraint will be illustrated in the following example. For this example, the original assembly configuration plays an important role, as is often the case.

2.4.2 Rolling Constraint Example

Consider a system involving a wheel rolling without slip on a plane, as shown in Figure 2.13. The wheel is driven by pulling on the connecting rod that passes through a pivoted eye block. The radii, R and r, are known. Figure 2.13(a) shows the system as originally assembled. In the original assembly configuration, the position of the contact point, S_o, is known; the subscript o is used to denote original assembly values. The connecting rod pivot is initially directly above the point of contact between the wheel and the supporting plane, and the connecting rod extends from that pivot to the eye block pivot. As the connecting rod is drawn through the eye block, the wheel rolls to the left as shown in Figure 2.13(b). Set up all equations required to determine the secondary kinematic variables.

First consider the information known for the original assembly position. There is enough information to determine the connecting rod length, B_o, and the initial inclination angle, C_o, as follows:

$$B_o = \sqrt{S_o^2 + (R + r)^2}$$

$$C_o = \text{Arctan}[(R + r)/S_o]$$

The original rotation angle is $A_o = 0$.

Now look at the displaced configuration. The system has one degree of freedom, here associated with the length of rod drawn through the eye

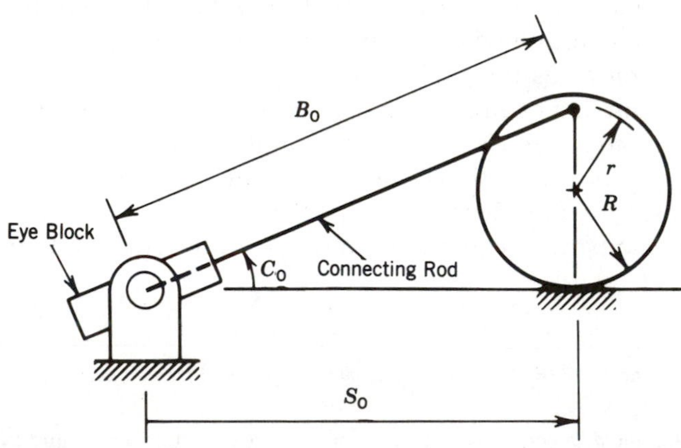

FIGURE 2.13 **(a)** Initial Configuration for Rolling Constraint Example

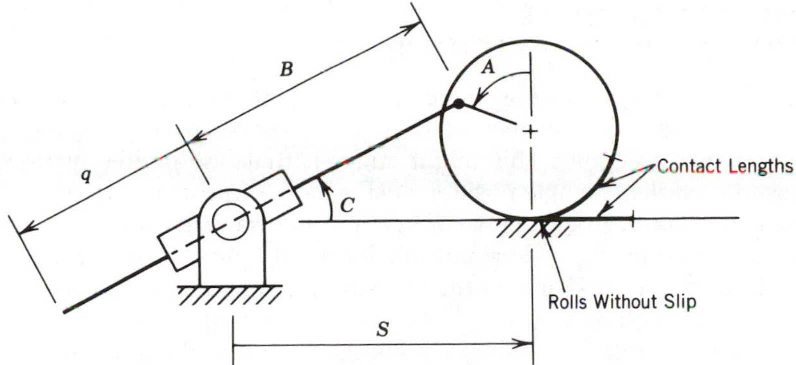

FIGURE 2.13 (b) Displaced Position for Rolling Constraint Example

block, q. There is one position loop, from which the position loop equations are

$$(B_o - q)\cos C + r\sin A - S = 0$$

$$(B_o - q)\sin C - r\cos A - R = 0$$

The unknowns appearing in the two loop equations are C, A, and S; it is evident that another equation is required. The remaining equation is the constraint relation expressing the fact that the wheel rolls without slipping. On the plane surface, the point of contact moves by the amount $(S_o - S)$ from the original configuration to that shown in Figure 2.13(b). On the surface of the wheel, the length of the arc through which the contact point moves is RA, and these two lengths must be equal for rolling without slip, so

$$S_o - S - RA = 0$$

This is the holonomic constraint expressing rolling without slipping for this problem. The full set of equations that must be solved for the secondary kinematic variables C, A, and S are

$$(B_o - q)\cos C + r\sin A - S = 0$$

$$(B_o - q)\sin C - r\cos A - R = 0$$

$$S_o - S - RA = 0$$

A closed-form solution may be sought for the three simultaneous, nonlinear equations, or a numerical solution may be determined using the Newton–Raphson method. For the Newton–Raphson solution, it is necessary to consider three unknowns; this is a simple matter when the program is written in matrix form, as in Appendix A2. The analysis of velocities and accelerations proceeds in the usual fashion, with the constraint relation differentiated along with the loop equations.

2.5 MULTILOOP SINGLE DEGREE OF FREEDOM MECHANISMS

All of the mechanisms considered to this point have been described in terms of a single position vector loop (two scalar equations) and have had one degree of freedom. This might suggest that one degree of freedom necessarily implies a single position vector loop, but that is not true. In this section, multiloop, single degree of freedom mechanisms will be addressed by means of an example. For multiple loops, the procedure for kinematic analysis is much the same as for the single loop case, except that more equations are required.

In writing the position loop equations, note that the loops used must be independent. In the example to follow, a kinematic analysis will be developed for a single degree of freedom mechanism involving two independent position vector loops. Additional loops can be identified, but they will not be independent. For independence, each loop should include some segment not a part of any other loop.

2.5.1 Four-Bar/Toggle Linkage

Figure 2.14 shows the kinematic skeleton for a mechanism sometimes used for a punch press. For this purpose, the input crank pivoted at the upper left is driven by a prime mover, usually with a large flywheel, while the punch tool is attached to the slider at the bottom of the diagram. For each revolution of the input crank, the tool executes one punch cycle. The link lengths R, C_2, C_3, and C_5 are known, as is the input crank pivot location, (X_1, Y_1). The single degree of freedom is associated with the input crank rotation angle, q. Set up the equations governing the secondary variable positions and velocity coefficients.

Position Analysis. The two loops to be used are identified in Figure 2.14. There is a third loop, obtained by tracing around the outside of the mechanism, but the equations describing the third loop are merely combinations of the equations determined for the first two loops; they contain no new information and are, therefore, *dependent*. The position loop equations are determined by summing displacements around each of the two independent loops. With two vector loops, four scalar position loop equations are obtained, as shown here:

$$f_1 = X_1 + R \cos q - C_2 \cos A_2 - C_3 \sin B_3 = 0$$

$$f_2 = Y_1 + R \sin q + C_2 \sin A_2 - C_3 \cos B_3 = 0$$

$$f_3 = C_3 \sin B_3 - C_5 \sin B_5 = 0$$

$$f_4 = Y - C_3 \cos B_3 - C_5 \cos B_5 = 0$$

These are four simultaneous, nonlinear equations in the four unknowns A_2, B_3, B_5 and Y. For a Newton–Raphson solution for this system of

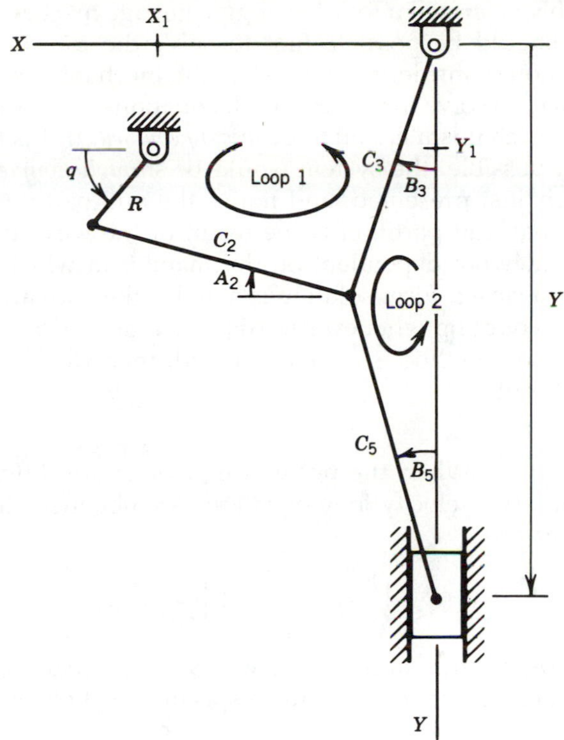

FIGURE 2.14 Four-Bar/Toggle Linkage

equations, the vector of secondary kinematic unknowns is

$$\{S\} = \text{Col}(A_2, B_3, B_5, Y)$$

and the Jacobian matrix is

$$\left[\frac{\partial f}{\partial s}\right] = \left[\begin{array}{cc|cc} C_2 \sin A_2 & -C_3 \cos B_3 & 0 & 0 \\ C_2 \cos A_2 & C_3 \sin B_3 & 0 & 0 \\ \hline 0 & C_3 \cos B_3 & -C_5 \cos B_5 & 0 \\ 0 & C_3 \sin B_3 & C_5 \sin B_5 & 1.0 \end{array}\right]$$

As discussed previously, the Newton–Raphson solution is achieved by an iterative refinement of the initial solution estimate.

Before leaving the position solution, there is one other noteworthy point. Look again at Figure 2.14. The upper loop is recognized as a four-bar linkage that is determined completely when q is specified. The second loop is a slider–crank mechanism for which B_5 and Y are determined once B_3 is known. (Because the slider–crank only works through much less than a full crank revolution, this is sometimes called a toggle linkage, and the whole assembly is termed a four-bar/toggle linkage mechanism.) This suggests that it should be possible first to solve the four-bar linkage loop equations without considering the rest of the mechanism; then, with that solution known, to solve the slider–crank equations. Because this is indeed possible, this mechanism is said to be *weakly coupled*. If this type of separation were not possible, the system would be *strongly coupled*. Look at the Jacobian matrix just presented, and notice the null partition in the upper right corner; that null partition is the result of the weak coupling. Weak coupling is somewhat dependent on the manner in which the equations are written. If a closed-form solution is sought, then advantage should be taken of weak coupling whenever it exists. If a numerical solution by the Newton–Raphson method is to be employed, then weak coupling is usually not significant.

Velocity Analysis. When the position equations are differentiated with respect to time, the velocity loop equations are obtained. In matrix form, these equations are

$$\left[\frac{\partial f}{\partial s}\right]\{\dot{S}\} = -\dot{q}\left\{\frac{\partial f}{\partial q}\right\}$$

Because the coefficient matrix on the left is the Jacobian matrix that was presented before, all that remains to be specified is the vector on the right side:

$$\left\{\frac{\partial f}{\partial q}\right\} = \begin{Bmatrix} -R\sin q \\ R\cos q \\ 0 \\ 0 \end{Bmatrix}$$

These equations may be solved readily for the secondary variable velocities or for the velocity coefficients. The preceding position and velocity coefficient solutions are demonstrated with the following numerical values.

Numerical Values. Consider a four-bar/toggle linkage with the following link lengths:

$$\begin{aligned} R &= 5 \text{ in.} \\ C_2 &= 14 \text{ in.} \\ C_3 &= 27 \text{ in.} \\ C_5 &= 44 \text{ in.} \end{aligned}$$

and that has the crank pivot at

$$\begin{aligned} X_1 &= 13.6 \text{ in.} \\ Y_1 &= 18 \text{ in.} \end{aligned}$$

For this system, the secondary kinematic variable values and the velocity coefficients will be determined when the input crank position is $q = 1.0$ radian.

The analysis described has been implemented in a computer program, a listing of which follows shortly. The program requests the input crank angle (line 1250) and initial estimates for A_2, B_3, B_5, and Y (lines 1270, 1290, 1310, and 1330). The Newton–Raphson solution logic is in lines 1400 to 1840, and the position solution results are in lines 1900 to 2050. Note that the basis for terminating the iteration successfully is printed by either line 2040 or 2050; if the termination occurs because of a failure to converge, this would cause line 1860 to print as a warning. The velocity coefficient analysis is implemented in lines 2100 through 2270, with results in lines 2290 through 2330. The problem specification, including the link lengths, residual evaluation, and evaluation of the Jacobian matrix, is given in lines 4000 through 4310.

The program has been executed for $q = 1.0$ radian, and the initial estimates are indicated below. The results are:

```
ENTER Q
? 1.0
ENTER INITIAL ESTIMATE FOR A2=S(1)
?.2
ENTER INITIAL ESTIMATE FOR B3=S(2)
?.1
ENTER INITIAL ESTIMATE FOR B5=S(3)
?.1
ENTER INITIAL ESTIMATE FOR Y=S(4)
?60
SOLUTION VALUES

S(1)=A2=.336015748
S(2)=B3=.114487188
S(3)=B5=.0701581542
S(4)=Y =70.7155059

FUNCTION VALUES
F(1)=4.35589391E-05
F(2)=3.06464644E-04
F(3)=-2.12726227E-05
F(4)=5.04734926E-04
NUMBER OF ITERATIONS =3
TERMINATION BASED ON MAGNITUDE OF RESIDUAL

VELOCITY COEFFICIENTS
K2=-.161312715
K3=-.184616221
K5=-.112823145
KY=.91742608
```

Note that the angles A_2, B_3, and B_5 all have small positive values, whereas Y is significantly larger than the initial estimate. As indicated by the signs of the velocity coefficients, all of the angles decrease as q increases, while the slider is moving down, $K_Y > 0$. These observations are in accord with an intuitive understanding of the working of the machine.

The program listing follows. The Matrix Operations Package from Appendix A1 must be added to the listing given.

```
1000 REM  FOUR BAR / TOGGLE LINKAGE
1010 REM  GOVERNING EQUATIONS AND
1020 REM  JACOBIAN MATRIX ELEMENTS ARE
1030 REM  PROGRAMMED AT LINE 4000
1040 REM
1050 DIM  F(4),S(4),DS(4),JM(4,4)
1060 DIM  O1(4),O2(4,5),O4(4,4)
1070 REM
1100 REM  SET CONTROLS
1110 REM  E1 = MAX ALLOWABLE RESIDUAL
1120 REM  E2 = MIN ALLOWABLE ADJUSTMENT
1130 REM  I9 = MAX NUMBER OF ITERATIONS
1140 E1=0.0001
1150 E2=0.0001
1160 I9=10
1170 REM
1200 REM  ENTER CRANK POSITION AND
1210 REM  INITIAL ESTIMATE FOR UNKNOWNS
1230 HOME
1240 PRINT"ENTER Q"
1250 INPUT Q
1260 PRINT"ENTER INITIAL ESTIMATE FOR A2=S(1)"
1270 INPUT S(1)
1280 PRINT "ENTER INITIAL ESTIMATE FOR B3=S(2)"
1290 INPUT S(2)
1300 PRINT"ENTER INITIAL ESTIMATE FOR B5=S(3)"
1310 INPUT S(3)
1320 PRINT"ENTER INITIAL ESTIMATE FOR Y=S(4)"
1330 INPUT S(4)
1340 REM
1400 REM BEGIN THE ITERATION
1410 FOR I=1 TO I9
1420 REM EVALUATE THE RESIDUAL AND THE JACOBIAN
1440 GOSUB 4000
1500 REM EVALUATE THE ADJUSTMENT
1510 FOR K=1 TO 4
1520 O2(K,5)=-F(K)
1530 FOR L=1 TO 4
1540 O2(K,L)=JM(K,L)
1550 NEXT L
1560 NEXT K
1570 Z1=4
1572 Z2=1
1574 Z3=1
1580 GOSUB 5400
1590 FOR K=1 TO 4
1600 DS(K)=O2(K,5)
```

```
1610 NEXT K
1620 REM
1700 REM  TEST FOR TERMINATION
1710 REM  FORM NORMS FOR RESIDUAL AND ADJUSTMENT
1720 N1=F(1)^2+F(2)^2+F(3)^2+F(4)^2
1730 N2=DS(1)^2+DS(2)^2+DS(3)^2+DS(4)^2
1740 IF N1>E1 THEN GOTO 1770
1750 V=1
1760 GOTO 1900
1770 IF N2>E2 THEN GOTO 1800
1780 V=2
1790 GOTO 1900
1800 REM  UPDATE SOLUTION ESTIMATES
1810 FOR K=1 TO 4
1820 S(K)=S(K)+DS(KS)
1830 NEXT K
1840 NEXT I
1850 REM
1860 PRINT"MAX NUMBER OF ITERATIONS"
1870 REM
1900 REM  OUTPUT
1910 PRINT"SOLUTION VALUES"
1920 PRINT
1930 PRINT"S(1)=A2=";S(1)
1940 PRINT"S(2)=B3=";S(2)
1950 PRINT"S(3)=B5=";S(3)
1960 PRINT"S(4)=Y =";S(4)
1970 PRINT
1980 PRINT"FUNCTION VALUES"
1990 PRINT"F(1)=";F(1)
2000 PRINT"F(2)=";F(2)
2010 PRINT"F(3)=";F(3)
2020 PRINT"F(4)=";F(4)
2030 PRINT"NUMBER OF ITERATIONS =";I
2040 IF V=1 THEN PRINT"TERMINATION BASED ON
     MAGNITUDE OF RESIDUAL
2050 IF V=2 THEN PRINT"TERMINATION BASED ON
     MAGNITUDE OF ADJUSTMENT
2060 REM
2100 REM  VELOCITY COEFFICIENT ANALYSIS
2110 REM
2120 REM  SET UP RIGHT SIDE VECTOR
2130 02(I,5)=R*SIN(Q)
2140 02(2,5)=-R*COS(Q)
2150 02(3,5)=0
2160 02(4,5)=0
2170 REM RE-EVALUATE JACOBIAN
2180 GOSUB 4000
2190 REM  SET COEF MATRIX EQUAL TO JACOBIAN
2200 FOR I=1 TO 4
2210 FOR J=1 TO 4
```

```
2220 O2(I,J)=JM(I,J)
2230 NEXT J
2240 NEXT I
2250 REM  SOLVE FOR VELOCITY COEFFICIENTS
2260 GOSUB 5400
2270 REM  PRINT VELOCITY COEFFICIENT RESULTS
2280 PRINT
2290 PRINT"VELOCITY COEFFICIENTS"
2300 PRINT"K2=";O2(1,5)
2310 PRINT"K3=";O2(2,5)
2320 PRINT"K5=";O2(3,5)
2330 PRINT"KY=";O2(4,5)
2340 END
2350 REM
4000 REM  PROBLEM GEOMETRY
4010 R=5
4020 C2=14
4030 C3=27
4040 C5=44
4050 X1=13.6
4060 Y1=18
4100 REM  FUNCTION EVALUATIONS
4110 F(1)=X1+R*COS(Q)-C2*COS(S(1))-C3*SIN(S(2))
4120 F(2)=Y1+R*SIN(Q)+C2*SIN(S(1))-C3*COS(S(2))
4130 F(3)=C3*SIN(S(2))-C5*SIN(S(3))
4140 F(4)=S(4)-C3*COS(S(2))-C5*COS(S(3))
4150 REM  JACOBIAN EVALUATION
4160 JM(1,1)=C2*SIN(S(1))
4170 JM(1,2)=-C3*COS(S(2))
4180 JM(1,3)=0
4190 JM(1,4)=0
4200 JM(2,1)=C2*COS(S(1))
4210 JM(2,2)=C3*SIN(S(2))
4220 JM(2,3)=0
4230 JM(2,4)=0
4240 JM(3,1)=0
4250 JM(3,2)=C3*COS(S(2))
4260 JM(3,3)=-C5*COS(S(3))
4270 JM(3,4)=0
4280 JM(4,1)=0
4290 JM(4,2)=C3*SIN(S(2))
4300 JM(4,3)=C5*SIN(S(3))
4310 JM(4,4)=1.0
4320 RETURN
4330 END
```

As illustrated here, the existence of multiple independent loops increases the number of equations required in each of the position, velocity, and acceleration solutions. Otherwise, it is exactly like any other single degree of freedom system analysis.

2.6 GENERAL KINEMATIC ANALYSIS FOR SINGLE DEGREE OF FREEDOM MECHANISMS

The kinematic analysis for a single degree of freedom mechanism has been presented by means of examples for several important cases. At this point a general analysis will be presented that will include all of the previous examples as special cases and will apply to all other single degree of freedom mechanisms as well. Recall that the idea of a *primary* kinematic variable or *generalized coordinate* associated with the degree of freedom was introduced in Section 1.2. The remaining kinematic variables were labeled as *secondary* variables. Consequently, the notation for this general analysis is

q = generalized coordinate

s_i = secondary coordinate, $i = 1, 2, \ldots N_2$

Position Analysis. The position vector loop equations, written in scalar form, and the constraint equations, if any are required, form a system of N_2, simultaneous, nonlinear equations:

$$f_1(q, s_1, s_2, \ldots s_{N2}) = 0$$
$$f_2(q, s_1, s_2, \ldots s_{N2}) = 0$$
$$\vdots \qquad\qquad \vdots$$

This system of equations must be solved either in closed form or by numerical means. For the numerical solution by the Newton–Raphson method, the solution vector $\{S\}$ and the residual vector $\{F\}$ each contain N_2 elements:

$$\{S\} = \mathrm{Col}(s_1, s_2, s_3, \ldots, s_{N2})$$
$$\{F\} = \mathrm{Col}(f_1, f_2, f_3, \ldots, f_{N2})$$

The Jacobian matrix is formed by differentiating the position loop functions, f_i, with respect to the secondary coordinates, s_i. Note the locations for the various derivatives in the Jacobian matrix:

$$\left[\frac{\partial f}{\partial s}\right] = \begin{bmatrix} \frac{\partial f_1}{\partial s_1} & \frac{\partial f_1}{\partial s_2} & \frac{\partial f_1}{\partial s_3} & \cdots & \frac{\partial f_1}{\partial s_{N2}} \\ \frac{\partial f_2}{\partial s_1} & \frac{\partial f_2}{\partial s_2} & \frac{\partial f_2}{\partial s_3} & \cdots & \frac{\partial f_2}{\partial s_{N2}} \\ \vdots & \vdots & \vdots & & \vdots \\ \frac{\partial f_{N2}}{\partial s_1} & \frac{\partial f_{N2}}{\partial s_2} & \frac{\partial f_{N2}}{\partial s_3} & \cdots & \frac{\partial f_{N2}}{\partial s_{N2}} \end{bmatrix}$$

As demonstrated previously, the Jacobian matrix plays a most significant role in the Newton–Raphson solution of the position equations, the velocity solution, and the acceleration solution. From this point on, the position solution is assumed to be complete, so that values are available for all position variables.

Velocity Analysis. The velocity loop equations are obtained by differentiating the position loop equations with respect to time, including the constraints if any were required. Expressed in matrix form, this is

$$\left[\begin{array}{c:c} \dfrac{\partial f}{\partial q} & \dfrac{\partial f}{\partial s} \end{array}\right] \left\{\begin{array}{c} \dot{q} \\ \hdashline \dot{S} \end{array}\right\} = \{0\}$$

$$(N_2 \times 1) \quad (N_2 \times N_2) \quad (1 \times 1)$$

$$(N_2 \times 1)$$

The elements of the coefficients matrix will usually be functions of the position variables, but these are known from the current position analysis. This partitioned matrix relation, however, may be separated into a system of linear simultaneous algebraic equations to be solved for the secondary velocities.

$$\left[\frac{\partial f}{\partial s}\right] \{\dot{S}\} = -\dot{q} \left\{\frac{\partial f}{\partial q}\right\}$$

Alternatively, the system may be rearranged to be solvable for the velocity coefficients.

$$\left[\frac{\partial f}{\partial s}\right] \{K_s\} = -\left\{\frac{\partial f}{\partial q}\right\}$$

Acceleration Analysis. A second time differentiation is required for the accelerations. In terms of the velocity coefficients (now known either numerically or in functional form), the velocity relation is

$$\{\dot{S}\} = \dot{q}\{K_s\}$$

which is readily differentiated using the chain rule:

$$\begin{aligned} \{\ddot{S}\} &= \ddot{q}\{K_s\} + \dot{q}^2 \frac{d(\{K_s\})}{dq} \\ &= \ddot{q}\{K_s\} + \dot{q}^2\{L_s\} \end{aligned}$$

If the velocity coefficients were determined in functional form, they can be differentiated to determine the velocity coefficient derivatives directly. If the velocity coefficients were determined numerically, this differentiation is not possible. For a numerical approach to evaluate the velocity coefficient derivatives, consider first the relation defining the column vector of velocity coefficients:

$$[J]\{K_s\} = -\left\{\frac{\partial f}{\partial q}\right\}$$

where $[J]$ is the Jacobian matrix. That matrix is known explicitly, as is the column vector on the right side. Taking an ordinary derivative of this equation with respect to q gives

$$\frac{d[J]}{dq}[K_s] + [J]\frac{d[K_s]}{dq} = -\frac{d}{dq}\left\{\frac{\partial f}{\partial q}\right\}$$

The velocity coefficient derivatives appear as a factor in the second term, so the equation is solved for that term,

$$[J][L_s] = -\frac{d[J]}{dq}[K_s] - \frac{d}{dq}\left\{\frac{\partial f}{\partial q}\right\}$$

Thus, the vector of velocity coefficient derivatives may be obtained as the solution of a system of linear equations, again with the Jacobian as the coefficient matrix. Correct interpretation of the right side of this equation requires careful observance of the rules for ordinary and partial differentiation.

Position, Velocity, and Acceleration for Other Points. Consider a point P on a mechanism member in curvilinear motion. Let (X_o, Y_o) be the coordinates of a point on the same body whose position is readily expressed in terms of the generalized coordinate and the secondary coordinates. Using this point as an origin, consider a body coordinate system (U, V) on the body containing P, in which (U_p, V_p) are the body coordinates of P. The base coordinates of P are (X_p, Y_p), which are readily expressed as

$$X_p = X_o + U_p \cos A - V_p \sin A$$

$$Y_p = Y_o + U_p \sin A + V_p \cos A$$

where A is the angle between the X–axis and the U–axis. The angle A is either the primary variable, a secondary coordinate, or can be determined from them. This expression is readily evaluated when the secondary coordinates are known.

The velocity of point P is determined by differentiating the two position equations with respect to time, and the acceleration requires yet another differentiation. Velocity coefficients and velocity coefficient derivatives are defined for the point P in the obvious manner. In this way, the position, velocity, and acceleration for any point P are readily expressed in terms of the generalized coordinate, q, and its velocity and acceleration, $\dot{q}$ and $\ddot{q}$.

The purpose of this demonstration is not to provide general formulae for use in problem solving. Rather, the reasons for the general analysis are (1) to show that the same processes, resulting in equations of the same form, apply for all closed-loop, single degree of freedom mechanisms and (2) to focus on the analytical process, independent of any particular system. If the final results for a particular problem do not have the forms shown in these general results, that suggests that an error has been made.

2.7 CONCLUSION

The basic approach to the kinematic analysis of single degree of freedom mechanism has been presented in this chapter. That process begins with the selection of a primary coordinate to be associated with the degree of freedom. The mechanism position is described by position equations that are the scalar components of the closed position vector loops and are written in terms of the primary coordinate and such secondary coordinates as may be required. These equations can be solved to provide values for the secondary variables after a value is assigned to the primary variable. The velocities and accelerations of the secondary variables are determined from the differentiated position loop equations.

The velocity of any secondary variable is directly proportional to the velocity of the primary variable, and this makes possible use of a position-dependent velocity coefficient to express the secondary velocity in terms of the primary velocity. The acceleration of any secondary variable can be expressed as the sum of two terms—the first, the product of the velocity coefficient with the acceleration of the primary variable, and the second, the product of the velocity coefficient derivative with the square of the primary velocity. The required velocity coefficient derivative is the derivative of the velocity coefficient with respect to the primary variable.

For any point on a moving body, base coordinates are readily expressed in terms of the primary variable, the secondary variables, and the body coordinates for that point of interest. When these relations are differentiated with respect to time, the base coordinate velocity and acceleration are obtained. For any such point in two-dimensional motion, there are two velocity coefficients and two velocity coefficient derivatives—one associated with each direction.

Writing the position loop equations is a process unique to each mechanism. After that, the solutions for the position loop equations, velocity coefficient equations, and equations for velocity coefficient derivatives may or may not be possible in closed form, although numerical solutions are generally available.

If the ideas summarized here have been mastered, there should be little difficulty in the kinematic analysis of any single degree of freedom mechanism. These ideas play a major role in the later static and dynamic analysis of such mechanisms, so kinematic analysis is only the beginning of their application.

REFERENCES

Crandall, S. H., Carnopp, D. C., Kurtz, E. F., and Pridmore–Brown, D. C., *Dynamics of Mechanical and Electromechanical Systems.* New York: McGraw–Hill, 1968.

Goldstein, H., *Classical Mechanics.* Reading, MA: Addison–Wesley, 1959.

Paul, B., *Kinematics and Dynamics of Planar Machinery*. Englewood Cliffs, NJ: Prentice–Hall, 1979.

PROBLEM SET

2.1 The device shown was used in early, mechanical analog computation. The dimension A and B are known constants, and q is an assigned position.

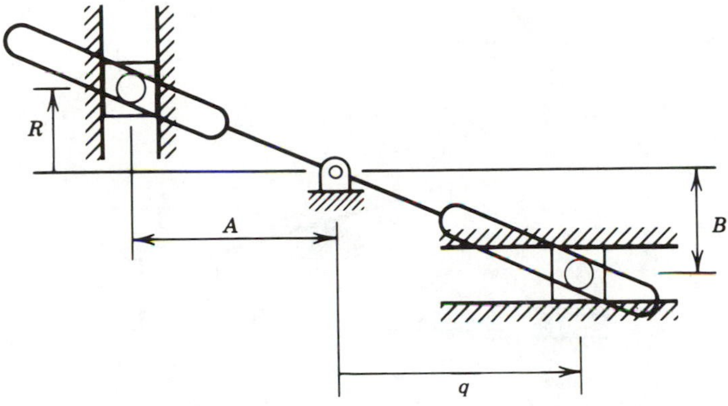

a. Express the output displacement, R, in terms of q, A, and B;

b. Determine the velocity coefficient, $K = dR/dq$;

c. Determine the velocity coefficient derivative, $L = dK/dq$;

d. Given that $A = 5$ in. and $B = 1.75$ in., evaluate R, K, and L for $q = 3.5$ in.;

e. Using the data from part d, evaluate $\dot{R}$ and $\ddot{R}$ for $\dot{q} = -2.5$ in./s, $\ddot{q} = 12.0$ in./s^2.

2.2 The figure shows a mechanism used on jet fighter aircraft to cause a rapid loss in air speed ("dive brakes"). A panel that is initially flush with the fuselage surface, view (a), is forced out into the airstream, view (b), by a hydraulic actuator. View (c) shows a schematic diagram of the system where S is the actuator displacement, B is the hinge angle, and A is the pushrod angle. The dimensions R, C_1, C_2, C_3, and C_4 are known constants. The panel is flush with the fuselage when $S = 0$.

a. Considering the actuator displacement S as the primary variable, write the position loop equations for this mechanism (do not solve);

b. Determine expressions for the velocity coefficients $K_a = dA/dS$ and $K_b = dB/dS$, remembering that S and $\dot{S}$ are assigned and that A and B are considered known from the solution for part a;

c. Determine the velocity coefficient derivatives $L_a = dK_a/dS$ and $L_b = dK_b/dS$.

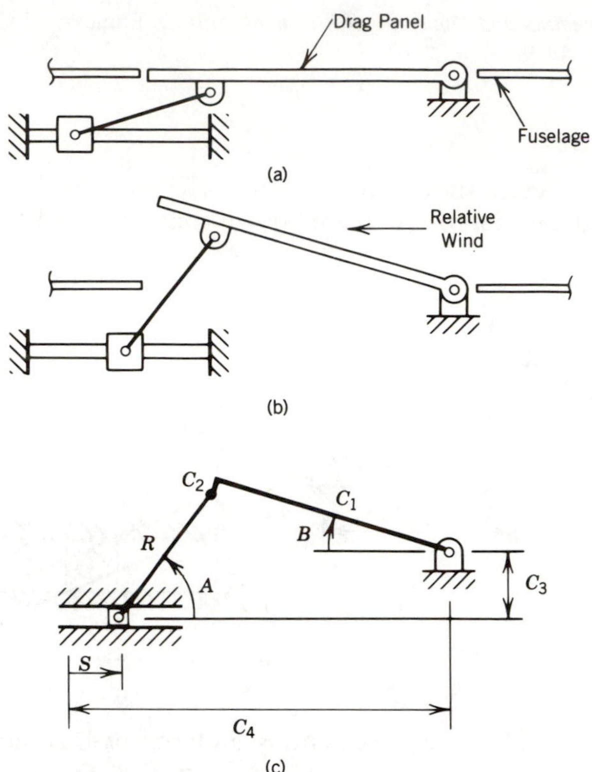

2.3 A retractable headlight mechanism for a sports car is shown in the figure. A motor driven nut at the left pivot causes the screw to move, raising or lowering the headlight. The dimensions C_1, C_2, and C_3 are known. Considering the screw position, S, as the primary variable, the secondary kinematic variables are the angles A and B.

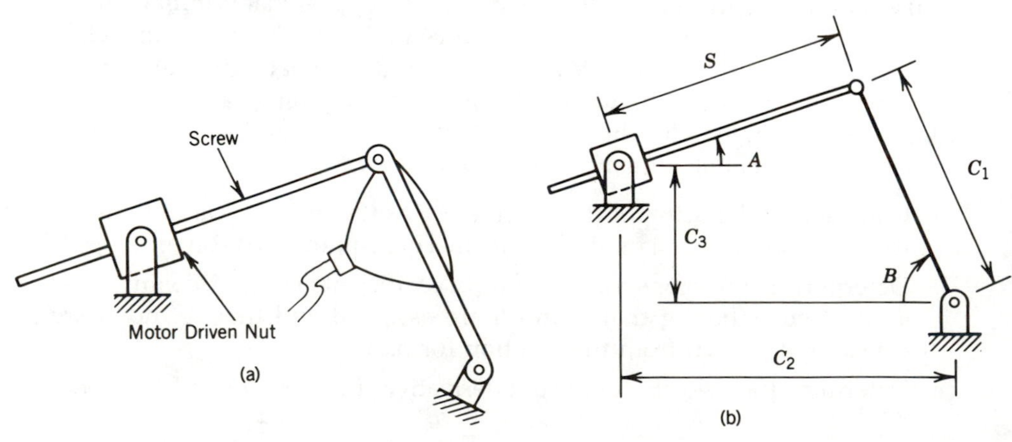

a. Write the position loop equations (do not solve);

b. Determine the velocity coefficients $K_a = dA/dS$ and $K_b = dB/dS$, remembering that S and $\dot{S}$ are considered assigned, and that A and B are considered known for the solution for part a;

c. Determine the velocity coefficient derivatives, $L_a = dK_a/dS$ and $L_b = dK_b/dS$.

2.4 The mechanism shown is called a "quick return mechanism." The flywheel is driven by a motor (not shown), causing the lever to oscillate to the left and the right. A tool is mounted on the horizontal slider and moves with it. The dimensions C, D, and R are known constants.

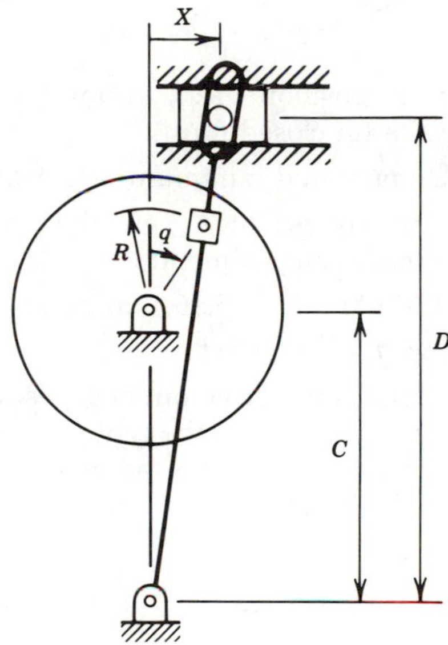

a. Determine, in closed form, X as a function of q;

b. Determine, in closed form, K_x as a function of q;

c. Determine, in closed form, L_x as a function of q;

d. Given the following dimensions, evaluate X, $\dot{X}$, and $\ddot{X}$ for $q = \pi/6$ rad, $\dot{q} = 22.5$ rad/s, and $\ddot{q} = 147$ rad/s^2.

$$C = 0.4 \text{ m} \qquad D = 0.75 \text{ m} \qquad R = 0.085 \text{ m}$$

2.5 The Geneva wheel is a mechanism that provides intermittent angular motion with a continuous angular input motion. The drive pin engages a slot in the output wheel, causing it to rotate until the drive pin disengages from the slot. At the initial engagement and again at separation, the velocity of the drive pin should be along the axis of the slot so as to avoid operating problems. For the Geneva mechanism shown, the lengths R_1, R_2, and D are known, and the input rotation is q.

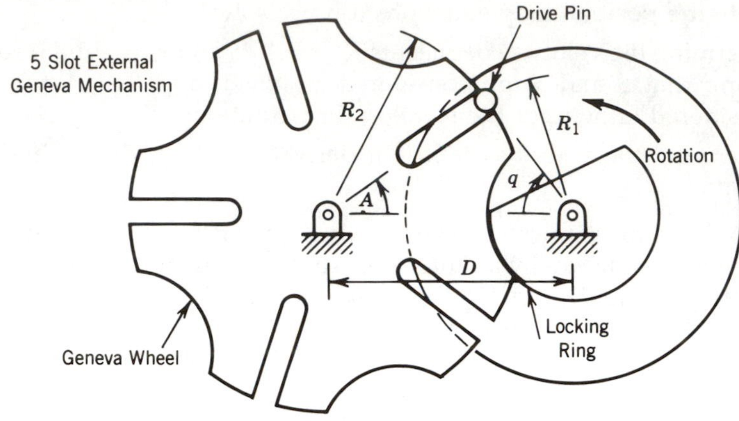

a. Determine the output position, $A(q)$, during the time the drive pin is engaged with the slot (in closed form);

b. Determine the maximum and minimum values of q for engagement;

c. Determine the velocity coefficient, K_a, and the velocity coefficient derivative, L_a, during pin-slot engagement;

d. Given that $R_1 = 4.335$ in., $R_2 = 5.9666$ in., and $D = 7.3751$ in., determine A, $\dot{A}$, and $\ddot{A}$ for $q = 0.15$ rad, $\dot{q} = 12.4$ rad/s, and $\ddot{q} = -29.0$ rad/s^2.

2.6 To avoid impact as the drive pin engages the slot in the Geneva wheel, the pin velocity must be along the centerline of the slot. For a drive pin of zero diameter that is located at a radius R_1, as shown in view (a), engaging a wheel with N slots,

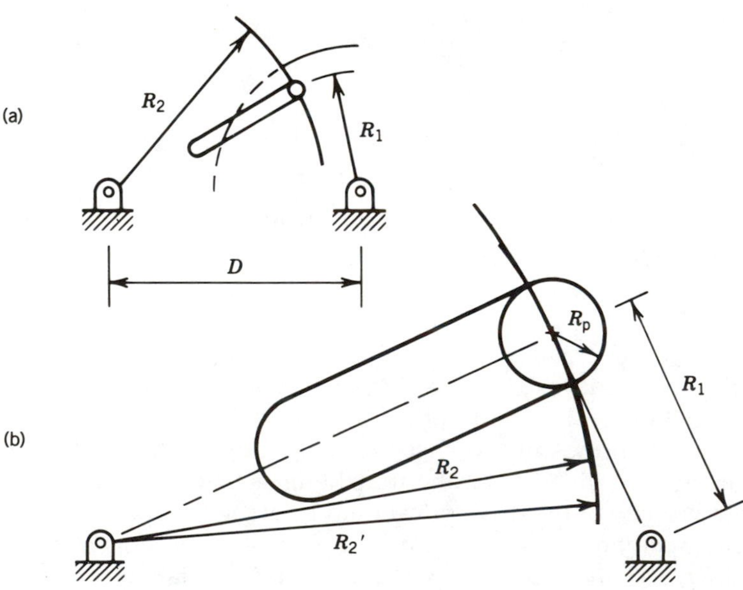

a. Determine the appropriate value for R_2;
b. Determine the appropriate value for D.

The value determined in part a for R_2 was based on a drive pin of zero diameter, which, of course, is not the actual case. It is necessary to use a slightly larger radius, R_2', as shown in view (b).

c. If the drive pin radius is R_p, determine the appropriate value of R_2';
d. If a drive pin diameter of 0.562 in. is to be used with a five-slot Geneva wheel that has the same dimensions as in part d of the preceding problem, evaluate R_2'.

2.7 A crank–lever drive is to be designed to meet the following criteria:

1. The lift at the tip of the slotted member is to be 7.85 in.;
2. The maximum speed at the tip of the slotted member is to be 120 in./s in absolute value with the crank speed of 10 rad/s;
3. The tip of the slotted member is to extend 1.45 in. beyond the outermost position of the slider pivot center point.

Determine the following three essential dimensions: C = distance between fixed pivots, R = crank radius, and L = length of the slotted member from the stationary pivot to the tip.

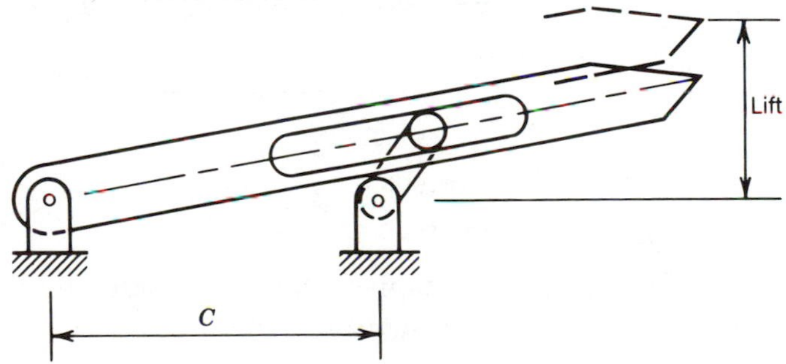

2.8 The figure shows the mechanism of a double-acting gas compression cylinder. A prime mover (not shown) turns the crank, causing the

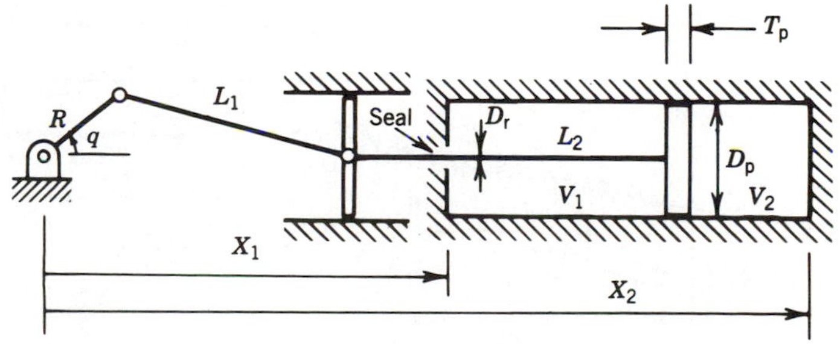

crosshead, piston rod, and piston assembly to move left and right. The piston rod, piston, and cylinder cavity are all assumed to be right circular cylinders, and the dimensions R, L_1, L_2, X_1, X_2, D_p, D_r, and T_p are all known.

a. Determine expressions for the chamber volumes, V_1 and V_2, in terms of the known parameters;
b. Determine the derivatives dV_1/dq and dV_2/dq.

2.9 For the double-acting gas compressor shown, the following dimensions apply:

$R = 5$ in. $\qquad D_p = 8$ in.
$L_1 = 22$ in. $\qquad D_r = 2.25$ in.
$L_2 = 20$ in. $\qquad T_p = 2$ in.
$X_1 = 35$ in.
$X_2 = 49$ in.

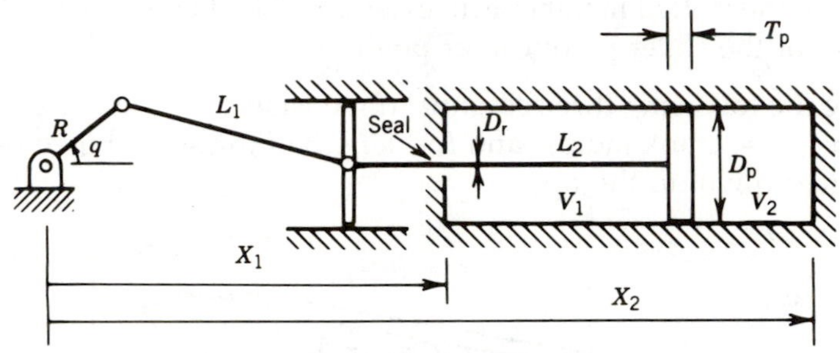

a. Make a plot of V_1 and dV_1/dq versus q;
b. Make a plot of V_2 and dV_2/dq versus q.

2.10 For the slider–crank mechanism shown, consider the slider position as the input variable, q. The dimensions of C_1, C_2, and L, and radii

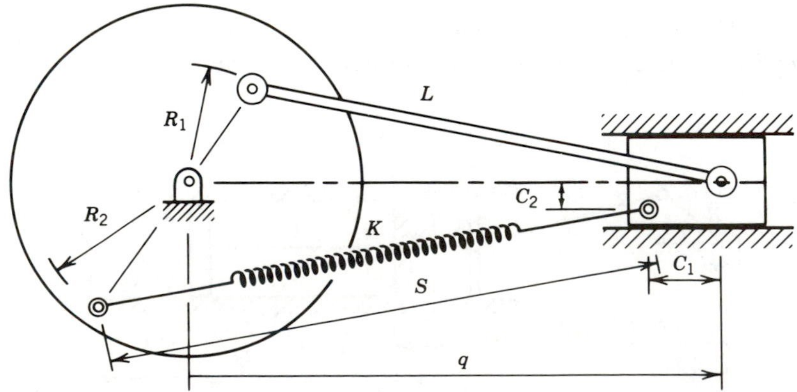

R_1 and R_2 are known. A spring of length S stretches from the slider to a point on the crankwheel directly opposite the connecting rod attachment point.

a. In closed form, determine S as a function of q;

b. In closed form, determine $K_s = dS/dq$;

c. Given the following data, evaluate S and K_s for $q = 105$ in.

$C_1 = 10$ in. $\quad C_2 = 4.5$ in. $\quad L = 82$ in.

$R_1 = 27$ in. $\quad R_2 = 30$ in.

2.11 The figure shows a slider–crank mechanism falling under the effect of gravity. The dimensions C_1, C_2, U_1, U_2, and D, and the angle C_o are known.

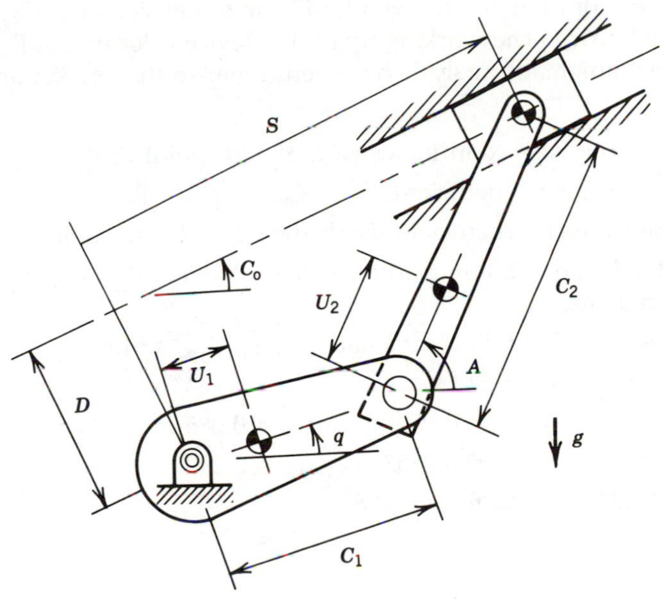

a. Determine $A(q)$ and $S(q)$;

b. Determine the velocity coefficients K_a and K_s;

c. Determine the velocity coefficient derivatives L_a and L_s;

d. Determine the center of mass velocity coefficients K_{1x}, K_{1y}, K_{2x}, K_{2y}.

2.12 The mechanism shown is used in a small, portable air compressor. Notice that the piston, wrist pin, and connecting rod are all one piece. The piston is spherical in shape, and the cavity is a right circular cylinder with a hemispherical end. The dimensions R_1, R_2, R_3, and D are known. If the piston is 0.010 inch from the end of the cavity at top-dead center, express the cylinder volume as a function of the crank angle, q.

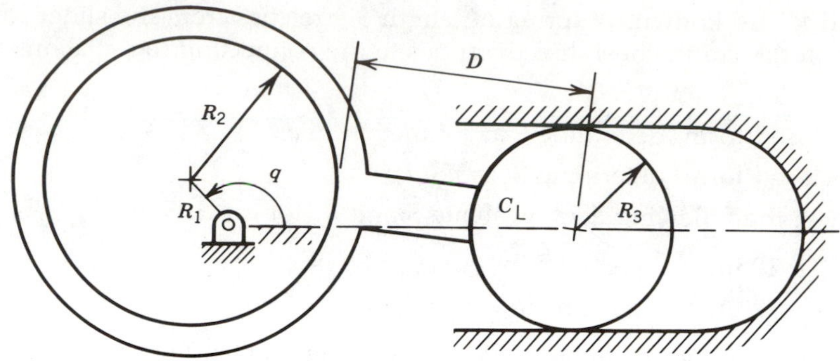

2.13 The mechanism shown is used to engage and advance a work piece one step for each crank revolution. The dimensions C_1, C_2, C_3, and radius R are known. The working tip of the device is located at P. Assume that an initial kinematic analysis has been done, so that A, K_a, and L_a are known.

a. Determine the base coordinates (X_p, Y_p) for point P;
b. Determine velocity coefficients K_{px}, K_{py} for point P;
c. Determine velocity coefficient derivatives L_{px}, L_{py} for point P;
d. Given the data that follow part e, evaluate X_p, Y_p, K_{px}, K_{py}, L_{px}, L_{py} for $q = 0.8$ radians;
e. Given the data that follow, evaluate $\dot{X}_p$, $\dot{Y}_p$, $\ddot{X}_p$, $\ddot{Y}_p$ for $q = 0.8$ rad, $\dot{q} = 14.5$ rad/s, and $\ddot{q} = 28$ rad/s^2.

$C_1 = 0.14$ m	$C_2 = 0.05$ m	$C_3 = 0.085$ m
$R = 0.055$ m	$A = 0.2857$	$K_a = 0.2937$
$L_a = -0.2599$	All for $q = 0.8$	

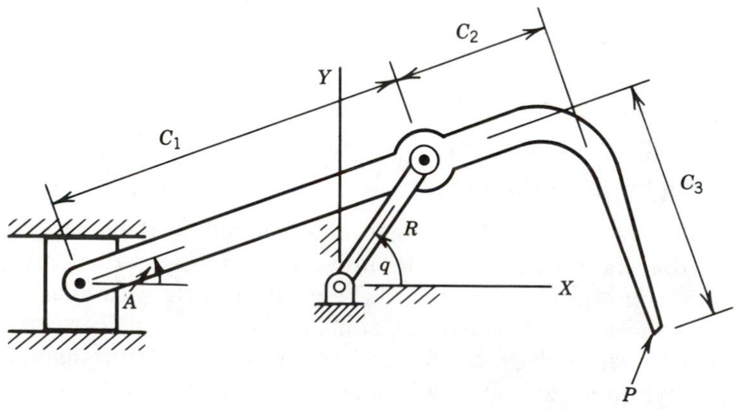

2.14 Some early steam engines used the mechanism shown, and it is still seen in some working model steam engines today. The cylinder oscillates about a fixed pivot as the piston moves back and forth. The oscillation of the cylinder is used to open and close inlet and exhaust ports to the cylinder, thus providing the necessary valve action.

a. Write the position loop equations and solve them in closed form;
b. Determine analytic expressions for the velocity coefficients and velocity coefficient derivatives associated with the secondary variables;
c. Determine analytic expressions for the velocity coefficients and velocity coefficient derivatives for the piston center of mass;
d. Using the data following part e, determine A, $\dot{A}$, and $\ddot{A}$ for $q = 0.75$ rad, $\dot{q} = 14$ rad/s, and $\ddot{q} = -27$ rad/s^2.
e. Using the following data, determine the magnitude of the velocity and the acceleration vectors for the piston center of mass under the conditions of part d.

$R = 14$ in. $\qquad L = 45$ in. $\qquad C = 30$ in.

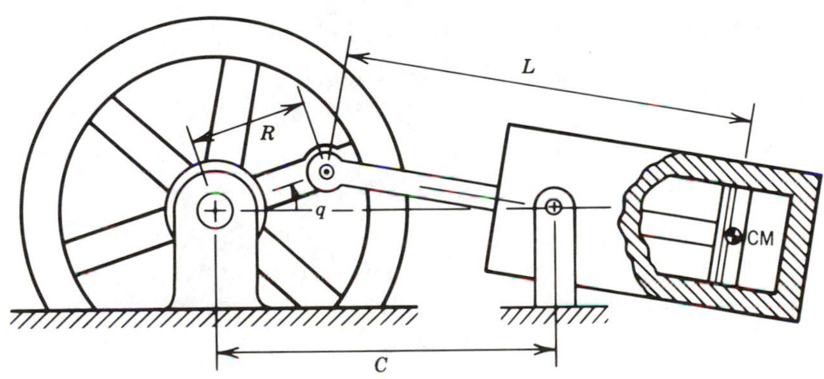

2.15 A hand-actuated pump mechanism is shown. The dimensions C_1, C_2, C_3, C_4, C_5, D_1, and D_2 are all known. Assume that the initial kinematic analysis for the secondary variables A and B has been completed and that A, B, K_a, K_b, L_a, and L_b are known. The condition of interest is when $q = 2.27$ in.

a. Using the data that follows part d, determine the numeric values for X_p, and Y_p, which are the base coordinates for point P;
b. Using the data that follows part d, determine numeric values for the velocity coefficients K_{px}, K_{py} for point P;
c. Using the data that follows part d, determine numeric values for the velocity coefficient derivatives, L_{px}, L_{py} for point P;

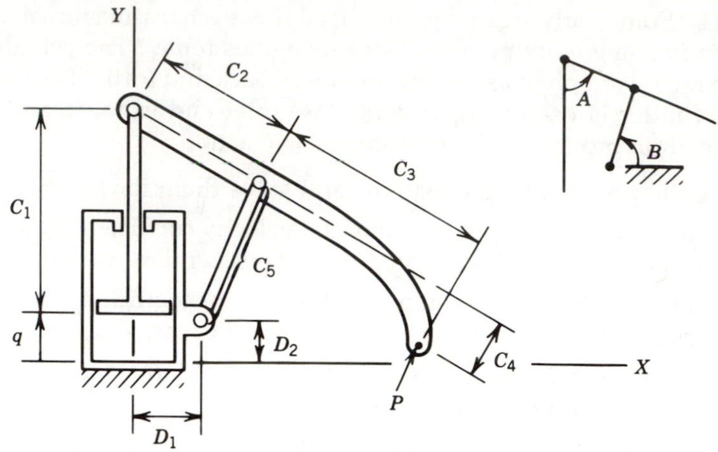

d. Given that $q = 2.27$ in., $\dot{q} = 13.6$ in./s and $\ddot{q} = -8.65$ in./s^2, determine the magnitudes for the velocity and acceleration of point P.

$C_1 = 10.5$ in.	$A = 1.2635$ radians
$C_2 = 8.32$ in.	$B = 1.0916$ radians
$C_3 = 26.22$ in.	$K_a = -0.10825$ 1/in.
$C_4 = 5.68$ in.	$K_b = 0.034691$ 1/in.
$C_5 = 8.85$ in.	$L_a = 7.3496\ (10^{-4})$ 1/in.^{2}
$D_1 = 3.85$ in.	$L_b = 0.010972$ 1/in.^{2}
$D_2 = 2.40$ in.	(this column only for $q = 2.27$ in.)

2.16 A window mechanism is shown in the figure, and the lengths C_1, C_2, C_3, C_4, and C_5 are known dimensional data. With the crank angle taken as the primary variable, the angles A and B are secondary variables.

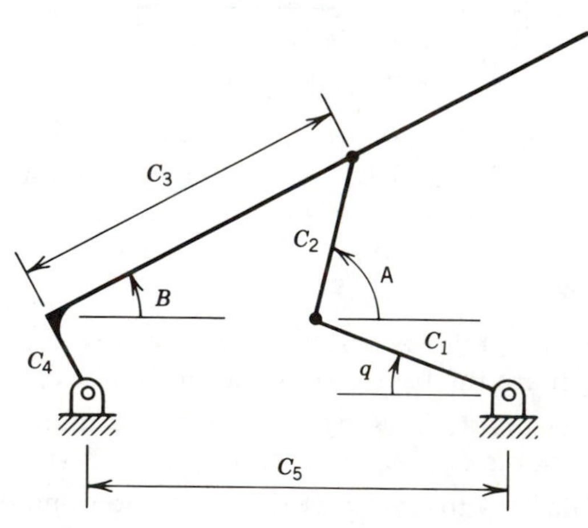

a. Write the equations required to determine A and B (do not solve);

b. Develop the equations required for numerical evaluation of the velocity coefficients K_a and K_b;

c. Develop the equations required for numerical evaluation of the velocity coefficient derivatives L_a and L_b (do not solve).

2.17 For the window mechanism of problem 2.16, the following data values are available:

$C_1 = 10.5$ in. $\quad C_4 = 1.2$ in.
$C_2 = 7.4$ in. $\quad C_5 = 16.5$ in.
$C_3 = 11.0$ in.

A numerical position solution is needed for $q = 0.15$ radians. As an initial estimate take $A = 1.2$ radians and $B = 0.8$ radians, and determine a second estimate for both A and B using the Newton–Raphson method.

2.18 The platen for a laminating press is supported on a four-bar hinge, as shown in the figure. The lengths C_1, C_2, C_3, and C_4 are known dimensions. Consider the angular position of the platen as the primary variable as indicated.

a. Write the equations required to determine the angles A and B (do not solve);

b. Write the equations required to determine the velocity coefficients K_a and K_b;

c. Write the equations required to determine the velocity coefficient derivatives L_a and L_b.

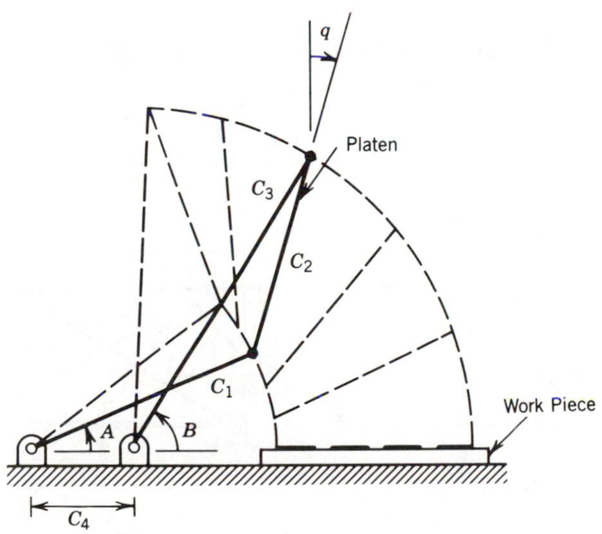

2.19 The four-bar linkage shown is the Galloway mechanism for which $\overline{OA} = \overline{OC}$ and $\overline{AB} = \overline{CB}$.

a. What is the path of point B?
b. What is the range of values for angle D as q moves from zero to 2π?
c. Write the equations required to determine D as a function of q (do not solve);
d. Determine the velocity coefficient K_d and the velocity coefficient derivative L_d, assuming that the equations for part c have been solved.

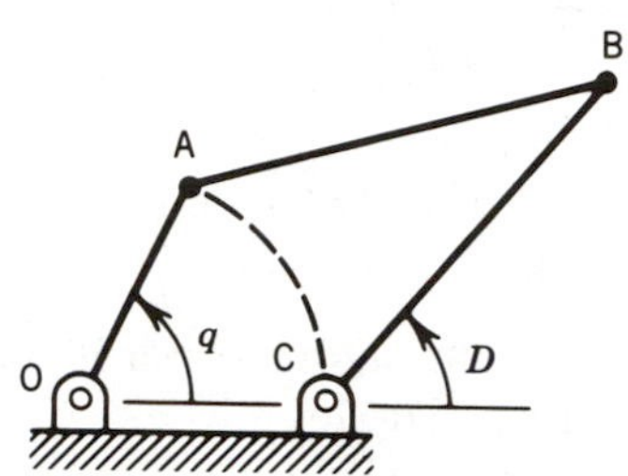

2.20 The disappearing platform mechanism shown has been used variously to support a typewriter, a sewing machine, and other equipment that is needed on the work surface only part of the time. When the machine is dropped to the storage position, a replacement panel fills the opening in the work surface. Consider the dimensions C_1, C_2, C_3, X_o, Y_o as known data, and take the horizontal position of the left corner of the work surface as the primary variable.

a. Determine the equations required to evaluate the angles B_1, B_2, and B_3 (do not solve);
b. Determine the equations required to evaluate the velocity coefficients K_{b1}, K_{b2}, and K_{b3}.

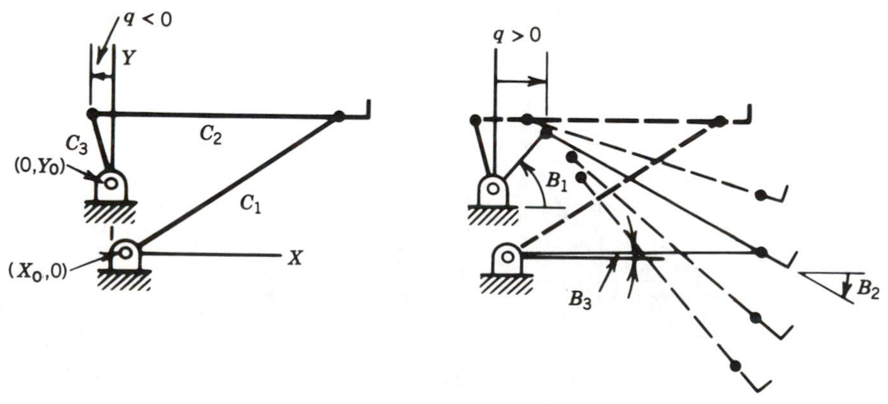

2.21 The mechanism shown in the figure is a toggle clamp, used to clamp a work piece between the table and the clamp foot (the space measured by Y_1). The force to close the clamp is applied on the coupler link, at a position described by body coordinates (U_o, V_o), at distance Y_o above the table surface. The dimensions C_1, C_2, C_3, C_4, C_5, C_6, C_7, C_8, and body coordinates (U_o, V_o) are all considered known. Determine a means to compute values numerically for the derivatives $K_{y1} = dY_1/dY_o$ for various positions of the clamp (this is a velocity coefficient of sorts, and is useful in the static force analysis).

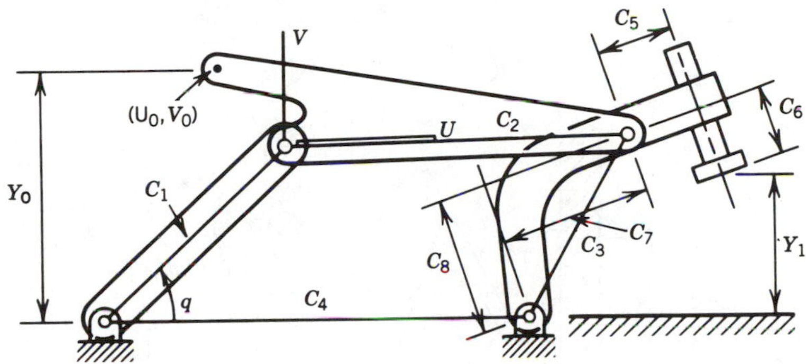

2.22 The linkage shown represents the front suspension of an automobile. The dimensions C_1, C_2, C_3, C_4, C_5, C_6, and S (S = spindle length) are known.

a. Set up the equations required for numerical determination of the secondary variables (do not solve);
b. Set up the equations for numerical evaluation of the velocity coefficients for the secondary variables;
c. Set up the equations for numerical evaluation of the velocity coefficient derivatives;
d. Determine the base coordinates for point P;
e. Express the velocity coefficients for point P;
f. Express the velocity coefficient derivatives for point P.

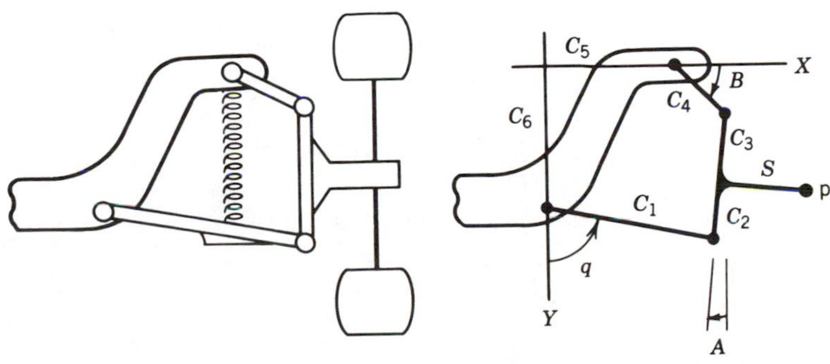

2.23 The mechanism shown is a type of garage door. The dimensional values C_1, C_2, . . . C_{11} are known data. A tension spring is to be attached between point E on the door and the stationary point designated as the spring anchor. The length of the spring is denoted as S.

a. Analyze the mechanism for all secondary variables (do not solve);
b. Express the velocity coefficients;
c. Express the spring length, S;
d. Express the spring length derivative dS/dq.

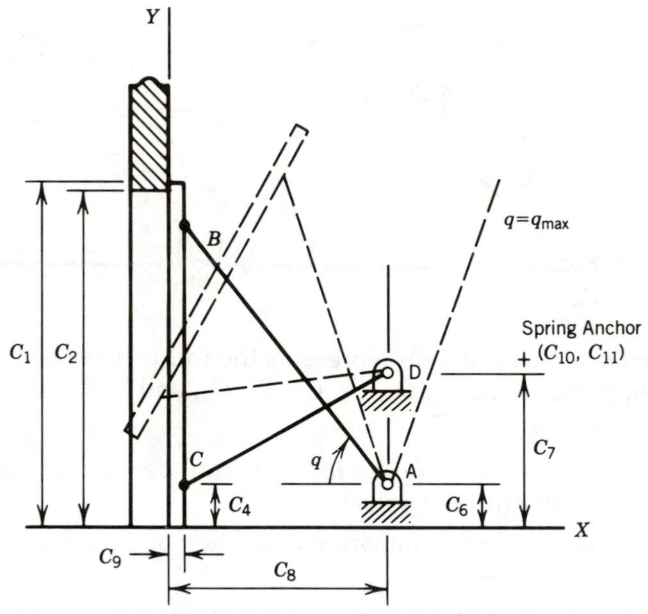

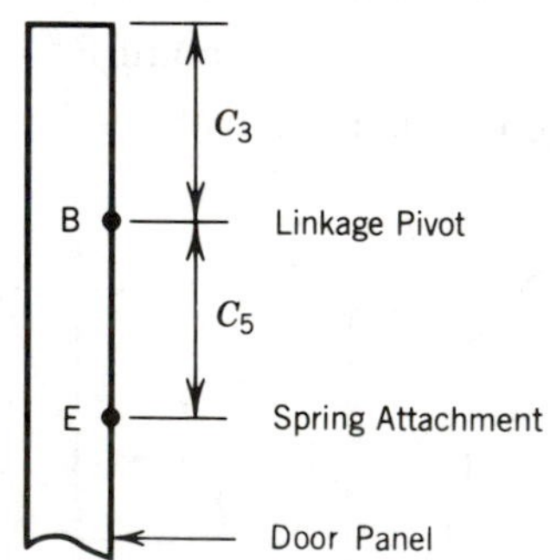

2.24 For the garage door mechanism of problem 2.23, the following dimensional data are available:

$C_1 = 7.00$ ft	$C_7 = 3.20$ ft
$C_2 = 6.82$ ft	$C_8 = 4.60$ ft
$C_3 = 0.90$ ft	$C_9 = 0.30$ ft
$C_4 = 0.90$ ft	$C_{10} = 7.50$ ft
$C_5 = 0.70$ ft	$C_{11} = 3.50$ ft
$C_6 = 0.60$ ft	

In this problem $q_{max} = 1.9199$ radians. Consider the door in nine positions, at evenly spaced intervals on q, including the $q = 0$ and $q = q_{max}$ positions.

a. Will the lower edge of the door clear the lintel?

b. Plot S versus q.

c. On the plot of part b, also plot dS/dq versus q.

2.25 The emergency main steam shut-off valve for a nuclear power plant is a massive, tapered plug that swings into a tapered seat to stop the flow of steam. When released, gravity causes the valve to close, with the impact reduced by the dashpot with coefficient B. The plug and its supporting links form a four-bar linkage with link lengths L_1, L_2, L_3, and L_4, as shown in the detail sketch. The center of mass of the plug is located by the body coordinates (U_c, V_c).

a. Set up the equations required for numerical evaluation of A_2 and A_3 as functions of q (do not solve);

b. Set up the equations required for numerical evaluation of the velocity coefficients K_{a2} and K_{a3};

c. Set up the equations required for numerical evaluation of the velocity coefficient derivatives L_{a2} and L_{a3}.

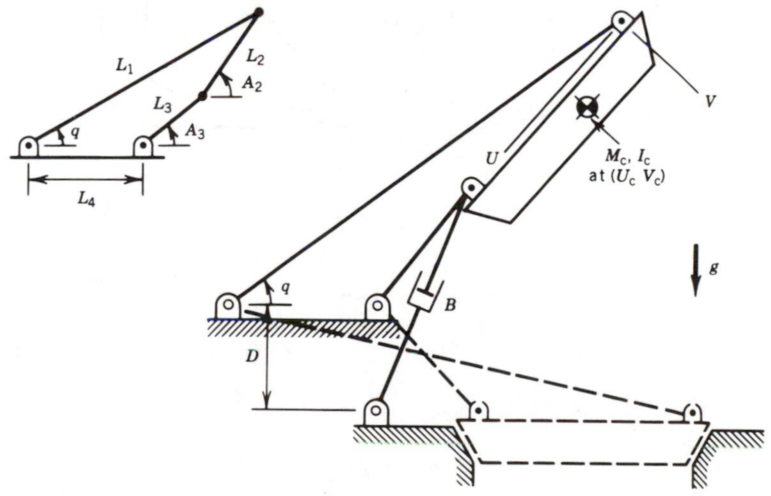

2.26 The mechanism shown is used to dress the semicircular edge of a grinding wheel. As the handle is displaced through the angle q, the diamond dressing tool moves in a circular arc about point C. Because the wheel spins about its rotation axis while the dressing tool moves, the result is that the surface of the wheel is reshaped to the desired semicircular contour. The dimensional data D_1, D_2, D_3, D_4, and D_5 are known. The left and right side linkage components are identical.

a. Show that the tool actually moves as described;
b. Express the radius R in terms of the adjustable parameter D_3 and the other fixed dimensions of the mechanism.

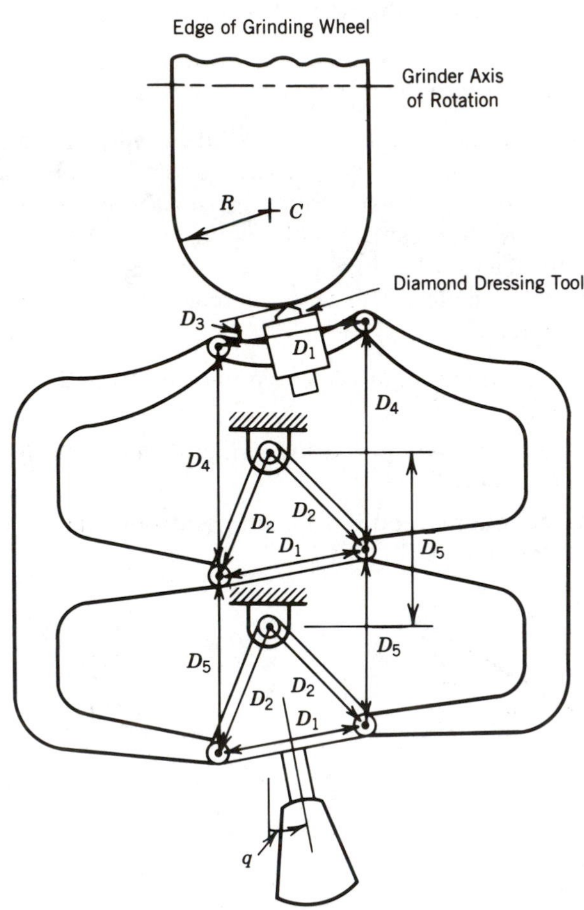

2.27 The wheel shown, while under the control of a crank and connecting rod, rolls without slipping on the horizontal surface. The system was assembled originally in the configuration shown by the broken lines, with the crank and connecting rod colinear and the wheel a distance S_o to the right of the crank pivot. The dimensional values C_1, C_2, r, R, and S_o are known.

a. Set up the equations required to determine the secondary variables (do not solve);
b. Set up the equations to be solved for the velocity coefficients;
c. Set up the equations required for numerical determination of the velocity coefficient derivatives.

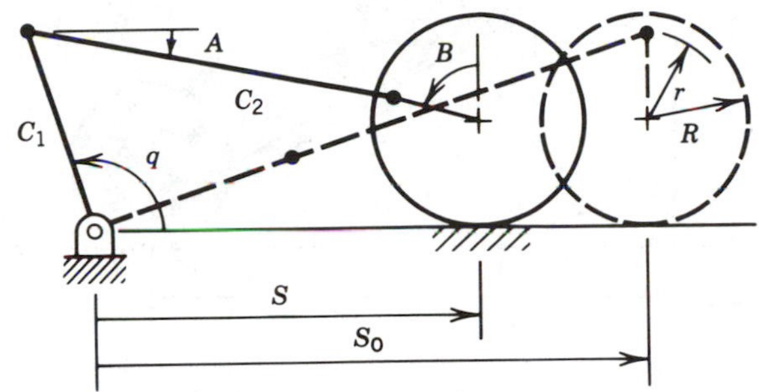

2.28 Under the control of a crank and connecting rod, the wheel rolls without slipping on a semicircular support of radius C_5. The original assembly configuration is shown in broken lines, with the crank and connecting rod colinear. The dimensions C_1, C_2, C_3, C_4, C_5, r, and R are known.

a. Set up the equations governing the secondary position variables (do not solve);
b. Set up the equations to be solved for the velocity coefficients;
c. Set up the equations for numerical determination of the velocity coefficient derivatives.

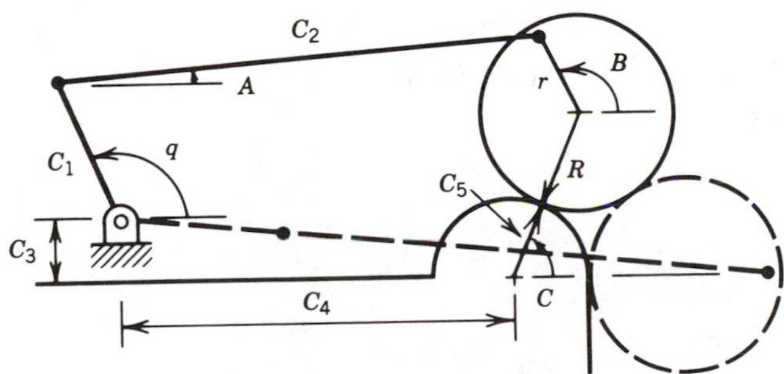

2.29 The wheel rolls to the right without slipping on the horizontal surface as shown. A slider on an overhead guide at height D is coupled to the wheel by a connecting rod of length L. The dimensions L, D, R, and r are known. When the process begins with $q = 0$, the wheel is at a known position $X_2 = X_{20}$.

a. Write equations solvable for the secondary position variables (do not solve);

b. Write the equations for numerical determination of the velocity coefficient derivatives.

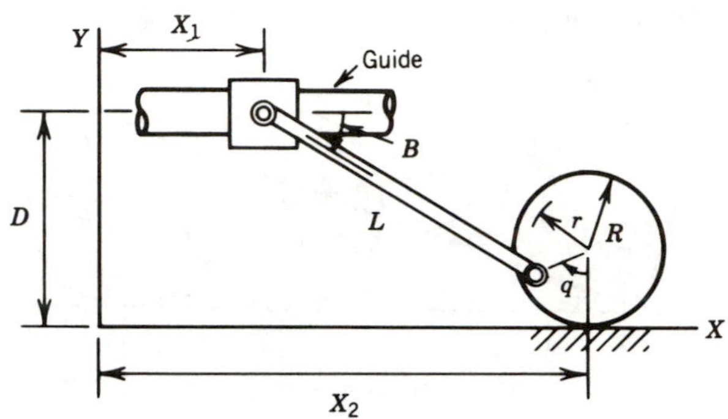

2.30 A right-triangular connecting link with sides L and H joins a slider to a wheel. As the slider moves to the right, the wheel rolls without slipping on the incline. The process begins with $q = 0$, $B = 0$, and $S = S_o$. Later, when $q = 5$ in., the secondary variables have the values $A = 0.7094$ rad, $B = 0.3642$ rad, $S = 14.27$ in. When $q = 5$ in., $\dot{q} = 16.5$ in./s, and $\ddot{q} = -27.7$ in./s^2, determine numerical values for

a. Base coordinates for point G;

b. Velocity coefficients for point G;

c. Velocity coefficient derivatives for point G;

d. Velocity magnitude and acceleration magnitude for point G.

$L = 22.0$ in.	$D = 18.214$ in.
$R_1 = 6.0$ in.	$C = 0.452$ in.
$R_2 = 7.0$ in.	$S_o = 10.9495$ in.
$H = 4.0$ in.	

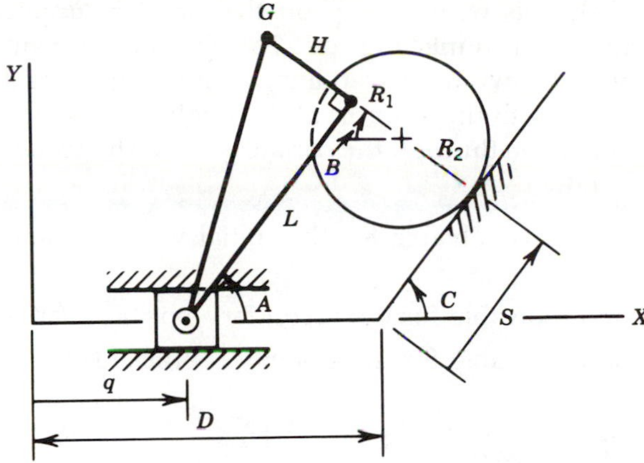

2.31 A disk rolls without slip on the circular arc, dragging the slider and connecting rod. When $q = 0$, the connecting rod is horizontal, with the pin joining the connecting rod to the disk located on the Y-axis, as shown in broken line. The dimensions L, R_0, R_1, and R_2 are known values.

a. Set up the equations governing the secondary position variables (do not solve);
b. Set up the equations for the velocity coefficients;
c. Set up the equations for numerical determination of the velocity coefficient derivatives.

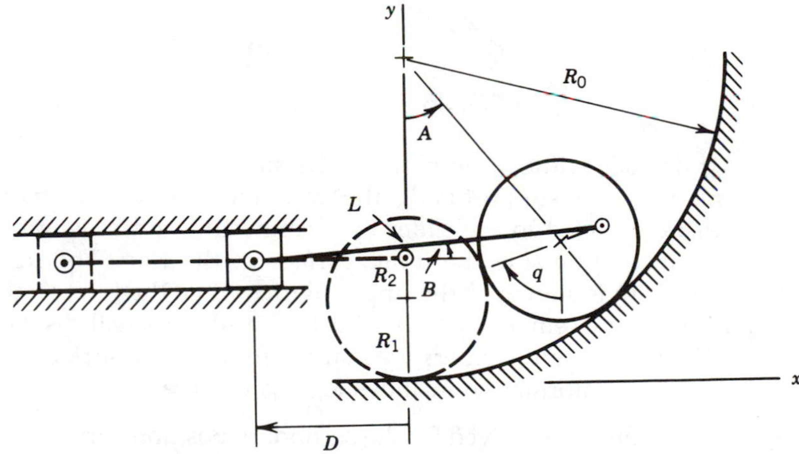

2.32 The disk rolls without slip on the circular support under the control of the crank and connecting rod. The system is assembled initially in the configuration shown in broken line, with the crank and connecting rod colinear, contact with the support at the angle B_o, and the connecting rod pin on the radial line through the contact point. The values R_1, R_2, R_3, L, a, and D are known.

a. Determine equations solvable for the initial values q_o and B_o (do not solve);
b. Set up equations solvable for the secondary position variables;
c. Set up equations solvable for the velocity coefficients.

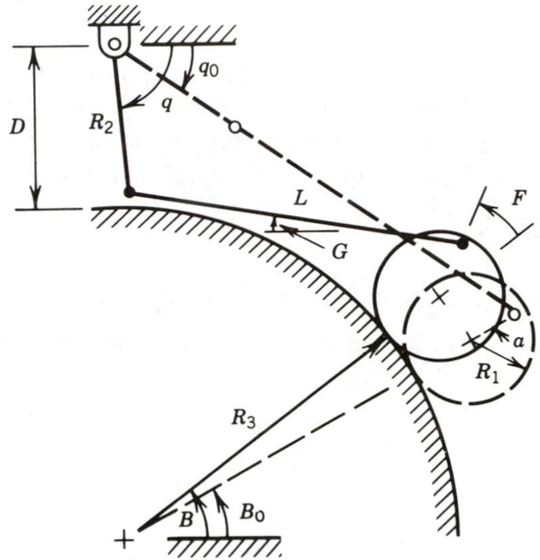

2.33 A ring rolls without slipping on the stationary support (cross-ruled). The radius of the support is R_s; the inner radius of the ring is R_r. The ring is also connected to the slider by a connecting rod of length C, attached a distance D outside the inner surface of the ring. The initial configuration ($q = 0$) is shown in the upper figure, while the lower figure shows a positive displacement ($q > 0$) with the resulting small displacement of the slider to the left. Centers are marked for both the support and the ring. The known information includes R_s, R_r, C, D, and X_{so}.

a. Write the equations to be solved for the secondary position variables (do not solve);
b. Write the equations to be solved for the velocity coefficients.

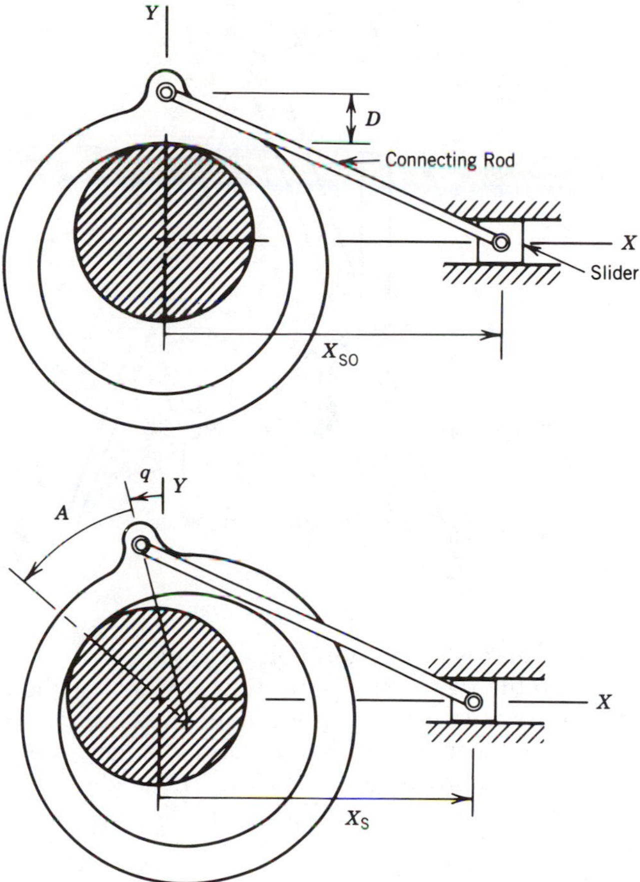

2.34 The clerestory windows found in an industrial shop are opened and closed with the mechanism shown. The disk shown is a pinion gear that is driven by a servo motor, and the arm shown tangent to the disk is a rack that is meshed with the gear and held in mesh by a guide that is not shown. As the pinion rotates through the angle q, both the rack angle A and the window angle B change. When the mechanism is first assembled, the window is closed and the pinion position is defined as $q = 0$.

a. Set up the equations required to determine the initial values for the secondary variables A and G (do not solve);

b. Formulate the equations to be solved for the secondary position variables;

c. Formulate the equations to be solved for the velocity coefficients.

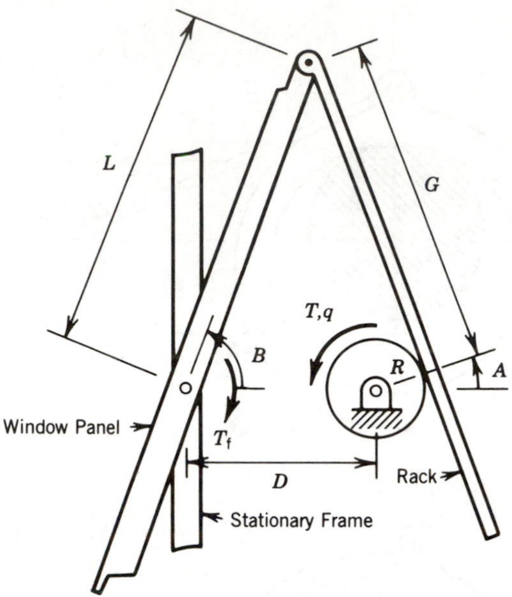

2.35 As the disk is turned through an angle $q > 0$, the slider moves along the track to the right. The connecting arm rolls without slip on the disk due to the force exerted by the guide, pressing the arm and disk together. As the angular position of the arm changes, the guide rotates with it about the axis of rotation of the disk. The angle q is zero when the arm is perpendicular to the track. The dimensions R and D are known values.

a. Set up the equations required to determine the secondary position variables (do not solve);
b. Set up the equations to be solved for the velocity coefficient.

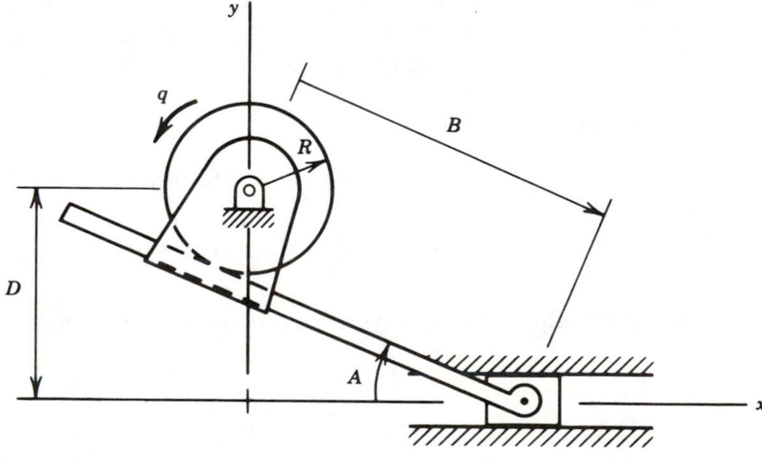

2.36 A light-duty clamshell bucket, which is actuated by a hydraulic cylinder, is shown in the figure. The dimensions D_1, D_2, D_3, and angle C are known parameters. The length of the hydraulic cylinder is taken as the input variable, q.

a. Set up the equations required to determine the secondary position variables (do not solve);

b. Set up the equations required to determine the velocity coefficients K_a and K_b.

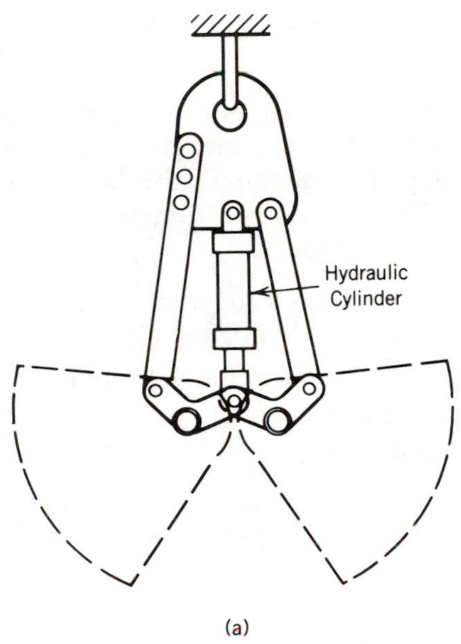

(a)

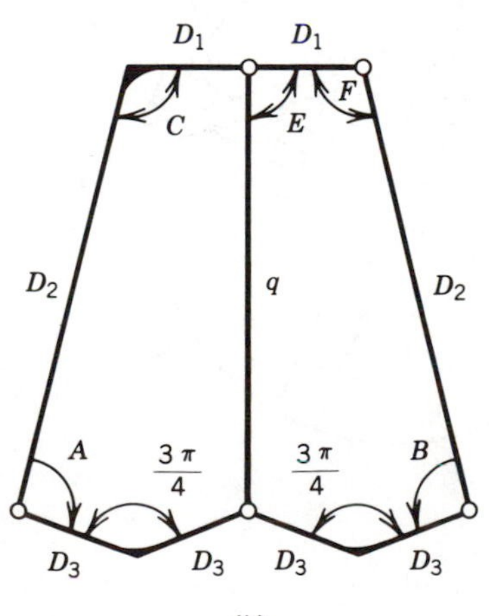

(b)

2.37 For the clamshell bucket of problem 2.36, the following data apply:

$D_1 = 5.0$ in. $\quad D_3 = 6.0$ in.
$D_2 = 22.0$ in. $\quad C = 1.7825$ radians

a. Solve the position equations numerically and make an overlay plot of the mechanism for one-inch increments on q from $q = 18$ inches to $q = 28$ inches;

b. Determine numerical values for the velocity coefficients K_a and K_b at each of the positions determined in part a;

c. Plot the velocity coefficients K_a and K_b versus q using the data from part b.

2.38 The mechanism shown is called a trammel crank drive. There are two sliders attached to the connecting rod. These sliders move in grooves on the crank plate. The same unbroken connecting rod extends to the external slider. The only known dimensions are L_1 and L_2, the two segments of the connecting rod.

a. Set up the equations required for numerical determination of the secondary position variables (do not solve);

b. Set up the equations governing the velocity coefficients;

c. Set up the equations to be solved for numerical determination of the velocity coefficient derivatives;

d. If the external slider is used in a punch action, how many cutting strokes does the punch make per revolution of the crank plate?

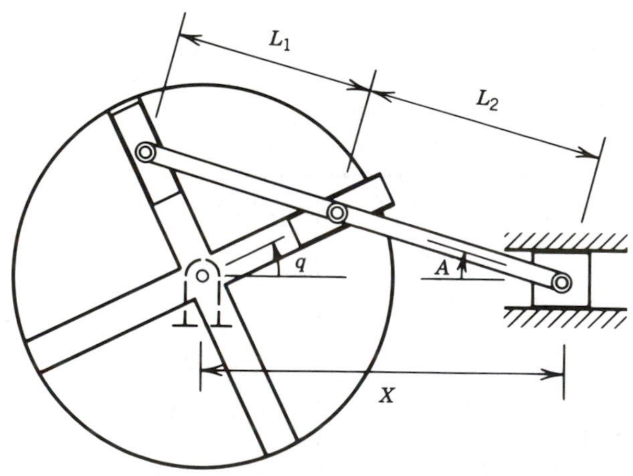

2.39 The mechanism shown is called a "quick return." As the crank rotates, as described by q, the two sliders must move in their respective guides with positions described by X and Y. The dimensions R and L are known values.

a. Formulate the equations that determine the secondary position variables (do not solve);

b. Set up the equations for determining the velocity coefficients;

c. If the crank rotates with constant velocity $\dot{q}$, for what crank position is the absolute value of the velocity of the horizontal slider at its maximum? Its minimum?

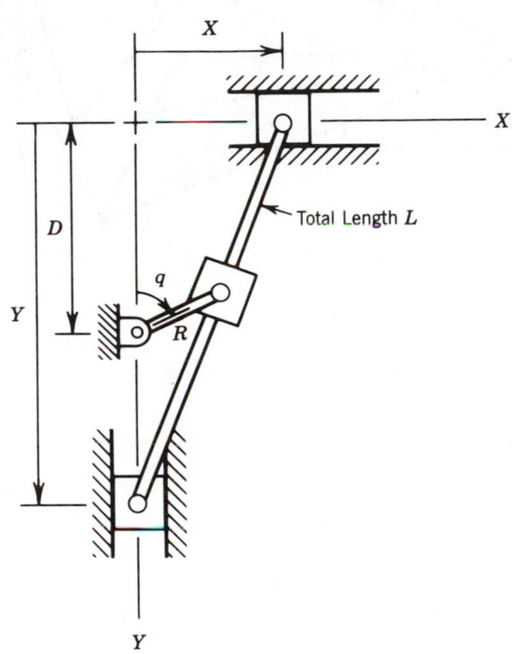

2.40 The mechanism shown is called Peaucellier's straight line mechanism. As the crank is moved through the angle q, the point P describes a straight line as shown in the short broken line. The lengths L_1, L_2, and R are all known values. Give an analytic proof that point P moves in a straight line.

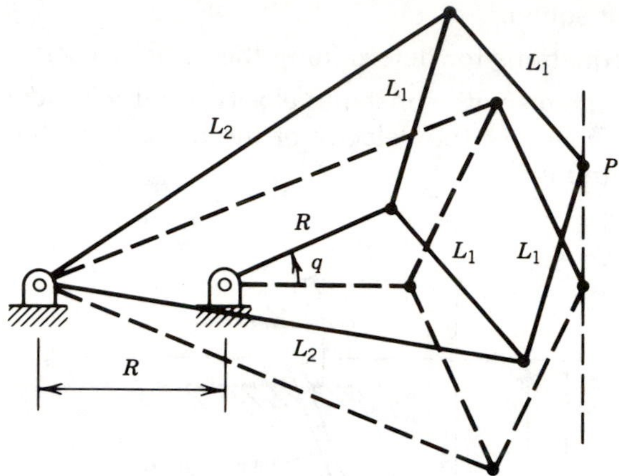

CHAPTER 3

Multidegree of Freedom Linkages

Linkages and other mechanisms with multiple degrees of freedom require multiple generalized coordinates that are equal in number to the number of degrees of freedom. The linkage configuration is not fully determined until all of the generalized coordinates are specified. For kinematic analysis, this is equivalent to saying that there will be multiple, assigned primary variables. This has relatively little effect on the position solution, but it complicates the velocity and acceleration analyses.

In Chapter 2, in the context of single degree of freedom mechanisms, most of the ideas and tools required for this chapter were introduced. These include primary and secondary variables, position loop equations, the Newton–Raphson numerical solution technique, the Jacobian matrix, velocity and acceleration equations, velocity coefficients, velocity coefficient derivatives, and body coordinates. With mostly minor modifications, all of these will be studied in this chapter as applied to multidegree of freedom systems. Because there are few new theoretical tools to be developed, most of the presentation will be by means of examples.

3.1 KINEMATIC ANALYSIS IN CLOSED FORM

As seen in the preceding chapter, the kinematic analysis for some mechanisms can be carried out in closed form. Although this is true for only a limited class of mechanisms, it is useful to consider this case for the insight it gives into the analysis process; sometimes the logic of the analysis is clouded by the use of numerical solutions. In the following two degree of freedom example, the position, velocity, and accelerations for the secondary variables are determined in closed form. The modifications required for multiple degrees of freedom become evident in the velocity and acceleration analyses. The motion of a particular point of interest is also developed.

3.1.1 Sliding Four-Bar Mechanism

A sliding four-bar mechanism is shown in Figure 3.1. It differs from the conventional four-bar linkage in that the pivot joining the second crank and the coupler is on a slider, so that the effective coupler length is variable. The system has two degrees of freedom that are associated with the two crank angles, q_1 and q_2. The secondary coordinates are the coupler inclination angle, A, and the effective length of the coupler, B. The point P is a point of interest on the coupler link, located by body coordinates (U_p, V_p). Determine the positions, velocities, and accelerations for the secondary variables, as well as the motion of the point P.

Position Analysis. The mechanism involves a single position loop that provides two scalar position loop equations:

$$f_1 = C_1 \cos q_1 + B \cos A - C_2 \cos q_2 - C_3 = 0$$

$$f_2 = C_1 \sin q_1 + B \sin A - C_2 \sin q_2 = 0$$

These equations are to be solved for the two secondary coordinates, A and B. Although numerical methods are certainly applicable, in this case a closed form solution is also available. When B is eliminated, the result is

$$\tan A = \frac{C_2 \sin q_2 - C_1 \sin q_1}{C_3 + C_2 \cos q_2 - C_1 \cos q_1}$$

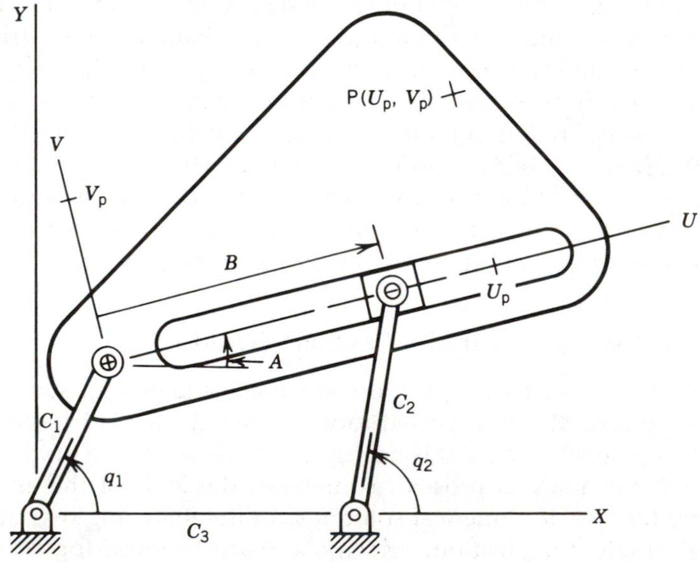

FIGURE 3.1 Sliding Four-Bar Linkage

For the linkage proportions indicated in the figure, $-\pi/2 < A < \pi/2$. Consequently, this equation can be properly solved for A using the principal value of the inverse tangent function. This value is then available for the solution for B:

$$B = \frac{C_3 + C_2 \cos q_2 - C_1 \cos q_1}{\cos A}$$

Velocity Analysis. The velocity loop equations are the result of time differentiation of the position loop equations; they are

$$-C_1\dot{q}_1 \sin q_1 + \dot{B} \cos A - B\dot{A} \sin A + C_2\dot{q}_2 \sin q_2 = 0$$

$$C_1\dot{q}_1 \cos q_1 + \dot{B} \sin A + B\dot{A} \cos A - C_2\dot{q}_2 \cos q_2 = 0$$

Recalling that q_1, q_2, $\dot{q}_1$, and $\dot{q}_2$ are to be assigned values, and that the values of A and B have already been determined, these are a pair of linear, simultaneous equations in the two unknown velocities, $\dot{A}$ and $\dot{B}$. When these are cast in matrix form, the coefficient matrix on the left side is recognized as the Jacobian matrix.

$$\begin{bmatrix} -B \sin A & \cos A \\ B \cos A & \sin A \end{bmatrix} \begin{Bmatrix} \dot{A} \\ \dot{B} \end{Bmatrix} = \begin{bmatrix} C_1 \sin q_1 & -C_2 \sin q_2 \\ -C_1 \cos q_1 & C_2 \cos q_2 \end{bmatrix} \begin{Bmatrix} \dot{q}_1 \\ \dot{q}_2 \end{Bmatrix}$$

For a closed-form solution, this equation is premultiplied by the inverse of the Jacobian matrix, and the result is algebraically simplified to give

$$\begin{Bmatrix} \dot{A} \\ \dot{B} \end{Bmatrix} = \begin{bmatrix} -(C_1/B) \cos (q_1 - A) & (C_2/B) \cos (A - q_2) \\ C_1 \sin (q_1 - A) & C_2 \sin (A - q_2) \end{bmatrix} \begin{Bmatrix} \dot{q}_1 \\ \dot{q}_2 \end{Bmatrix}$$

This shows that the secondary velocities, $\dot{A}$ and $\dot{B}$, can be expressed as linear combinations of the two primary velocities, $\dot{q}_1$ and $\dot{q}_2$. The velocity coefficients for this case form a (2×2) matrix, rather than simply a column vector (as in the previous chapter):

$$[K] = \begin{bmatrix} K_{a1} & K_{a2} \\ K_{b1} & K_{b2} \end{bmatrix}$$

$$= \begin{bmatrix} -(C_1/B) \cos (q_1 - A) & (C_2/B) \cos (A - q_2) \\ C_1 \sin (q_1 - A) & C_2 \sin (A - q_2) \end{bmatrix}$$

Then the relation between the secondary velocities, velocity coefficients and primary velocities is

$$\begin{Bmatrix} \dot{A} \\ \dot{B} \end{Bmatrix} = \begin{bmatrix} K_{a1} & K_{a2} \\ K_{b1} & K_{b2} \end{bmatrix} \begin{Bmatrix} \dot{q}_1 \\ \dot{q}_2 \end{Bmatrix}$$

In general, for the multidegree of freedom situation, the velocity coefficients will form a rectangular matrix, rather than the column vector associated with the velocity coefficients for a single degree of freedom. The fact that the velocity coefficient matrix is square for this example is just coincidence.

Acceleration Analysis. The acceleration will be obtained by differentiating the expression for the secondary velocities in terms of the velocity coefficient matrix and the primary velocities. The alternative is to differentiate the velocity loop equations to obtain the acceleration loop equations, and those must then be solved for the accelerations. Following the route indicated, the process is as follows:

$$\begin{Bmatrix} \ddot{A} \\ \ddot{B} \end{Bmatrix} = \begin{bmatrix} K_{a1} & K_{a2} \\ K_{b1} & K_{b2} \end{bmatrix} \begin{Bmatrix} \ddot{q}_1 \\ \ddot{q}_2 \end{Bmatrix} + \dot{q}_1 \frac{\partial}{\partial q_1} \begin{bmatrix} K_{a1} & K_{a2} \\ K_{b1} & K_{b2} \end{bmatrix} \begin{Bmatrix} \dot{q}_1 \\ \dot{q}_2 \end{Bmatrix} + \dot{q}_2 \frac{\partial}{\partial q_2} \begin{bmatrix} K_{a1} & K_{a2} \\ K_{b1} & K_{b2} \end{bmatrix} \begin{Bmatrix} \dot{q}_1 \\ \dot{q}_2 \end{Bmatrix}$$

$$\begin{Bmatrix} \ddot{A} \\ \ddot{B} \end{Bmatrix} = [K] \begin{Bmatrix} \ddot{q}_1 \\ \ddot{q}_2 \end{Bmatrix} + \dot{q}_1 [L_1] \begin{Bmatrix} \dot{q}_1 \\ \dot{q}_2 \end{Bmatrix} + \dot{q}_2 [L_2] \begin{Bmatrix} \dot{q}_1 \\ \dot{q}_2 \end{Bmatrix}$$

where the velocity coefficient partial derivative matrices are

$$[L_1] = \begin{bmatrix} (C_1/B)(1 - K_{a1}) \sin (q_1 - A) & -(C_2/B) K_{a1} \sin (A - q_2) \\ +(C_1/B^2) K_{b1} \cos (q_1 - A) & -(C_2/B^2) K_{b1} \cos (A - q_2) \\ C_1(1 - K_{a1}) \cos (q_1 - A) & C_2 K_{a1} \cos (A - q_2) \end{bmatrix}$$

$$[L_2] = \begin{bmatrix} -(C_1/B) K_{a2} \sin (q_1 - A) & -(C_2/B)(K_{a2} - 1) \sin (A - q_2) \\ +(C_1/B^2) K_{b2} \cos (q_1 - A) & -(C_2/B^2) K_{b2} \cos (A - q_2) \\ -C_1 K_{a2} \cos (q_1 - A) & C_2(K_{a2} - 1) \cos (A - q_2) \end{bmatrix}$$

Notice the role of the individual velocity coefficients in the differentiation to express the velocity coefficient partial derivative matrices.

In the case of multiple degrees of freedom, a family of velocity coefficient derivative matrices appears, equal in number to the number of degrees of freedom. This is emphasized when they are described as velocity coefficient *partial* derivative matrices.

Position, Velocity, and Acceleration for Point P

The base coordinates of point P are readily determined after the secondary variables have been determined:

$$X_p = C_1 \cos q_1 + U_p \cos A - V_p \sin A$$
$$Y_p = C_1 \sin q_1 + U_p \sin A + V_p \cos A$$

With one time differentiation, the velocity components for point P are

$$\begin{Bmatrix} \dot{X}_p \\ \dot{Y}_p \end{Bmatrix} = \begin{bmatrix} -C_1 \sin q_1 & 0 \\ C_1 \cos q_1 & 0 \end{bmatrix} \begin{Bmatrix} \dot{q}_1 \\ \dot{q}_2 \end{Bmatrix} + \begin{bmatrix} -U_p \sin A + V_p \cos A & 0 \\ U_p \cos A - V_p \sin A & 0 \end{bmatrix} \begin{Bmatrix} \dot{A} \\ \dot{B} \end{Bmatrix}$$

$$= \begin{bmatrix} -C_1 \sin q_1 & 0 \\ C_1 \cos q_1 & 0 \end{bmatrix} \begin{Bmatrix} \dot{q}_1 \\ \dot{q}_2 \end{Bmatrix} + \begin{bmatrix} -U_p \sin A + V_p \cos A & 0 \\ U_p \cos A - V_p \sin A & 0 \end{bmatrix} \begin{bmatrix} K_{a1} & K_{a2} \\ K_{b1} & K_{b2} \end{bmatrix} \begin{Bmatrix} \dot{q}_1 \\ \dot{q}_2 \end{Bmatrix}$$

$$= \begin{bmatrix} -C_1 \sin q_1 - K_{a1}(U_p \sin A + V_p \cos A) & -K_{a2}(U_p \sin A + V_p \cos A) \\ C_1 \cos q_1 + K_{a1}(U_p \cos A - V_p \sin A) & K_{a2}(U_p \cos A - V_p \sin A) \end{bmatrix} \begin{Bmatrix} \dot{q}_1 \\ \dot{q}_2 \end{Bmatrix}$$

$$= \begin{bmatrix} K_{px1} & K_{px2} \\ K_{py1} & K_{py2} \end{bmatrix} \begin{Bmatrix} \dot{q}_1 \\ \dot{q}_2 \end{Bmatrix}$$

In the final result shown, the coefficient matrix is the velocity coefficient matrix for point P. Notice again that a rectangular array, not a column vector, is obtained.

With $\ddot{A}$ and $\ddot{B}$ already determined, the quickest way to obtain the acceleration of point P is simply to differentiate the velocity equations with respect to time, bearing in mind that the secondary accelerations are already known. Beginning with the first expression given for the velocities $\dot{X}_p$ and $\dot{Y}_p$, and differentiating, the acceleration expressions are:

$$\ddot{X}_p = -C_1\ddot{q}_1 \sin q_1 - C_1\dot{q}_1^2 \cos q_1 - \ddot{A}(U_p \sin A + V_p \cos A) - \dot{A}^2(U_p \cos A - V_p \sin A)$$

$$\ddot{Y}_p = C_1\ddot{q}_1 \cos q_1 - C_1\dot{q}_1^2 \sin q_1 + \ddot{A}(U_p \cos A - V_p \sin A) - \dot{A}^2(U_p \sin A + V_p \cos A)$$

Numerical Values. The position and velocity calculations just described have been carried out numerically with the following system parameters:

$C_1 = 1.5$ in. $\quad U_p = 9.0$ in.
$C_2 = 2.2$ in. $\quad V_p = 0.0$ in.
$C_3 = 3.5$ in.
$L = 9.0$ in.

With the input crank angles at $q_1 = 0.85$ radians and $q_2 = 0.25$ radians, the position solution gives the following results:

$$A = -0.125 \text{ radians}$$
$$B = 4.678 \text{ in.}$$
$$X_p = 9.920 \text{ in.}$$
$$Y_p = 6(10^{-3}) \text{ in.}$$

Purely by coincidence, point P is very close to the X–axis for this example. This is consistent with the crank angles and the fact that A has a negative value. In this configuration, and with input rotation rates $\dot{q}_1 = -2.6$ rad/s and $\dot{q}_2 = 3.5$ rad/s, the system velocities are

$$\dot{A} = 1.581 \text{ rad/s}$$
$$\dot{B} = -2.573 \text{ in./s}$$

$$\dot{X}_P = 4.703 \text{ in./s}$$
$$\dot{Y}_P = 11.548 \text{ in./s}$$

The cranks are rotating in opposite senses, which results in a relatively large velocity magnitude for point P, $|\mathbf{V}_P| = 12.469$ in./s $(= \sqrt{(4.703)^2 + (11.548)^2})$. For these positions and velocities, and with the input crank accelerations $\ddot{q}_1 = 0.42$ rad/s^2 and $\ddot{q}_2 = 0.68$ rad/s^2, the system accelerations are

$$\ddot{A} = 2.701 \text{ rad/s}^2$$
$$\ddot{B} = -3.946 \text{ in./s}^2$$
$$\ddot{X}_P = -26.470 \text{ in./s}^2$$
$$\ddot{Y}_P = 19.722 \text{ in./s}^2$$

The acceleration magnitude at point P is $|\mathbf{A}_P| = \sqrt{(-26.470)^2 + (19.722)^2} = 33.009$ in./s^2.

3.2 KINEMATIC ANALYSIS WITH NUMERICAL SOLUTION

Multidegree of freedom mechanisms that lend themselves to complete, closed-form analysis such as in the preceding section are relatively rare; mechanisms requiring numerical solutions for the describing kinematic equations are much more common.

Position and Velocity Analysis. The position solution process is exactly as it was before, beginning with the position loop equations. The solution is obtained by the Newton–Raphson method; its implementation requires the Jacobian matrix, which may be obtained directly or as a part of the velocity analysis. The position loop equations are differentiated with respect to time to give the velocity equations. In preparation to determine the unknown velocities, the velocity equations can be written in matrix form as

$$\underset{(N_2 \times N_2)}{[J]} \; \underset{(N_2 \times 1)}{\{\dot{S}\}} = \underset{(N_2 \times N_1)}{[B]} \; \underset{(N_1 \times 1)}{\{\dot{q}\}}$$

where N_1 is the number of primary variables, and N_2 is the number of secondary variables. The matrices in this equation are $[J]$, the Jacobian, $\{\dot{S}\}$, the column vector of unknown, secondary velocities, $[B]$, a rectangular matrix of coefficients, and $\{\dot{q}\}$, the column vector of primary velocities. The velocity coefficient matrix, $[K]$, is defined by

$$\{\dot{S}\} = [K]\{\dot{q}\}$$

Consequently, it is evident that $[K]$ can be written as

$$[K] = [J]^{-1}[B]$$

For computational efficiency and reduced roundoff error, the velocity coefficient matrix is obtained as the solution of a system of simultaneous linear equations:

$$\underset{(N_2 \times N_2)}{[J]} \quad \underset{(N_2 \times N_1)}{[K]} \quad = \quad \underset{(N_2 \times N_1)}{[B]}$$

For the solution of such a system, see the discussion in Appendix A1.4.

Acceleration Analysis. Secondary accelerations are expressed by differentiating the expression for the secondary velocities in terms of the velocity coefficient matrix and the column vector of primary velocities. In general, the elements of the velocity coefficient matrix are functions of all of the primary variables, so the chain rule is required for this differentiation. The resulting expression for the secondary accelerations is as follows:

$$\begin{aligned}\{\ddot{S}\} &= \dot{q}_1 \frac{\partial[K]}{\partial q_1}\{\dot{q}\} + \dot{q}_2 \frac{\partial[K]}{\partial q_2}\{\dot{q}\} + \cdots + [K]\{\ddot{q}\} \\ &= \dot{q}_1[L_1]\{\dot{q}\} + \dot{q}_2[L_2]\{\dot{q}\} + \cdots + [K]\{\ddot{q}\}\end{aligned}$$

where the notation $[L_i]$ represents $\partial[K]/\partial q_i$. As introduced in the preceding section, the $[L_i]$ matrices are the velocity coefficient partial derivative matrices, analogous to the velocity coefficient derivative vectors found in single degree of freedom systems. Differentiation is with respect to the primary variable denoted by the subscript i. The next question is how to obtain the $[L_i]$ matrices numerically.

The velocity coefficient matrix is defined by the relation

$$[J][K] = [B]$$

where the elements of the Jacobian matrix, $[J]$, and the right-side matrix, $[B]$, are known explicitly. This expression is differentiated with respect to a typical primary variable, q_i, with the result:

$$\frac{\partial[J]}{\partial q_i}[K] + [J]\frac{\partial[K]}{\partial q_i} = \frac{\partial[B]}{\partial q_i}$$

The expression is solved for the term containing the required derivative matrix as a factor:

$$[J]\frac{\partial[K]}{\partial q_i} = \frac{\partial[B]}{\partial q_i} - \frac{\partial[J]}{\partial q_i}[K]$$

This is simply a system of linear equations in the elements of the required derivative matrix. Note that the coefficient is the Jacobian matrix, as would be expected. The formal solution for the velocity coefficient derivative is as follows:

$$[L_i] = \frac{\partial[K]}{\partial q_i} = [J]^{-1}\left(\frac{\partial[B]}{\partial q_i} - \frac{\partial[J]}{\partial q_i}[K]\right)$$

However, the numerical evaluation is obtained better as the solution of a system of linear simultaneous equations. With the velocity coefficient derivative matrices determined, the secondary accelerations can be evaluated from the expression just developed. For the following example, secondary positions, velocities, and accelerations are determined as well as center of mass positions and velocities.

3.2.1 Four-Bar Mechanism with Translating Crank Pivot

The mechanism shown in Figure 3.2 is called a "four-bar mechanism" with translating crank pivot. In reality, it is not a four-bar mechanism at all because there is no fourth bar of fixed length. The terminology only suggests the association. The two primary variables are the crank pivot location and the crank rotation, q_0 and q_1, respectively. The two secondary variables are the angles A_2 and A_3. The second crank pivot is stationary at the point (X_4, Y_4). Position, velocity, and acceleration are to be determined for the secondary variables, and the position and velocity are required for each center of mass.

Position Analysis. The position loop equations depend on both of the primary variables, as follows:

$$q_0 + C_1 \cos q_1 + C_2 \cos A_2 + C_3 \cos A_3 - X_4 = 0$$

$$C_1 \sin q_1 + C_2 \sin A_2 + C_3 \sin A_3 - Y_4 = 0$$

For specified values of q_0 and q_1, these two equations are to be solved for A_2 and A_3. A closed-form solution may be possible, but a numerical solution will certainly be simpler to obtain. Such a solution is assumed to have been completed so as to proceed to the next stage of analysis.

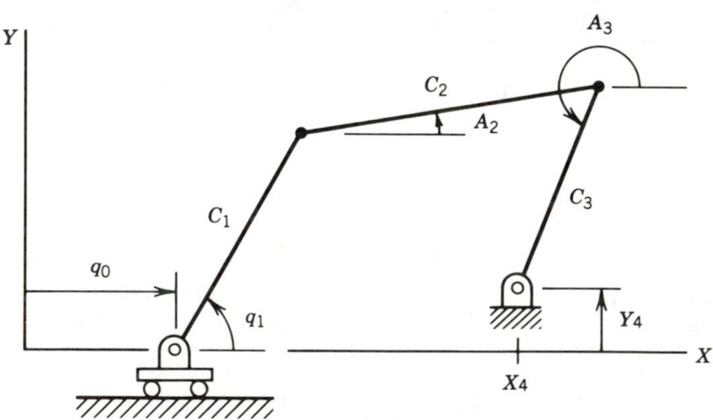

FIGURE 3.2 Kinematic Skeleton Four-Bar Mechanism with Translating Crank Pivot

Velocity Analysis. The position loop equations are differentiated to give the velocity relations:

$$\begin{bmatrix} -C_2 \sin A_2 & -C_3 \sin A_3 \\ C_2 \cos A_2 & C_3 \cos A_3 \end{bmatrix} \begin{Bmatrix} \dot{A}_2 \\ \dot{A}_3 \end{Bmatrix} = \begin{bmatrix} -1 & C_1 \sin q_1 \\ 0 & -C_2 \cos q_1 \end{bmatrix} \begin{Bmatrix} \dot{q}_0 \\ \dot{q}_1 \end{Bmatrix}$$

The left side of this equation consists of the Jacobian matrix that premultiplies the column vector of secondary velocities. The right side has the coefficient matrix $[B]$ multiplying the column vector of primary velocities. A numerical solution of this system of linear equations is accomplished readily, with the result expressing the secondary velocities $\dot{A}_2$ and $\dot{A}_3$ as linear combinations of $\dot{q}_0$ and $\dot{q}_1$:

$$\begin{Bmatrix} \dot{A}_2 \\ \dot{A}_3 \end{Bmatrix} = \begin{bmatrix} K_{a20} & K_{a21} \\ K_{a30} & K_{a31} \end{bmatrix} \begin{Bmatrix} \dot{q}_0 \\ \dot{q}_1 \end{Bmatrix}$$

Acceleration Analysis. Differentiating the velocity equation above gives the secondary accelerations:

$$\begin{Bmatrix} \ddot{A}_2 \\ \ddot{A}_3 \end{Bmatrix} = \dot{q}_0[L_0] \begin{Bmatrix} \dot{q}_0 \\ \dot{q}_1 \end{Bmatrix} + \dot{q}_1[L_1] \begin{Bmatrix} \dot{q}_0 \\ \dot{q}_1 \end{Bmatrix} + [K] \begin{Bmatrix} \ddot{q}_0 \\ \ddot{q}_1 \end{Bmatrix}$$

The necessary velocity coefficient derivative matrices are determined as solutions of the following two equations:

$$[J][L_0] = \begin{bmatrix} 0 & 0 \\ 0 & 0 \end{bmatrix} + \begin{bmatrix} C_2K_{a20} \cos A_2 & C_3K_{a30} \cos A_3 \\ C_2K_{a20} \sin A_2 & C_3K_{a30} \sin A_3 \end{bmatrix} [K]$$

$$[J][L_1] = \begin{bmatrix} 0 & C_1 \cos q_1 \\ 0 & C_1 \sin q_1 \end{bmatrix} + \begin{bmatrix} C_2K_{a21} \cos A_2 & C_3K_{a31} \cos A_3 \\ C_2K_{a21} \sin A_2 & C_3K_{a31} \sin A_3 \end{bmatrix} [K]$$

The secondary accelerations are now readily evaluated in terms of the primary velocities and accelerations and the coefficient matrices $[K]$, $[L_0]$, and $[L_1]$.

Base Coordinates and Velocities. Consider next the matter of base coordinates and base coordinate velocities for the centers of mass for the three links as shown in Figure 3.3. The centers of mass are located in their respective body coordinate systems by the body coordinate pairs (U_{c1}, V_{c1}), (U_{c2}, V_{c2}), and (U_{c3}, V_{c3}). The center of mass position vectors are written by inspection:

$$\begin{Bmatrix} X_{c1} \\ Y_{c1} \end{Bmatrix} = \begin{Bmatrix} q_0 + U_{c1} \cos q_1 - V_{c1} \sin q_1 \\ U_{c1} \sin q_1 + V_{c1} \cos q_1 \end{Bmatrix}$$

$$\begin{Bmatrix} X_{c2} \\ Y_{c2} \end{Bmatrix} = \begin{Bmatrix} q_0 + C_1 \cos q_1 + U_{c2} \cos A_2 - V_{c2} \sin A_2 \\ C_1 \sin q_1 + U_{c2} \sin A_2 + V_{c2} \cos A_2 \end{Bmatrix}$$

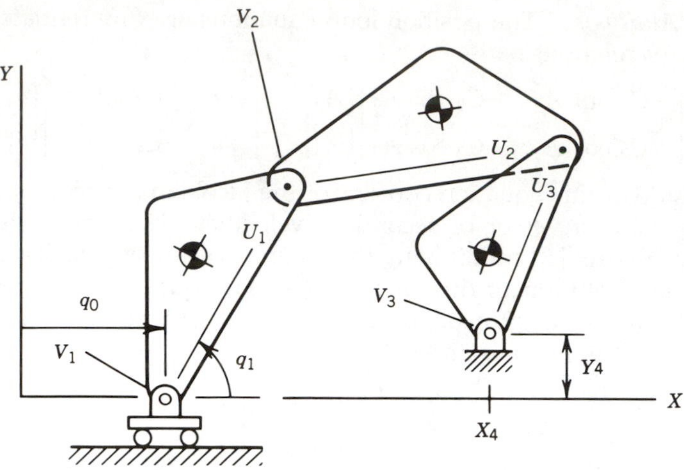

FIGURE 3.3 Pictorial Representation for a Four-Bar Mechanism with Translating Crank Pivot Showing CM Locations

$$\begin{Bmatrix} X_{c3} \\ Y_{c3} \end{Bmatrix} = \begin{Bmatrix} X_4 - U_{c3} \cos A_3 + V_{c3} \sin A_3 \\ Y_4 - U_{c3} \sin A_3 - V_{c3} \cos A_3 \end{Bmatrix}$$

When these expressions are differentiated with respect to time, in addition to terms involving $\dot{q}_0$ and $\dot{q}_1$, there will also be terms involving $\dot{A}_2$ and $\dot{A}_3$. The elements of the rectangular velocity coefficient matrix will be required to re-express the latter in terms of $\dot{q}_0$ and $\dot{q}_1$. With the required substitutions made, the center of mass velocity components are

$$\begin{Bmatrix} \dot{X}_{c1} \\ \dot{Y}_{c1} \end{Bmatrix} = \begin{bmatrix} 1 & -U_{c1} \sin q_1 - V_{c1} \cos q_1 \\ 0 & U_{c1} \cos q_1 - V_{c1} \sin q_1 \end{bmatrix} \begin{Bmatrix} \dot{q}_0 \\ \dot{q}_1 \end{Bmatrix}$$

$$= [K_1] \begin{Bmatrix} \dot{q}_0 \\ \dot{q}_1 \end{Bmatrix}$$

$$\begin{Bmatrix} \dot{X}_{c2} \\ \dot{Y}_{c2} \end{Bmatrix} = \begin{bmatrix} K_{x20} & K_{x21} \\ K_{y20} & K_{y21} \end{bmatrix} \begin{Bmatrix} \dot{q}_0 \\ \dot{q}_1 \end{Bmatrix} = [K_2] \begin{Bmatrix} \dot{q}_0 \\ \dot{q}_1 \end{Bmatrix}$$

where

$$K_{x20} = 1 - K_{a20}(U_{c2} \sin A_2 + V_{c2} \cos A_2)$$
$$K_{x21} = -C_1 \sin q_1 - K_{a21}(U_{c2} \sin A_2 + V_{c2} \cos A_2)$$
$$K_{y20} = K_{a20}(U_{c2} \sin A_2 - V_{c2} \sin A_2)$$
$$K_{y21} = C_1 \cos q_1 + K_{a21}(U_{c2} \cos A_2 - V_{c2} \sin A_2)$$

$$\begin{Bmatrix} \dot{X}_{c3} \\ \dot{Y}_{c3} \end{Bmatrix} = \begin{bmatrix} K_{x30} & K_{x31} \\ K_{y30} & K_{y31} \end{bmatrix} \begin{Bmatrix} \dot{q}_0 \\ \dot{q}_1 \end{Bmatrix} = [K_3] \begin{Bmatrix} \dot{q}_0 \\ \dot{q}_1 \end{Bmatrix}$$

where

$$K_{x30} = K_{a30}(U_{c3} \sin A_3 + V_{c3} \cos A_3)$$
$$K_{x31} = K_{a31}(U_{c3} \sin A_3 + V_{c3} \cos A_3)$$
$$K_{y30} = K_{a30}(-U_{c3} \cos A_3 + V_{c3} \sin A_3)$$
$$K_{y31} = K_{a31}(-U_{c3} \cos A_3 + V_{c3} \sin A_3)$$

For some purposes (see Section 8.4 for details), it will be useful to combine the foregoing into a single velocity coefficient matrix, which gives all of the center-of-mass velocity components and the rotation rates for all links. Such a velocity coefficient matrix will be (9×2), and may be defined as

$$\begin{Bmatrix} \dot{X}_{c1} \\ \dot{Y}_{c1} \\ \dot{X}_{c2} \\ \dot{Y}_{c2} \\ \dot{X}_{c3} \\ \dot{Y}_{c3} \\ \dot{q}_1 \\ \dot{A}_2 \\ \dot{A}_3 \end{Bmatrix} = \begin{bmatrix} & [K_1] & \\ & [K_2] & \\ & [K_3] & \\ 0 & & 1 \\ K_{a20} & & K_{a21} \\ K_{a30} & & K_{a31} \end{bmatrix} \begin{Bmatrix} \dot{q}_0 \\ \dot{q}_1 \end{Bmatrix} = [K] \begin{Bmatrix} \dot{q}_0 \\ \dot{q}_1 \end{Bmatrix}$$

It is a simple matter to evaluate this (9×2) velocity coefficient matrix numerically. Note that for this particular problem, it is also possible to obtain analytic expressions for all of the elements of the velocity coefficient matrix.

3.3 NUMERICAL SOLUTION FOR MULTILOOP MECHANISMS

Each of the examples considered to this point have involved only a single position loop, resulting in two scalar loop equations. However, this is by no means the only case. As with single degree of freedom mechanisms, there are multiloop, multiple degree of freedom systems that result in one pair of position loop equations for each independent loop. This increases the number of equations to be dealt with simultaneously, but through matrix notation this is not significantly more difficult. The following example will demonstrate the process.

3.3.1 Power Shovel

A typical crawler-mounted power shovel is shown pictorially in Figure 3.4. The three hydraulic cylinders control the motion of the boom and the bucket, and immediately suggest that the mechanism has three degrees of freedom.

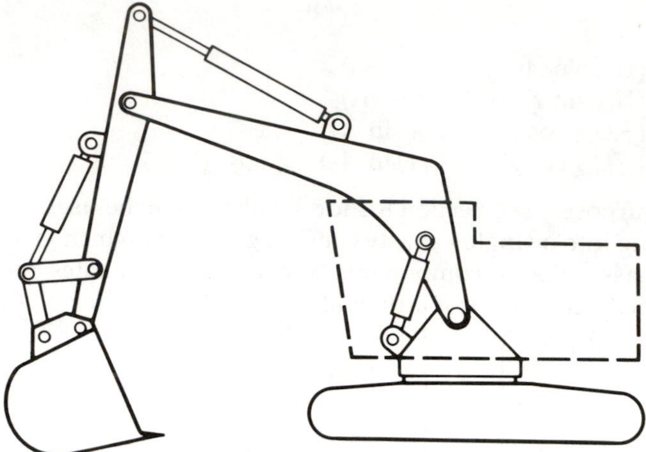

FIGURE 3.4 Pictorial Representation of a Power Shovel

The operator, by means of the hydraulic control valves in the cab, controls the lengths of these three components. Consequently, these may be considered as the three input variables. The dimensions and kinematic variables are shown on Figure 3.5, where the three hydraulic cylinder lengths are identified as q_1, q_2, and q_3. The dimensions $C_1, C_2, \ldots C_{15}$ are lengths from a pivot or welded intersection to the next such point, without regard for the fact that a member may continue beyond the end of a specific dimension. The angular dimensions C_{16} and C_{17} are fixed and known. The secondary variables are A_1, B_1, A_2, B_2, A_3, B_3, A_4, and B_4; E is also an unknown, but it is readily expressible

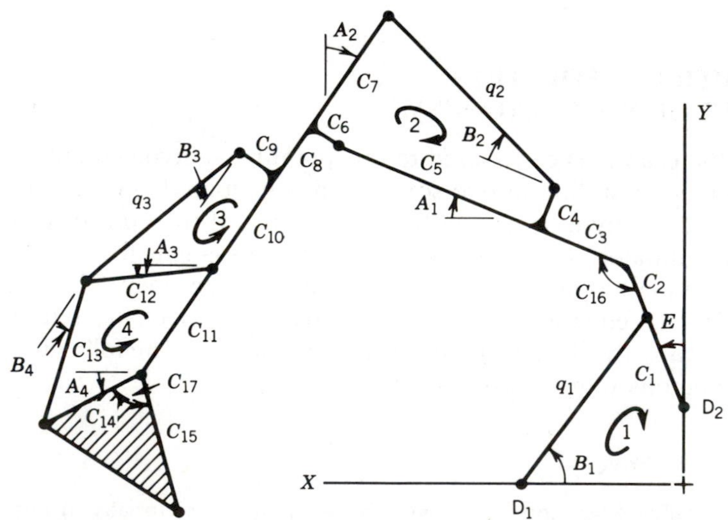

FIGURE 3.5 Kinematic Skeleton for the Power Shovel Showing Segment Lengths, Angles, and Loops

in terms of A_1. Note that all of the As are measured from either a horizontal or vertical reference to one of the machine members. The Bs, on the other hand, measure the relative angle from a major member to a nearby member.

Position Analysis. For the position analysis, four loops must be described. They are identified on Figure 3.5. Before writing the loop equations, note that the angle E is dependent on A_1:

$$E = C_{16} - A_1 - \pi/2$$

The loop equations are

Loop 1

$$D_1 - q_1 \cos B_1 - C_1 \sin E = 0$$
$$D_2 - q_1 \sin B_1 + C_1 \cos E = 0$$

Loop 2

$$C_5 \cos A_1 + C_4 \sin A_1 + C_6 \cos A_2 - C_7 \sin A_2 - q_2 \cos (A_1 + B_2) = 0$$
$$C_5 \sin A_1 - C_4 \cos A_1 + C_6 \sin A_2 + C_7 \cos A_2 - q_2 \sin (A_1 + B_2) = 0$$

Loop 3

$$C_9 \cos A_2 + q_3 \sin(A_2 + B_3) - C_{12} \cos A_3 - C_{10} \sin A_2 = 0$$
$$C_9 \sin A_2 - q_3 \cos(A_2 + B_3) + C_{12} \sin A_3 + C_{10} \cos A_2 = 0$$

Loop 4

$$C_{12} \cos A_3 + C_{13} \sin(A_2 - B_4) - C_{14} \cos A_4 - C_{11} \sin A_2 = 0$$
$$-C_{12} \sin A_3 - C_{13} \cos(A_2 - B_4) + C_{14} \sin A_4 + C_{11} \cos A_2 = 0$$

One approach to this problem would be to treat this simply as a system of nine nonlinear equations to be solved simultaneously. Alternatively, because the equations above are weakly coupled, they can be considered as four pairs of equations, each pair to be solved in turn. If this alternate approach is adopted, the equations from the first loop give values for B_1 and E; these are used to evaluate A_1, and E is not required again. Thereafter, the equations of the second loop give A_2 and B_2, and so forth.

Velocity Analysis. The velocity analysis is based on time differentiation of the position loop equations. With E eliminated and the remaining eight equations differentiated, the resulting system of equations can be written in matrix form as

$$\underset{(8 \times 8)}{[J]} \quad \underset{(8 \times 1)}{\{\dot{S}\}} = \underset{(8 \times 3)}{[B]} \quad \underset{(3 \times 1)}{\{\dot{q}\}}$$

where

$[J]$ is the Jacobian matrix
$\{\dot{S}\} = \text{Col}(\dot{A}_1, \dot{B}_1, \dot{A}_2, \dot{B}_2, \dot{A}_3, \dot{B}_3, \dot{A}_4, \dot{B}_4)$, the vector of secondary velocities
$[B]$ is the matrix of coefficients for the input velocities
$\{\dot{q}\} = \text{Col}(\dot{q}_1, \dot{q}_2, \dot{q}_3)$, the vector of input velocities

The non-zero entries in the Jacobian matrix are

$$J(1,1) = C_1(-\cos C_{16} \sin A_1 + \sin C_{16} \cos A_1)$$
$$J(1,2) = q_1 \sin B_1$$
$$J(2,1) = -C_1(\sin C_{16} \sin A_1 + \cos C_{16} \cos A_1)$$
$$J(2,2) = -q_1 \cos B_1$$
$$J(3,1) = C_4 \cos A_1 - C_5 \sin A_1 + q_2 \sin (A_1 + B_2)$$
$$J(3,3) = -(C_6 \sin A_2 + C_7 \cos A_2)$$
$$J(3,4) = q_2 \sin (A_1 + B_2)$$
$$J(4,1) = C_4 \sin A_1 + C_5 \cos A_1 - q_2 \cos (A_1 + B_2)$$
$$J(4,3) = C_6 \cos A_2 - C_7 \sin A_2$$
$$J(4,4) = -q_2 \cos (A_1 + B_2)$$
$$J(5,3) = -(C_9 \sin A_2 + C_{10} \cos A_2) + q_3 \cos (A_2 + B_3)$$
$$J(5,5) = C_{12} \sin A_3$$
$$J(5,6) = q_3 \cos (A_2 + B_3)$$
$$J(6,3) = C_9 \cos A_2 - C_{10} \sin A_2 + q_3 \sin (A_2 + B_3)$$
$$J(6,5) = C_{12} \cos A_3$$
$$J(6,6) = q_3 \sin (A_2 + B_3)$$
$$J(7,3) = C_{13} \cos (A_2 - B_4) - C_{11} \cos A_2$$
$$J(7,5) = -C_{12} \sin A_3$$
$$J(7,7) = C_{14} \sin A_4$$
$$J(7,8) = -C_{13} \cos (A_2 - B_4)$$
$$J(8,3) = C_{13} \sin (A_2 - B_4) - C_{11} \sin A_2$$
$$J(8,5) = -C_{12} \cos A_3$$
$$J(8,7) = C_{14} \cos A_4$$
$$J(8,8) = -C_{13} \sin (A_2 - B_4)$$

The non-zero entries in the $[B]$ matrix are

$$B(1,1) = \cos B_1$$
$$B(2,1) = \sin B_1$$
$$B(3,2) = \cos(A_1 + B_2)$$
$$B(4,2) = \sin(A_1 + B_2)$$
$$B(5,3) = -\sin(A_2 + B_3)$$
$$B(6,3) = \cos(A_2 + B_3)$$

The formal solution for the velocities is then

$$\underset{(8\times 1)}{\{\dot S\}} = \underset{(8\times 8)}{[J]^{-1}}\ \underset{(8\times 3)}{[B]}\ \underset{(3\times 1)}{\{\dot q\}}$$

The velocity coefficient matrix, $[K]$, is again defined such that

$$\underset{(8\times 1)}{\{\dot S\}} = \underset{(8\times 3)}{[K]}\ \underset{(3\times 1)}{\{\dot q\}}$$

and, as before, $[K]$ is the solution of

$$[J][K] = [B]$$

Acceleration Analysis. The velocity relation is differentiated with respect to time to express the secondary accelerations:

$$\{\ddot S\} = \dot q_1 \frac{\partial[K]}{\partial q_1}\{\dot q\} + \dot q_2 \frac{\partial[K]}{\partial q_2}\{\dot q\} + \dot q_3 \frac{\partial[K]}{\partial q_3}\{\dot q\} + [K]\{\ddot q\}$$

or, in the abbreviated notation for the velocity coefficient derivative,

$$\{\ddot S\} = \dot q_1[L_1]\{\dot q\} + \dot q_2[L_2]\{\dot q\} + \dot q_3[L_3]\{\dot q\} + [K]\{\ddot q\}$$

For numerical determination of the velocity coefficient derivatives, the defining relation for the velocity coefficient matrix is differentiated with respect to one of the primary variables,

$$\frac{\partial[J]}{\partial q_i}[K] + [J]\frac{\partial[K]}{\partial q_i} = \frac{\partial[B]}{\partial q_i}$$

The second factor of the second term is $[L_i]$, the required velocity coefficient derivative, and the equation can be cast as a system of simultaneous equations solvable for the elements of $[L_i]$:

$$\underset{(8\times 8)}{[J]}\ \underset{(8\times 3)}{[L_i]} = -\underset{(8\times 8)}{\frac{\partial[J]}{\partial q_i}}\ \underset{(8\times 3)}{[K]} + \underset{(8\times 3)}{\frac{\partial[B]}{\partial q_i}}$$

The elements of $[J]$ and $[B]$ are all known, so these partial derivatives can be evaluated directly. Let the first of the three required partial derivatives of $[J]$ be denoted $[J_1]$, where

$$[J_1] = \frac{\partial[J]}{\partial q_1}$$

The non-zero elements of $[J_1]$ are as follows:

$$J_1(1,1) = -C_1(\cos C_{16}\cos A_1 + \sin C_{16}\sin A_1)K(1,1)$$
$$J_1(1,2) = \sin B_1 + q_1\cos B_1\, K(2,1)$$
$$J_1(2,1) = C_1(-\sin C_{16}\cos A_1 + \cos C_{16}\sin A_1)\, K(1,1)$$
$$J_1(2,2) = -\cos B_1 + q_1\sin B_1 K(2,1)$$

$$J_1(3,1) = -[C_4 \sin A_1 + C_5 \cos A_1 - q_2 \cos (A_1 + B_2)]K(1,1)$$
$$J_1(3,3) = (-C_6 \cos A_2 + C_7 \sin A_2)K(3,1)$$
$$J_1(3,4) = q_2 \cos (A_1 + B_2)\ K(1,1)$$
$$J_1(4,1) = [C_4 \cos A_1 - C_5 \sin A_1 + q_2 \sin (A_1 + B_2)]K(1,1)$$
$$J_1(4,3) = -(C_6 \sin A_2 + C_7 \cos A_2)K(3,1)$$
$$J_1(4,4) = q_2 \sin (A_1 + B_2)\ K(1,1)$$
$$J_1(5,3) = -[C_9 \cos A_2 - C_{10} \sin A_2 + q_3 \sin (A_2 + B_3)]K(3,1)$$
$$J_1(5,5) = C_{12} \cos A_3\ K(5,1)$$
$$J_1(5,6) = -q_3 \sin (A_2 + B_3)\ K(3,1)$$
$$J_1(6,3) = -[C_9 \sin A_2 + C_{10} \cos A_2 - q_3 \cos (A_2 + B_3)]K(3,1)$$
$$J_1(6,5) = -C_{12} \sin A_3\ K(5,1)$$
$$J_1(6,6) = q_3 \cos (A_2 + B_3)\ K(3,1)$$
$$J_1(7,3) = -[C_{13} \sin (A_2 - B_4) + C_{11} \sin A_2]K(3,1)$$
$$J_1(7,5) = -C_{12} \cos A_3\ K(5,1)$$
$$J_1(7,7) = C_{14} \cos A_4\ K(7,1)$$
$$J_1(7,8) = C_{13} \sin (A_2 - B_4)\ K(3,1)$$
$$J_1(8,3) = [C_{13} \cos (A_2 - B_4) - C_{11} \cos A_2]K(3,1)$$
$$J_1(8,5) = C_{12} \sin A_3\ K(5,1)$$
$$J_1(8,7) = -C_4 \sin A_4\ K(7,1)$$
$$J_1(8,8) = -C_{13} \cos (A_2 - B_4)\ K(3,1)$$

In writing out the elements of $[J_1]$, some terms appear to have been omitted; however, these terms involve a zero factor from the velocity coefficient matrix, $[K]$. These zero factors also will be taken into account in expressing the derivative of $[B]$, which follows. In a similar manner, the partial derivative of $[B]$ with respect to q_1 is denoted $[B_1]$, for which the non-zero elements are the following:

$$B_1(1,1) = -\sin B_1\ K(2,1)$$
$$B_1(2,1) = \cos B_1\ K(2,1)$$
$$B_1(3,2) = -\sin(A_1 + B_2)\ K(1,1)$$
$$B_1(4,2) = \cos(A_1 + B_2)\ K(1,1)$$
$$B_1(5,3) = -\cos(A_2 + B_3)\ K(3,1)$$
$$B_1(6,3) = -\sin(A_2 + B_3)\ K(3,1)$$

With these two matrices and the previously determined velocity coefficient matrix, the matrix $[L_1]$ can be evaluated now. Similar steps are required to evaluate the two remaining velocity coefficient matrices, $[L_2]$ and $[L_3]$. Once the velocity coefficient matrix and the three velocity coefficient partial derivative matrices are known, the secondary accelerations are readily evaluated, as just indicated.

Numerical Values. The kinematic analysis of the power shovel has been evaluated numerically based on the following dimensions, all in inches except where noted:

$$
\begin{array}{ll}
C_1 = 30 & C_{10} = 60 \\
C_2 = 14 & C_{11} = 14 \\
C_3 = 30 & C_{12} = 20 \\
C_4 = 6 & C_{13} = 18 \\
C_5 = 70 & C_{14} = 10 \\
C_6 = 3 & C_{15} = 21 \\
C_7 = 25 & C_{16} = 2.12 \text{ radians} \\
C_8 = 12 & C_{17} = 1.42 \text{ radians} \\
C_9 = 6 &
\end{array}
$$

For the assigned values $q_1 = 46$ in., $q_2 = 76$ in., and $q_3 = 70$ in., the position solution angles are

$$
\begin{array}{ll}
A_1 = 0.219570 \text{ radians} & B_1 = 1.231979 \text{ radians} \\
A_2 = 0.195236 & B_2 = 0.251588 \\
A_3 = 0.270786 & B_3 = 0.170355 \\
A_4 = 0.771711 & B_4 = 0.743913
\end{array}
$$

This position is shown in the computer-generated Figure 3.6. The tick marks are at 10-inch intervals on both axes; note that there is some distortion, as seen in the slightly different spacing of the tick marks on the two axes. For the machine in this position, the velocity coefficient matrix is

$$
[K] = \begin{bmatrix}
0.053783 & 0.0 & 0.0 \\
-0.027527 & 0.0 & 0.0 \\
0.053783 & -0.043035 & 0.0 \\
0.0 & -0.005491 & 0.0 \\
-0.053783 & 0.043035 & 0.062169 \\
0.0 & 0.0 & -0.010556 \\
-0.053783 & 0.043035 & 0.122606 \\
0.0 & 0.0 & 0.034016
\end{bmatrix}
$$

Certain patterns in the velocity coefficient matrix are of interest. Notice that $K_{a11}(K_{11})$, $K_{a21}(K_{31})$, $K_{a31}(K_{51})$, and $K_{a41}(K_{71})$ all have the same numerical value, although they have different signs. Why does this happen? Notice that $K_{a22}(K_{32})$, $K_{a32}(K_{52})$, and $K_{a42}(K_{72})$ all have the same numerical value, although again different signs. Again, why? What is the significance of the zeroes in the first column? In the second column? In the third

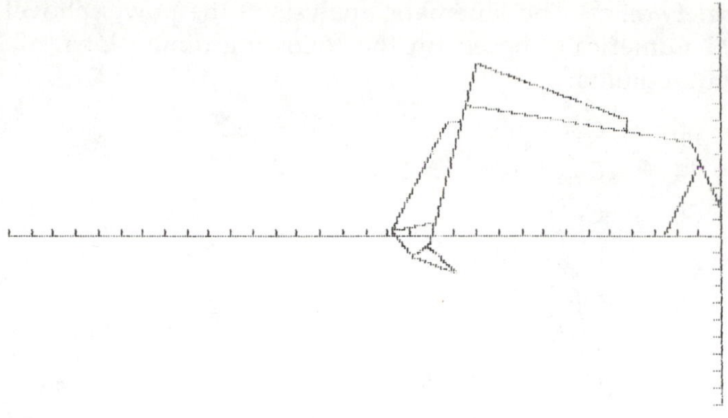

FIGURE 3.6 Computer Generated Configuration Plot for the Power Shovel

column? Do the signs of the various elements appear reasonable for the configuration shown? All of these questions should be readily answerable, if the material presented has been understood.

With the machine in this same position, the velocity coefficient matrix $[L_1]$ is evaluated as described earlier. The result is

$$[L_1] = (10^{-2})\begin{bmatrix} 0.178812 & 0.0 & 0.0 \\ -0.088756 & 0.0 & 0.0 \\ 0.178812 & 0.0 & 0.0 \\ 0.0 & 0.0 & 0.0 \\ -0.178812 & 0.0 & 0.0 \\ 0.0 & 0.0 & 0.0 \\ -0.165373 & -0.010753 & 0.160926 \\ -0.096208 & 0.076981 & -0.073073 \end{bmatrix}$$

At this point, evaluation of the secondary accelerations must wait, pending numerical evaluation of the two remaining velocity coefficient partial derivative matrices.

3.4 GENERAL ANALYSIS FOR MULTIDEGREE OF FREEDOM MECHANISMS

As shown by the preceding examples, there are mechanisms for which more than one generalized coordinate is required to specify the system configuration; these are multidegree of freedom mechanisms. Such mechanisms may involve one or more position vector loops, as has been demonstrated. In this section, the analyses demonstrated in the preceding example problems are summarized in general form. The notation is as follows:

N_1 = number of generalized coordinates
N_2 = number of secondary coordinates
q_i = generalized coordinates, i = 1, 2, . . . N_1
s_i = secondary coordinates, i = 1, 1, . . . N_2

Position Analysis. The position analysis begins by formulating the position vector loop equations in scalar form, just as for a single degree of freedom case. Any necessary constraint equations are also included with the position loop equations. This gives rise to a system of simultaneous, nonlinear equations of the form

$$f_1(q_1, q_2, \ldots q_{N1}, s_1, s_2, s_3, \ldots s_{N2}) = 0$$
$$f_2(q_1, q_2, \ldots q_{N1}, s_1, s_2, s_3, \ldots s_{N2}) = 0$$
$$\vdots \qquad\qquad\qquad \vdots$$

With all of the generalized coordinates specified, this is a system of N_2 equations to be solved for the N_2 unknown secondary coordinates. These may be solved in closed form or numerically using the Newton–Raphson method. The fact that there are multiple generalized coordinates has no effect on the solution for the secondary coordinates because all of the generalized coordinate values are specified. The position solution vector, {S}, and the residual vector, {F}, are defined as before:

$$\{S\} = \mathrm{Col}(s_1, s_2, s_3, \ldots s_{N2})$$
$$\{F\} = \mathrm{Col}(f_1, f_2, f_3, \ldots f_{N2})$$

The Jacobian matrix is

$$\left[\frac{\partial f}{\partial s}\right] = \begin{bmatrix} \frac{\partial f_1}{\partial s_1} & \frac{\partial f_1}{\partial s_2} & \frac{\partial f_1}{\partial s_3} & \cdots & \frac{\partial f_1}{\partial s_{N2}} \\ \frac{\partial f_2}{\partial s_1} & \frac{\partial f_2}{\partial s_2} & \frac{\partial f_2}{\partial s_3} & \cdots & \frac{\partial f_2}{\partial s_{N2}} \\ \vdots & \vdots & \vdots & & \vdots \\ \frac{\partial f_{N2}}{\partial s_1} & \frac{\partial f_{N2}}{\partial s_2} & \frac{\partial f_{N2}}{\partial s_3} & \cdots & \frac{\partial f_{N2}}{\partial s_{N2}} \end{bmatrix}$$

The significance of the Jacobian matrix in the Newton–Raphson solution of the position equations, the velocity solution, the acceleration solution, and the numerical evaluation of the velocity coefficient derivatives has been discussed previously. From this point on, it will be assumed that the position solution is complete, and that values are available for all position variables.

Velocity Analysis. The velocity loop equations are obtained by time differentiation of the position loop equations (and any necessary constraint equations). Assuming no explicit time dependence, this differentiation may be written in matrix form as

$$\left[\begin{array}{c:c}\dfrac{\partial f}{\partial q} & \dfrac{\partial f}{\partial s}\end{array}\right]\left\{\begin{array}{c}\dot{q}\\ \hdashline \dot{S}\end{array}\right\} = \{0\}$$

$$(N_2 \times N_1) \quad (N_2 \times N_2) \quad \begin{array}{c}(N_1 \times 1)\\(N_2 \times 1)\end{array}$$

When the partitions are separated, the result is a system of linear simultaneous algebraic equations, to be solved for the secondary velocities; that is,

$$\left[\frac{\partial f}{\partial s}\right]\{\dot{S}\} = -\left[\frac{\partial f}{\partial q}\right]\{\dot{q}\}$$

The coefficient matrix on the left side is the Jacobian matrix, as seen before. The formal solution, in terms of the inverse of the Jacobian matrix, shows that the secondary velocities can each be written as a linear combination of the generalized velocities,

$$\{\dot{S}\} = -\left[\frac{\partial f}{\partial s}\right]^{-1}\left[\frac{\partial f}{\partial q}\right]\{\dot{q}\}$$

$$\{\dot{S}\} = [K]\{\dot{q}\}$$

where the velocity coefficient matrix, $[K]$, is

$$[K] = -\left[\frac{\partial f}{\partial s}\right]^{-1}\left[\frac{\partial f}{\partial q}\right]$$

$$(N_2 \times N_1) \qquad (N_2 \times N_2) \qquad (N_2 \times N_1)$$

Acceleration Analysis. Matrix notation is awkward for a completely general analysis of the secondary accelerations. Matrix notation is limited to one- and two-dimensional arrays, but a three-dimensional array would be most convenient. Nevertheless, the following approach works well. The secondary velocities are expressed in terms of the velocity coefficients and the primary velocities as

$$\{\dot{S}\} = [K]\{\dot{q}\}$$

When this relation is differentiated with respect to time, the chain rule is required to reflect the changes in $[K]$ as the generalized coordinates change with time:

$$\begin{aligned}\{\ddot{S}\} &= \dot{q}_1\frac{\partial[K]}{\partial q_1}\{\dot{q}\} + \dot{q}_2\frac{\partial[K]}{\partial q_2}\{\dot{q}\} + \cdots + [K]\{\ddot{q}\}\\ &= \dot{q}_1[L_1]\{\dot{q}\} + \dot{q}_2[L_2]\{\dot{q}\} + \cdots + [K]\{\ddot{q}\}\end{aligned}$$

If $[K]$ is known in explicit analytical form, the required derivatives can be evaluated. This is a direct approach to evaluating the secondary accelerations.

If $[K]$ has only been determined numerically, then a numerical evaluation of the velocity coefficient derivative matrices is also required. For this purpose, consider again the relation defining the velocity coefficient matrix:

$$\left[\frac{\partial f_i}{\partial s_k}\right][K] = -\left[\frac{\partial f_i}{\partial q_j}\right]$$

When this equation is differentiated with respect to a typical primary variable, q_n, the result includes the required velocity coefficient derivative matrix as a factor in the second term, as follows:

$$\left[\frac{\partial^2 f_i}{\partial q_n \partial s_k}\right][K] + \left[\frac{\partial f_i}{\partial s_k}\right][L_n] = -\left[\frac{\partial^2 f_i}{\partial q_n \partial q_j}\right]$$

The velocity coefficient derivative matrix $[L_n]$ is then evaluated by the numerical solution of the system of equations:

$$\left[\frac{\partial f_i}{\partial s_k}\right][L_n] = -\left[\frac{\partial^2 f_i}{\partial q_n \partial s_k}\right][K] - \left[\frac{\partial^2 f_i}{\partial q_n \partial q_j}\right]$$

The coefficient matrix on the left side is the familiar Jacobian matrix that is known explicitly. The first matrix on the right is the partial derivative of the Jacobian matrix with respect to the primary variables, q_n; this can be written explicitly. The second term on the right is the partial derivative of the previous righthand-side matrix $[B]$, also differentiated with respect to the same primary variable, q_n; this is also known explicitly. (Recall that the definition of $[B]$ included a negative sign, here shown explicitly.) Thus, all of the matrices required for the numerical evaluation of $[L_n]$ are known. As usual, this should be treated as a system of simultaneous linear equations to be solved.

As discussed in Section 2.6 (dealing with the general analysis of single degree of freedom systems), the purpose here is not to provide general formulae for problem solving. Instead, the intent is (1) to show the form of the equations to be expected in all multidegree of freedom kinematic analyses and (2) to focus on the analytical process without reference to a specific problem. For problem solving, each problem should be formulated from fundamentals and carried through to the final results. If the results do not follow the forms indicated here, it is likely that an error has been made.

3.5 CONCLUSION

For multidegree of freedom mechanisms, the kinematic analysis is an extension of the ideas introduced previously for the analysis of single degree of freedom mechanisms. Although the number of primary coordinates increases in the case of multiple degrees of freedom, the position solution process scarcely changes at all. The solution for secondary velocities or velocity coefficients is more complicated because the velocity coefficient matrix for multiple degrees of freedom is rectangular, rather than a column vector. The complication extends further to the velocity coefficient derivatives, where partial derivatives of the rectangular velocity coefficient matrix are required.

The use of body coordinates and base coordinates for points of interest remains exactly as it was established previously for a single degree of freedom. The velocity coefficients and velocity coefficient partial derivatives are, again, rectangular matrices, as noted previously.

Despite the increased complexity usually associated with multidegree of freedom problems, for some cases closed-form solutions are available for every step. In general, however, the required solutions will need to be obtained by numerical procedures. The position solution, velocity coefficients, and velocity coefficient partial derivatives—all useful concepts as presented for kinematic analysis—will also be very significant in the multidegree of freedom static and dynamic analyses presented in Chapters 6 and 8.

REFERENCES

Paul, B., *Kinematics and Dynamics of Planar Machinery*. Englewood Cliffs, N.J.: Prentice–Hall, 1979.

Svoboda, A., *Computing Mechanisms and Linkages*. New York: Dover, 1969.

Tuttle, S. B., *Mechanisms for Engineering Design*. New York: John Wiley, 1967.

PROBLEM SET

3.1 The mechanism shown may be considered a "floating" slider–crank mechanism because the crank pivot is not stationary. The dimensions R, L, and C are known.

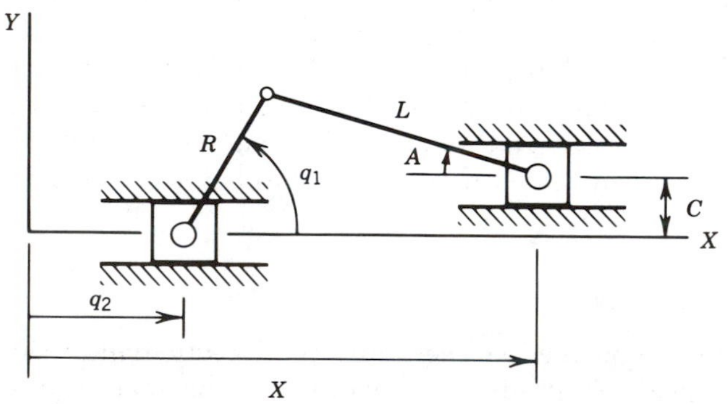

a. Set up the necessary position equations and solve in closed form for the secondary variables A and X;
b. Obtain a closed-form solution for the velocity coefficient matrix;
c. Obtain a closed-form solution for the velocity coefficient derivative matrices $[L_1]$ and $[L_2]$;
d. Using the dimensions given in part f, evaluate A and X for $q_1 = 0.85$ radians and $q_2 = 0.095$ m;
e. For q_1 and q_2 as in part d, and $\dot{q}_1 = 1.85$ rad/s, $\dot{q}_2 = -2.45$ m/s, evaluate the velocity coefficient matrix and the velocities $\dot{A}$ and $\dot{X}$;
f. For $q_1, q_2, \dot{q}_1$, and $\dot{q}_2$ as in parts d and e, and $\ddot{q}_1 = 11.66$ rad/s^2, $\ddot{q}_2 = 4.25$ m/s^2, evaluate the velocity coefficient derivative matrices $[L_1]$ and $[L_2]$, and the accelerations $\ddot{A}$ and $\ddot{X}$.

$R = 0.125$ m $\quad L = 0.287$ m $\quad C = 0.008$ m

3.2 In the modified slider–crank mechanism shown, the plunger operates through a guide bushing that itself moves in a vertical guide. The crank radius R and the connecting rod length L are known.

a. Set up the position loop equations and solve in closed form for A and X as functions of q_1 and q_2;
b. Obtain a closed-form solution for the velocity coefficient matrix;
c. Obtain a closed-form solution for the velocity coefficient derivative matrices $[L_1]$ and $[L_2]$;
d. Using the data following part f, evaluate A and X for $q_1 = 1.55$ radians and $q_2 = 2.27$ in.
e. For q_1 and q_2 as in part d, and $\dot{q}_1 = 26.83$ rad/s and $\dot{q}_2 = -14.09$ in./s, evaluate the velocity coefficient matrix and the velocities $\dot{A}$ and $\dot{X}$;
f. For $q_1, q_2, \dot{q}_1$, and $\dot{q}_2$ as in parts d and e, and $\ddot{q}_1 = -57.27$ rad/s^2 and $\ddot{q}_2 = -8.25$ in./s^2, evaluate the velocity coefficient derivative matrices $[L_1]$ and $[L_2]$, and the accelerations $\ddot{A}$ and $\ddot{X}$.

$R = 7.38$ in. $\quad L = 22.85$ in.

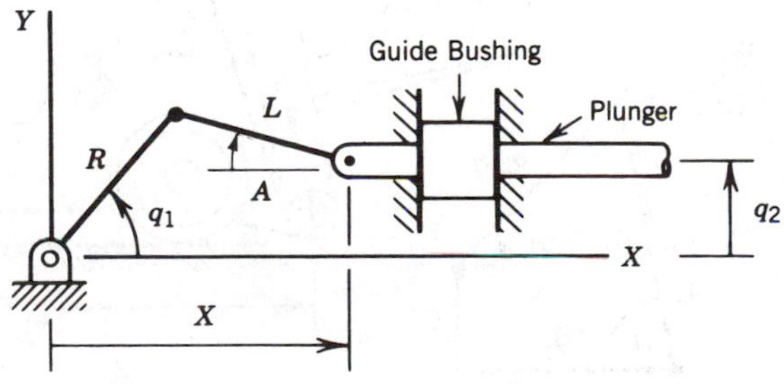

3.3 Sometimes it is desirable to vary the compression ratio of a slider–crank machine—either for use as an engine or as a compressor. One way to achieve a variable compression ratio while maintaining a fixed chamber geometry is shown in the figure. The vertical position of the crosshead guide is described by q_2, whereas q_1 is the crank rotation, as usual. The lengths R, L_1, L_2, and C are known.

a. Determine closed-form solutions for the secondary positions variables A_1, A_2, X_1, and X_2;
b. Determine a closed-form solution for the velocity coefficient matrix;
c. Using the following data, evaluate A_1, A_2, X_1, X_2, and the velocity coefficient matrix for $q_1 = 1.35$ rad and $q_2 = 3.22$ in.

$R = 4.25$ in. $\quad L_1 = 15.5$ in.
$C = 2.15$ in. $\quad L_2 = 22.1$ in.

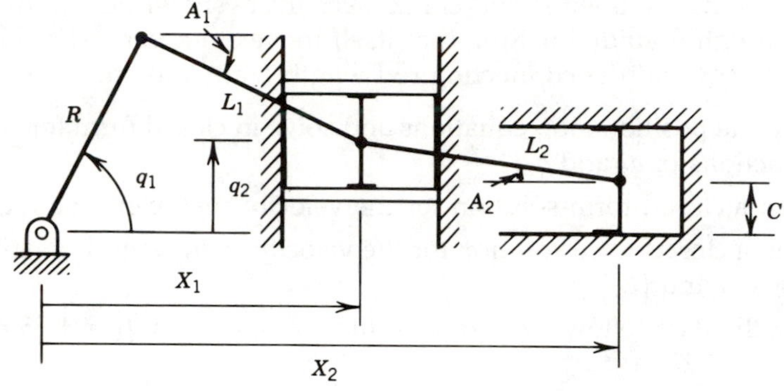

3.4 The figure shows a fork lift carrying a crate. There are three degrees of freedom: (1) the forward motion of the vehicle, (2) the extension

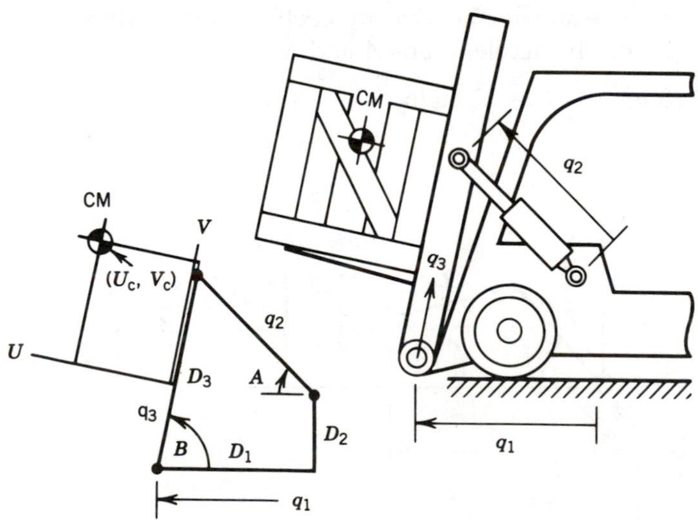

of the hydraulic cylinder controlling the angle of the lift, and (3) the elevation of the forks. The dimensions D_1, D_2, D_3, U_c, and V_c are known.

a. Set up equations solvable for the secondary position variables (do not solve);
b. Set up equations solvable for the velocity coefficients (do not solve);
c. Set up equations solvable for the velocity coefficient derivative matrices $[L_1]$, $[L_2]$, and $[L_3]$;
d. Determine base coordinates for the center of mass of the crate;
e. Determine the (2×3) velocity coefficient matrix for the center of mass;
f. Determine the velocity coefficient derivative matrices for the center of mass.

3.5 The mechanism shown consists of two links pinned to each other and to sliders operating along the X– and Y–axes, respectively. The centers of mass for the two links are located at distances C_1 and C_2 away from the sliders. The dimensions L_1, L_2, C_1, and C_2 are known.

a. Set up the position loop equations (do not solve);
b. Determine the velocity coefficient matrix in closed form, assuming that the position loop equations have been solved previously;
c. Determine base coordinates for the link centers of mass, (X_1, Y_1) and (X_2, Y_2);
d. Determine the velocity coefficient matrix relating $\mathrm{Col}(\dot{X}_1, \dot{Y}_1, \dot{X}_2, \dot{Y}_2)$ to the vector $\mathrm{Col}(\dot{q}_1, \dot{q}_2)$.

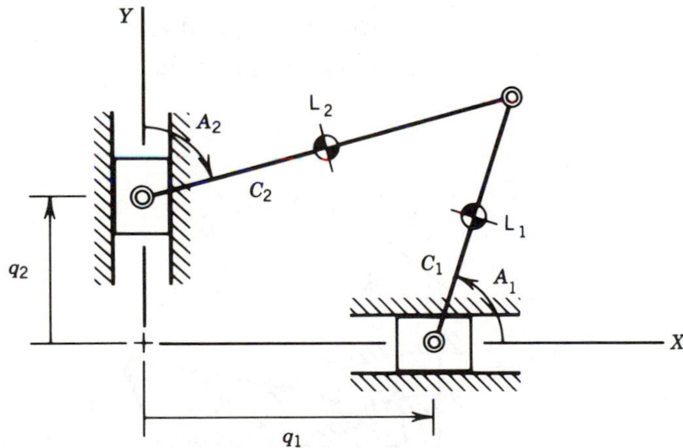

3.6 The figure shows a mechanism consisting of two links pinned to each other and to a pair of guided sliders. One of the sliders operates along the X–axis, while the second slider is guided along a line passing through the origin at an angle C to the X–axis. The dimensions D_1 and D_2, and angle C are known.

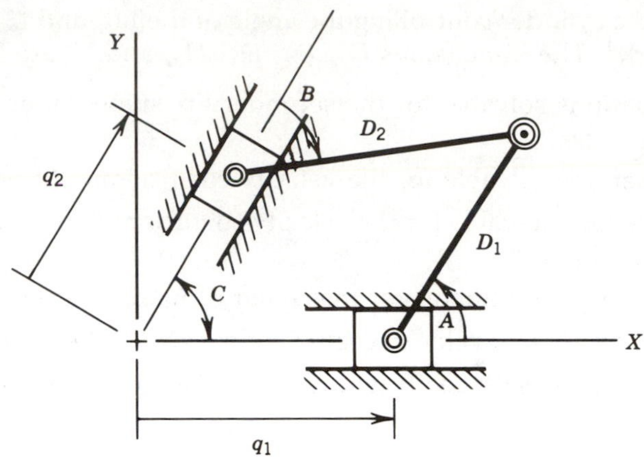

a. Set up the position loop equations (do not solve);
b. Set up the equations required to determine the velocity coefficient matrix (do not solve);
c. Set up the equations required for numerical determination of the velocity coefficient derivative matrices $[L_1]$ and $[L_2]$ (do not solve).

3.7 Two links, each of length L, are pinned to each other and to sliders guided along a circular path of radius R_1. The positions of the two sliders are described by the angles q_1 and q_2. The dimensions R_1 and L are known.

a. Determine B_1, B_2, and B_3 as functions of q_1 and q_2 in closed form;
b. Determine $\dot{B}_1$, $\dot{B}_2$, and $\dot{B}_3$ as functions of $\dot{q}_1$ and $\dot{q}_2$ in closed form;
c. Determine R_2 and $\dot{R}_2$ in terms of q_1 and q_2 and their derivatives.

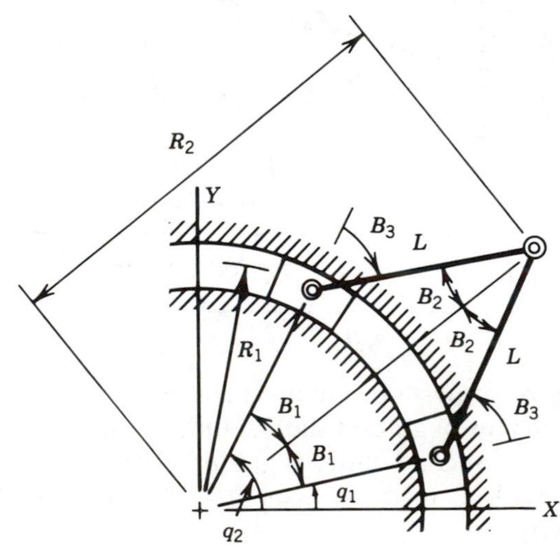

3.8 The figure shows what could be called a "floating" four-bar linkage—floating in the sense that there are no fixed points (and clearly no fourth bar). The crank pivots that would normally be fixed are mounted on sliders, moving parallel to the X– and Y–axes, respectively. The dimensions C_1, C_2, C_3, C_4, and C_5 are known.

a. Write the position loop equations (do not solve);
b. Set up the equations to determine the (2×3) velocity coefficient matrix;
c. Set up the equations required to numerically evaluate $[L_1]$, $[L_2]$, and $[L_3]$, assuming that the equations developed in parts a and b have been solved.

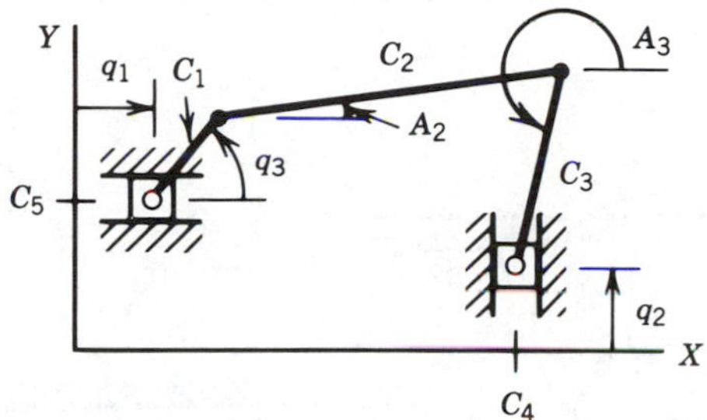

3.9 The figure shows a reel wound with a tape. The free end of the tape is initially at the point (R, R), so that the unsupported length is $B_o = R$; this is the reference orientation for the reel, $G = 0$. The free end of the tape is then moved to (q_1, q_2), causing the reel to rotate through the angle G and increasing the unsupported length to B.

a. Set up the position equations solvable for B and G for given values of q_1 and q_2, assuming that the tape remains taut (do not solve);
b. Set up the equations to be solved for the velocity coefficient matrix relating $\dot{B}$ and $\dot{G}$ to $\dot{q}_1$ and $\dot{q}_2$;
c. Set up the equations to numerically determine the velocity coefficient derivative matrices $[L_1]$ and $[L_2]$.

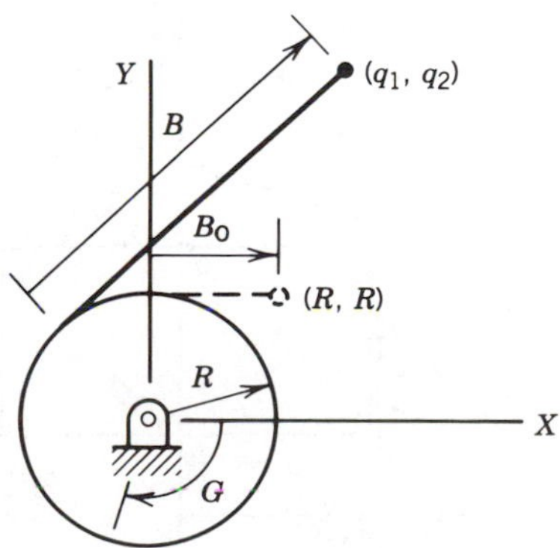

3.10 A disk and linkage assembly is initially in the configuration shown by the broken lines, with the length D_1 vertical and the length D_2 horizontal. The horizontal slider is then displaced a distance q_1 to the right, while the disk rolls through the angle q_2, where there is no slip between the disk and the supporting surface. The dimensions R_1, R_2, D_1, and D_2 are known.

a. Set up position equations solvable for secondary variables A and B (do not solve);
b. Set up the matrix equation solvable for the velocity coefficient matrix;
c. Set up the equations required for numerical determination of the velocity coefficient derivative matrices $[L_1]$ and $[L_2]$.

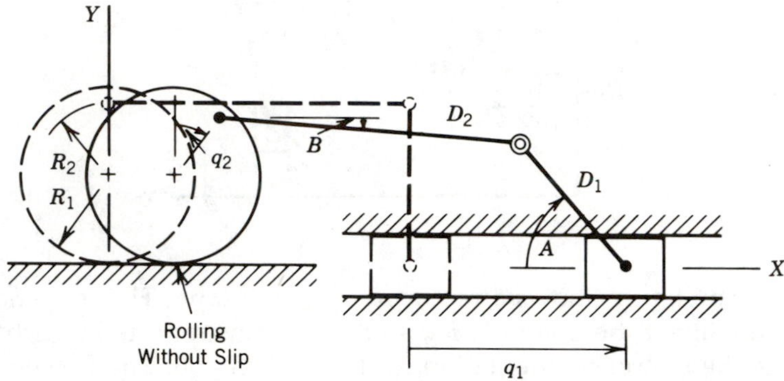

3.11 The motion of the disk is controlled by the slider position, q_1, and the elevation of the table, q_2, through the connecting rod of length L. The disk rolls on the table surface without slipping. When $q_1 = q_2 = 0$, the disk is centered on the table, $X = A = 0$. The dimensions L and D are known.

a. Write the position equations for the system;
b. Set up the matrix equation to determine the velocity coefficient matrix;

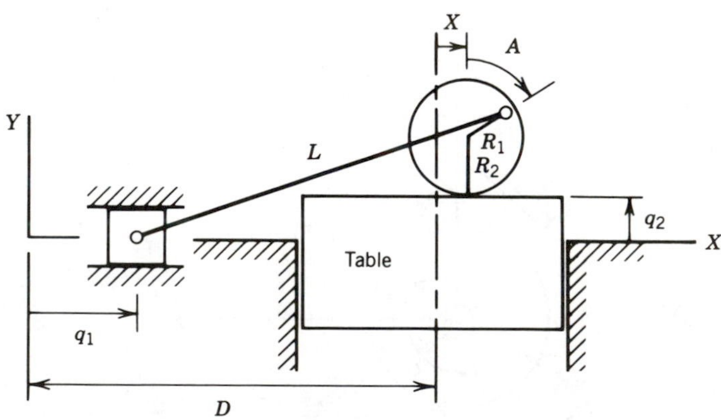

c. Set up the equations to numerically determine the velocity coefficient derivative matrices $[L_1]$ and $[L_2]$.

3.12 For the mechanism shown, the arm is held in contact with the disk at all times by a guide (not shown), and the disk rolls without slipping on the arm. The initial state is indicated in broken line, with the slider Ⓘ at the origin and the arm along the X–axis. As q_1 increases, the slider moves to the right and the arm moves past the underside of the disk, thereby increasing A. As q_2 increases, slider Ⓘ Ⓘ, carrying the disk pivot, moves upward, causing the angle A to decrease. The dimensions C, D, E, R, and W are known.

a. Formulate the position equations (do not solve);
b. Develop equations solvable for the velocity coefficient matrix (in matrix form);
c. Set up the equations to solve for the velocity coefficient derivatives $[L_1]$ and $[L_2]$;
d. Determine the base coordinates for the point M;
e. Determine the velocity coefficients for point M;
f. Determine the velocity coefficient derivative matrices for point M.

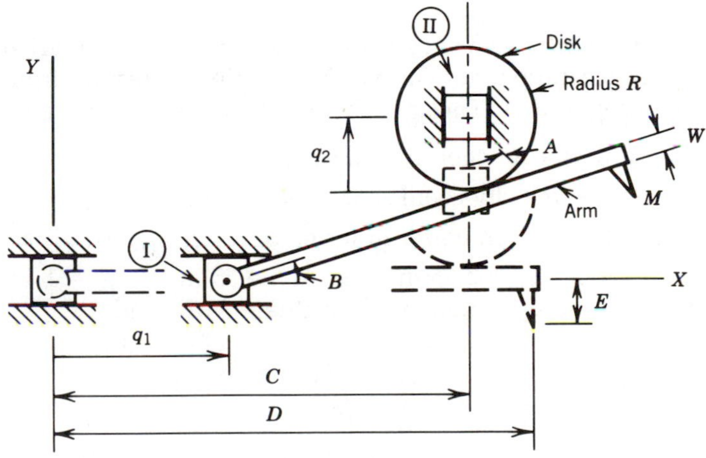

3.13 A small lathe is driven by an electric motor with a gravity-tensioning arrangement as shown. At rest, the system configuration is as shown in the broken lines. When power is applied, the motor torque puts tension in the lower side of the belt, causing the motor to rise and the upper side of the belt to slacken. The rotations of the lathe spindle and the motor pulley are q_1 and q_2, respectively, and the length of the tight side of the belt is D. The dimensions C_1, C_2, C_3, C_4, R_1, and R_2 are known.

a. Set up the position equations to be solved for A, B, and D (do not solve);

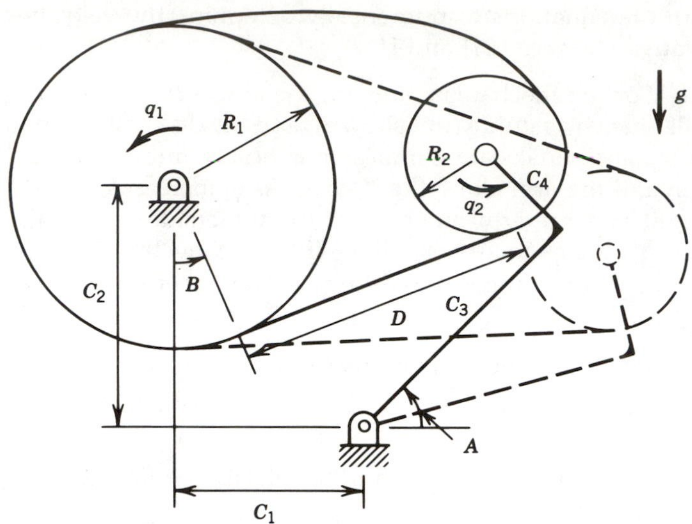

b. Set up the equations to be solved for the velocity coefficient matrix (in matrix form);

c. Set up the equations to be solved for the velocity coefficient derivative matrices $[L_1]$ and $[L_2]$.

3.14 The mechanism shown was used formerly as a mechanical analog computing component. The vertical motion R results from the two input motions q_1 and q_2. The angle between the two arms is a right angle.

a. Determine R as a function of q_1 and q_2;

b. Determine the velocity coefficient (row) vector (K_{r1}, K_{r2});

c. Determine the acceleration $\ddot{R}$ in terms of q_1, $\dot{q}_1$, $\ddot{q}_1$, q_2, $\dot{q}_2$, and $\ddot{q}_2$;

d. For $q_1 = 3.5$ in., $\dot{q}_1 = 2.75$ in./s, $\ddot{q}_1 = -4.55$ in./s^2, $q_2 = 1.13$ in., $\dot{q}_2 = -6.29$ in./s, and $\ddot{q}_2 = 3.86$ in./s^2, evaluate R, $\dot{R}$, and $\ddot{R}$.

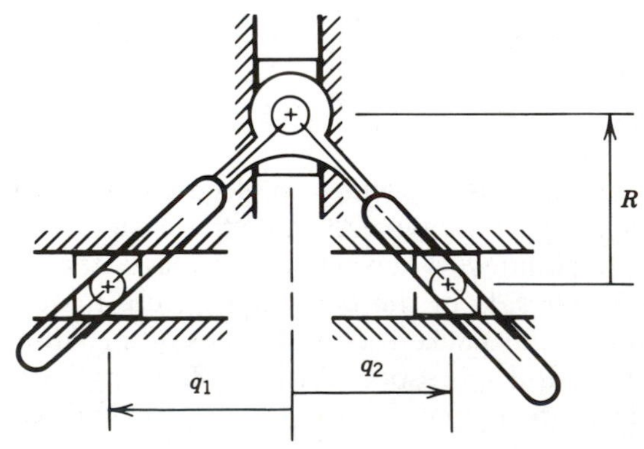

3.15 The figure shows a variation of the mechanism in Problem 3.14. The difference is that the arms no longer form a right angle; the included angle is now C.

a. Determine the response R as a function of q_1 and q_2;
b. Determine the velocity coefficient (row) vector (K_{r1}, K_{r2});
c. Express the acceleration $\ddot{R}$ in terms of q_1, $\dot{q}_1$, $\ddot{q}_1$, q_2, $\dot{q}_2$, and $\ddot{q}_1$.

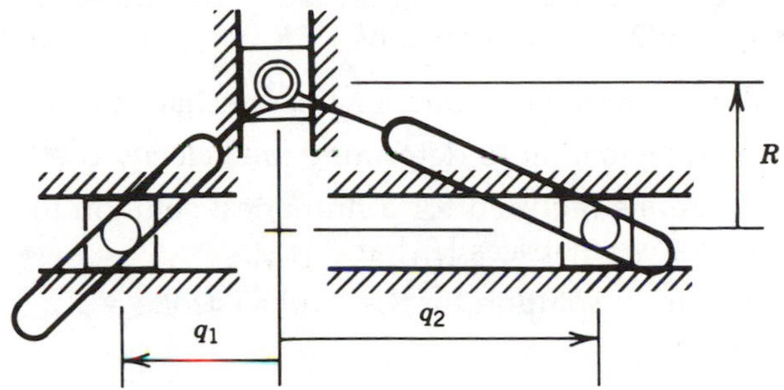

3.16 In mechanical analog computation, the mechanism shown was called a *summing linkage*. The distances B and C are known. Do not assume that the angle A is small.

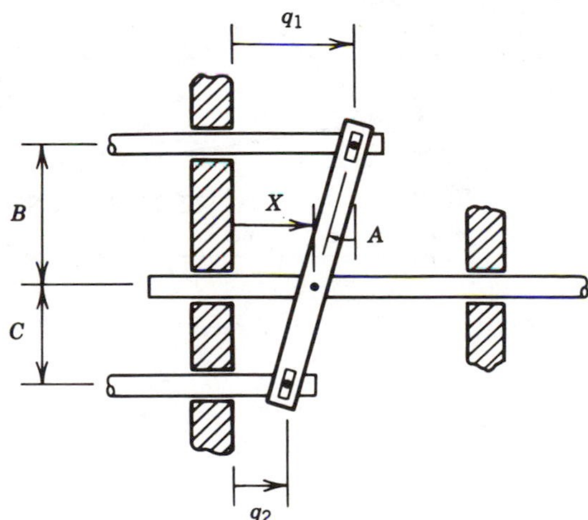

a. Determine the response A and X in terms of q_1 and q_2;
b. Determine the velocity coefficient matrix in closed form;
c. Determine the velocity coefficient derivative matrices $[L_1]$ and $[L_2]$ in closed form;

d. For the dimensions and state described, evaluate X, $\dot{X}$, and $\ddot{X}$.

$B = 0.15$ m	$C = 0.08$ m
$q_1 = 0.135$ m	$q_2 = 0.029$ m
$\dot{q}_1 = 0.95$ m/s	$\dot{q}_2 = 0.83$ m/s
$\ddot{q}_1 = -0.15$ m/s^2	$\ddot{q}_2 = -0.54$ m/s^2

3.17 The mechanism shown provides a variable stroke length, depending on the position of the control arm, q_2, as the input crank rotation, q_1, moves through a revolution. The stroke is minimum for $q_2 = 0$ and maximum for $q_2 = \pi/2$. The dimensions L, R_1, R_2, and S are known.

a. Determine the position equations for this mechanism (do not solve);
b. Set up the matrix equation to determine the velocity coefficient matrix;
c. Set up the equations required for a numerical solution for the velocity coefficient derivative matrices $[L_1]$ and $[L_2]$;
d. Obtain a closed-form solution for the stroke (stroke = $X_{max} - X_{min}$) as a function of q_2 only.

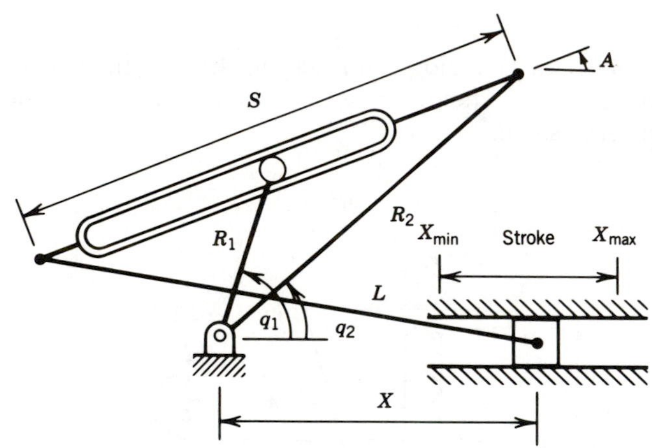

CHAPTER 4

Cam Systems

4.1 INTRODUCTION

A *cam* is a solid body shaped such that its motion imparts a prescribed motion to a second body, called the *follower,* that is maintained in contact with the cam. The shape of the cam and the physical relation between the cam and the follower define a particular functional relation between the cam position and the follower position over a range of cam positions. Using cams is one of the simplest ways to generate complex motions with high repeatability and reliability, and with reasonable costs. For these reasons, cams are widely used in many types of modern machinery.

There are many types of cam and follower systems, some of which are illustrated in Figure 4.1 and Figure 4.2. The cams shown in Figure 4.1 are of the type known as *disk* or *plate cams;* this type of cam will be the principal subject of this chapter. Disk cams are commonly found in internal combustion engines, timer mechanisms, machine tools, and numerous other devices. Other types of cams are shown in Figure 4.2, including the *wedge cam,* the *barrel* or *cylinder cam,* and the *face cam.* The wedge cam is used in some vending mechanisms, whereas the barrel cam can be found in various machine tools. The cam types shown in Figure 4.2 are less common; they will not be considered further here.

The four cam–follower systems shown in Figure 4.1 all use the same disk cam, but each illustrates a different follower type. These followers may be classified according to the type of surface that contacts the cam and to the type of motion given to the follower. The followers shown in the top row, Figure 4.1(a) and (b) are both flat-faced; roller followers are shown in Figure 4.1(c) and (d). The follower motions in the left column are transla-

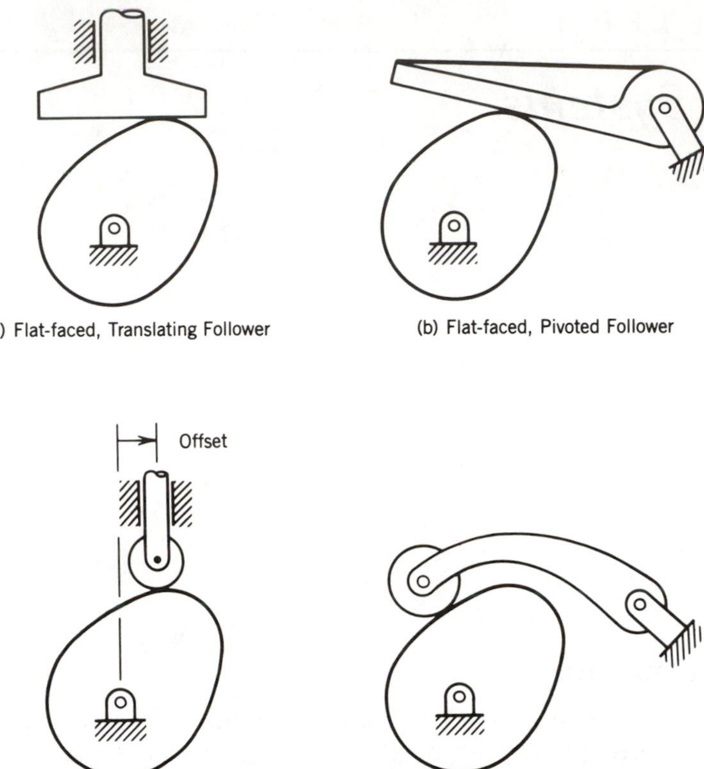

(a) Flat-faced, Translating Follower

(b) Flat-faced, Pivoted Follower

(c) Offset Roller Follower

(d) Pivoted Roller Follower

FIGURE 4.1 Common Disk Cam and Follower Systems

tional (Figure 4.1(a) and (c)), whereas those in the right column rotate about a fixed pivot (Figure 4.1(b) and (d)). Roller followers reduce friction and wear, but they are more expensive than flat-faced followers. Alternatively, the flat-faced followers have the benefit of simplicity and low cost. Finally, for each case shown in Figure 4.1, the driving motion is rotation of the disk cam. The response motion of the follower may be either translation or rotation.

For a cam–follower system, the response motion is the motion of a particular point in the follower. That point is called the *trace point* because it traces the response motion. In the case of a flat-faced, translating follower, the trace point is chosen on the center line of the follower guide at the contact surface. For a roller follower, the trace point is taken as the roller axis of rotation. As long as a flat-faced follower is considered rigid, the definition of the trace point is not particularly critical. For a rigid, translating flat-faced follower, however, all points in the body have the same translational motion, so no one point has unique motion. For a rigid,

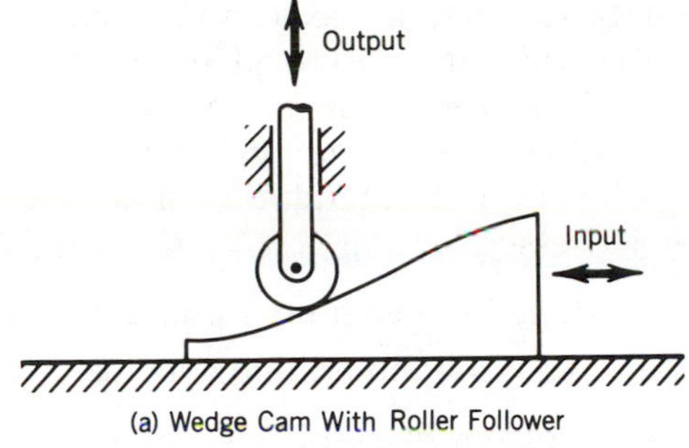

(a) Wedge Cam With Roller Follower

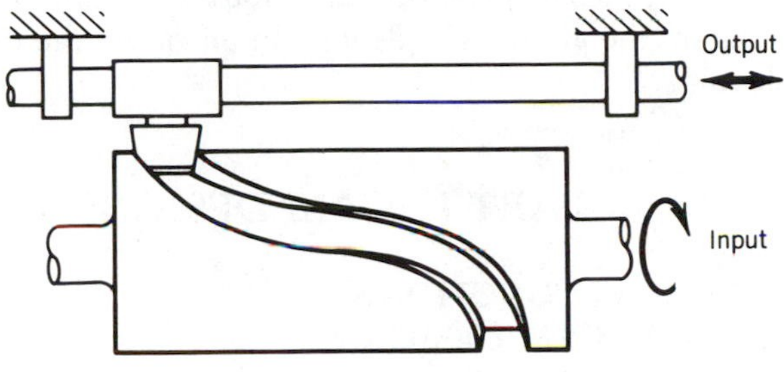

(b) Barrel Cam With Translating Roller Follower

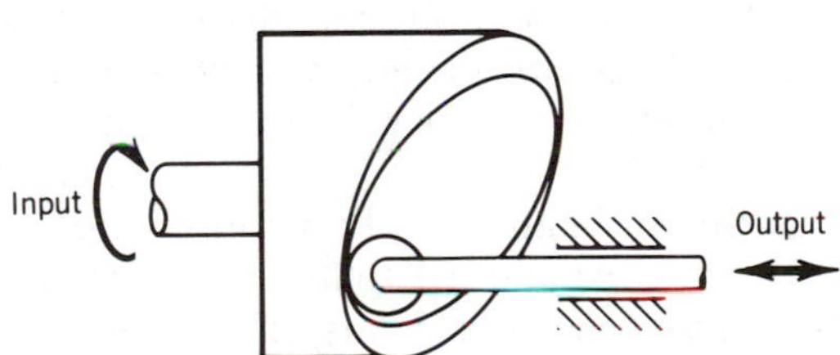

(c) Face Cam With Roller Follower

FIGURE 4.2 Various Other Cam and Follower Systems

pivoted flat-faced follower, all parts of the body have the same rotation and, again, no location has unique motion. The following distinctions are very important:

1. If a roller follower is used, the trace point is not on the roller, but is the roller axis, fixed in the translating or rotating component;
2. All bodies are assumed rigid, and consequently, no deformations are considered in either the follower or cam.

The *cam profile* is that part of the cam surface that comes into contact with the follower. Inside the plate cam profile, the *base circle* is the smallest circle centered on the cam rotation axis and tangent to the cam profile. The *pitch curve* is the closed path described by the trace point as it moves relative to the cam. The *prime circle* is the smallest circle centered on the cam rotation axis and tangent to the pitch curve.

Considering cams from the standpoint of kinematics, two major questions need to be addressed:

1. How is a cam designed to generate a prescribed response motion?
2. For a given cam, what motion will be generated?

The first issue is a problem of synthesis; the second is one of analysis. Correspondingly, this chapter is divided in two parts: Part I—Cam Design, addresses the synthesis question, and Part II—Cam Mechanism Kinematics, considers the analysis question. In all cases, the discussion is limited to disk cams.

PART I CAM DESIGN

4.2 DISPLACEMENT FUNCTIONS AND GRAPHICAL CAM DESIGNS

A mathematical description that relates the follower displacement to the angular position of the cam is called the *displacement function* for that cam–follower system, $f(A)$, where A is the rotational displacement of the cam, measured in radians. The follower displacement may be either translation or rotation with the corresponding linear or angular units. A plot of the displacement function over one cycle of cam rotation is called a *displacement diagram;* a typical displacement diagram is shown in Figure 4.3.

The full range of the linear or angular follower displacement is called the *lift* and denoted L. The cam rotation is shown divided into four

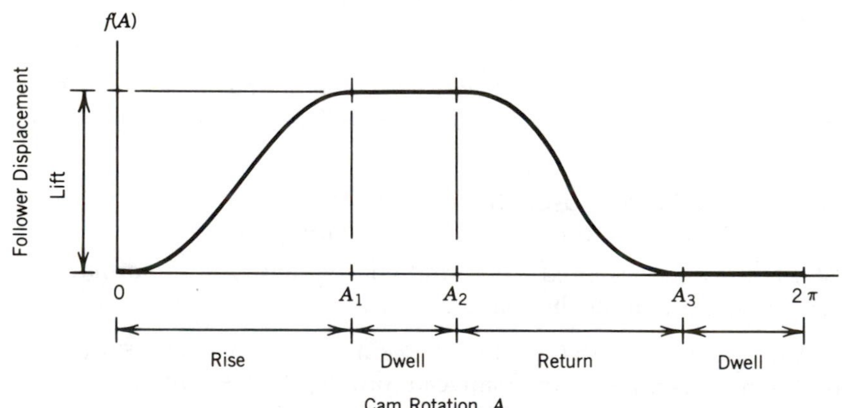

FIGURE 4.3 Typical Displacement Diagram

phases—*rise,* during which the displacement goes from zero to the maximum value; *dwell,* a period of sustained maximum displacement; *return,* in which the displacement diminishes from the maximum to zero value; and finally, a second *dwell* at zero displacement. These four phases are separated by four break points—the beginning (or end) of the cycle, A_1, A_2, and A_3 (the beginning and end of the cycle are the same point because this is a periodic process). The description of the displacement function is given piecewise:

$$\begin{aligned} f(A) &= f_1(A) && 0 \le A < A_1 \\ &= L && A_1 \le A < A_2 \\ &= f_3(A) && A_2 \le A < A_3 \\ &= 0 && A_3 \le A < 2\pi \end{aligned}$$

The dwell at full lift is sometimes reduced to zero length, bringing the break points A_1 and A_2 into coincidence. In this case, it may or may not be possible to use a single function to describe the lift over the range zero to A_3.

From an operational viewpoint, the two dwells are usually the functionally important parts of the cycle, whereas the rise and return are merely transitions. From the standpoint of cam design, the dwells require very little attention. However, the rise and return intervals require detailed consideration because of the transient dynamic phenomena occurring in these intervals.

There is no limit to the number of functions that can be used to describe the rise and return intervals. At the same time, there are several criteria that must be met in choosing these functions. Clearly, because these curves represent the physical motion of the follower, they must be continuous and differentiable. For a smooth transition to the adjacent dwells, the functions $f_1(A)$ and $f_3(A)$ must each have zero derivatives at the end of their respective intervals. As will be shown shortly, there is also cause for concern regarding the second derivatives of these functions, particularly at the transitions to the dwell intervals. To see how the displacement function is related to the time history of the follower, let $H(A)$ denote the position of the cam follower:

$$H(A) = f(A) + \mathit{Constant}$$

The velocity of the follower is determined by differentiation of this relation, using the chain rule for differentiation:

$$\dot{H}(A) = \frac{df(A)}{dA}\frac{dA}{dt} = f'(A)\,\dot{A}$$

where the prime notation denotes differentiation of a function *with respect to its argument*, in this case, A. This notation is important and will be used repeatedly in later discussions. As usual, the dot denotes differentiation with respect to time. The acceleration is obtained by another differentiation,

$$\ddot{H}(A) = \ddot{A}\ f'(A) + \dot{A}^2\ f''(A)$$

If $f''(A)$ is very large, this indicates very high acceleration of the cam follower. Discontinuities in the second derivative are likely to occur at the transition from one displacement description to the next; $f''(A)$ is often zero on one side of the junction and non-zero on the other side. If $f''(A)$ is discontinuous, then the cam–follower acceleration is also discontinuous. When this happens, the contact force between the cam and follower is discontinuous, which is an impact situation. This leads to surface damage and accelerated wear of the cam and follower, as well as exciting vibration throughout the remainder of the system. Evidently, in choosing the displacement function, it is necessary to consider carefully the first and second derivatives as well as the function itself.

Consider the construction of a very simple rise curve, constructed from parabolic segments. The rise interval extends from zero to A_1, and the lift is L. A single parabola cannot be adjusted to fit smoothly over this entire interval, but using two parabolas, one can be fitted at each end and the two can be joined in the middle. Thus, consider the two parabolas

$$F_1 = K_1\ A^2$$

$$F_2 = L - K_2\ (A - A_1)^2$$

with derivatives

$$F_1' = 2K_1A \qquad F_1'' = 2K_1$$

$$F_2' = -2K_2(A - A_1) \qquad F_2'' = -2K_2$$

It is readily evident that $F_1(0) = F_1'(0) = 0$, $F_2(A_1) = L$, and $F_2'(A_1) = 0$; this assures a smooth transition from the rise curve to the adjacent dwells. What remains is to make the curve continuous over the rise interval. If the two functions F_1 and F_2 are to meet at the center of the interval with a common slope,

$$F_1(A_1/2) = F_2(A_1/2)$$

$$F_1'(A_1/2) = F_2'(A_1/2)$$

From these two conditions, the values for K_1 and K_2 must be

$$K_1 = K_2 = 2L/A_1^2$$

Thus, the complete description of the rise curve for this example is as follows:

$$f(A) = 2L(A/A_1)^2 \qquad 0 \le A < A_1/2$$

$$= L - 2L\,\frac{(A_1 - A)^2}{A_1^2} \qquad A_1/2 \le A < A_1$$

This function and its first derivative are continuous throughout the entire interval, including the end points. The second derivative, $f''(A)$, is continuous everywhere except for a finite discontinuity at the ends and the middle of the interval where the two parabolas meet. Developing a similar return curve appropriate to the interval $A_2 \leq A < A_3$ is a useful exercise; the process is illustrated in the next example.

Consider next the development of a sinusoidal displacement function. The required function will provide a sinusoidal rise over the interval zero to A_1, a dwell at full lift from A_1 to A_2, a return over the interval A_2 to A_3, and a dwell at zero from A_3 to the end of the cycle. The required sinusoid is of the form $(1 - \cos X)$, and it must move through one half cycle in the interval zero to A_1 and again in the interval A_2 to A_3. To express the return curve, given the form for a rise curve, two substitutions are required: (1) replace the argument A with $A_3 - A$ and (2) replace the phase duration A_1 with $A_3 - A_2$. Note that the range of the function $(1 - \cos X)$ is from zero to two, so the coefficient is $L/2$ rather than L. The required displacement function is given piecewise as follows:

$$
\begin{aligned}
f(A) &= (L/2)[1 - \cos(\pi A/A_1)] && 0 \leq A < A_1 \\
&= L && A_1 \leq A < A_2 \\
&= (L/2)\{1 - \cos[\pi(A_3 - A)/(A_3 - A_2)]\} && A_2 \leq A < A_3 \\
&= 0 && A_3 \leq A < 2\pi
\end{aligned}
$$

Notice that although the function and the first derivative are continuous over the entire cycle, the second derivative has a finite discontinuity at each of the break points.

The matter of choosing a suitable displacement function for a particular design assignment is an art, and has been explored extensively for many years. Many types of displacement functions have been studied, including sinusoidal, cycloidal, and polynominal functions. Several elementary displacement functions are listed in Table 4.1. Of the functions tabulated, only the cycloidal displacement function is not restricted to low-speed operations. The parabolic and sinusoidal functions have finite discontinuities in the second derivative at both ends of the rise interval, and the parabolic and cubic curves have second derivative discontinuities at the middle of the rise interval. An extensive presentation of the topic is beyond the scope of this book, but further information can be obtained from the references at the end of this chapter. The volume of published work on this matter attests to the significance of cams as useful machine elements and to their often difficult design requirements. The displacement function fully specifies the required motion for a particular cam. As such, it provides all of the kinematic information required to design the cam and, in the past, was used directly in graphical cam design techniques. Although such graphical methods are no longer recommended, they are useful in clarifying the relationship between the displacement function and the cam pro-

TABLE 4.1 Elementary Rise Curves

Name	*Functional Form*	
Parabolic	$f(A) = 2L(A/A_1)^2$	$0 \leq A < A_1/2$
	$= L\{1 - 2[(A_1 - A)/A_1]^2\}$	$A_1/2 \leq A < A_1$
Cubic	$f(A) = 4L(A/A_1)^3$	$0 \leq A < A_1/2$
	$= L\{1 - 4[(A_1 - A)/A_1]^3\}$	$A_1/2 \leq A < A_1$
Sinusoidal	$f(A) = \frac{L}{2}\left[1 - \cos\left(\frac{\pi A}{A_1}\right)\right]$	$0 \leq A < A_1$
Cycloidal	$f(A) = \frac{L}{\pi}\left[\frac{\pi A}{A_1} - \frac{1}{2}\sin\left(\frac{2\pi A}{A_1}\right)\right]$	$0 \leq A < A_1$

where

$f(A)$ = cam–follower displacement
L = lift, maximum follower displacement
A = cam rotation
A_1 = angular duration of the rise

file. For this reason only, two graphical cam design procedures are discussed in the following sections and several graphical design problems are also included at the end of the chapter.

Graphical Design of a Cam with a Flat-Faced Translating Follower

The graphical design process for this case is shown in Figure 4.4. The development of the cam profile is done in two parts:

1. The radial position of the cam follower is determined for each of a sequence of cam angular positions;
2. The cam profile is drawn as a smooth curve, tangent to each of the follower face positions in the sequence.

The design begins with a layout of the previously chosen displacement function in the upper right corner of the Figure 4.4 (the displacement diagram). Note that this particular displacement function has no dwell at full lift; the return begins immediately after the rise. The abscissa of this plot is divided into a number of points over the cam rotation cycle (12 points are shown in Figure 4.4). Construction then moves to a point chosen for the cam axis of rotation (there is a comment later about the choice of this point) where radial lines are constructed at angles equal to the angular divisions of the cam cycle. These radial lines are numbered corresponding to the angular division points. The base line of the displacement diagram is then projected to intersect a vertical line through the cam axis of rotation, and the cam follower is outlined in this position, as shown. A circle, cen-

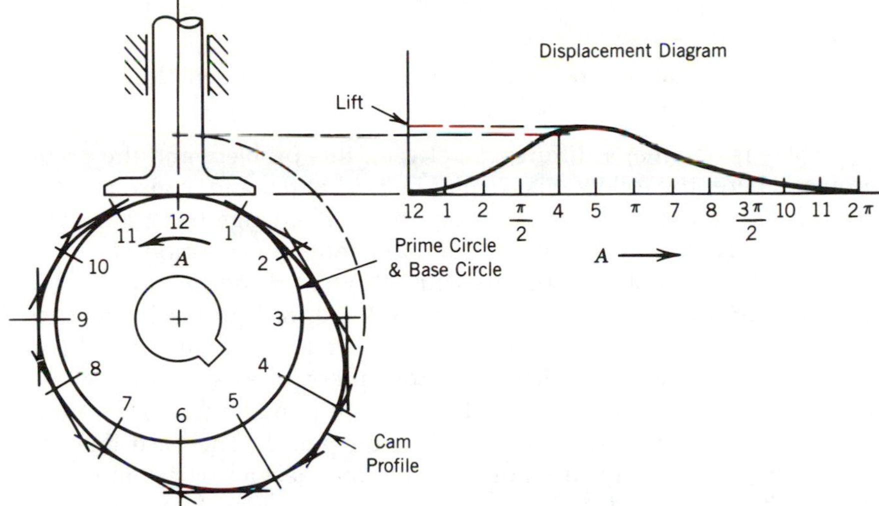

FIGURE 4.4 Graphical Design of a Cam with Flat-Faced Translating Follower

tered at the cam axis of rotation and tangent to the follower face lines, is drawn next; this is the base circle for this cam–follower system.

The ordinate corresponding to each of the cycle division points is then laid off radially outside the base circle. This process is performed graphically by projection; the broken lines show the transfer of the ordinate from point 4 on the displacement diagram to the vertical line representing zero cam displacement, then by rotation about the cam axis of rotation to the intersection with the radial line 4. At this point a perpendicular to the radial line is erected, *representing the follower face*. This process is carried out for each division point, resulting in a sequence of perpendiculars that bound the cam profile. The cam profile is then drawn as a smooth curve that is tangent to each of the follower face images.

The process just described should be reviewed carefully to make clear the manner in which the cam profile is determined from the displacement function. Note that the cam follower is also involved in this determination because a sequence of images of the flat follower face define the boundaries of the cam profile. However, note that the vertical location of the point representing the cam axis of rotation is actually a design decision because this determines the base circle radius. Finally, the size of the base circle controls the overall size of the cam because the cam profile is developed outside the base circle.

Graphical Design of a Cam with a Radial, Translating Roller Follower

Figure 4.5 describes this graphical design process. The same two-stage development applies here as was used for the cam with flat-faced translating follower:

1. The radial position of the cam follower is determined for each of a sequence of cam angular positions;
2. The cam profile is drawn as a smooth curve tangent to each of the follower outlines in the sequence.

There is one major difference between this problem and the previous problem; here, the follower is circular instead of flat and, consequently, its images will be circular arcs. Although this will change the graphical process distinctly, the two solutions still have much in common.

As before, the design begins with a layout of the previously chosen displacement function in the upper right corner of Figure 4.5. The cycle is divided as before, with 12 divisions also used for this example. This is followed by erection of radial lines from the cam axis of rotation at the angular positions represented by the division points; these lines are then numbered to correspond with the division points. The base line of the displacement diagram is then projected to intersect a vertical line through the previously selected cam axis of rotation; this defines the trace point (the roller axis) for zero cam rotation. The follower outline is then constructed about this trace point, as shown. A circle is drawn, centered on the cam axis of rotation and passing through the roller axis. This circle is the prime circle. The base circle is also drawn about the same center but is tangent to the roller in the follower outline at zero cam rotation. The trace point positions are then measured on each radial line, outward from the prime circle. As previously, this is performed graphically by projection. The broken lines in this case show the projection of ordinate 3, first horizontally to the vertical line through the cam axis of rotation, then by rotation about that center to intersect the radial line 3. This defines the trace point when

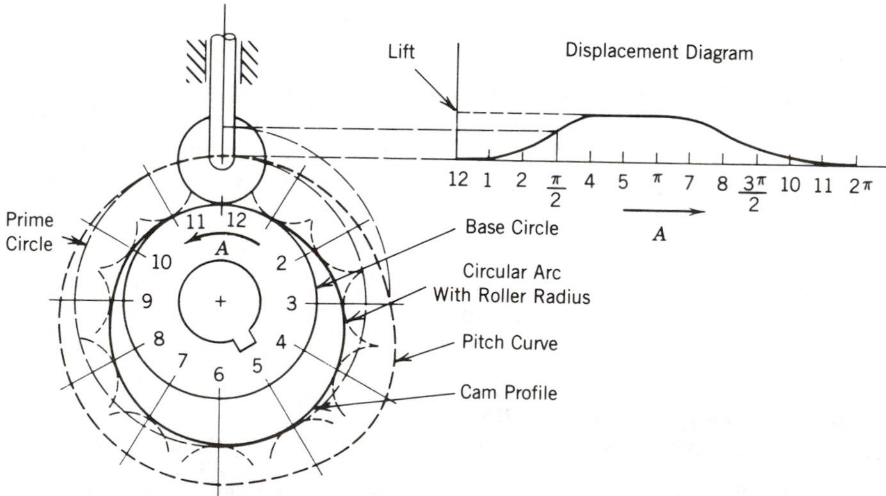

FIGURE 4.5 Graphical Design of a Cam With Radial, Translating Follower

the cam rotation is $\pi/2$. The roller–follower image is then drawn at this position, shown as an arc with short broken lines. The process is performed for the ordinate associated with each division point. In this case it results in a sequence of circular arcs that bound the cam profile. The cam profile is then drawn as a smooth curve, tangent to each of the arcs representing the roller–follower images.

The process for the roller–follower design problem should be compared to that for the flat-faced follower, particularly to identify steps common to both. The role of the cam–follower shape in each should also be compared. Two design decisions were made in this process, (1) the assignment of a follower roller radius and (2) the assignment of the prime circle radius. Both of these choices were made arbitrarily in the present development, yet they have a significant impact on the size, stress conditions, and function of the final design. There will be more discussion on these matters in the later sections that deal with analytical design methods.

4.3 A KINEMATIC THEOREM FOR RIGID BODIES

Consider a rigid body rotating about a fixed axis at point O, as shown in Figure 4.6. There are two intersecting lines inscribed on the body, O–O′ and P–P′. The rotation of the body is described by the angle $A(t)$, measured from the stationary X–axis to the line O–O′. At the instant shown, the line P–P′ makes an angle B with the X–axis. At this instant, a unit vector along the line P–P′ is $\mathbf{E}_b$, which is defined as

$$\mathbf{E}_b = \mathbf{i} \cos B + \mathbf{j} \sin B$$

The intersection of lines O–O′ and P–P′ defines a point denoted by 1. The position of that point is $\mathbf{R}_1$, where

$$\mathbf{R}_1 = |\mathbf{R}_1| \, (\mathbf{i} \cos A + \mathbf{j} \sin A)$$

and the position of any other point 2 on the line P–P′ is $\mathbf{R}_2$, where

$$\mathbf{R}_2 = \mathbf{R}_1 + \mathbf{E}_b D$$

The velocity of point 2 is $\mathbf{V}_2$, which is expressed as

$$\begin{aligned} \mathbf{V}_2 &= \mathbf{k}\,\dot{A} \times \mathbf{R}_2 \\ &= \mathbf{k}\,\dot{A} \times (\mathbf{R}_1 + \mathbf{E}_b\, D) \\ &= \mathbf{V}_1 + \mathbf{k} \times \mathbf{E}_b\, \dot{A}\, D \end{aligned}$$

where $\mathbf{V}_1$ is the velocity of point 1. The component of $\mathbf{V}_2$ in the direction of the line P–P′ is

$$\begin{aligned} \mathbf{V}_2 \cdot \mathbf{E}_b &= \mathbf{E}_b \cdot (\mathbf{V}_1 + \mathbf{k} \times \mathbf{E}_b\, \dot{A}\, D) \\ &= \mathbf{E}_b \cdot \mathbf{V}_1 + \mathbf{E}_b \cdot (\mathbf{k} \times \mathbf{E}_b)\, \dot{A}\, D \\ &= \mathbf{E}_b \cdot \mathbf{V}_1 \end{aligned}$$

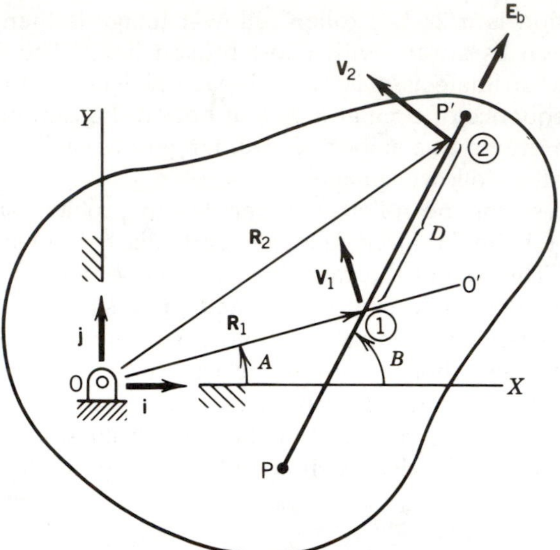

FIGURE 4.6 Velocities at Points Along a Line Inscribed on a Rigid Body

The vector product $\mathbf{k} \times \mathbf{E}_b$ is perpendicular to $\mathbf{E}_b$; the dot product of $\mathbf{E}_b$ with a vector perpendicular to itself is zero, so the last term vanishes. The result is that the component of $\mathbf{V}_2$ along the line P–P′ is the same as the component of $\mathbf{V}_1$ *along that line.* Note that if the axis of rotation was not fixed but instead, was translating, this would contribute the same additional velocity component to each of the velocities $\mathbf{V}_1$ and $\mathbf{V}_2$. When the velocity relation is dotted with $\mathbf{E}_b$, the equality just shown would still be valid. These results are summarized in the following theorem:

> All points along a line inscribed on a rotating body have the same component of velocity *along that line.*

This theorem becomes completely obvious if considered on physical grounds when the definition of a rigid body is recalled. That definition includes the fact that the separations between all particles of the rigid body are constants. This requires that if one particle has a component of velocity in a particular direction, all other particles along that line must move at the same speed in that direction to maintain constant the separations between particles. This theorem will be used frequently in the discussions on cams and in later discussions on gears.

4.4 ANALYTICAL DESIGN OF A CAM WITH A FLAT-FACED, TRANSLATING FOLLOWER

The term "analytical design" in this context refers to an analytical process to determine a cam that will have a specified displacement function. This

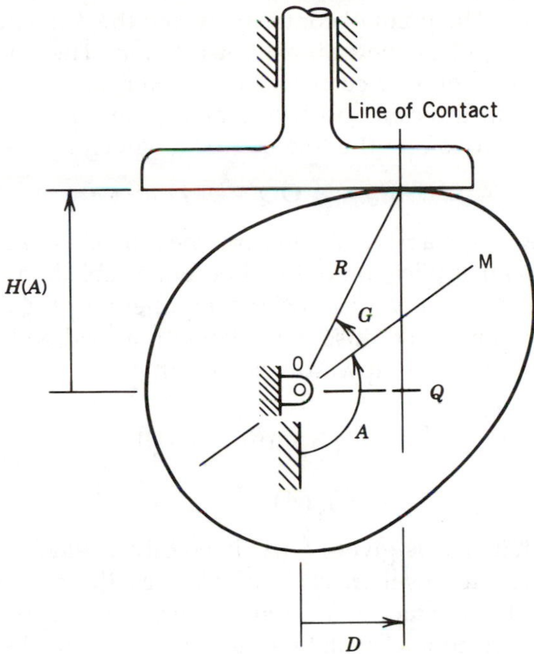

FIGURE 4.7 Cam With Flat-Faced Translating Follower

section describes such a process for a cam with a flat-faced, translating follower, as shown in Figure 4.7. The displacement function is assumed to be chosen previously, and the primary problem is to determine a corresponding cam profile. Much of the analysis presented for the four cam–follower system types considered here is based on the work of Raven.[1]

The cam axis of rotation is point O, and the cam rotation is the angle A, measured from a stationary line parallel to the follower motion around to the line OM; as the cam rotates CCW (counterclockwise), the angle A increases. Line OM is called the *body reference line,* a line fixed on the cam surface. The body reference line is very basic to the description of the cam profile; it may be thought of as existing before the cam profile, and that the profile is laid out *with respect to the body reference line.* The position of the follower is given by the function $H(A)$:

$$H(A) = R_o + f(A)$$

where $f(A)$ is the previously chosen displacement function, and R_o is the base circle radius. The distance from O to the cam profile, measured along OM from O and away from M, is R_o; this is the point of contact when $A=0$. The assignment of R_o is a design decision, and there will be more discussion regarding this choice later; for the present, consider it assigned.

[1] F. H. Raven, "Analytical Design of Disk Cams and Three-Dimensional Cams by Independent Position Equations," *Journal of Applied Mechanics, Trans. ASME,* pp. 18–24, March, 1959.

Contact Location. The point of contact between the cam and follower is a distance D to the right of the cam axis of rotation. The line through the point of contact and perpendicular to the follower face is called the *line of contact*. All points in the cam and laying along the line of contact have a component of velocity along the line of contact given by

$$V_c = \dot{A}\ \widehat{OQ} = \dot{A}\ D$$

where Q is at the foot of a normal line from the line of contact to the center of rotation; this is an application of the theorem established in Section 4.3. In particular, this is the upward velocity component for the cam particle in contact with the follower. Because the follower is a rigid body, all points in the follower have the same upward velocity, that is,

$$V_f = \dot{H} = \frac{d}{dt}[R_o + f(A)]$$

$$= \dot{A}\ f'(A)$$

Due to the fact that $f(A)$ is given, $f'(A)$ is readily available. Furthermore, because the cam and follower are each solid bodies, there can be no overlap or penetration at the contact point. On the other hand, the cam and follower must remain in contact at all times to function correctly. These conditions are expressed analytically by requiring that the velocity components V_c and V_f be equal; note that this requires only that the upward components be equal. Nothing is said about the components parallel to the follower face. Equating the velocity components leads to an equation readily solvable for the contact position D, with the result:

$$D = f'(A)$$

Contact is to the right of the cam axis of rotation when D is greater than zero; this must occur in the rise phase of the cam cycle. Similarly, D is less than zero for contact to the left of the axis of rotation, and this occurs during the return phase of the motion. Note that this last equation is dimensionally correct; D and f each have length units, and the differentiation is with respect to the angle A (measured in radians). This relation leads to the limits of the wear surface because

$$D_{max} = f'_{max}$$

$$D_{min} = f'_{min}$$

and the total length of the wear surface is $D_{max} - D_{min}$.

Cam Profile. The cam profile is specified by giving the polar coordinates (R, G) of the contact point with respect to the body reference line OM. To determine these values, consider two position equations:

$$R \sin(A + G) = D = f'(A)$$

$$-R \cos(A + G) = H(A) = R_o + f(A)$$

Note that the angle $(A + G)$ is a second quadrant angle, and that the signs attached to the sine and cosine functions make the left members of both equations positive. From these equations, the angle $(A + G)$ is

$$A + G = \text{Atan}_2\{ f'(A), -[R_o + f(A)] \}$$

where Atan_2 is the two-argument arctangent function that places the result in the appropriate quadrant, depending on the signs of both arguments (see Appendix A4). The polar coordinates are

$$G = -A + \text{Atan}_2\{ f'(A), -[R_0 + f(A)] \}$$
$$R = \sqrt{f'^2 + (R_o + f)^2}$$

Note that the coordinates are given in terms of the parameter A, the cam rotation angle. Using these equations, the polar coordinates for the contact point can be calculated for any cam rotation angle. As the rotation angle sweeps through the range 0 to 2π, all points on the cam profile are generated.

Hertz Contact Stresses. One of the important considerations in the design of a disk cam is the state of stress in the cam. If the contact stresses are too large, there will be spalling and other signs of surface fatigue in addition to excessive wear; all of these modify the cam profile. Stresses resulting from direct bearing of one curved body on another are usually referred to as *Hertz contact stresses,* after Heinrich Hertz who developed the theory describing them. Hertz stresses are usually studied using two-dimensional elastic theory, either *plane stress* or *plane strain.* It is important to be clear as to which approximation is relevant for disk cams.

Plane Stress. Consider a thin plate in the X–Y plane with stresses uniformly distributed over the thickness. The top and bottom surfaces are stress free. The stresses σ_{zz}, σ_{xz}, and σ_{yz} are zero on both boundaries, and may reasonably be assumed small throughout the thickness. Then, the only non-zero stresses are σ_{xx}, σ_{yy}, and σ_{xy}; this is the two-dimensional plane stress system.

Plane Strain. Along the Z–axis of coordinates, consider a long cylindrical body with loading perpendicular to the cylindrical surface. This suggests that there will be no axial displacements of the particles within the body, from which the strains ε_{zz}, ε_{xz}, ε_{yz} are zero. The remaining nonzero strains are ε_{xx}, ε_{yy}, ε_{xy}, and these make up a two-dimensional plane strain system. In this situation, the state of stress is the same at all cross sections, and the problem can be completely analyzed in terms of a typical cross section.

Generally, the disk cam is thin relative to its radial dimensions, so plane stress is the appropriate approximation. The plane stress analysis for edge contact of two circular disks is given by Budynas.[2] For the contact stress calculation, consider the following definitions:

- P_1 and P_2 are the radii of curvature of the two disks at the point of contact;
- E_1 and E_2 are the Young's modulii for the two disk materials;
- t is the thickness of the disks;
- F is the total force transferred between the bodies.

Then, the maximum normal stress on the contact surface is

$$\sigma_o = \sqrt{\frac{F\, E_1\, E_2\, (P_1 + P_2)}{\pi\, t\, P_1\, P_2\, (E_1 + E_2)}}$$

If one of the surfaces is flat, then the associated radius of curvature is infinite, and the contact surface maximum normal stress simplifies to

$$\sigma_o = \sqrt{\frac{F\, E_1\, E_2}{\pi\, t\, P\, (E_1 + E_2)}}$$

where P is the radius of curvature of the curved body in contact with the flat surface. The most damaging stress state occurs a short distance below the contact surface; at that point the shear stress is a maximum at a value approximately $0.3\sigma_0$.

Radius of Curvature. For proper operation, the cam profile must be smooth and have a continuously turning tangent. If this condition is not met, the cam could have a cusp or "corner." At a cusp, the tangent line is undefined, and the radius of curvature is zero. This would certainly lead to stress difficulties because the normal contact stress varies with the reciprocal of the square root of the radius of curvature. For a typical cam profile, the radius of curvature varies from point to point along the profile, so that for finite rotations, the radius of curvature must clearly be considered a variable. However, for infinitesimal rotations, the radius of curvature may be considered constant and the location of the center of curvature to be fixed on the cam. Thus, for an infinitesimal rotation, the contact zone may be replaced locally with a circular arc. Furthermore, because the rest of the cam profile has no function for such a rotation, it may have any form whatever. This type of replacement is shown in Figure 4.8, where a short circular arc replaces the actual profile near the point of contact, and an irregular boundary replaces the remainder of the cam profile. This figure emphasizes the fact that this part of the boundary makes no difference. The center of curvature is on the line of contact at point C*, and the distance from C* to the point of contact is P, the radius of curvature. The

[2] R. G. Budynas, *Advanced Strength and Applied Stress Analysis,* pp. 154–158, McGraw–Hill, New York, 1977.

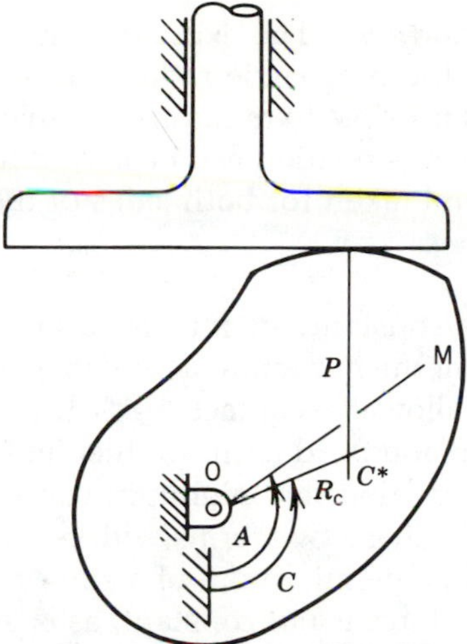

FIGURE 4.8 Determination of the Radius of Curvature

distance from the cam rotation axis to C* is R_c, an unknown. Each of the angles C and A are measured from the common vertical reference line to a line fixed on the cam. Thus, they differ by at most a constant, and

$$\frac{dC}{dA} = +1$$

Consider now the position equations:

$$D(A) = f'(A) = R_c \sin C$$

$$H(A) = R_o + f(A) = -R_c \cos C + P$$

The negative sign in the vertical equation is required because, as shown, C is a second quadrant angle. These equations involve three unknowns: R_c, C, and P. To eliminate two of them in a single substitution, the horizontal position equation is differentiated with respect to A, as follows:

$$\begin{aligned} \frac{dD}{dA} = f''(A) &= R_c \cos C \frac{dC}{dA} \\ &= R_c \cos C \end{aligned}$$

When this is applied in the vertical position equation, the result may be solved for the radius of curvature; that is,

$$P = R_o + f(A) + f''(A)$$

Presented in this form, the equation provides a means to evaluate the radius of curvature at any point on the cam profile in terms of the base radius and the displacement function.

Base Radius Determination. The expression just developed can be used as a guide to choose the base circle radius. In most cases, the base circle should be as small as possible to minimize the material and space required for the cam. The previous relation can be solved for the base circle radius, and then the minimum taken for both sides of the equation, as follows:

$$R_{o\,min} = (P - f - f'')_{min}$$

If the contact force is constant, such as for a gravity-loaded follower in a slowly moving system, the minimum acceptable radius of curvature can be determined from the allowable contact stress; this fixes the first term on the right side. The second and third terms on the right side are functions of the angle A, so it is evident that the minimum value for R_o is determined by minimizing the sum of these two terms with respect to A. Note, however, that what is required is the minimum of a sum, not the sum of individual minima! If the contact force is not constant, as is usually the case, then the minimum value of R_o must be determined iteratively to assure satisfactory contact stress for all cam positions.

Milling Cutter Coordinates. In the manufacture of cams, it is often desirable to cut the cam profile using a milling machine. The milling cutter is a rotating, cylindrical tool with cutting edges evenly spaced around the circumference. To manufacture a specified cam, it is necessary to know the required location of the milling cutter axis with respect to the work piece to generate each point on the cam profile. This information is given in terms of body coordinates fixed on the cam blank material. The following discussion describes the calculation of these body coordinates for manufacture of a cam to function with a flat-faced, translating follower.

The cam profile to be manufactured is shown in Figure 4.9, along with the outline of the milling cutter. The body reference line OM serves as a basis for a full set of body coordinates (U, V), with U along OM, and V normal to U such that the angle G rotates from U toward V. The objective is to specify body coordinates for the milling cutter, (U_m, V_m) as functions of the cam rotation angle, A. Then, as A sweeps from 0 to 2π, all of the required positions for the milling cutter are generated. It should be noted that these calculations are for a specific value of the milling cutter radius, R_m, and that use of a different diameter cutter will necessitate calculation of different milling cutter coordinates.

In the position shown in Figure 4.9, the milling cutter axis is located at (X_m, Y_m) relative to the X–Y coordinates. These values are:

$$X_m = f'(A)$$

$$Y_m = H(A) + R_m = R_o + f(A) + R_m$$

The transformation from X–Y coordinates to U–V coordinates requires a rotation through the angle $(A - \pi/2)$. When this is done, the milling cutter

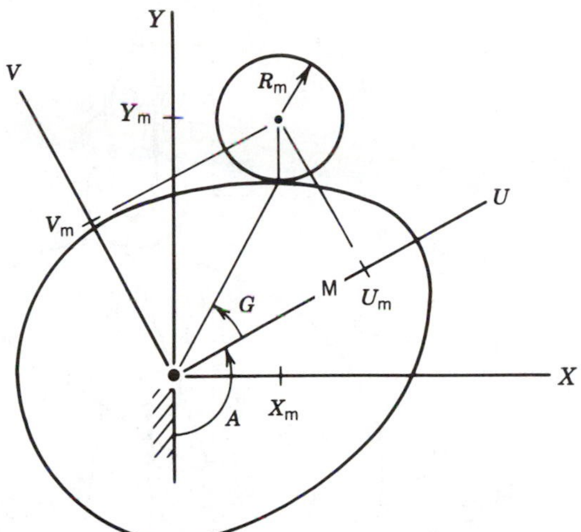

FIGURE 4.9 Milling Cutter Coordinate Determination for Cam With a Flat-faced Translating Follower

coordinates are:

$$U_m = X_m \sin A - Y_m \cos A$$

$$V_m = X_m \cos A + Y_m \sin A$$

These coordinates describe the required location of the cam cutter axis to generate the point in contact with the follower at the rotation angle A.

4.5 ANALYTICAL DESIGN OF A CAM WITH AN OFFSET, TRANSLATING ROLLER FOLLOWER

To maintain the translational response motion of the previous system, but with reduced friction and wear, a cam with a translating roller follower can be used. The follower motion need not be radial with respect to the cam axis of rotation; the follower may be offset as shown in Figure 4.10.

As before, the cam axis of rotation is point O and the cam rotation angle is A. The rotation is measured from a stationary line parallel to the follower motion, around to the body reference line OM. For zero cam rotation, the cam follower is at its lowest position, as shown in Figure 4.10(b); the vertical position of the trace point in this condition is a function of the radius of the prime circle, R_{po}, and the offset, E, that is,

$$H_o = \sqrt{R_{po}^2 - E^2}$$

The prime circle radius, R_{po}, and the roller radius, R_f, are design decisions; for the present they are considered to have been assigned (more discussion

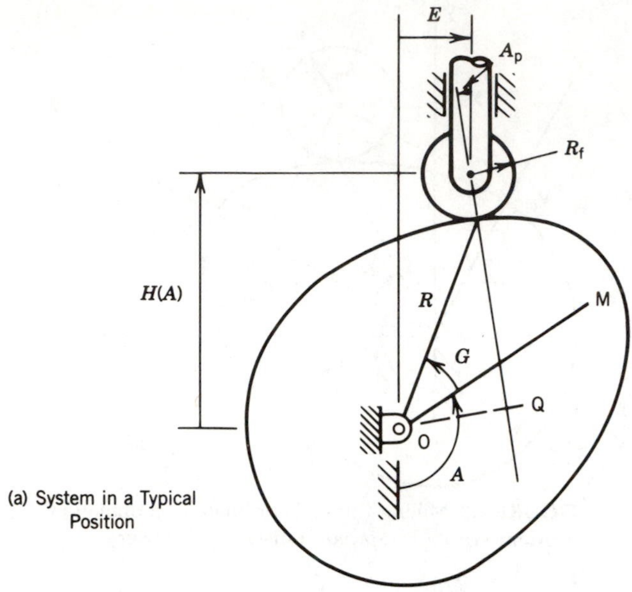

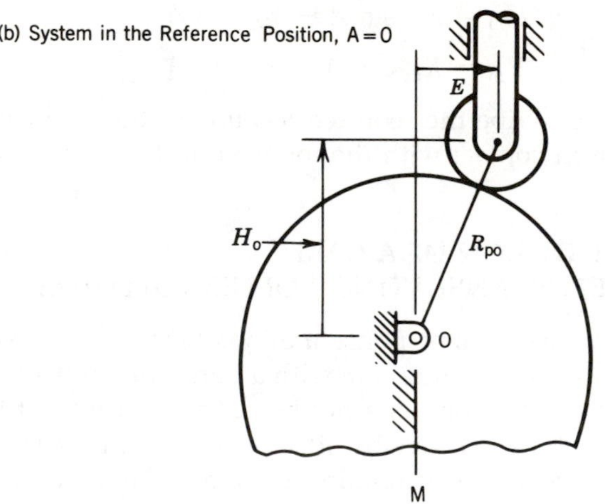

FIGURE 4.10 Cam With Offset, Translating, Roller Follower

on that choice follows later). The position of the trace point as a function of cam rotation is

$$H(A) = H_o + f(A) = \sqrt{R_{po}^2 - E^2} + f(A)$$

Pressure Angle. The line of contact is the normal to the tangent at the point of contact between the cam and the follower. This line necessarily also passes through the roller axis, which is the trace point. The angle between the line of contact and the axis of motion for the follower is called the *pressure angle*, A_p, as shown in Figure 4.10(a). It is necessary to express the pressure angle, which is useful in itself and as a step toward determining the required cam profile. The following analysis relates the pressure angle to the displacement function.

The cam follower must neither penetrate the cam profile nor separate from it. As before, this condition is expressed in terms of velocities by requiring that the velocity components along the line of contact for the particles in contact (one in the cam and the other in the follower) must be the same. The velocity of the contact point in the cam follower may be considered to consist of two parts: the rotational velocity term, and the vertical motion term, $\dot{H}$, which is superimposed on all points in the roller body. The rotational velocity term is normal to the line of contact. Therefore, it contributes nothing to the component of velocity along the line of contact. The component of $\dot{H}$ that is along the line of contact is $\dot{H} \cos A_p$, so the velocity along the line of contact for the contact particle in the follower is

$$V_f = \dot{H} \cos A_p$$

For the contact particle in the cam, the velocity along the line of contact is

$$V_c = \dot{A}\, \widehat{OQ}$$

To determine the length $\widehat{OQ}$, refer to Figure 4.11, where it is evident that

$$\widehat{OQ} = H \sin A_p + E \cos A_p$$

which makes the cam velocity component

$$V_c = \dot{A}(H \sin A_p + E \cos A_p)$$

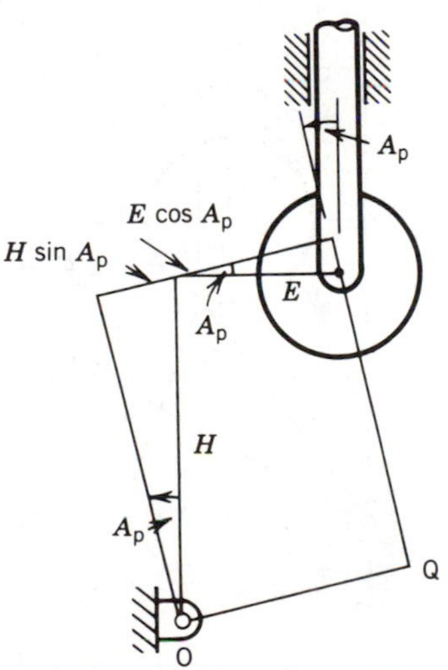

FIGURE 4.11 Evaluation of $\widehat{OQ}$

When the two velocity components are equated, the result may be solved for tan A_p,

$$\tan A_p = \frac{1}{H}\left(\frac{\dot{H}}{\dot{A}} - E\right) = \frac{1}{H}\left(\frac{dH}{dA} - E\right)$$

From the expression for $H(A)$, it is evident that dH/dA is $f'(A)$, so that the final expression for the pressure angle is

$$A_p = \text{Arctan}\left[\frac{f'(A) - E}{\sqrt{R_{po}^2 - E^2} + f(A)}\right]$$

The pressure angle is a measure of the lateral component of load exerted on the follower, a force that will tend to bind the follower, and increase friction and wear. Experience indicates that for satisfactory cam performance, the maximum pressure angle should usually not exceed $\pi/6$ radians, although there are reports of satisfactory operation in special cases for pressure angles as high as 0.83 radians. From the preceding equation, it is evident that an increase in the prime circle radius will decrease the pressure angle.

Cam Profile. Using the pressure angle just determined, the polar coordinates for the cam profile, (R, G), with respect to the body reference line OM can now be developed. Referring to Figure 4.12, a new angle, B, is introduced, where:

$$B = \text{Arctan}\left[\frac{E + R_f \sin A_p}{H(A) - R_f \cos A_p}\right]$$

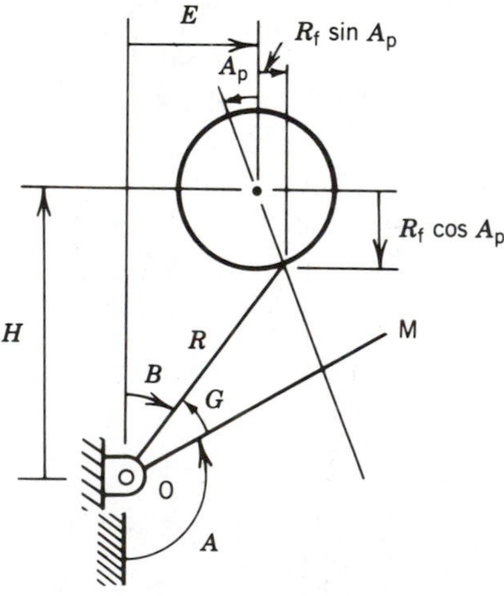

FIGURE 4.12 Determination of Cam Profile

The angle B, the cam rotation angle A, and the polar coordinate angle G are related such that

$$A + G + B = \pi$$

This equation is readily solvable for the polar coordinate angle G, while the radius R is evaluated through the Pythagorean theorem:

$$G = \pi - A - \text{Arctan}\left[\frac{E + R_f \sin A_p}{H(A) - R_f \cos A_p}\right]$$

$$R = \sqrt{[H(A) - R_f \cos A_p]^2 + (E + R_f \sin A_p)^2}$$

As in the previous case, the polar coordinates for the cam profile are expressed in terms of the parameter A, the cam rotation.

Radius of Curvature. To determine the radius of curvature, the actual cam profile is again considered locally replaced by a circular arc, as shown in Figure 4.13. The cam profile center of curvature is C*, a point on the line of contact a distance P below the contact point, where P is the radius of curvature for the cam profile. The point C* is also the center of curvature for the pitch curve and the radius of curvature for that pitch curve is P_p. To find the cam profile radius of curvature, it will be easiest first to determine

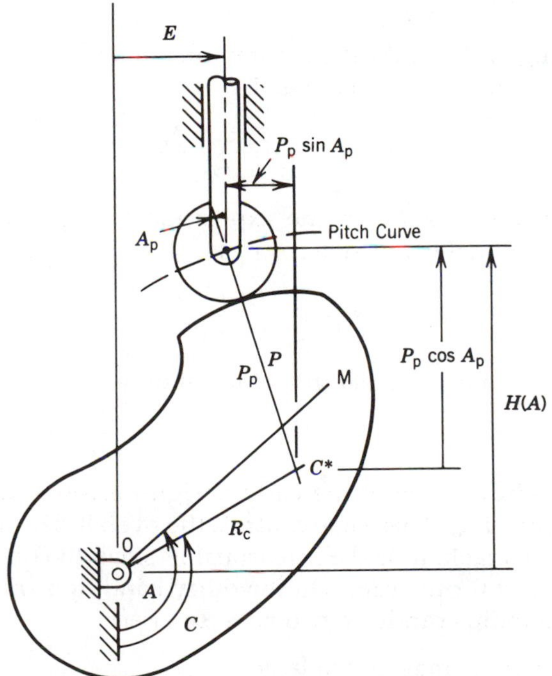

FIGURE 4.13 Radius of Curvature Determination

the pitch curve radius of curvature, then to subtract the roller radius. The center of curvature is fixed on the cam at an unknown distance R_c from the axis of rotation. The angles C and A are each measured to a line in the cam from a common reference; thus, they differ by at most a constant, and

$$\frac{dC}{dA} = +1$$

Two position equations are available involving the pitch curve radius of curvature, P_p, as follow:

$$P_p \cos A_p = H + R_c \cos C$$

and

$$P_p \sin A_p = R_c \sin C - E$$

Note that angle C, as shown, is a second quadrant angle; this explains the sign attached to the cos C term. These two equations involve three unknowns: R_c, C, and P_p. Two of these unknowns may be eliminated by using the derivative of the horizontal position equation,

$$P_p \cos A_p \, A_p' = R_c \cos C \, C'$$

where $C' = +1$. With this substitution in the vertical position equation, the only remaining unknown is P_p,

$$P_p = \frac{H \sec A_p}{1 - A_p'}$$

The quantity A_p' is determined by differentiating the expression for the pressure angle, with the eventual result

$$A_p' = \frac{f'' - f' \tan A_p}{H \sec^2 A_p}$$

This expression replaces A_p' in the expression for P_p; after substantial manipulation, the final expression for P_p is

$$P_p = \frac{[H^2 + (f' - E)^2]^{3/2}}{H^2 - Hf'' + (f' - E)(2f' - E)}$$

The cam profile radius of curvature is then simply

$$P = P_p - R_f$$

Prime Circle Radius. One of the early design decisions required is the prime circle radius, R_{po}. This value controls the overall size of the cam, and therefore, it is desirable to make it reasonably small to conserve material and space required for the cam. On the other hand, too small a value for the prime circle radius can have two adverse effects:

1. The pressure angle may be too large;
2. The state of stress in the cam may be unacceptable.

The state of stress depends on the roller radius and the cam profile radius of curvature. The profile radius of curvature, in turn, depends on the pitch curve radius of curvature and the roller radius. The prime circle describes the minimum radial position of the pitch curve, and hence, affects the stress through its effect on the profile radius of curvature.

Roller Radius. There are two considerations that constrain the roller radius:

1. The effect on stresses, both in the cam and in the roller;
2. Kinematic response at a point of minimum cam profile radius of curvature.

The first of these considerations makes it desirable to increase the roller radius, so as to reduce stresses. The calculation of Hertz contact stresses was discussed in Section 4.4. The second consideration limits the maximum size in relation to the minimum radius of curvature of the cam profile. This limitation is discussed below.

With the prime circle radius chosen, and before any choice has been made regarding a roller radius, the pitch curve is determined fully by the displacement function. The cam profile is then defined inside the pitch curve as the surface required to support the roller while the trace point moves along the pitch curve. This is shown in Figure 4.14, where the pitch curve is shown in broken line, and three possible roller sizes are consid-

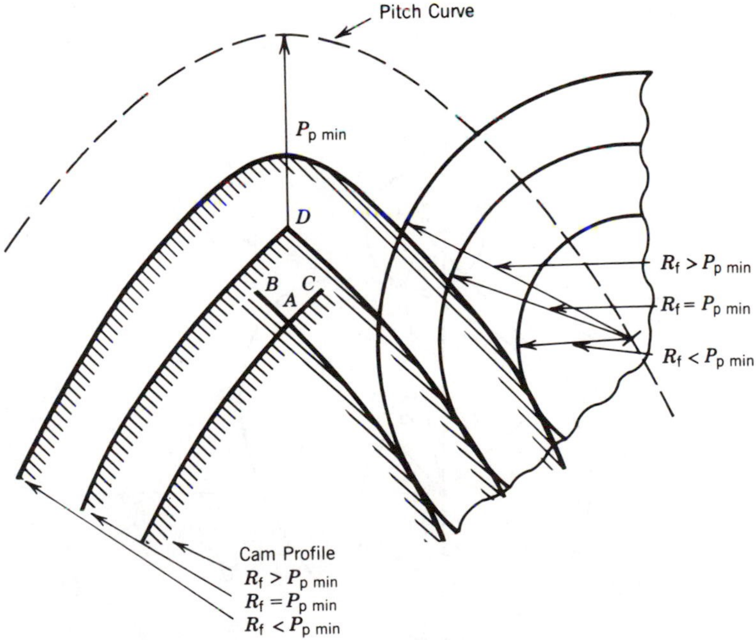

FIGURE 4.14 Relation of Pitch Curve and Roller Radius to Cam Profile

ered. The smallest roller radius results in the outermost cam profile, smooth and apparently kinematically acceptable as shown. Consider the largest roller shown; the cam surface $\widehat{AB}$ is required to support the roller as it approaches the position of minimum pitch curve radius of curvature. Then, as the roller moves away, having passed that point, the follower must pass through the profile segment $\widehat{AB}$ while being supported on the segment $\widehat{AC}$. This is a contradictory situation; each part of the motion requires material for support that must be cut away to allow another part of the motion to occur. This situation is called *undercutting*, and it results in a cusp on the cam profile. The distinguishing features among these three rollers are their radii compared to the pitch curve minimum radius of curvature. For any roller radius less than the minimum pitch radius of curvature, a smooth cam profile with a continuously turning tangent line results. If the roller radius is greater than the pitch curve minimum radius of curvature, undercutting will always result and the trace point will not follow the pitch curve. For the intermediate case, where the roller radius is exactly equal to the pitch curve minimum radius of curvature, the trace point follows the pitch curve but the cam profile has an unacceptable cusp.

Milling Cutter Coordinates. As discussed in Section 4.5, it is often desirable to determine milling cutter coordinates for manufacturing the cam profile, this time for a cam intended to function with a translating, roller follower. The required cam profile and the milling cutter outline are shown in Figure 4.15. The milling cutter radius is R_m, whereas the radius of the

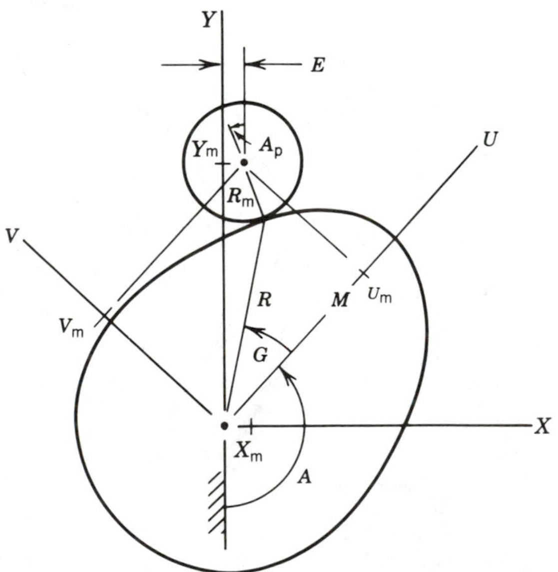

FIGURE 4.15 Milling Cutter Coordinate Determination for Cam with Translating, Roller Follower

intended roller follower is R_f. With respect to the X–Y coordinate system, the milling cutter axis coordinates are

$$X_m = E + (R_f - R_m) \sin A_p$$

$$Y_m = H(A) - (R_f - R_m) \cos A_p$$

With the necessary rotation by $(A - \pi/2)$ to the body coordinate system, the body coordinates for the milling cutter axis are

$$U_m = X_m \sin A - Y_m \cos A$$

$$V_m = X_m \cos A + Y_m \sin A$$

As the angle A sweeps through the range 0 to 2π, these functions determine all locations for the milling cutter axis required to generate the specified cam profile.

4.6 ANALYTICAL DESIGN OF A CAM WITH A PIVOTED, FLAT-FACED FOLLOWER

A cam system with a pivoted, flat-faced follower is called for when an angular motion is required for the cam follower response. A typical system of this type is shown in Figure 4.16. The displacement function, $f(A)$, for this type of system describes the variation in the angular position of the follower, and consequently, is measured in radians. In the analysis to follow, the displacement function and the dimensions C_1, C_2, and C_3 are assumed to have been chosen previously.

The cam rotates about a fixed axis at O, with the rotation measured by the angle A from a stationary reference line to the body reference line OM. When the cam rotation is zero, the angular position of the follower is B_o; the value B_o depends on the base circle radius and the previously chosen dimensions C_1, C_2, and C_3. As the cam rotates, the follower is required to move according to the relation

$$B(A) = B_o + f(A)$$

Contact Location. The line of contact is the normal to the tangent line at the point of contact; in this case the normal to the follower face at the contact point. On the follower, the point of contact is located by the distance D, as shown in Figure 4.16. For sustained contact without penetration, the velocity components along the line of contact for the two points in contact—one on the cam and one on the follower—must be the same. For the contact point on the follower, the component of velocity along the line of contact is

$$V_f = \dot{B}\, D$$

Applying the theorem presented in Section 4.3, the velocity component along the line of contact for the contact point on the cam is

$$V_c = \dot{A}\, \widehat{OQ}$$

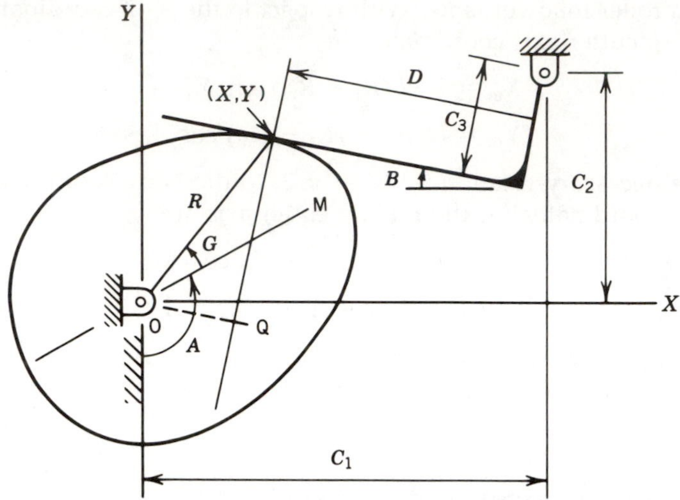

FIGURE 4.16 Cam With Pivoted, Flat-faced Follower

Equating these two velocity components leads to

$$\frac{\overset{\frown}{OQ}}{D} = \frac{\dot{B}}{\dot{A}} = \frac{dB}{dA} = B'(A) = f'(A)$$

The next steps are to eliminate the distance $\overset{\frown}{OQ}$ and to solve for the contact position, D. To this end, consider an equation expressing position measured parallel to the follower face:

$$\overset{\frown}{OQ} + D + C_2 \sin B = C_1 \cos B$$

Using these two relations, the length $\overset{\frown}{OQ}$ is eliminated and the contact position is determined:

$$D = \frac{C_1 \cos B - C_2 \sin B}{1 + f'(A)}$$

Cam Profile. With the distance D known, the rectangular coordinates (X, Y) of the contact point are readily determined, as follows:

$$X = C_1 - C_3 \sin B - D \cos B$$

$$Y = C_2 - C_3 \cos B + D \sin B$$

The cam rotation, A, the cam profile polar coordinate angle, G, and the angular position of the contact point are related, such that

$$A + G - \pi/2 = \text{Atan}_2(Y, X)$$

The polar coordinate angle, G, is determined from the preceding relation, whereas the cam profile radius, R, is determined by the Pythagorean theorem:

$$G = \pi/2 - A + \text{Atan}_2(Y, X)$$
$$R = \sqrt{X^2 + Y^2}$$

As in the previous cases, these equations give the polar coordinates of the cam profile in parametric form, where the parameter is the cam rotation angle, A.

Radius of Curvature. Following the now familiar pattern, the actual cam profile at the contact point is replaced locally with a circular arc, while the rest of the profile is of no consequence; this is shown in Figure 4.17. The center of curvature is on the line of contact at point C*, a distance P below the point of contact, where P is the cam profile radius of curvature. The center of curvature is fixed on the cam at an unknown distance R_c from the cam center of rotation. The angles C and A each measure the angular position of a line in the cam from a common reference. Thus, they differ by at most a constant, and

$$\frac{dC}{dA} = +1$$

Consider now the two position loop equations:

$$R_c \sin C + P \sin B + D \cos B + C_3 \sin B - C_1 = 0$$
$$-R_c \cos C + P \cos B - D \sin B + C_3 \cos B - C_2 = 0$$

These involve the three unknowns R_c, C, and P. To eliminate R_c and C in a single substitution, the derivative of the vertical equation is required to eliminate the term $R_c \sin C$ in the horizontal equation:

$$R_c \sin C\, C' - P \sin B\, B' - D' \sin B - D \cos B\, B' - C_3 \sin B\, B' = 0$$

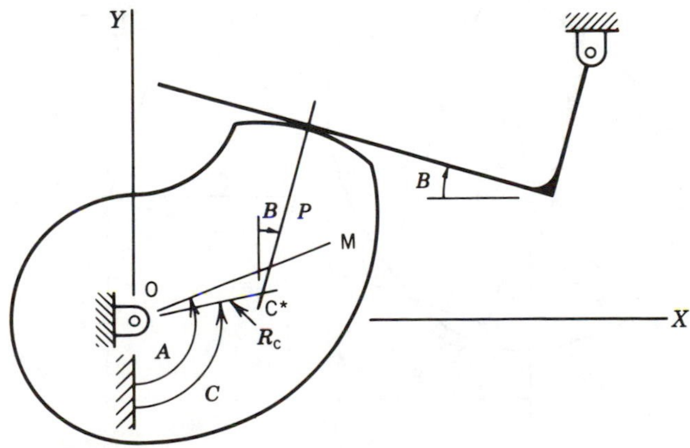

FIGURE 4.17 Radius of Curvature Determination

It is also necessary to differentiate the expression for the contact location to express D':

$$D' = \frac{-f'(C_1 \sin B + C_2 \cos B) - f''D}{1 + f'}$$

With these substitutions, and recalling that $C' = +1$ and $B' = f'$, the quantity $R_c \sin C$ is eliminated and the result solved for the radius of curvature, P. After much algebraic manipulation, the cam profile radius of curvature is

$$P = \frac{(1 + 2f')(C_1 \sin B + C_2 \cos B) + f''D}{(1 + f')^2} - C_3$$

4.7 ANALYTICAL DESIGN OF A CAM WITH A PIVOTED, ROLLER FOLLOWER

A cam with a pivoted, roller follower provides angular follower response combined with low friction and wear. A system of this type is shown in Figure 4.18. The displacement function is assumed to have been chosen previously, as are the dimensions C_1, C_2, and C_3.

The cam rotates about a fixed pivot at O, while the follower rotates about the fixed pivot at E. The cam rotation is measured by the angle A from a fixed reference line to the body reference line OM. The follower position is described by the angle B, measured as shown in the figure. The angle B is given by

$$B(A) = B_o + f(A)$$

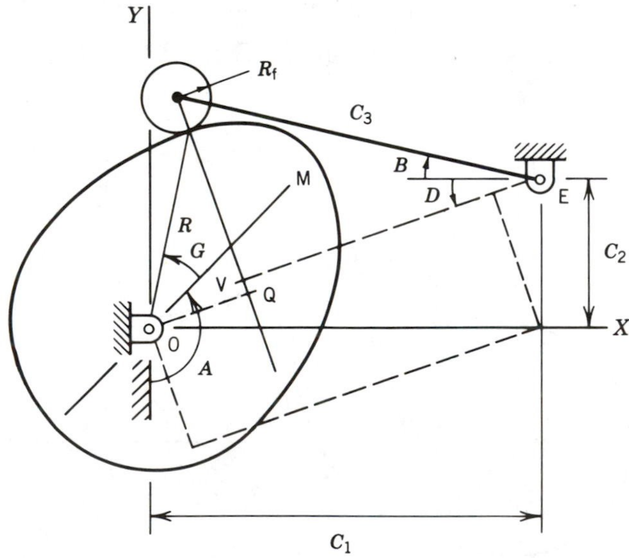

FIGURE 4.18 Cam With Pivoted Roller Follower

where B_o is the angular position of the follower arm when the cam rotation angle A is zero. The value B_o, the base radius for the cam, the roller follower radius, and C_1, C_2, and C_3 are related as follows:

$$(R_o + R_f)^2 = (C_1 - C_3 \cos B_o)^2 + (C_2 + C_3 \sin B_o)^2$$

Orientation of the Line of Contact. The angle D is introduced to express the orientation of the line of contact. In some respects, this angle is comparable to the pressure angle of Section 4.5. As shown in Figure 4.18, the angle D is measured from a horizontal reference through point E, down to the normal from the line of contact through point E (the latter shown as the broken line VE). Note that this angle also appears several other places in the figure, and in particular, it is the deviation of the line of contact from the vertical. For the point of contact in the cam, the velocity along the line of contact is

$$V_c = \dot{A}\,\widehat{OQ}$$

For the velocity of the contact point in the follower, recall that the rotational velocity component is normal to the line of contact, so that the component along the line of contact is

$$V_f = \dot{B}\,\widehat{VE}$$

(If this is unclear, review the discussion in Section 4.5 regarding determination of the pressure angle.) As before, the cam and follower must remain in contact without penetration, which is expressed by requiring that the preceding velocity components be equal. This leads to a relation between $\widehat{OQ}$ and $\widehat{VE}$:

$$\widehat{OQ} = \frac{\dot{B}}{\dot{A}}\,\widehat{VE} = \frac{dB}{dA}\,\widehat{VE} = B'\,\widehat{VE} = f'\,\widehat{VE}$$

Evidently, it is necessary to express the length $\widehat{VE}$ from the right triangle involving the follower arm and the side $\widehat{VE}$, as follows:

$$\begin{aligned}\widehat{VE} &= C_3 \cos (B + D)\\ &= C_3 (\cos B \cos D - \sin B \sin D)\end{aligned}$$

The position equation in terms of lengths measured parallel to the line VE is

$$\widehat{OQ} + \widehat{VE} = C_1 \cos D + C_2 \sin D$$

This is a system of three equations in the three unknowns $\widehat{OQ}$, $\widehat{VE}$, and D. When $\widehat{OQ}$ and $\widehat{VE}$ are eliminated, the result can be solved for D:

$$D = \text{Arctan}\left[\frac{C_3 (1 + f') \cos B - C_1}{C_3 (1 + f') \sin B + C_2}\right]$$

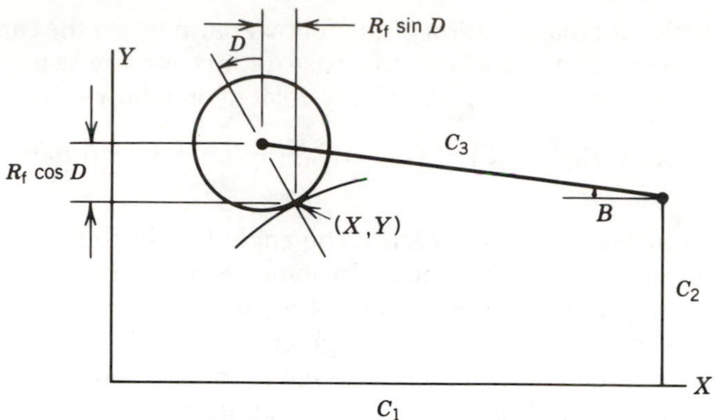

FIGURE 4.19 Cam Profile Determination

Cam Profile. The rectangular coordinates (X, Y) of the point of contact can now be determined easily. Referring to Figure 4.19, it is evident that they are

$$X = C_1 - C_3 \cos B + R_f \sin D$$

$$Y = C_2 + C_3 \sin B - R_f \cos D$$

The angular position of the contact point is related to the cam rotation A and the profile polar coordinate angle G such that

$$A + G - \pi/2 = \text{Atan}_2(Y, X)$$

Solving this relation for G and using the Pythagorean theorem for the length of the radius gives the polar coordinates for the contact point measured from the body reference line:

$$G = \pi/2 - A + \text{Atan}_2(Y, X)$$

$$R = \sqrt{X^2 + Y^2}$$

These are the polar coordinates of the cam profile, in terms of the parameter A.

Radius of Curvature. As in previous sections, the cam profile is considered to be replaced locally by a circular arc, whereas the remainder of the profile is completely arbitrary. This is shown in Figure 4.20. The center of curvature is the point C*, a point on the line of contact at an unknown distance R_c from the cam axis of rotation. The point C* is a distance P from the point of contact between the cam and follower, and a distance P_p from the cam follower axis. The length P is the cam profile radius of curvature; P_p is the radius of curvature for the pitch curve. The angle C is measured from the stationary vertical reference line to the radial line on the cam

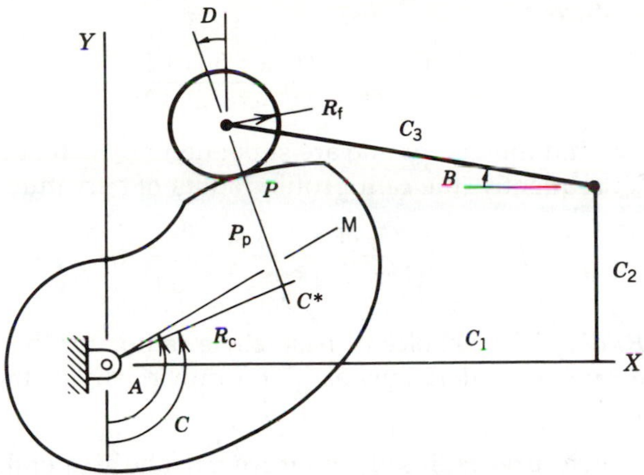

FIGURE 4.20 Radius of Curvature Determination

surface passing through C^*. Evidently C and A differ by, at most, a constant, so that

$$\frac{dC}{dA} = +1$$

Consider next the horizontal and vertical loop equations:

$$R_c \sin C - P_p \sin D + C_3 \cos B - C_1 = 0$$

$$-R_c \cos C + P_p \cos D - C_3 \sin B - C_2 = 0$$

This gives two equations in the three unknowns R_c, C, and P_p. As before, it is possible to eliminate two of the three in a single substitution by using the derivative of the horizontal equation in the vertical equation. Thus,

$$R_c C' \cos C - P_p D' \cos D - C_3 B' \sin B = 0$$

To determine D', the previous expression for D can be written as

$$\tan D = \frac{C_3 (1 + f') \cos B - C_1}{C_3 (1 + f') \sin B + C_2}$$

which is a convenient place to begin the differentiation. After much manipulation, the final result is

$$D' = \frac{C_3 f''(C_2 \cos B + C_1 \sin B) - C_3^2 (1 + f')^2 + C_3 f'(1 + f') (C_1 \cos B - C_2 \sin B)}{C_1^2 + C_2^2 + C_3^2 (1 + f')^2 + 2C_3 (1 + f') (C_2 \sin B - C_1 \cos B)}$$

With D' known and available for substitution into the differentiated horizontal loop equation, the system can then be solved for the pitch curve radius of curvature:

$$P_p = \frac{C_2 + C_3\,(1 + f')\sin B}{(1 - D')\cos D}$$

The pitch curve and the cam profile are separated by the follower radius of curvature. Consequently, the cam profile radius of curvature is

$$P = \frac{C_2 + C_3\,(1 + f')\sin B}{(1 - D')\cos D} - R_f$$

Base Circle Radius. The choice of base circle radius for this case is governed by the same considerations as given previously for the translating roller follower:

1. Minimum pitch curve radius of curvature must be greater than the roller radius;
2. Stresses must be acceptable.

In addition, the equation given at the beginning of this section relating B_o, R_o, R_f, C_1, C_2, and C_3 applies.

PART II CAM MECHANISM KINEMATICS

In this part, the second of the two major questions is addressed: "For a given cam, what motion will be generated?" There are many situations in which the freedom to use a cam designed from the specific application does not apply. This may be due to the need to work within available manufacturing capabilities or to make use of existing part stocks. Consequently, the questions to be considered here are, for a specified cam profile,

1. What is the displacement function?
2. What is the velocity coefficient function?
3. What is the velocity coefficient derivative?

In Section 4.2, the follower position, velocity, and acceleration were observed to be related to the cam rotation according to the following equations:

$$H(A) = f(A) + \mathit{Constant}$$

$$\dot{H}(A) = \dot{A}f'(A)$$

$$\ddot{H}(A) = \ddot{A}f'(A) + \dot{A}^2 f''(A)$$

In a sense, the answers to questions 2 and 3 are evident: the velocity coefficient is $f'(A)$, and the velocity coefficient derivative is $f''(A)$. It will be useful, however, to look further into this matter, as well as to return to the first question. The special case of a circular profile is considered first, before looking at a more general shape.

4.8 ECCENTRIC CIRCLE CAM

Figure 4.21 shows several cam profiles, each in the form of one or more circular arcs pivoted about a point other than the center of the arc. This type of cam is called an *eccentric,* which means "not having the same center." The cam profile may be only a limited arc, as shown in Figure 4.21(b), or it may be a full circle, as in Figure 4.21(a). The third cam profile shown is a composite of four circular arcs, each of different radius. The

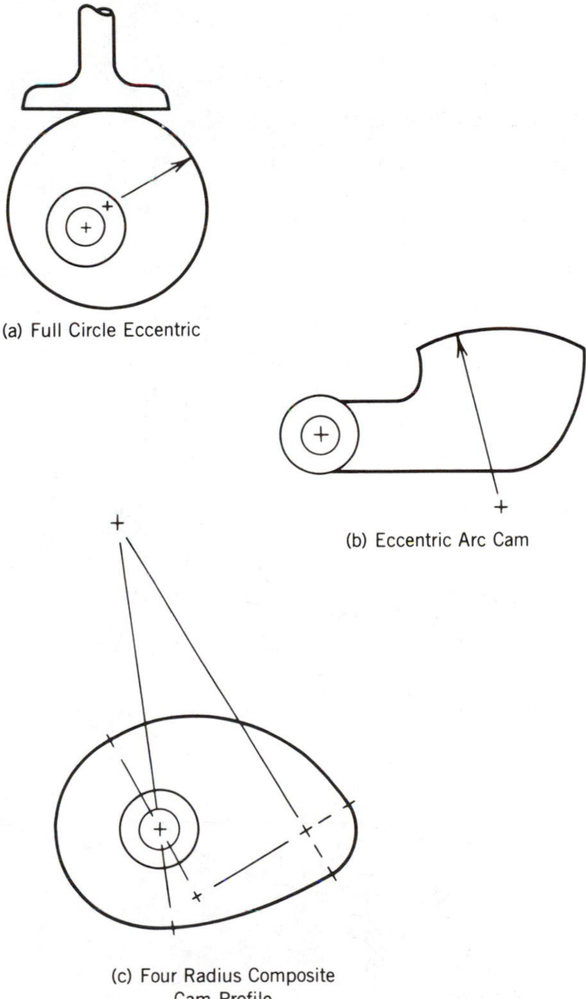

FIGURE 4.21 Various Types of Eccentric Circle Cams

process of "splicing" several circular arcs together, as illustrated in Figure 4.21(c), provides the means to generate rather complex motion based entirely on circular arcs. Furthermore, it is relatively inexpensive to manufacture cam profiles made up of circular arcs. The following discussions apply to any situation where contact between the follower and the cam profile occurs on an eccentric circular arc.

Flat-Faced Translating Follower. A flat-faced translating follower is shown with an eccentric arc cam in Figure 4.22. For convenience, let the body reference line OM pass through the center of curvature for the contact surface, the point C*. The length R_c is called the *eccentricity*. The radius of the circular arc is P, the radius of curvature. For this system, the follower position is

$$\begin{aligned} H(A) &= -R_c \cos A + P \\ &= R_c (1 - \cos A) + P - R_c \end{aligned}$$

By inspection, the displacement function is

$$f(A) = R_c (1 - \cos A)$$

from which the first two derivatives are

$$K_h = f'(A) = R_c \sin A$$

$$L_h = f''(A) = R_c \cos A$$

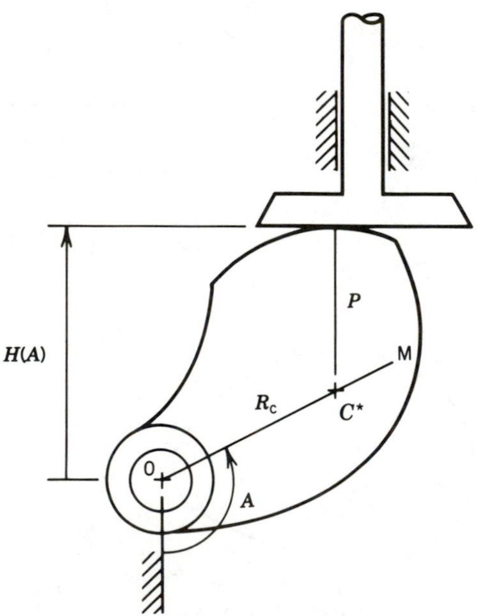

FIGURE 4.22 Eccentric Arc Cam with Flat-faced Translating Follower

The velocity and acceleration of the follower, expressed in terms of the cam rotation angle A and its derivatives, are

$$\dot{H}(A) = \dot{A}f'(A) = \dot{A}\,R_c \sin A$$

$$\ddot{H}(A) = \ddot{A}f'(A) + \dot{A}^2 f''(A)$$

$$= \ddot{A}\,R_c \sin A + \dot{A}^2\,R_c \cos A$$

Offset, Translating Roller Follower. A full-circle eccentric cam with an offset, translating roller follower is shown in Figure 4.23. The body reference line OM is again taken through the center of curvature for the contact surface, point C*. The known data include the following:

R_c: eccentricity
P: cam profile radius of curvature
R_f: follower radius
E: follower offset

To determine the position of the cam follower, consider the horizontal and vertical loop equations:

$$R_c \sin A - (R_f + P) \sin A_p - E = 0$$

$$-R_c \cos A + (R_f + P) \cos A_p - H = 0$$

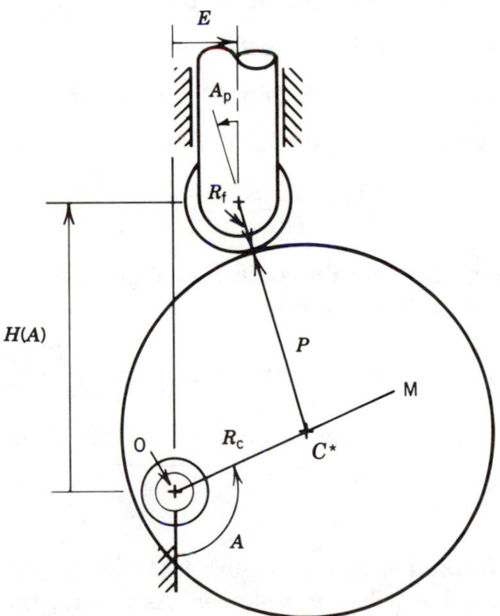

FIGURE 4.23 Full Circle Eccentric Cam with Offset, Translating, Roller Follower

In writing these equations, it has been necessary to introduce the unknown pressure angle, A_p. These equations can be solved explicitly for the pressure angle and the follower position:

$$A_p = \text{Arcsin}\left(\frac{R_c \sin A - E}{R_f + P}\right)$$

$$H = (R_f + P) \cos A_p - R_c \cos A$$

When the cam rotation angle A is zero, let the pressure angle be denoted A_{po}, and the follower position is

$$H(0) = (R_f + P) \cos A_{po} - R_c$$

This last amount is the constant difference between $H(A)$ and $f(A)$, so that

$$f(A) = (R_f + P)(\cos A_p - \cos A_{po}) - R_c(\cos A - 1)$$

To determine the velocity relations, the preceding loop equations are differentiated, with the result

$$\dot{A}\, R_c \cos A - \dot{A}_p\, (R_f + P) \cos A_p = 0$$

$$\dot{A}\, R_c \sin A - \dot{A}_p\, (R_f + P) \sin A_p - \dot{H} = 0$$

These may be solved in closed form to give the velocities $\dot{A}_p$ and $\dot{H}$:

$$\dot{A}_p = \frac{\dot{A}\, R_c \cos A}{(R_f + P) \cos A_p}$$

$$\dot{H} = \dot{A}\, R_c\, (\sin A - \cos A \tan A_p)$$

From this last result, the velocity coefficient K_h can be identified:

$$K_h = f'(A) = R_c\, (\sin A - \cos A \tan A_p)$$

Also, the derivative dA_p/dA is available; that is,

$$\frac{dA_p}{dA} = \frac{R_c \cos A}{(R_f + P) \cos A_p}$$

With $f'(A)$ known, the velocity coefficient derivative $L_h = f''(A)$ can be obtained directly by differentiation:

$$L_h = f''(A) = R_c \left(\cos A + \sin A \tan A_p - \frac{R_c \cos^2 A}{(R_f + P) \cos^3 A_p}\right)$$

The acceleration of the follower is then written in the familiar form

$$\ddot{H} = \ddot{A}\, K_h + \dot{A}^2\, L_h$$

Flat-Faced, Oscillating Follower. Figure 4.24 shows a full-circle, eccentric cam with a flat-faced, oscillating follower. As before, the body reference line OM is chosen to pass through the center of curvature, point C*. The known data include the dimensions C_1, C_2, C_3, P, and R_c. Horizontal and

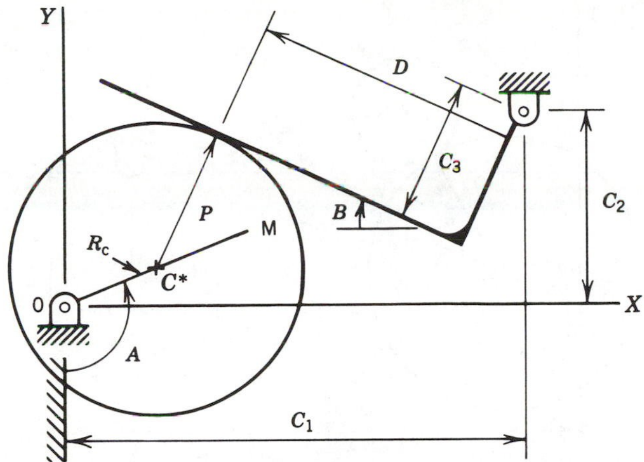

FIGURE 4.24 Full Circle Eccentric Cam with Oscillating Flat-faced Follower

vertical loop equations can be written involving the two unknowns B and D. When these are solved for the follower angle B, the result is

$$B = -\text{Arccos}[(P + C_3)/W(A)] + E$$

where

$$E = \text{Atan}_2(C_1 - R_c \sin A,\ C_2 + R_c \cos A)$$

$$W(A) = \sqrt{(C_1 - R_c \sin A)^2 + (C_2 + R_c \cos A)^2}$$

With B known, the distance D may be determined from the two loop equations with the result

$$D = (C_1 - R_c \sin A) \cos B - (C_2 + R_c \cos A) \sin B$$
$$= W(A) \sin (E - B)$$

Differentiating the loop equations with respect to time gives the velocity loop equations. With some manipulation, the velocity coefficient for the follower angle is

$$K_b = R_c \sin(A + B)$$

Further differentiation of the velocity coefficient with respect to A gives the velocity coefficient derivative:

$$L_b = R_c\ (1 + K_b) \cos(A + B)$$

Pivoted Roller Follower. An oscillating follower with roller contact is shown with an eccentric circle cam in Figure 4.25. Once again, the body reference line OM has been chosen to pass through the center of curvature, point C*. The dimensions C_1, C_2, C_3, P, R_c, and R_f are considered known. The combined effect of the cam disk and the follower disk is to maintain a

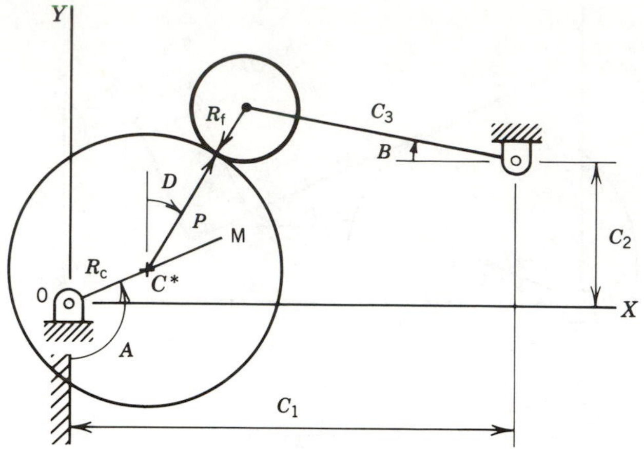

FIGURE 4.25 Full Circle Eccentric Cam with Pivoted, Roller Follower

constant distance between the center of the cam disk and the trace point at the tip of the follower arm. The same result could be obtained with a link of length $(P + R_f)$. It is evident then, that this cam–follower system is equivalent to a four-bar linkage. As suggested previously in Chapter 2, a numerical solution of the position loop equations is recommended. That solution will provide values for both of the angles B and D as functions of the input rotation A.

When the position loop equations are differentiated with respect to time, the resulting pair of equations is linear in the unknown velocities, $\dot{B}$ and $\dot{D}$. These can be solved explicitly for the unknown velocities and then further rearranged to give the velocity coefficients, as follows:

$$K_b = \frac{R_c \sin(A + D)}{C_3 \cos(B - D)}$$

$$K_d = \frac{-R_c \cos(A + B)}{(P + R_f)\cos(B - D)}$$

Finally, the velocity coefficient derivative, L_b, can be obtained by differentiating K_b with respect to A. The result is

$$L_b = \left(\frac{R_c}{C_3}\right) \frac{[\cos(B - D)\cos(A + D)(1 + K_d) + \sin(B - D)\sin(A + D)(K_b - K_d)]}{\cos^2(B - D)}$$

The foregoing results for the eccentric circle cam suggest an approach that could be applied to analyse the follower motion that results from a cam of arbitrary shape. This is possible because the arbitrary shape can be approximated locally at each point as a circular arc. However, it is also an awkward approach, so a more direct alternative is used in the following discussion for cams of arbitrary shape.

4.9 ANGLE BETWEEN RADIUS AND NORMAL

Before discussing the follower response for a cam of arbitrary shape, it is useful to establish a means to calculate the angle between a radial line and the outward normal at a point on the cam profile as shown in Figure 4.26. The cam profile is assumed to be known in polar form

$$R = R(G)$$

and the angle of interest is denoted V. The unit vectors $\mathbf{E}_r$ and $\mathbf{E}_g$ are along and perpendicular to the radial line, respectively. The vector $\mathbf{R}$ is along the radial line,

$$\mathbf{R} = R(G)\,\mathbf{E}_r$$

The first step is to determine a unit vector along the outward normal at the point considered. When the position vector is differentiated with respect to G, the result is a vector tangent to the cam profile:

$$\frac{d\mathbf{R}}{dG} = R'\,\mathbf{E}_r + R\,\mathbf{E}_g$$

A unit tangent vector may be constructed simply by normalizing this last vector,

$$\mathbf{E}_t = \frac{R'\,\mathbf{E}_r + R\,\mathbf{E}_g}{\sqrt{R^2 + R'^2}}$$

Let $\mathbf{E}_n$ denote the unit normal vector, which can be written as a linear combination of $\mathbf{E}_r$ and $\mathbf{E}_g$,

$$\mathbf{E}_n = a\mathbf{E}_r + b\mathbf{E}_g$$

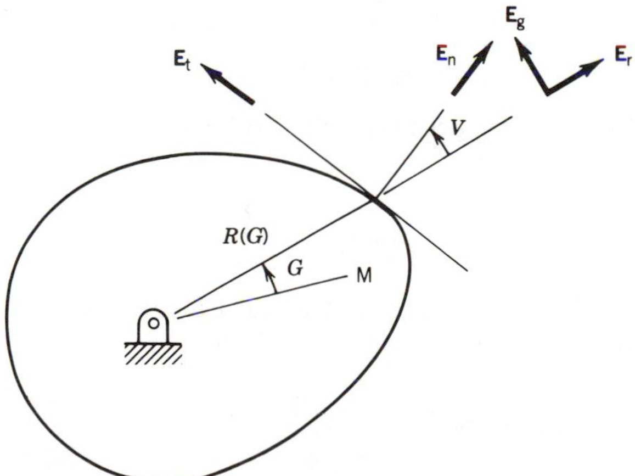

FIGURE 4.26 Angle Between a Radial Line and the Normal to the Cam Profile

where a and b are unknown coefficients. Because $\mathbf{E}_n$ is a unit vector, the coefficients a and b must be related, such that

$$a^2 + b^2 = 1$$

Furthermore, by definition $\mathbf{E}_n$ is normal to the unit tangent vector $\mathbf{E}_t$, so that

$$\mathbf{E}_n \cdot \mathbf{E}_t = 0$$

from which

$$a\,R' + b\,R = 0$$

The unit vector $\mathbf{E}_n$ then is determined by solving these relations for a and b, with the following results:

$$a = \frac{R}{\sqrt{R^2 + R'^2}}$$

$$b = \frac{-R'}{\sqrt{R^2 + R'^2}}$$

and

$$\mathbf{E}_n = \frac{R\,\mathbf{E}_r - R'\,\mathbf{E}_g}{\sqrt{R^2 + R'^2}}$$

The cosine of V is obtained through a dot product of $\mathbf{E}_r$ with $\mathbf{E}_n$, whereas the sine of V is determined by the cross-product of $\mathbf{E}_r \times \mathbf{E}_n$:

$$\cos V = \frac{R}{\sqrt{R^2 + R'^2}}$$

$$\sin V = \frac{-R'}{\sqrt{R^2 + R'^2}}$$

Finally, the angle between the radius and the normal to the cam profile is

$$V = \mathrm{Atan}_2(-R', R)$$

Derivatives of V with respect to G will also be required, and these are best obtained by starting with

$$\tan V = -\frac{R'}{R}$$

Performing two differentiations with respect to the angle G, the first and second derivatives of V with respect to G are

$$V' = \frac{R'^2 - RR''}{R^2}\cos^2 V$$

$$V'' = -\cos^2 V\,\frac{(R^2\,R''' - 3R\,R'\,R'' + 2\,R'^3)}{R^3} - 2\,V'^2 \tan V$$

These expressions for V, V', and V'' are rather cumbersome for further analytical work, but they are evaluated readily for numerical work using a computer.

4.10 FOLLOWER RESPONSE FOR ARBITRARY CAM PROFILE

In this section, the follower response for an arbitrary cam profile is considered. The cam is assumed to be known in polar form, $R(G)$, where R is the radius and G is the polar angle, measured from a reference line OM fixed in the cam. The objective is to describe the follower position, velocity, and acceleration by determining the functions $f(A)$, $f'(A)$, and $f''(A)$.

Flat-Faced, Translating Follower. Figure 4.27 shows a typical cam profile with a flat-faced, translating follower. The point of contact between the cam and follower has polar coordinates (R, G). The normal to the cam profile at the point of contact is parallel to the displacement of the follower. From the geometry, it is evident that

$$A + G + V = \pi$$

where the angle V is a known function of the angle G, as discussed in Section 4.9. The vertical position loop equation is

$$R(G) \cos V - H = 0$$

For a direct approach, these two equations may be solved for the two unknowns G and H in terms of A. This requires an iterative solution of the

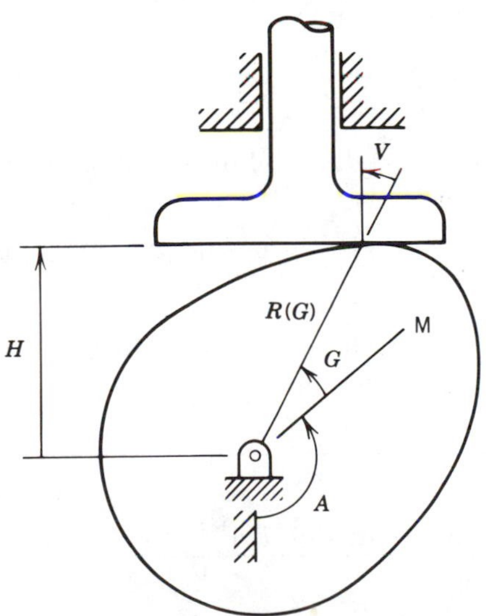

FIGURE 4.27 Cam With Specified Profile and Flat-faced, Translating Follower

first equation to evaluate G. Then, H is readily evaluated from the second equation. As an alternative, consider G as the independent variable (instead of A), with A and H as the unknowns. Because $V(G)$ is a known function, the first equation is readily solvable for A and the second is for H. Then, the displacement function is

$$f(A) = H - R_o = R(G) \cos V - R_o$$

If H and A are tabulated as functions of G, then for a prescribed value of A, the associated values of G and H can be determined using inverse interpolation. This is the most satisfactory approach for many purposes, and will be assumed for further work in this text.

For the velocity coefficient and the velocity coefficient derivative, it is necessary to differentiate f or H with respect to A, but H and A are known from preceding work as functions of G. Then, the first derivative, $f'(A)$, is obtained by the chain rule for differentiation:

$$f'(A) = \frac{df}{dA} = \frac{df}{dG}\frac{dG}{dA}$$

$$= \frac{\dfrac{df}{dG}}{\dfrac{dA}{dG}}$$

The two required derivatives are determined by differentiating the position equations just established:

$$\frac{df}{dG} = R' \cos V - R\, V' \sin V$$

$$\frac{dA}{dG} = -1 - V'$$

where $R' = dR/dG$ and $V' = dV/dG$. Putting all of the pieces together gives the final expression for $f'(A)$:

$$f'(A) = \frac{df}{dA} = \frac{R\, V' \sin V - R' \cos V}{1 + V'}$$

Carrying this type of analysis one step further, the second derivative is given by the following expression:

$$f''(A) = \frac{d^2f}{d^2A}$$

$$= \frac{-(R'V' \sin V + R\, V'' \sin V + R\, V'^2 \cos V - R'' \cos V + R'V' \sin V)}{(1 + V')^2}$$

$$+ \frac{V''(RV' \sin V - R' \cos V)}{(1 + V')^3}$$

where the indicated derivatives of R and V are with respect to G. The derivatives of V with respect to G were expressed in Section 4.9 in terms of the derivatives of $R(G)$ with respect to G. Alternatively, if the functions A, H, R, and V have been tabulated as functions of G, the necessary derivatives for the evaluation of f' and f'' can be evaluated numerically using the interpolation process described in Appendix A5. As in previous cases, the velocity and acceleration of the follower can be written as

$$\dot{H} = \dot{A}\ f'(A)$$

$$\ddot{H} = \ddot{A}\ f' + \dot{A}^2\ f''$$

Translating, Roller Follower. A cam with translating, roller follower is shown in Figure 4.28. The follower is offset by an amount E, and the radius of the follower is R_f. The cam profile is assumed to be given in polar form, $R = R(G)$, and the angle between the radius and the normal at any point is the known function $V(G)$. The horizontal and vertical position loop equations are

$$R\sin(A + G) + R_f \sin(A + G + V) - E = 0$$

$$-R\cos(A + G) - R_f \cos(A + G + V) - H = 0$$

As in the preceding case, it is convenient to consider G as the independent variable with A and H as the unknowns. This avoids the numerical solution

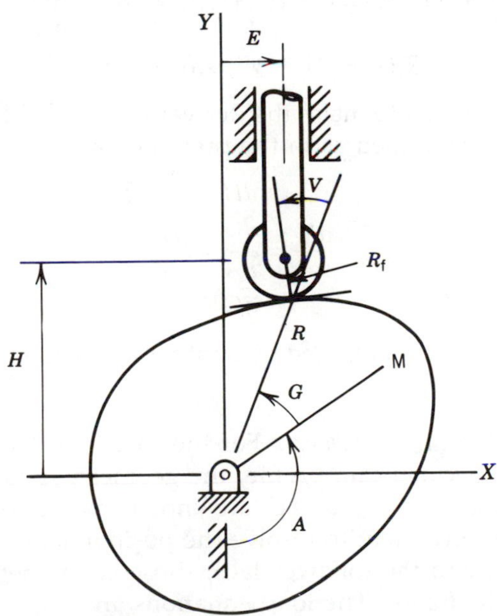

FIGURE 4.28 Cam with Specified Profile and Offset Translating Roller Follower

required for G and H if A is considered as the independent variable. Taking G as independent, these equations may be solved explicitly for A and H as follows:

$$A = -G - B - \text{Arcsin}(E/D)$$
$$H = -R\cos(A + G) - R_f \cos(A + G + V)$$

where

$$B = \text{Atan}_2(R_f \sin V, R + R_f \cos V)$$
$$D = \sqrt{(R + R_f \cos V)^2 + (R_f \sin V)^2}$$

The value D is the distance from the fixed pivot point of the cam to the trace point on the follower. The angle B is between the line defined by D and the radial line through the point of contact.

Because G is considered as the independent variable, the chain rule is required again to evaluate the derivative $f'(A) = df/dA$. The derivative of A with respect to G is determined by differentiating the horizontal loop equation with respect to G, with the following result:

$$\frac{dA}{dG} = \frac{-R' \sin(A + G) - R\cos(A + G) - R_f (1 + V')\cos(A + G + V)}{R\cos(A + G) + R_f \cos(A + G + V)}$$

In a similar manner, the derivative of H with respect to G is determined from the vertical loop equation:

$$\frac{dH}{dG} = -R' \cos(A + G) + R(1 + A')\sin(A + G)$$
$$+ R_f(1 + A' + V')\sin(A + G + V)$$

where A' is understood to mean the derivative dA/dG. The derivative of H with respect to A is formed as in the previous case:

$$\frac{dH}{dA} = \frac{\frac{dH}{dG}}{\frac{dA}{dG}} = \frac{H'}{A'}$$

The second derivative is obtained by another application of the chain rule.

Flat-Faced, Oscillating Follower. For the case of a flat-faced, oscillating follower with a specified cam profile, the geometry is as shown in Figure 4.29. The dimensions C_1, C_2, and C_3 are known, as is the cam profile $R(G)$. In this case, it is convenient to resolve the position loop equations parallel and perpendicular to the follower face; those components are shown in broken line on the figure. The loop equations are

$$R\sin V + D + C_2 \sin B - C_1 \cos B = 0$$
$$R\cos V + C_3 - C_2 \cos B - C_1 \sin B = 0$$

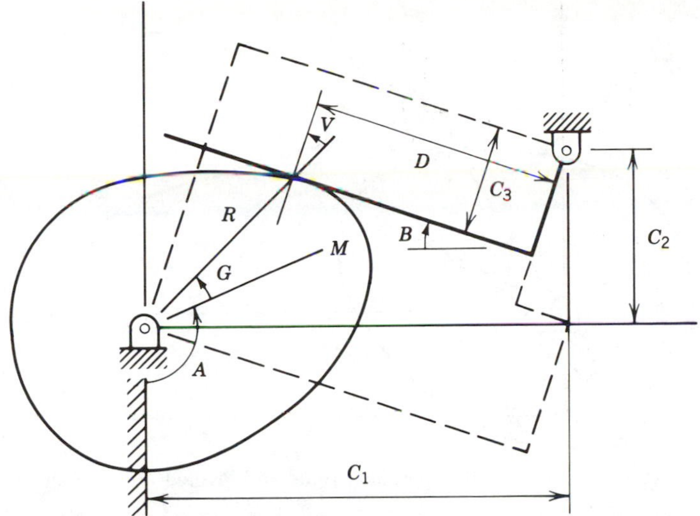

FIGURE 4.29 Cam With Specified Profile and Flat-faced, Oscillating Follower

As in the two preceding cases, it is convenient to take the angle G as the independent variable, and to determine B and D. The follower angle B is determined from the second loop equation, with the result

$$B = -E + \text{Arcsin}\left(\frac{R \cos V + C_3}{\sqrt{C_1^2 + C_2^2}}\right)$$

where

$$E = \text{Atan}_2(C_2, C_1)$$

With B known, the first loop equation can be used to evaluate D. The angle A is then determined from the geometric relation.

$$A + G + V + B = \pi$$

As before, the solutions for R, V, B, D, and A can be tabulated readily as functions of G. When a particular value of A is specified, inverse interpolation can be used to determine the associated value of G as well as the other variables. If the loop equations are differentiated with respect to time, the result is a pair of equations that are linear in the unknown velocities $\dot{B}$ and $\dot{D}$. The equations can then be rearranged to give the velocity coefficients K_b and K_d, and be further differentiated to yield L_b.

Pivoted, Roller Follower. The final case to be considered is that of a pivoted follower with roller contact, as shown in Figure 4.30. The horizontal and vertical position loop equations for that system are as follows:

$$R \sin(A + G) + R_f \sin(A + G + V) + C_3 \cos B - C_1 = 0$$

$$-R \cos(A + G) - R_f \cos(A + G + V) - C_3 \sin B - C_2 = 0$$

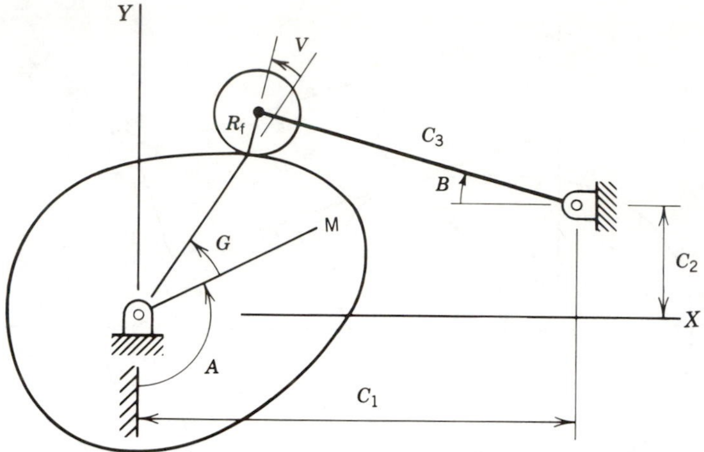

FIGURE 4.30 Cam With Specified Profile and Pivoted, Roller Follower

There is no advantage to using G as the independent variable in place of A because a numerical solution will be required in any case. Following that numerical position solution, the velocity coefficients and velocity coefficient derivatives will be obtained as the solutions of the systems of linear equations resulting from successive time differentiations of the position equations.

4.11 CONCLUSION

The subject of cams is very large, and the discussion of this chapter has been limited necessarily only to disk cams and their four common follower types. An elementary discussion of displacement function has also been presented, but for most purposes the displacement function is assumed to have been previously specified. Displacement functions have been studied extensively in the literature, and the interested reader is encouraged to look elsewhere for more information on the art of choosing an appropriate displacement function.

The first part of the chapter dealt with the analytical design of a cam profile to generate a specified follower response when employed with a particular follower type. In addition to purely kinematic considerations, the Hertz contact stress analysis applicable to cams was presented also. In the second part of the chapter, the follower motion generated by a specified cam profile was analyzed. The special case of response to eccentric circle cams was developed before going to the general discussion for cams of arbitrary profile.

Because of their high reliability, accuracy, and relatively low cost in large volume, cams will continue to be used widely for many years to come. The incorporation of computerized synthesis and analysis methods with computerized manufacturing will make cams more easily incorporated into various types of machinery.

REFERENCES

Boresi, A. P., Sidebottom, O. M., Seely, F. B., and Smith, J. O., *Advanced Mechanics of Materials.* 3rd ed. New York: Wiley, 1978.

Juvinall, R. C., *Fundamentals of Machine Component Design.* New York: Wiley, 1983.

Mabie, H. H. and Ocvirk, F. W., *Mechanisms and Dynamics of Machinery,* 3rd ed. New York: Wiley, 1975.

Poritsky, H., "Stresses and Deflections of Cylindrical Bodies in Contact With Application to Contact of Gears and Locomotive Wheels," *J. of Applied Mechanics, Trans. ASME,* June 1950, pp. 191–201.

Rothbart, H. A., *Cams.* New York: Wiley, 1956.

Shigley, J. E. and Uicker, J. J., *Theory of Machines and Mechanisms.* New York: McGraw–Hill, 1980.

PROBLEM SET

4.1 Consider a cam for which the rise extends over 0.82 radian of cam rotation for a lift of 2.3 in. The rise is in the form of two identical cubic arcs ($Y = aX^3$), each tangent to the adjoining dwell and tangent to each other at the midpoint of the rotation interval. Determine 11 equally spaced points on this displacement curve, including both end points.

4.2 The rise portion of a particular cam displacement curve is to be described as two identical hyperbolic sine-squared arcs. The lower arc is of the form

$$f(A) = B \sinh^2(CA)$$

with appropriate values of B and C, and the upper arc is similar. Each arc is tangent to the adjacent dwell, and the two arcs are tangent to each other at the midpoint of the rotation interval. Let L denote the lift and T the cam rotation during the rise phase. The value of C is 2.0.

a. Determine an expression for the coefficient B as a function of L and T;
b. For the case $L = 1.2$ in. and $T = 0.5$ radians, evaluate $f(T/4)$;
c. For L and T as in part (b), evaluate $f(5T/8)$.

4.3 The return phase of a cam cycle spans 0.70 radian of cam rotation, beginning at $A_2 = 4.22$ radians, and removes a lift of 1.65 in. The displacement function during the return is in the form of a sine-squared arc, taken over one-quarter period of the sine function. (Remember that the displacement function must be tangent to the dwells occurring both before and after the return.) Determine nine equally spaced points on the displacement function over the return phase, including both end points.

4.4 The rise phase of the two cam profiles are to be compared. For the first cam, the displacement function is sinusoidal; for the second, the displacement function is cycloidal. For each cam, the rise extends from zero to $A_1 = 0.6\pi$ with the lift $L = 0.45$ in.

a. Tabulate and plot both displacement functions at 11 evenly spaced points over the rise interval, including both end points;
b. Tabulate and plot both velocity coefficient functions, $f'(A)$, at the same 11 points;
c. Tabulate and plot both velocity coefficient derivative functions, $f''(A)$, at the same 11 points;
d. Which of the two displacement functions will provide the most desirable operating characteristics? Why?

4.5 A particular cam is intended to provide sinusoidal motion during the return phase of the cam cycle. The return occurs in the interval $A_2 = 1.15\pi$ to $A_3 = 1.78\pi$, and a lift of 0.375 in. is removed. Tabulate and plot the following functions at 11 evenly spaced points over the return interval:

a. Displacement function, $f(A)$;
b. Velocity coefficient, $f'(A)$;
c. Velocity coefficient derivative, $f''(A)$;
d. Velocity coefficient second derivative, $f'''(A)$;
e. What is the largest absolute value for the velocity coefficient second derivative? Where does it occur?

4.6 Use the graphical approach to design a plate cam for a flat-faced, translating follower to meet the following kinematic specification:

Lift:	$L = 15$ mm	
Rise:	sinusoidal	$0 \le A < 1.35$ radians
Dwell:		$1.35 \le A < 1.75$ radians
Return:	parabolic	$1.75 \le A < 4.22$ radians
Dwell:		$4.22 \le A < 2\pi$ radians

Take the base circle radius as 50 mm (this is a design decision). Make a full-size engineering drawing that shows the following:

a. Displacement function;
b. Follower in contact with the cam profile at $A = 0$;
c. Development of the cam profile, based on 12 divisions of the circle.

4.7 Using the graphical approach, design a plate cam for use with a radial, translating roller follower that meets the kinematic specification given in Problem 4.6. Take the roller radius as 9.5 mm, and the prime circle radius as 65 mm (these are design decisions). Make a full-size engineering drawing that shows the following:

a. Displacement function;
b. Follower in contact with the cam profile at $A = 0$;
c. Development of the cam profile, based on 12 divisions of the circle.

4.8 Consider a cam for use with a flat-faced, translating follower having a rise and immediate return spanning 3.0 radians of cam rotation and described by

$$f(A) = L \sin^2(\pi A/3) \qquad 0 \le A < 3.0$$
$$= 0 \qquad 3.0 \le A < 2\pi$$

and where L is the lift, 0.887 in. Both the cam and the follower are made of steel, $E = 3(10^7)$ lb/in.^{2} Assume the base radius, R_o is 2.35 in., and

a. Determine the length of the cam follower wear surface;
b. Determine polar coordinates (R, G) for points on the cam profile for $A = 0.0, 0.5, 1.0, 1.5, 2.0, 2.5,$ and 3.0 radians;
c. Make a sketch of the cam profile;
d. The minimum radius of curvature for this cam profile occurs at $A = 1.5$ radians. If the contact force is 20 lb and the contact stress is to be limited to 7800 lb/in.^{2}, determine the minimum thickness for the cam.

4.9 Consider a cam, working with a flat-faced, translating follower, where the follower guide is radial with respect to the cam axis of rotation. The system is designed to meet the following kinematic specification:

Lift:	$L = 0.45$ in.	
Rise:	half-cycle of sinusoid	$0 \le A < 0.4722$
Dwell:		$0.4722 \le A < 2.3030$
Return:	half-cycle of sinusoid	$2.3030 \le A < 3.7525$
Dwell:		$3.7525 \le A < 2\pi$

Base Circle Radius = 3.365 in.

a. What is the minimum follower face length to the left of the guide? To the right? Minimum total face length?
b. Determine polar coordinates (R, G) for the profile point in contact with the follower for each of the following positions: $A = 0, \pi/6, \pi/3, \pi/2, 2\pi/3, 5\pi/6, \pi,$ and $7\pi/6$ radians.

4.10 Consider a cam, operating with a flat-faced, translating follower, and having the following kinematic specification:

Lift:	$L = 0.54$ in.	
Rise:	cycloidal	$0 \le A < 1.34$
Dwell:		$1.34 \le A < 1.62$
Return:	sinusoidal	$1.62 \le A < 4.38$
Dwell:		$4.38 \le A < 2\pi$

Base Radius: $R_o = 3.25$ in.

The follower guide is radial with respect to the cam axis of rotation.

a. What is the minimum follower face length to the left of the guide? To the right? Minimum total face length?

b. Tabulate polar coordinates (R, G) for the profile point in contact with the follower for the following positions: $A = 0, \pi/6, \pi/3, \pi/2, 2\pi/3, 5\pi/6, \pi, 7\pi/6, 4\pi/3, 3\pi/2, 5\pi/3, 11\pi/6$, and 2π;

c. Use the tabulated data of part (b) to make a sketch of the cam profile.

4.11 Working with an offset translating roller follower, a particular cam satisfies the following displacement function:

$$f(A) = 0.377\,[1 - \cos(A/0.6)] \qquad 0 \le A < 1.2\pi$$
$$= 0.0 \qquad 1.2\pi \le A < 2\pi$$

The offset is 0.35 in., the prime circle radius is 2.27 in., and the follower radius is 0.65 in. For $A = 0.68\pi$,

a. Determine the polar coordinates for the contact point;

b. Determine the pressure angle;

c. Determine the radius of curvature.

4.12 For a particular cam with a pivoted, flat-faced follower, the follower rotational displacement is given by

$$f(A) = 0.08\,[1 - \cos(2A/1.65)] \qquad 0 \le A < 1.65\pi$$
$$= 0.0 \qquad 1.65\pi \le A < 2\pi$$

The dimensional parameters are $C_1 = 9.77$ in., $C_2 = 1.33$ in., and $C_3 = 0.93$ in. The angle B_o is zero.

a. Determine the minimum radius for the cam profile;

b. For $A = 1.23\pi$, determine the polar coordinates for the contact point on the cam profile;

c. For $A = 1.23\pi$, determine the radius of curvature for the contact point on the cam profile.

4.13 Consider a cam operating with a translating follower and having the displacement function

$$f(A) = L \sin^2[\pi A/(2A_1)]$$

during the rise motion, and a similar functional form for the return phase. The return phase occurs over the rotation interval $A_2 = 1.27\pi$ to $A_3 = 1.82\pi$, and a lift of 0.87 in. is removed. The cam rotates at constant angular speed, $\dot{A} = 22$ rad/s. At 11 equally spaced points over the return interval, tabulate and plot the answers for parts a, b, and c; answer the questions posed in parts d and e.

a. Displacement during the return phase;
b. Velocity coefficient, $f'(A)$, during the return phase;
c. Velocity coefficient derivative, $f''(A)$, during the return phase.
d. At the point $A = 4.85$ radians, what are the velocity and acceleration of the cam follower?
e. Is the acceleration discontinuous at any point?

4.14 The cam shown functions only through a partial revolution, $\pi/2 \leq A \leq \pi$. During that interval, the roller follower contacts an involute surface, as indicated. The offset of the follower axis is R_b, which is also the base circle radius used to generate the involute. The follower radius is R_f. In terms of these known quantities (R_b, R_f, and A, which is assigned), determine an expression for the position of the tracepoint, $H(A)$, over the interval of motion.

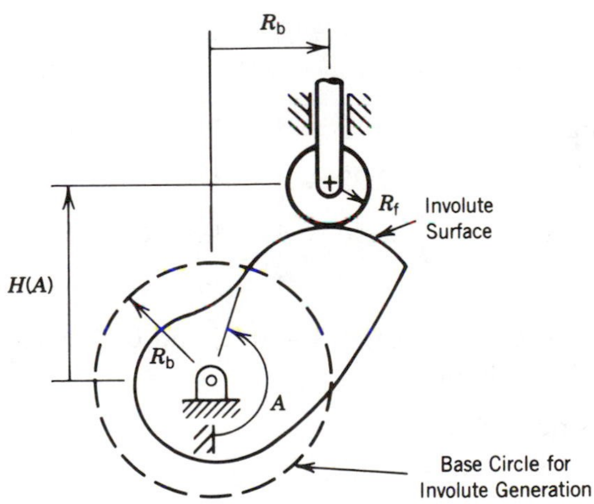

4.15 For the slider–crank cam–follower system shown, the dimensions D, R, C, and R_o (base circle radius) are known. The cam and follower are described by the known displacement function $f(A)$.

a. Express the follower position, H, as a function of q;
b. Express the velocity coefficient of $K_h = dH/dq$;
c. Express the velocity coefficient derivative, L_h.

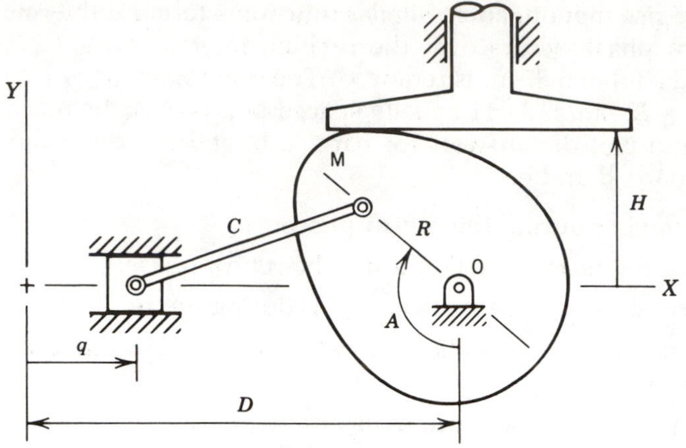

4.16 For the system described in the preceding problem, the known dimensions are

$$D = 0.405 \text{ m}$$
$$R = 0.055 \text{ m}$$
$$C = 0.350 \text{ m}$$
$$R_o = 0.220 \text{ m}$$

The displacement function for the cam is

Lift:	$L = 0.012$ m	
Rise:	sinusoidal	$0 \leq A < 2.33$ radians
Dwell:		$2.33 \leq A < 3.40$
Return:	sinusoidal	$3.40 \leq A < 4.87$
Dwell:		$4.87 \leq A < 2\pi$

The motion begins with $q = 0$ and $A = \pi/2$, and both q and A are increasing. For $q = 0.075$, determine

a. Follower position, H;
b. Velocity coefficient, $K_h = dH/dq$;
c. Velocity coefficient derivative, L_h.

4.17 The mechanism shown uses a pivoted flat-faced follower to move a slider through the distance S when the cam rotation is $q = A$. The slider centerline passes through the cam axis of rotation. When $A = 0$, the slider position is $S = 0$. The dimensions C_1, C_2, C_3, C_4, C_5, and L are known, as are the angles E_1 and E_2. The profile satisfies the rotational displacement function

$$B(A) = B_o + f(A)$$

where B_o and $f(A)$ are known.

a. Determine a sequence of closed-form calculations to evaluate $S = S(A)$ (it is not necessary to combine all of the calculations into one);
b. Determine an expression for K_s;
c. Determine an expression for L_s. (Hint: First determine angle D, then S).

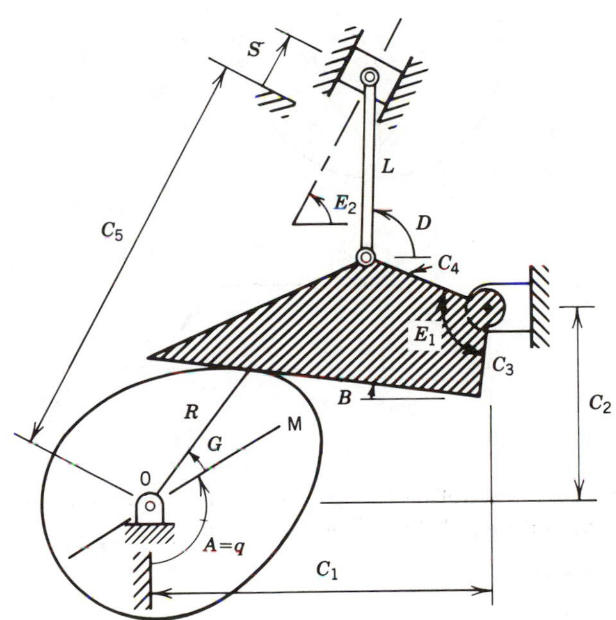

4.18 For the system shown, the cam shaft speed is given by

$\dot{A}(t) = C_1\,(1 - e^{-C_2 t})$ where $A(0) = 0$

$C_1 = 80$ rad/s

$C_2 = 0.38$ 1/s

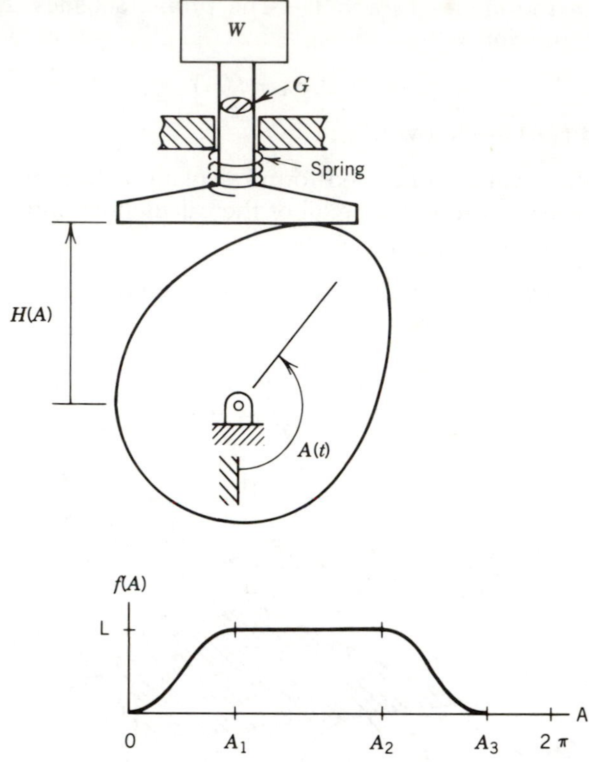

The displacement function is as shown—a sinusoidal rise, dwell, sinusoidal return, and a dwell at zero. The break points for this curve are

$$A_1 = 2.55 \text{ rad}$$
$$A_2 = 3.45 \text{ rad}$$
$$A_3 = 5.65 \text{ rad}$$

while the lift is $L = 0.85$ in. The follower is spring loaded against the cam, with a preload of 12 lb when the follower is in a static condition at its lowest position. The spring rate is $K = 5$ lb/in. The spring, the follower, and the shaft above may all be considered massless, but they drive a block weighing 4.73 lb. The follower, the shaft, and the block are all rigid. There is no gravitational effect to be considered. At the time $t = 2.0$ s, determine numerical values for the following:

a. Acceleration of the weight;

b. Force in the spring;

c. Contact force between the cam and follower;

d. Force in the shaft at the section indicated G.

4.19 A cam with a pivoted, flat-faced follower is connected to a sliding link as shown. The angular displacement of the follower is given by

$$B(A) = B_o + f(A)$$

where $f(A)$ is the displacement function and B_o is the minimum rotation. The distance D and the angle C are known constants.

a. Obtain an expression for the velocity coefficient, $K_x = dX/dA$, in terms of B_o, C, D, f, and f';
b. If the displacement function is

$$f(A) = 0.32\,[1 - \cos(2\pi A/A_1)] \qquad 0 \le A < A_1$$
$$= 0.0 \qquad A_1 \le A < 2\pi$$

where

$A_1 = 1.65\pi$ radians
$B_o = 0.21$ radians
$C = 1.02$ radians
$D = 14.5$ in.

then, evaluate K_x for $A = 1.22$ radians.

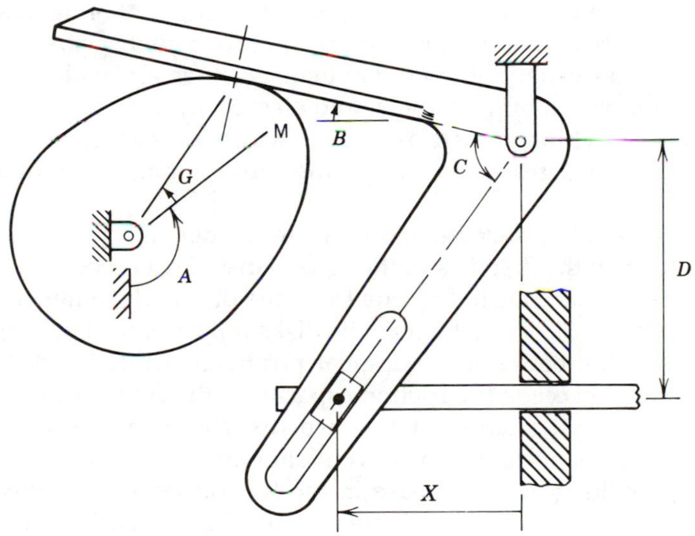

CHAPTER 5

Gears

5.1 INTRODUCTION

The transmission of power from one shaft to another, often with a change in speed of rotation, is frequently accomplished with gears. Gears are found in a wide variety of everyday items such as alarm clocks, kitchen mixers, automobiles, office copiers, and electric drills. Gears are also used in industrial machinery such as lathes, winches, tractors, and printing presses. Certainly, gears are among the most commonly used machine elements.

If two disks are pressed together, edge-to-edge, it is possible to transmit rotation from the first disk to the second disk. Friction between the two contact areas acts to transmit torque from one disk to the other; such disks are called *friction wheels.* As long as the disks roll without slipping, there is a definite relation between the angular positions of the two disks. If the torque required exceeds the friction capacity of the contact, the disks will slip. The positional relation of the two disks is then changed and the required torque is not met. To prevent slipping, the contact force can be increased, resulting in an increase in the friction capacity of the contact. This occurs at the expense of increased bearing loads, bearing friction, stress and deflection in the shafting, and stresses within the disks. A better way to overcome slipping is to increase the coefficient of friction through a more rough surface. The rougher surface involves larger surface irregularities, which are more difficult to shear away and, thus, more effectively resist slippage. The extreme extension of this idea occurs when teeth are cut on the two wheels, in which case they are called *gears.*

If the teeth are cut straight across the edge of a disk, in a direction parallel to the axis of rotation, the resulting gear is called a *spur gear.*

Alternatively, teeth may be cut along a helix generated around the axis of the gear. In this latter case, the teeth will cross the edge of the disk at an angle, and these gears are called *helical gears.* Spur gears or helical gears are used where power is to be transmitted between parallel shafts. If the two shafts intersect, power may be transmitted from one to the other using *bevel gears,* for which each gear is cut from a cone. There are also situations where the two shaft axes are non-intersecting, non-parallel lines; these situations are served with *skew gears, hypoid gears,* or *worm gears.* However, in this chapter only spur gears will be discussed; for more information on other types, see the references listed at the end of the chapter.

5.2 VELOCITY RATIO

The *velocity ratio* is one of the most important aspects of a pair of gears. It is the ratio of the angular velocity of one gear to that of a second gear that is in mesh with the first. It is important that this ratio be constant, not only on the average, but instantaneously. Consider first a situation in which the average velocity ratio is constant but the instantaneous value is variable.

Nonconstant Velocity Ratio. Figure 5.1 shows a gear pair such that the teeth of one gear consist of M_1 radial pins of uniform width that are equally spaced around the circumference of the disk; this is called a *spoked gear.* The teeth of the second gear consist of M_2 circular pins evenly spaced between two disks, each pin with its axis perpendicular to the gear disks and parallel to their axis of rotation. This gear type is called a *lantern.* Gearing of this sort was used in medieval Europe in the construction of grain mills and other machines.

Referring to Figure 5.2, the loop equations describing the position relation are:

$$R_1 \cos q + (R_2 + W) \sin A + D \cos A - C = 0$$

$$R_1 \sin q + (R_2 + W) \cos A - D \sin A = 0$$

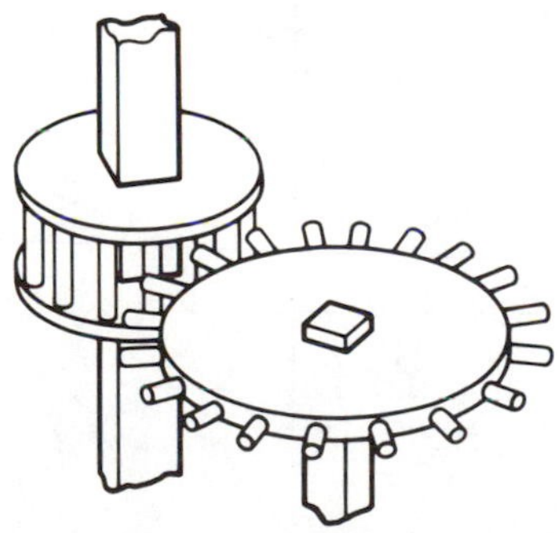

FIGURE 5.1 Medieval Lantern and Spoked Gears in Mesh

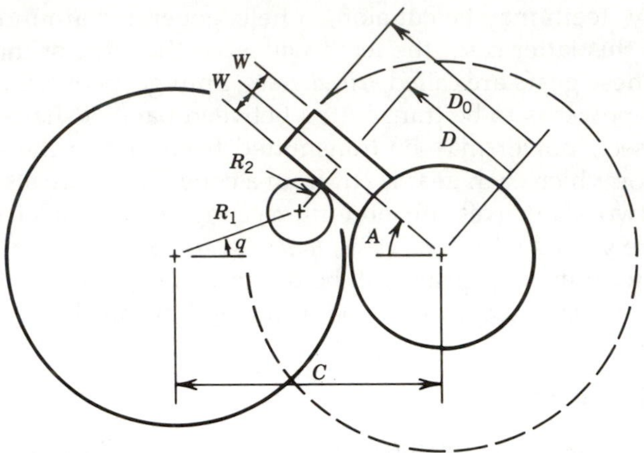

FIGURE 5.2 Schematic Diagram for Lantern and Spoked Gears

These relations apply over an interval $q_1 \leq q \leq q_2$ during the engagement of a single pair of teeth, with the corresponding range $A_1 \leq A \leq A_2$. The dimensions R_1, R_2, W, D_o, and C are presumed known, and the loop equations are solvable for A and D as functions of the lantern rotation q. The velocity loop equations are the result of differentiating the position equations, and can be used to determine velocity coefficients for the system. The solutions for A and K_a as functions of q for typical proportions are shown in Figure 5.3. The range of values for q shown there spans the duration of one tooth-pair engagement, and thus presents one cycle of the periodic function K_a. However, the angle A is not periodic because it is increasing continuously. It is evident that K_a is not a constant but, instead, varies with position. This indicates that the velocity ratio, $\dot{A}/\dot{q}$, is not con-

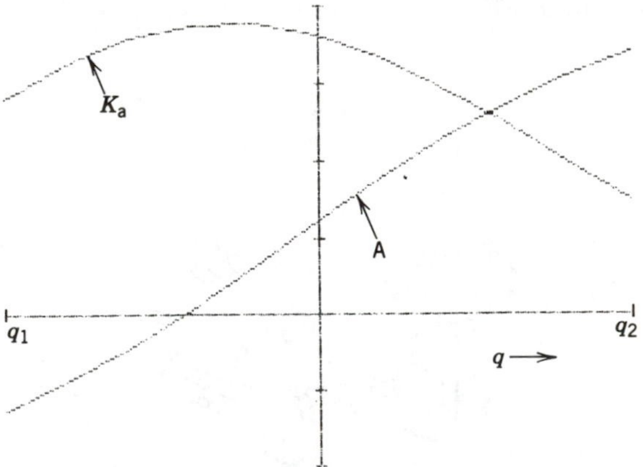

FIGURE 5.3 Position and Velocity Coefficient Curves For Lantern and Spoked Gear Pair

stant, but varies with position through the mesh. If the input angular velocity, $\dot{q}$, is constant, the output angular velocity, $\dot{A}$, will vary periodically during the engagement of each tooth pair. It should be noted that the average output shaft angular velocity, taken over a full revolution of the output gear, will be constant for constant input velocity even though the instantaneous value varies.

What difference does it make if the angular velocity ratio varies? Consider again the case where the input speed is constant, perhaps because of the use of a very large flywheel. The output gear is connected through a shaft of finite stiffness to a rotating machine component with finite inertia. As the angular velocity of the output gear varies, the shaft will be alternately twisted and relaxed in an effort to cause the driven component to follow the speed variations of the output gear. This alternating torque in the shaft will excite torsional vibrations in the driven machine, with resulting noise, vibration, and fatigue damage. For crude, low-speed, low-torque machinery this situation may be tolerable, but for modern high-speed, high-power, machinery it can not be allowed.

Condition for Constant Velocity Ratio. Because it is important to have a constant velocity ratio, under what conditions does it occur? In Figure 5.4, two cams are shown in contact, representing teeth on two gears. The two cams rotate about fixed pivots at C_1 and C_2, separated by the center distance C. The cams have instantaneous angular velocities $\dot{A}_1$ and $\dot{A}_2$, respectively. At the point of contact, point H, the tangent line is B_1–B_2, and N_1–N_2 is the normal line. The normal is also called the "line of action" or the "line of contact." The cam on the left is the driver, and the cam on the right is driven. As discussed previously, the cam contact surfaces should never

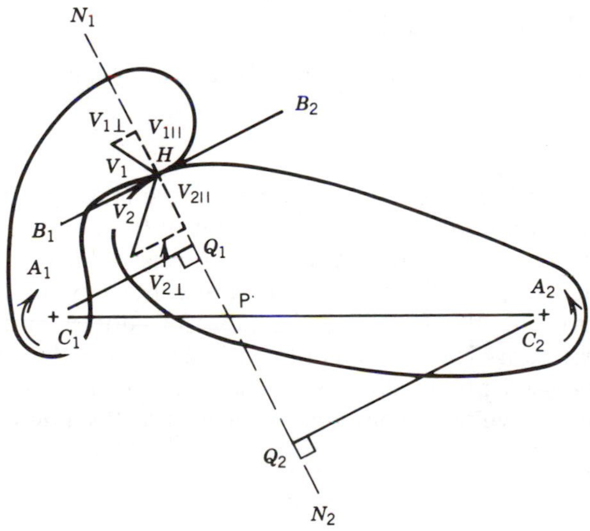

FIGURE 5.4 Two Cams Representing Teeth on Two Gears

separate or overlap. This is expressed by requiring that the velocity components *along the line of contact* be the same for the two points in contact. Each cam rotates about a fixed pivot, so that every point in the cam moves on a circle about the fixed pivot for that cam. The velocity vectors, $\mathbf{V}_1$ for the contact point on body 1 and $\mathbf{V}_2$ on body 2, are shown in Figure 5.4. Let Q_1 denote the foot of a normal line from C_1 to the line of contact, and let Q_2 be the foot of a normal line from C_2 to the line of contact. According to the theorem developed in Section 4.3, the components of $\mathbf{V}_1$ and $\mathbf{V}_2$ along the line of contact are

$$V_{1_{\|}} = \dot{A}_1 \, \overset{\frown}{C_1Q_1}$$

$$V_{2_{\|}} = \dot{A}_2 \, \overset{\frown}{C_2Q_2}$$

Note that these two must be equal to each other for the cams to remain in continued contact. When these two expressions are equated and solved for the velocity ratio $\dot{A}_2/\dot{A}_1$, the result is

$$\frac{\dot{A}_2}{\dot{A}_1} = \frac{\overset{\frown}{C_1Q_1}}{\overset{\frown}{C_2Q_2}}$$

The intersection of the line of contact with the line of centers defines the point P, called the *pitch point*. The two triangles C_1Q_1P and C_2Q_2P are geometrically similar. For similar triangles, corresponding parts are proportional to each other. Consequently, the velocity ratio is also

$$\frac{\dot{A}_2}{\dot{A}_1} = \frac{\overset{\frown}{C_1P}}{\overset{\frown}{C_2P}}$$

If the velocity ratio is to remain constant, the ratio $\overset{\frown}{C_1P}/\overset{\frown}{C_2P}$ must also be constant, which says that P must always divide the center distance in the same ratio. The condition for a constant velocity ratio, sometimes called the "Fundamental Law of Gearing," is:

> *For a constant angular velocity ratio, the location of the pitch point must be constant.*

Sliding Velocity of the Point of Contact. Before leaving Figure 5.4, it is useful to also determine the velocity of sliding at the point of contact. This velocity is significant in terms of friction and wear on gear teeth. For each cam, the magnitude of the velocity for the contact point is readily expressed in terms of the radius to the point of contact and the angular velocity for the cam. The component of velocity along the line of contact has been expressed previously. Using that component and the Pythagorean theorem, the velocity components normal to the line of contact can also be expressed

$$\begin{aligned} V_{1_{\perp}}{}^2 &= |\mathbf{V}_1|^2 - V_{1_{\|}}{}^2 \\ &= (\overset{\frown}{C_1H_1}{}^2 - \overset{\frown}{C_1Q_1}{}^2)\, \dot{A}_1{}^2 \end{aligned}$$

$$= \widehat{Q_1H_1}^2 \dot{A}_1^2$$

$$V_{1\perp} = \widehat{Q_1H_1}\, \dot{A}_1$$

and similarly,

$$V_{2\perp} = \widehat{Q_2H_2}\, \dot{A}_2$$

In terms of $\widehat{PH}$, $\widehat{Q_1H_1}$, $\widehat{Q_2H_2}$, $\widehat{Q_1P}$, and $\widehat{Q_2P}$, all considered as positive lengths, the total sliding velocity is the sum of these two magnitudes,

$$\begin{aligned} V_s &= V_{1\perp} + V_{2\perp} \\ &= \widehat{Q_1H_1}\, \dot{A}_1 + \widehat{Q_2H_2}\, \dot{A}_2 \\ &= (\widehat{PH} - \widehat{Q_1P})\, \dot{A}_1 + (\widehat{PH} + \widehat{Q_2P})\, \dot{A}_2 \\ &= \widehat{PH}\, (\dot{A}_1 + \dot{A}_2) - \widehat{Q_1P}\, \dot{A}_1 \left(1 - \frac{\widehat{Q_2P}\, \dot{A}_2}{\widehat{Q_1P}\, \dot{A}_1}\right) \\ &= \widehat{PH}\, (\dot{A}_1 + \dot{A}_2) \end{aligned}$$

The second term on the right is zero because $\widehat{Q_2P}/\widehat{Q_1P}$ is yet another way to express the velocity ratio—again, based on similar triangles. The result is that the velocity of sliding is proportional to the distance from the point of contact to the pitch point, and also to the sum of the gear speeds.

5.3 CONJUGATE PROFILES

In the previous section, the Fundamental Law of Gearing was established which states that for a constant angular velocity ratio, the pitch point must be stationary. The position of the pitch point is determined by the gear profiles at the point of contact. For the pitch point to remain stationary, the normal at the contact point (the line of contact) must continue to intersect the line of centers at the same location, even though the point of contact varies as the two teeth pass through engagement. If the tooth profiles of the two gears are such that the pitch point remains stationary, the two gear tooth profiles are said to be *conjugate* to each other. Thus, conjugate tooth profiles satisfy the law of gearing, and gears that have conjugate gear tooth profiles will have a constant angular velocity ratio. In the example given in Section 5.2, in which the tooth on one gear was simply a circular pin and the tooth on another gear was a spoke of uniform width, the pitch point was nonstationary, and the gear pair did not have a constant angular velocity ratio.

Notice that the property of being conjugate (or nonconjugate) is a property of the *pair* of gear tooth forms, and not a property of either tooth form individually. Given any tooth form for the first gear it is, in principle, possible to determine a tooth form for the second gear such that the pair will be conjugate; a procedure for this is given by Beggs.[1] Thus, the circular

[1] J. S. Beggs, *Mechanism*, pp. 68–72, New York: McGraw–Hill, 1955.

pin and the uniform spoke are a nonconjugate pair. There is, however, a tooth form conjugate to the circular pin (or a different form that can be used with the uniform spoke) such that the pair will satisfy the Fundamental Law of Gearing. In the case of the circular pin tooth, the conjugate tooth profile can not be physically realized because it requires material bodies to pass through one another during a portion of the motion.

The desirability of conjugate tooth forms has been established, and the ability (in principle) to form a conjugate pair given any tooth form for the first gear has been stated. These facts would seem to suggest that a great variety of gear tooth forms would be in use, but this is not the case. In fact, the vast majority of gears use what is called an *involute* tooth profile. The one significant exception is the *cyclodial* tooth profile used in the mechanical watch and clock industry. A few other tooth forms may be used, such as the Novikov tooth forms (which are a nonconjugate pair), but these are quite uncommon. For this reason, the present discussion will be limited to the kinematics of involute gear teeth. Before discussing the involute gear tooth itself, some properties of the involute must be developed.

5.4 PROPERTIES OF THE INVOLUTE

The involute of a circle is the path traced by the tip of a taut string unwound from a circular body. In Figure 5.5, consider the string to be wound initially around a circle, called the *base circle,* with the free end at the intersection of the base circle with the vertical axis. The string is inextensi-

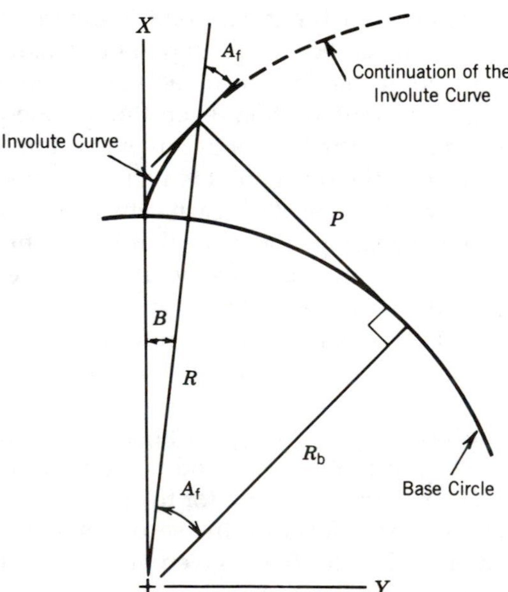

FIGURE 5.5 Geometry of the Involute Curve

ble and must remain taut, so the initial motion must be perpendicular to the circle in the upward direction. As more string is unwound, the path traced by the tip bends to the right as shown. Figure 5.5 displays the situation when the string has been unwound from an arc of $\pi/4$ radians. The broken line at the top of the figure indicates the further extention of the involute curve. The length of the straight part of the string is P, which is the instantaneous radius of curvature of the involute. The radius of the circular form (the base circle) on which the string is originally wound is R_b, called the *base radius*. The polar coordinates for the top of the string are (R, B), the radius and the polar angle for the point on the involute curve.

The polar angle to the point where the string is tangent to the base circle is the sum of two angles, $B + A_f$, where the second angle, A_f, is called the *flank angle*. Notice that the flank angle is identified a second time in Figure 5.5 as the angle between the tangent to the involute and the radial line to the point on the involute curve; these angles are identical.

The radius of curvature, P, is expressed in two ways, once in terms of the arc length and again using the trigonometry of a right triangle:

$$\begin{aligned} P &= R_b\,(B + A_f) \\ &= R_b \tan A_f \end{aligned}$$

If P is eliminated between these two expressions, the result can be solved for the angle B:

$$\begin{aligned} B &= \tan A_f - A_f \\ &= \mathrm{inv}(A_f) \end{aligned}$$

where the *involute function*, $\mathrm{inv}(X)$, is defined as $\tan(X) - X$. Another useful relation from the right triangle expresses the radius to the involute in terms of the flank angle; that is,

$$R = R_b / \cos A_f$$

The preceding two expressions give the polar coordinates of any point on the involute curve, (R, B) in terms of the parameter, A_f, the flank angle.

If a point on the involute is specified, either by giving the radius R or the polar angle B, one of these relations must be solved for A_f to determine the associated value of the flank angle. For a given value of R, the determination of A_f is straightforward; if the value of B is given, the determination of A_f is more involved. In this case, the relation defining the involute function is written as an equation to be solved for A_f:

$$\tan A_f - A_f - B = 0$$

A numerical solution is required, and Newton's method (the Newton–Raphson method for the case of one unknown) is quite workable. Using Newton's method with a starting estimate of $A_f = 0.4$, solution estimates for which the absolute value of the residuals are less than 10^{-6} are generally obtained in no more than six iterations. This process is readily programmed for a computer, but it is also suited for use in a programmable calculator or for manual iteration.

When the involute is used as a gear tooth form, two involute curves are employed as shown in Figure 5.6. The length along the circular arc from one involute to the other is called the circular thickness of the tooth. At times it is of interest to determine the circular thickness at some radius, given the circular thickness at another radius. Consider R_b, R_p, T_p, and R as known, with the circular thickness at R, denoted $T(R)$, to be determined.

The flank angle at R is $A_f(R)$, and this may be evaluated from the right triangle relation between R and R_b:

$$A_f(R) = \mathrm{Arccos}(R_b/R)$$

With $A_f(R)$ known, the involute function can be evaluated as follows:

$$\mathrm{inv}[A_f(R)] = \tan[A_f(R)] - A_f(R)$$

Similarly, the flank angle at R_p is determined, and the involute function evaluated for that argument.

Looking at the difference between the angles subtended by $T(R)/2$ and $T_p/2$, the following relation is evident:

$$\frac{T_p}{2R_p} - \frac{T(R)}{2R} = B(R) - B_p$$

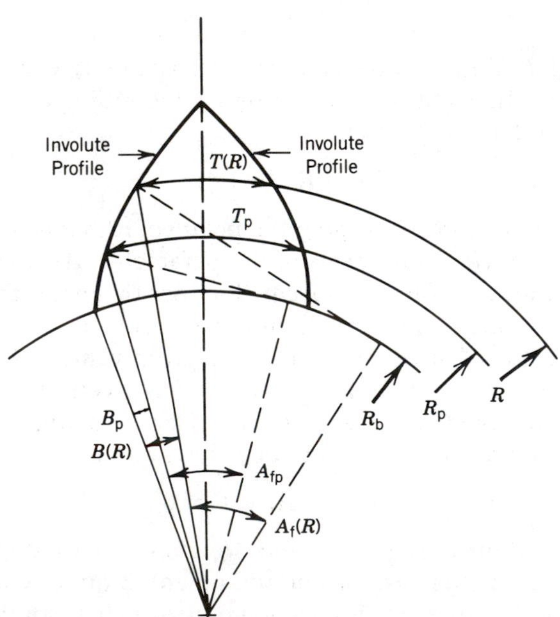

FIGURE 5.6 Determination of Gear Tooth Circular Thickness

This can be solved for the required circular thickness, $T(R)$:

$$T(R) = 2R\left\{\frac{T_p}{2R_p} + \text{inv}(A_{fp}) - \text{inv}[A_f(R)]\right\}$$

Furthermore, if it is necessary to find the radius at which the circular thickness has a specified value, $T(R)$ is then known, and this equation can be solved numerically for R. Again, Newton's method of solution is recommended.

5.5 INVOLUTE AS A GEAR TOOTH

To understand the action of involute gear teeth, it is useful to consider first the process of winding a thin, inextensible tape from one disk to another, keeping the tape taut at all times. The two disks concerned are mounted on fixed pivots, with the constant center distance C as shown in Figure 5.7(a). The radii of the disks are R_{b1} and R_{b2}, respectively. The unsupported length of tape is straight and tangent to both disks. The points of tangency are T_1 and T_2, which are fixed in space.

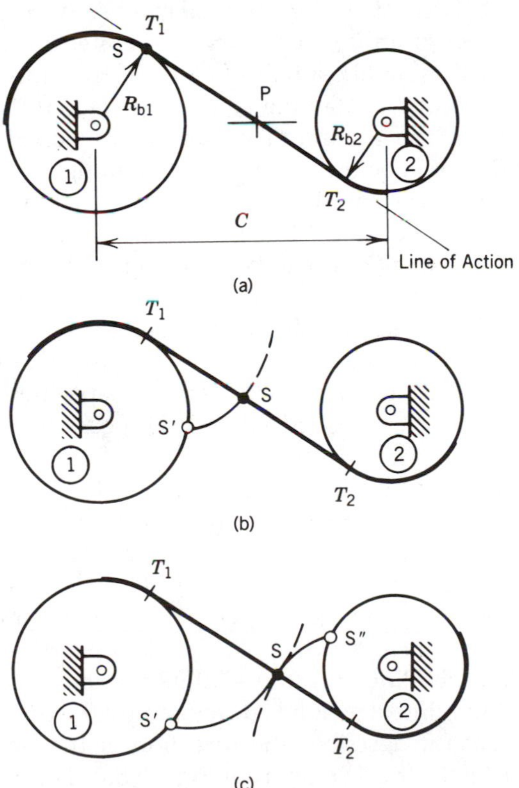

FIGURE 5.7 Involute Tooth Action: Winding a Tape from One Disk to Another

As shown in Figure 5.7, the process begins with the tape wrapped over the top of disk 1 and around the bottom of disk 2. The point of particular interest on the tape is labeled S, and S coincides with T_1 as the process begins. Counterclockwise rotation of disk 2 pulls the tape, causing a clockwise rotation of disk 1. After some rotation, the system appears as shown in Figure 5.7(b), where point S is on the straight part of the tape and is not in contact with either disk. The point on disk 1 previously supporting point S on the tape is labeled S′, and the path of point S *relative to disk 1* is the involute from S′ to the present position; the short broken line indicates the continuation of that involute curve. Disk 1 is the base circle from which this involute curve is generated.

The process of winding on a tape is simply the reverse of winding it off, and the tape considered here is in the process of being wound onto the second disk. The point S must then be moving along an involute curve *relative to the second disk,* as shown in Figure 5.7(c). Thus, at any point along the transition from disk 1 to disk 2, the point S moves simultaneously along two involute curves, the first generated as it unwinds from disk 1 (base circle radius R_{b1}) and the second generated as it winds onto disk 2 (base circle radius R_{b2}). When the point S reaches disk 2, it will coincide with the position marked S″. Next, consider physical extensions to both disks, each extension in the form of an involute curve generated from the associated base circle and functioning like a pair of cams. If the tape is removed and the motion controlled by maintaining contact between the two cams, there will be no change in the motion. Two important observations must be noted here, and they will be true at every point along the transition. First, for S to be on both curves simultaneously, the curves must be tangent to each other at point S. Secondly, the line tangent at S to the two involute curves is perpendicular to the unsupported length of tape. Therefore, the unsupported length of tape defines the line of action. The unsupported length of tape (the line of action) always intersects the line of centers at the same location, P, and thus the pitch point is stationary. Consequently, the condition for constant angular velocity ratio, as specified by the Fundamental Law of Gearing, is satisfied when both gears have involute tooth forms.

All contact between gear teeth occurs along the line of action. Recall that the line of action is the path of S during the transition from the first base circle to the second, and that the two gear tooth forms are always tangent to each other at S. In the absence of friction, the contact force between two bodies pressed together will be along the normal to the surface at the point of contact. Thus, for frictionless gears, the contact force will always be along the line of action—a line that has a fixed orientation. The orientation of the line of action is described by the angle A_p, called the *pressure angle* because it describes the direction of the "pressure" (force) between the gear teeth; this is shown in Figure 5.8. For a pair of involute teeth that are *in mesh at the pitch point,* the flank angle at the point of contact is equal to the pressure angle.

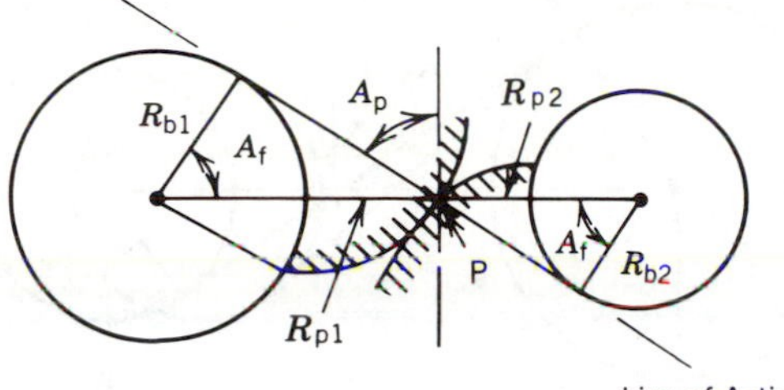

FIGURE 5.8 Definition of the Pressure Angle

If disk 1 is rotated through an angle A_1, the tape length $R_{b1}A_1$ is transferred to the second disk, requiring a rotation A_2 for the second disk. The two rotations are related such that

$$R_{b1}\,A_1 = R_{b2}\,A_2$$

The ratio of the rotations, which is also the ratio of angular velocities, is

$$\frac{A_2}{A_1} = \frac{R_{b1}}{R_{b2}}$$

The *pitch radius* is defined to be the distance from the gear center to the pitch point; the pitch radii are denoted R_{p1} and R_{p2} in Figure 5.8. The pitch radii are related to the base radii by

$$R_b = R_p \cos A_p$$

Therefore, the ratio of rotations can also be written as the inverse ratio of the pitch radii:

$$\frac{A_2}{A_1} = \frac{R_{p1}}{R_{p2}}$$

Because the sum of the pitch radii is equal to the (constant) center distance, the pitch radii can be interpreted as the radii of a pair of friction wheels kinematically equivalent to the two gears. For gear-train calculations, this is often a convenient way to think about the gears.

Some properties are intrinsic to an individual gear, such as the base circle, whereas other properties belong to a gear pair, such as the pitch point and the pressure angle. The center distance and the line of action play critical roles in defining the actual pitch point and the actual pressure angle. Neither the pitch point nor the pressure angle exists until both gears are mounted; only then are the actual centers established, and the center distance and the line of action defined. The line of action is defined by the base circles of both gears, at the actual center distance because the line of action is tangent to both base circles.

The result of changing the center distance, as it affects both the pitch point location and the pressure angle, is shown in Figure 5.9. With the nominal center distance, C, the pitch point is P and the pressure angle is

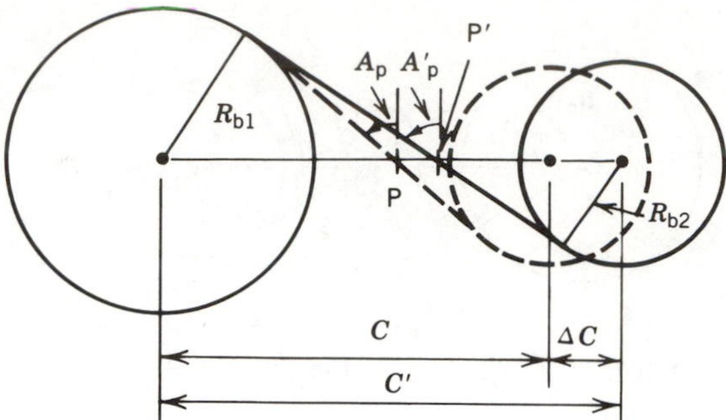

FIGURE 5.9 Effect of Changing Center Distance

A_p. When the center distance is increased to $C' = C + \Delta C$, the pitch point moves to the right to P′, and the pressure angle increases to $A_p{}'$. The most important observation is that conjugate action is preserved, and that the velocity ratio remains unchanged. This means that extremely accurate center distances are not required for conjugate action with involute teeth. In contrast, with cyclodial teeth conjugate action is achieved only at the design center distance. This result, and ease of manufacture, are major factors that make the involute tooth form used almost universally. For conjugate action, the contact surfaces of gear teeth are of involute form, but there are many other parts of the gear and the gear teeth that also must be described. A list of common gear terms follows, some of which are identified in Figure 5.10.

Pitch Circle: A circle with pitch radius; this circle rolls without slip on the pitch circle of the mating gear when the two are in mesh.

Addendum: The radial distance from the pitch circle to the top land.

Dedendum: The radial distance from the pitch circle to the bottom land.

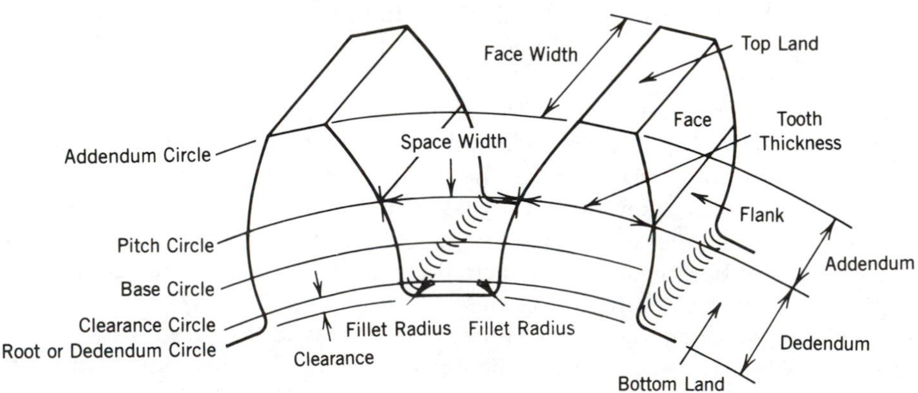

FIGURE 5.10 Gear Terminology

Addendum Circle: The circle tangent to the top land of the teeth.

Root or Dedendum Circle: The circle tangent to the bottom land of the teeth.

Whole Depth: The sum of the addendum and the dedendum dimensions.

Working Depth: The sum of addendum dimensions for two gears in mesh.

Clearance: The amount by which the dedendum of the gear in question exceeds the addendum of the gear with which it will mesh.

Circular Thickness: The arc length between the two sides of a gear tooth, measured along the pitch circle, unless otherwise specified.

Backlash: The amount by which the space width exceeds the thickness of the engaging tooth measured along the pitch circles.

Circular Pitch: The arc length along the pitch circle between corresponding profiles on adjacent teeth; the spatial period of the gear teeth as measured along the pitch circle.

Base Pitch: The arc length along the base circle between corresponding points on adjacent teeth; the spatial period as measured on the base circle. This is also the distance along the line of action from one tooth to the next on the same gear.

Diametral Pitch: The number of teeth per inch of pitch circle diameter; used with US customary units only, where the units for diametral pitch are *teeth per inch.*

Module: The pitch circle diameter per tooth, usually used with SI units where the unit for the module is *millimeters* (per tooth, the latter not expressed).

Pinion: The smaller of two gears in mesh.

5.6 AGMA STANDARDS AND TOOTH PROPORTIONS

To assure interchangability and to minimize the variety of cutting tools required to make gears, many gears are made according to pre-set standards. A gear tooth standard specifies the tooth proportions—such as addendum and dedendum dimensions and tooth thickness—in terms of a parameter such as the diametral pitch (P) for a standard pressure angle. The United States' standard is supported jointly by the American Gear Manufacturers Association (AGMA) and the American National Standards Institute (ANSI). For equal addendum gearing, as described in Table 5.1, the gear and pinion both have the same addendum. Nonstandard gear sets are also built, for which the gear and pinion have different addendum dimensions (long and short addendum gears), usually to eliminate interference (see Section 5.7). Important features from the AGMA Standard are summarized in Table 5.1, "AGMA Standard Tooth System for Spur Gears."

TABLE 5.1 AGMA Standard Tooth System for Spur Gears[a]

(Letter P Denotes the Diametral Pitch)

Item	*Coarse Pitch* $P < 20$		*Fine Pitch* $P \geq 20$
Pressure Angle	20 deg	25 deg	20 deg
Addendum	1/P	1/P	1/P
Dedendum	1.25/P	1.25/P	1.2/P + .002 in. (min)
Clearance	0.25/P	0.25/P	0.2/P + .002 in. (min)
Working Depth	2.0/P	2.0/P	2.0/P
Whole Depth	2.25/P	2.25/P	2.2/P + .002 in. (min)
Circular Thickness	$\pi/2P$	$\pi/2P$	1.5708/P
Minimum Number of Teeth			
Pinion	18	12	10
per Pair	36	24	
Source:			
AGMA Standard	201.02	201.02	207.05

[a] This data is extracted from USA Standard Tooth Proportions for Coarse-Pitch Involute Spur Gears (USAS B6.1–1968) and USA Standard System—Tooth Proportions for Fine-Pitch Involute Spur and Helical Gears (USAS B6.7–1967) with the permission of the publisher, the American Gear Manufacturers Association, 1330 Massachusetts Avenue, N.W., Washington, D.C. 20005.

For some tooth forms, the dedendum circle is outside the base circle. In this case, all of the tooth profile is an involute curve, except for the fillet at the bottom land. There are other cases where the dedendum circle is inside the base circle, such as is shown in Figure 5.10. Because the involute is undefined inside the base circle, that part of the tooth flank inside the base circle must be something other than an involute curve. For this case, in drawings it is customary to show the tooth flank below the base circle as a radial line. However, the tooth flank is usually determined by the manufacturing method used. Various methods of cutting gear teeth leave different forms for the flank below the base circle, but this is of more concern with regard to strength than kinematics. This is true because there should be no contact on the non-involute surface.

5.7 CONTACT RATIO, INTERFERENCE, AND UNDERCUTTING

As gear tooth pairs engage, the point of contact moves along the line of action from the point of initial engagement to the point where the teeth separate. This distance is called the *length of action*. The *arc of approach* is the gear rotation angle from the point of initial contact to the pitch point. The *arc of recess* is the rotation angle from tooth contact at the pitch point to the point where the teeth separate. During the arc of approach, tooth engagement becomes deeper; the teeth move toward disengagement during the arc of recess.

As a pair of gear teeth pass through engagement, for proper operation a second pair of teeth must begin to mesh before contact is lost between the first pair. This means that at some positions two pairs of teeth are in mesh, while at other positions only one pair is engaged. The ratio of the total length of action to the distance between adjacent tooth contacts is the average number of tooth pairs engaged; this ratio is called the *contact ratio.* The contact ratio is one measure of the quality of gear design: a higher contact ratio indicates a better design. In all cases, the contact ratio must be greater than 1.0, with minimum values in the range of 1.2 to 1.4.

As shown in Figure 5.11, the line of action is defined by the line tangent to the two base circles at points T_1 and T_2. For the indicated rotations, the first contact occurs when the pinion tooth first crosses the line of action at A_1. The tooth pair remain in contact as the point of contact moves from A_1 to A_2. At A_2, contact ends because the gear tooth no longer intersects the line of action. The distance $\widehat{A_1A_2}$ is the length of action, and is evaluated in the following manner:

$$\begin{aligned} L &= \widehat{A_1A_2} \\ &= \widehat{A_1T_2} + \widehat{T_1A_2} - \widehat{T_1T_2} \\ &= \sqrt{R_{a2}^2 - R_{b2}^2} + \sqrt{R_{a1}^2 - R_{b1}^2} - C \sin A_p \end{aligned}$$

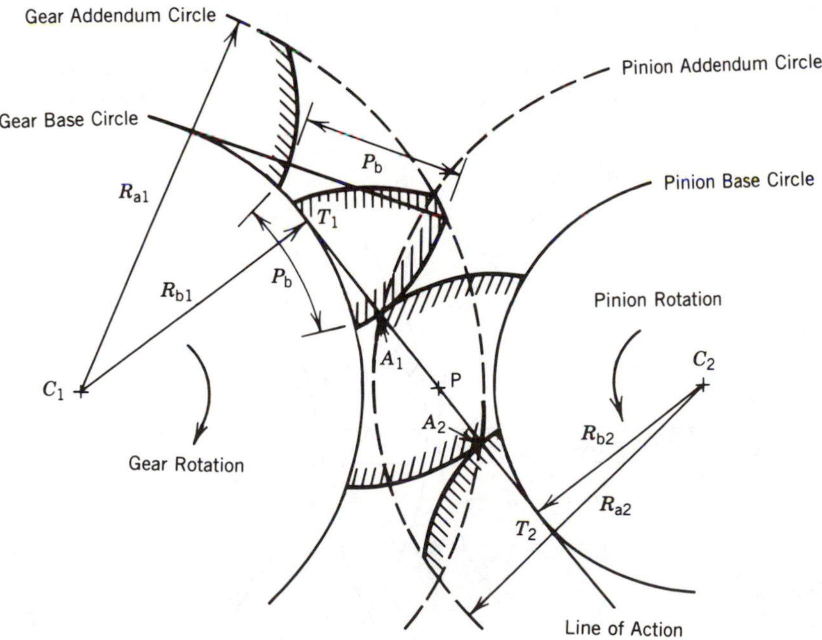

FIGURE 5.11 Contact Ratio and Interference Considerations

Here, R_{a1}, R_{a2} are addendum radii, R_{b1}, R_{b2} are base radii, and C is the center distance.

Consider two gear teeth simultaneously in mesh with two pinion teeth, as shown in Figure 5.11. The distance along the line of action from one gear tooth to the next is the base pitch, P_b. In Figure 5.11, this distance is dimensioned between adjacent teeth, and it is also identified as a length measured along the base circle. If there are N_1 teeth on the gear and N_2 teeth on the pinion, the base pitch is

$$P_b = 2\pi R_{b1}/N_1 = 2\pi R_{b2}/N_2$$

The contact ratio, M_c, is the ratio of the length of action to the base pitch:

$$M_c = L/P_b$$

For the gear and pinion shown in Figure 5.11, consider increasing the pinion addendum radius while holding all other dimensions constant. As R_{a2} is increased, the point of initial contact moves toward T_1. If A_1 reaches T_1, contact occurs at the very beginning of the involute; further increase of R_{a2} will cause contact to occur inside the base circle of the gear. Because the involute curve is not defined inside the base circle, such contact occurs necessarily on a non-involute surface and results in non-conjugate action. Contact on a non-involute surface is referred to as *interference*. To avoid interference, the gear and pinion addendum radii must be less than their respective critical values:

$$R_{a1\,crit} = \sqrt{R_{b1}{}^2 + C^2 \sin^2 A_p}$$

$$R_{a2\,crit} = \sqrt{R_{b2}{}^2 + C^2 \sin^2 A_p}$$

These critical values are identified in Figure 5.12. In making gears, one of the cutting methods involves using a cutting tool shaped like a pinion and

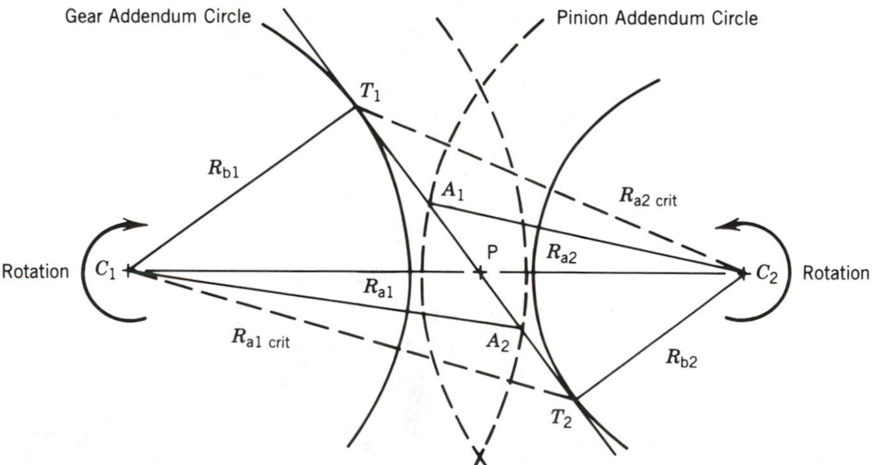

FIGURE 5.12 Critical Radii for Interference

moved parallel to the gear axis of revolution. If interference develops between the cutting tool and the gear being cut, the cutter simply removes the interfering material from the flank of the tooth inside the base circle, often taking away some material outside the base circle as well; this is called *undercutting*. Other methods of gear cutting can also result in undercutting. Undercutting seriously weakens the tooth and introduces the possibility of interference and resulting non-conjugate action when the gear is mounted with its working pinion. Obviously, undercutting is to be avoided.

5.8 SIMPLE AND COMPOUND GEAR TRAINS

The remainder of this chapter will address combinations of gears running together to accomplish useful motions. Such combinations are called *gear trains*. The essential requirement for two gears to run in mesh is that the circular pitch be the same for both gears. Recall that the circular pitch is the arc length, measured along the pitch circle, from one tooth profile to the corresponding point on the next tooth profile. If two gears have the same circular pitch, then the teeth on one will arrive in mesh at the same rate as those on the second gear. This requirement is expressed analytically in terms of the circular pitch, P_c:

$$P_{c1} = P_{c2}$$

$$\frac{2\pi R_{p1}}{N_1} = \frac{2\pi R_{p2}}{N_2}$$

where R_{p1} and R_{p2} are the pitch radii and N_1 and N_2 are the number of teeth on the two gears. If the π factors are eliminated and the fractions are inverted, the result shows that the dimetral pitch values must be equal for the two gears to run together.

Consider several gears running together in mesh, each gear mounted on a shaft supported in bearings fixed with respect to an external frame. This arrangement is called a *simple gear train*, an example of which is shown in Figure 5.13. Simple gear trains are found in applications where several shafts must rotate with a constant speed and phase relation between them, such as in a printing press or the hour-, minute-, and second-hand shafts for a mechanical clock. When constructed with spur gears, a simple gear train is necessarily coplanar, and the diametral pitch of all gears must be the same.

As shown in Figure 5.13, the gears are represented by their pitch circles, drawn tangent to each other at the pitch point on each centerline. Conjugate action assures that the pitch circles roll on each other without slipping; this allows the actual gear to be replaced with a friction wheel of diameter equal to twice the pitch radius of the gear. Let A_i represent the rotation of the ith gear, where all gear rotations are counted positive in the same sense, either clockwise (CW) or counterclockwise (CCW). The arc

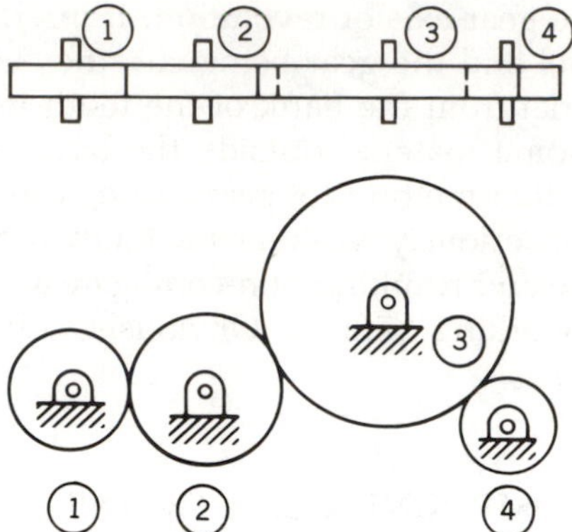

FIGURE 5.13 Simple Gear Train

length, counted positive when passing upward through mesh, must be the same for each pair of adjacent gears:

$$A_1 R_{p1} = -A_2 R_{p2}$$
$$A_2 R_{p2} = -A_3 R_{p3}$$
$$A_3 R_{p3} = -A_4 R_{p4}$$

where counterclockwise rotations have been taken as positive. If gear 1 and gear 4 are considered as "input" and "output," respectively, the intervening gears 2 and 3 are called *idler gears,* or simply *idlers.* The ratio of output rotation to input rotation is called the *train ratio;* this is also the velocity coefficient for the gear train, expressing the linear relation between input and output rotation rates. For the train shown, the train ratio is determined by successive substitutions:

$$A_4 = \left(-\frac{R_{p3}}{R_{p4}}\right)\left(-\frac{R_{p2}}{R_{p3}}\right)\left(-\frac{R_{p1}}{R_{p2}}\right) A_1$$

$$\frac{A_4}{A_1} = -\frac{R_{p1}}{R_{p4}} = -\frac{N_1}{N_4}$$

where N_1 and N_4 are the numbers of teeth on gears 1 and 4, respectively. The tooth numbers are in the same proportion as the pitch radii because both gears have the same diametral pitch. Notice that the idler radii simply disappear, and only the input and output gear radii are important to the train ratio. Idler gears simply fill the gap between the input and output gears and control the direction of rotation. This second function is particularly significant because each idler reverses the sense of the output rotation, as compared to the sense of the output if that idler were omitted. If driven machinery is attached to each of the gears 2, 3, and 4, then each becomes an "output," with a corresponding train ratio relating its relation to that of the input gear 1:

$$\frac{A_2}{A_1} = -\frac{R_{p1}}{R_{p2}} \qquad \frac{A_3}{A_1} = \frac{R_{p1}}{R_{p3}}$$

For large train ratios, simple gear trains are often not practical because of the large size required for one gear; a compound gear train is usually much more compact. Consider a gear train built with several shafts mounted in bearings that are fixed with respect to an external frame. However, some of the shafts carry two gears that are joined so that they must rotate together; this system is a *compound gear train.* Figure 5.14 shows a typical compound gear train where two of the shafts carry two gears each. The only gear engaged by gear 1 is gear 2; therefore, those two gears must have the same diametral pitch. Gear 3 engages only gear 4, so this pair must have a common diametral pitch. There is no requirement, however, for the diametral pitch of gears 1 and 2 to be the same as the diametral pitch of gears 3 and 4; they are often different because of their different strength requirements. If a compound train is constructed using spur gears, all gears cannot be coplanar, but instead must occupy two or more parallel planes.

Referring to Figure 5.14, consider gears 1 and 6 as the input and output gears, respectively. As for the simple train, consider all gear rotations positive in the same sense (CCW) and all arc lengths positive when passing upward through mesh. The constraints for rolling without slipping are

$$A_1 R_{p1} = -A_2 R_{p2}$$

$$A_3 R_{p3} = -A_4 R_{p4}$$

$$A_5 R_{p5} = -A_6 R_{p6}$$

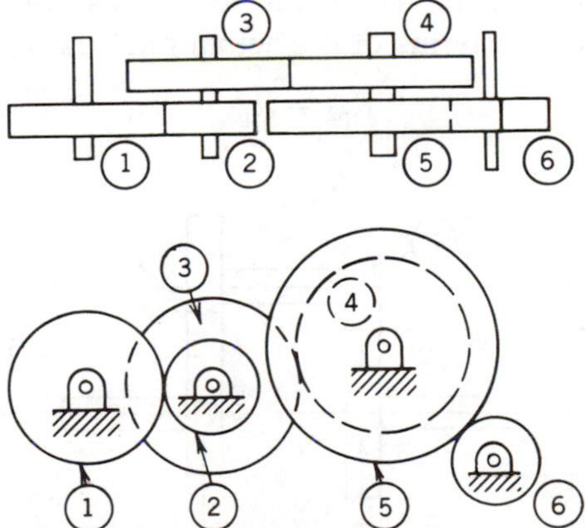

FIGURE 5.14 Compound Gear Train

In addition, there are constraints expressing the fact that gears 2 and 3 and gears 4 and 5 must rotate together:

$$A_2 = A_3$$

$$A_4 = A_5$$

Rotation A_6 is expressed in terms of rotation A_1 by multiple substitutions:

$$A_6 = \left(-\frac{R_{p5}}{R_{p6}}\right)\left(-\frac{R_{p3}}{R_{p4}}\right)\left(-\frac{R_{p1}}{R_{p2}}\right) A_1$$

and the train ratio is

$$\frac{A_6}{A_1} = -\frac{R_{p5}}{R_{p6}} \times \frac{R_{p3}}{R_{p4}} \times \frac{R_{p1}}{R_{p2}}$$

$$= -\frac{N_5}{N_6} \times \frac{N_3}{N_4} \times \frac{N_1}{N_2}$$

where N_1, N_2, N_3, N_4, N_5, and N_6 are the numbers of teeth on the six gears. The replacement of radius ratios with tooth number ratios is correct because, for each such ratio, the gears involved must have the same diametral pitch. As power flows from gear 1 to gear 6, at each gear contact the odd-numbered gear is the driving gear and the even-numbered gear is the driven gear. The train ratio result for compound gear sets may be generalized as follows:

$$\begin{matrix}\text{Train} \\ \text{Ratio}\end{matrix} = \pm \frac{\text{Product of tooth numbers for driving gears}}{\text{Product of tooth numbers for driven gears}}$$

where the positive sign is taken when there are an odd number of intermediate shafts and the negative sign is used for an even number of intermediate shafts.

In certain situations the input and output shafts for a compound train must be aligned; this is called a *reverted compound train,* and a typical example is shown in Figure 5.15. The center distances are the same for both pairs of gears so that, in terms of the actual pitch radii,

$$R_{p1} + R_{p2} = R_{p3} + R_{p4}$$

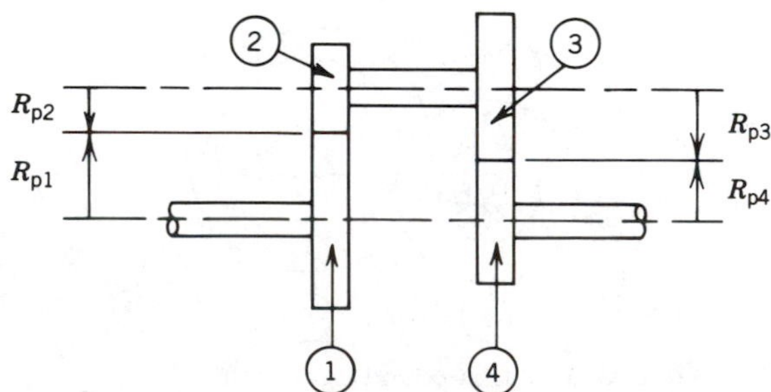

FIGURE 5.15 Reverted Compound Train

In many cases, this condition can not be met using standard gears mounted at nominal center distances. Allowing the center distance to increase provides for some design flexibility at the expense of increased pressure angles and decreased contact ratios.

5.9 PLANETARY GEAR TRAINS

All the gear trains discussed in the preceding section involve gear shaft centerlines that are stationary with respect to an external frame. A second important class of gear trains, called *planetary gear trains,* also exists in which some of the gear shaft centerlines are not stationary. Typical simple and compound planetary gear trains are shown in Figures 5.16 and 5.17. The name stems from a rather obvious analogy with the solar system. The central gear is called the *sun gear,* and the gears circulating around the sun are termed *planets*. However, there is no astronomical analog to the *ring gear,* an internal gear that engages the outside of the planets. The planet shafts are located by the *planet carrier,* as shown in the figure; the planet carrier is also called the *arm* or *spider*. Planetary gear trains are alternatively called *epicyclic gear trains* because points on the planet gears trace epicycloidal curves.

Figures 5.16 and 5.17 show pictorial views in the axial direction (on the left) along with schematic section views (on the right). In the schematics, the planets are shown rotated into the picture plane, even though in the case of three evenly spaced planets there are never two in the picture plane at one time. The gear shafts appear as lines in the schematic view, and the schematic is symmetric about the centerline. Commonly, the schematic is abbreviated by showing only the top half.

Notice that there are three shafts out of the gear train, one for each of the sun gear, the planet carrier, and the ring gear. In many cases, one of the shafts will be fixed to the frame so that only two shafts need to be brought out. If all three can rotate, then at least two must be brought out coaxially. If the planet carrier is fixed, the assembly is no different from the gear trains considered in the previous section. The more common situation

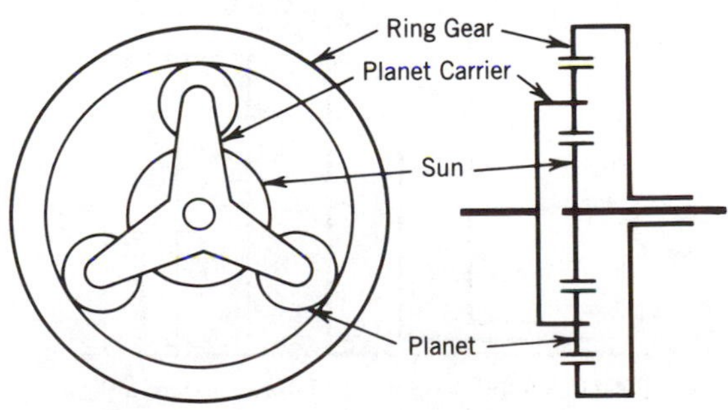

FIGURE 5.16 Simple Planetary Train

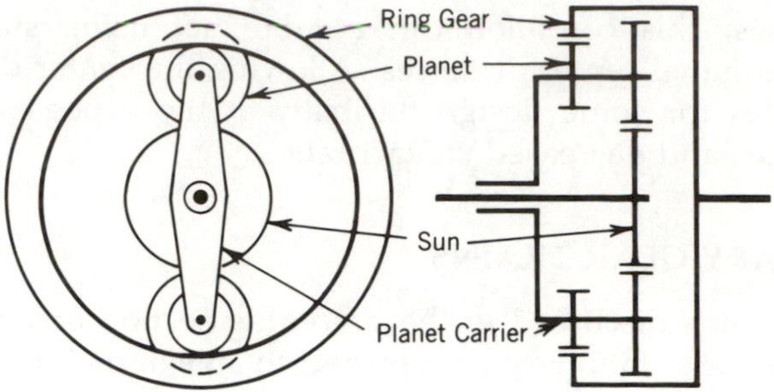

FIGURE 5.17 Compound Planetary Train

is to fix either the sun or the ring gear, in which case the planets circulate around the sun gear. Note that the planet rotation *is not* available for an external connection; there is no external shaft that directly drives, or is driven by, the individual planets. The sun, the planet carrier, and the ring gear are the only input or output rotations from the planetary gear train.

With three connecting shafts, the planetary gear train must have two assigned rotations to determine an output; this is a two-degree-of-freedom mechanism. If one of the shafts is fixed, that is an assigned zero rotation. Then, one additional rotation will determine the output. Fixing one shaft reduces the mechanism to a single degree of freedom.

Consider the simple planetary train shown in Figure 5.18, where the three concentric shafts to the left connect to the sun, the planet carrier, and the ring gear. Pitch radii are indicated for all gears, and the planet carrier radius is also provided. Let A_s, A_c, A_p, and A_r denote the rotations of the sun, planet carrier, planet, and the ring gear, all measured relative to a stationary reference frame, and positive in the sense that will move the upper edge of the body out of the plane of the figure. Two equations are required to express the condition of rolling without slip at the sun–planet contact and at the planet–ring contact. It is useful to think of these equa-

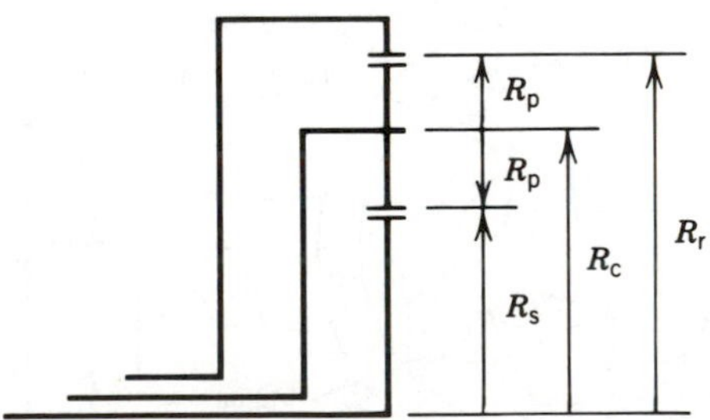

FIGURE 5.18 Schematic Half Section of a Simple Planetary Train

tions as expressing the equality of arc length passing through engagement on each of the gears at the contact:

$$R_s A_s = R_c A_c - R_p A_p \qquad \text{Sun–Planet}$$

$$R_c A_c + R_p A_p = R_r A_r \qquad \text{Planet–Ring}$$

The planet rotation may be eliminated by adding these two equations:

$$R_s A_s + R_r A_r = 2 R_c A_c$$

This is the kinematic relation between the three shaft rotations, A_s, A_c, and A_r. If only the tooth numbers are available for each gear, instead of pitch radii, it is useful to rewrite the preceding relation in those terms. For the simple planetary train, all components must have the same diametral pitch. Recall that the gear pitch radius is the number of teeth divided by twice the diametral pitch. On this basis, the pitch radii may be replaced and the common factor removed from the equation, with the result

$$N_s A_s + N_r A_r = 2 (N_s + N_p) A_c$$

where the Ns are the gear tooth numbers. If any of the components are fixed, the associated rotation is zero and that term drops from the preceding equation. This relation may be differentiated with respect to time to obtain the relation between the shaft rotation rates and the angular accelerations.

For the compound planetary train shown in Figure 5.19, the analysis is only slightly more involved. The constraints describing rolling without slipping are

$$R_s A_s = R_c A_c - R_{p1} A_p \qquad \text{Sun–Planet}$$

$$R_r A_r = R_c A_c + R_{p2} A_p \qquad \text{Planet–Ring}$$

When the planet rotation is eliminated, the remaining equation relating the rotations of the sun gear, planet carrier, and the ring gear is

$$R_{p2} R_s A_s + R_{p1} R_r A_r = (R_{p1} R_c + R_{p2} R_c) A_c$$

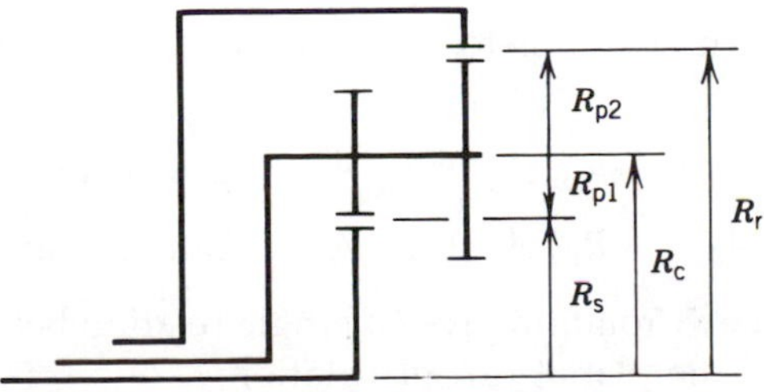

FIGURE 5.19 Schematic Half Section of a Compound Planetary Train

As before, it is usually preferable to express this relation in terms of tooth numbers. To that end, the pitch radius of each gear is expressed in terms of the number of teeth and the diametral pitch of the gear. Let the diametral pitch for the sun and planet 1 be denoted P_1, and let the diametral pitch for the second planet and the ring gear be P_2. Then, the required radii are

$$R_s = \frac{N_s}{2P_1} \qquad R_r = \frac{N_r}{2P_2}$$

$$R_{p1} = \frac{N_{p1}}{2P_1} \qquad R_{p2} = \frac{N_{p2}}{2P_2}$$

$$R_c = R_s + R_{p1} \qquad R_c = R_r - R_{p2}$$

$$= \frac{N_s + N_{p1}}{2P_1} \qquad = \frac{N_r - N_{p2}}{2P_2}$$

The two expressions for the planet carrier radius will be used in rewriting the kinematic relation in terms of the tooth numbers. When these substitutions have been made and common factors removed, the three rotations are related by

$$N_{p2}\, N_s\, A_s + N_{p1}\, N_r\, A_r = (N_{p1}\, N_r + N_{p2}\, N_s)\, A_c$$

If this expression is solved for any of the rotations, the resulting tooth number ratios in each case will involve ratios of gears having a common diametral pitch. For two gears with common diametral pitch, the tooth number ratio is equivalent to the corresponding radius ratio. The process of describing the kinematics of a compound planetary train and determining the train ratio is demonstrated in the following example.

5.9.1 Compound Planetary Train

Consider the planetary train shown in Figure 5.20, in which there is no ring gear, but rather a second sun gear. The pitch radii and tooth numbers are indicated in the figure, but only the tooth numbers are known. The input is the rotation of the sun gear 1, and the planet carrier is the output member. What is the train ratio? Note that there is no requirement that the two planet gears be of the same circular pitch, and they will usually be different.

As before, the first step is to express the rolling constraint conditions for each gear mesh:

$$R_{s1}\, A_1 = -R_{p1}\, A_p + R_c\, A_c \qquad \text{Sun 1–Planet 1}$$

$$R_{s2}\, A_2 = -R_{p2}\, A_p + R_c\, A_c \qquad \text{Sun 2–Planet 2}$$

Eliminating the planet rotation gives a single relation between the two sun gear rotations and the planet carrier rotation:

$$R_{p2}\, R_{s1}\, A_1 - R_{p1}\, R_{s2}\, A_2 = (R_c\, R_{p2} - R_c\, R_{p1})\, A_c$$

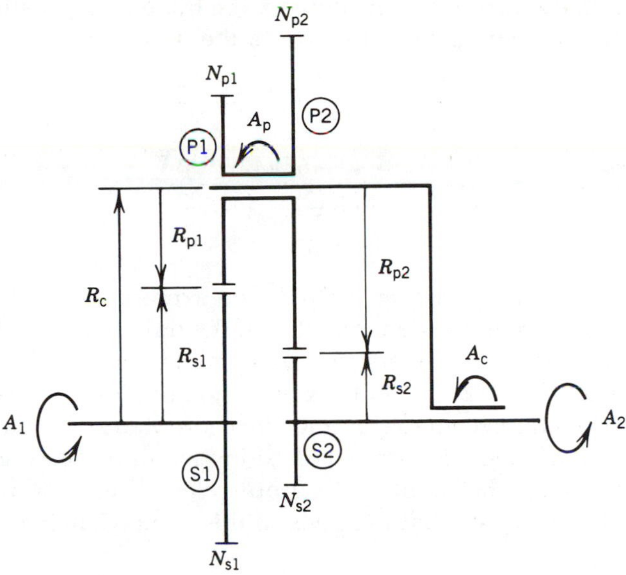

FIGURE 5.20 Planetary Gear Train For Example of Section 5.9.1

Let P_1 denote the (unknown) diametral pitch of the first sun and planet pair, while P_2 is the (unknown) diametral pitch of the second sun and planet pair. To eliminate the gear radii in favor of the tooth numbers, the following substitutions are needed:

$$R_{s1} = \frac{N_{s1}}{2P_1} \qquad R_{s2} = \frac{N_{s2}}{2P_2}$$

$$R_{p1} = \frac{N_{p1}}{2P_1} \qquad R_{p2} = \frac{N_{p2}}{2P_2}$$

$$R_c = R_{s1} + R_{p1} \qquad R_c = R_{s2} + R_{p2}$$

$$= \frac{N_{s1} + N_{p1}}{2P_1} \qquad = \frac{N_{s2} + N_{p2}}{2P_2}$$

When the planet rotation is eliminated and common factors are removed, the kinematic relation between the two sun gear rotations and the planet carrier rotation is

$$N_{p2}\, N_{s1}\, A_1 - N_{p1}\, N_{s2}\, A_2 = (N_{p2}\, N_{s1} - N_{p1}\, N_{s2})\, A_c$$

Note that the diametral pitch values are not required because they are eliminated in the common factor removal.

Now consider sun gear 2 to be fixed, and let the tooth numbers be as follows:

$$N_{s1} = 42 \qquad N_{p1} = 53$$

$$N_{s2} = 22 \qquad N_{p2} = 67$$

Because the planet carrier is considered to be the output rotation with the rotation of the first sun gear as the input, the train ratio is

$$\frac{A_c}{A_1} = \frac{N_{p2}\,N_{s1}}{N_{p2}\,N_{s1} - N_{p1}\,N_{s2}}$$

$$= \frac{(67)(42)}{(67)(42) - (53)(22)}$$

$$= 1.70752$$

For this system, with sun gear 2 fixed, the expression in the train ratio involves a difference in the denominator. If the train had two identical sun gears and two identical planets, the tooth numbers would be such that $N_{s1} = N_{s2}$ and $N_{p1} = N_{p2}$. Under this condition, the difference in the denominator would become zero, and the computed train ratio would be undefined. What does this mean? For an insight into this condition, consider interchanging the input and output roles. Then, the train ratio is zero. Does that suggest what happens in this somewhat special gear set?

5.10 GEAR TRAIN DESIGN

Choosing tooth numbers to achieve a specified train ratio is a problem in synthesis, and is usually much more difficult than analyzing a given gear train. For gears that turn through multiple revolutions, an integer number of teeth on the circumference of each gear is required. This requirement limits the possible velocity ratios for a gear pair to the ratios of relatively small integers, which may not offer a satisfactory approximation to the required train ratio. For a gear that does not turn a full revolution, an integer number of teeth on the circumference is not necessary; this offers greater flexibility in achieving a required train ratio.

The train ratio for a set of gears is the ratio of two integer products. Consequently, the train ratio is necessarily a rational fraction. This means that if the required train ratio is an irrational number, the ratio can only be approximated. If gears of unlimited size could be used, any rational train ratio value could be obtained, but unlimited size is unrealistic. Most gear train design is constrained to have no less than 18 or 20, and no more than 100 or 120 teeth on any gear. With this constraint, many rational values for the train ratio can not be obtained exactly, and must be approximated. Even so, the number of combinations possible in a train of four, six, or even eight gears is enormous, so some sort of guidance is needed to approximate a specified train ratio. Three tools are presented below, but their value depends on the skill and artistry of the user. They are (1) continued fraction approximation, (2) the Brocot Table, and (3) an algorithm for the prime factors of any positive integer.

Another consideration in the design of gear trains is the need for a condition called "hunting." In heavily loaded gear trains, where wear is a major concern, each pair of gears in mesh should have no common factor. With a common factor, the same pairs of teeth will engage repeatedly, so that a few pairs of teeth repeatedly will carry the maximum load and wear more rapidly. The lack of a common factor, however, tends to distribute the wear more uniformly around the gear circumference, a phenomena that is called "hunting." On occasion, the ratio may be modified by the addition of one extra tooth (called a "hunting tooth") to eliminate the existence of a common factor. Note that the problem is associated with the existence of a common factor *in a single gear mesh.* Consequently, for compound trains, the difficulty can usually be overcome by regrouping the factors of the train ratio calculation used to arrive at the individual tooth numbers.

Continued Fraction Approximation

Consider a required train ratio, in the interval 0 to 1.0, expressed as a finite decimal fraction. This is expressible immediately as a rational fraction because there is some power of 10, say n, such that the gear ratio is expressible as an integer, say N, divided by 10^n, that is, $N/10^n$. Clearly, both numerator and denominator are integers. These two integers would then be the appropriate number of teeth for the required gear ratio to be accomplished in a single gear pair. Unfortunately, the integers involved are often far too large. The required ratio can be written as

$$\frac{N}{10^n} = \frac{1}{Q + R/N}$$

where Q is the quotient and R is the remainder. As a first approximation, the remainder can be dropped:

$$\frac{N}{10^n} \doteq \frac{1}{Q} = \frac{N_1}{D_1}$$

where N_1 and D_1 are the integers used as numerator and denominator for this first approximation. This is usually quite crude, so consider the process carried to a second step:

$$\frac{N}{10^n} = \cfrac{1}{Q_1 + \cfrac{1}{Q_2 + R_2/R_1}} \doteq \frac{Q_2}{Q_2 Q_1 + 1} = \frac{N_2}{D_2}$$

where the second remainder, R_2, is dropped. Again, N_2 and D_2 are the integers used for the numerator and denominator of the second approximation. By dropping the remainder at any stage, the result is expressible as a pair of integers, N_i/D_i. The process will eventually reach the point where the numerator is the original value N and the denominator is the original divisor, 10^n (although it is rarely useful to carry it that far). The sequence of fractions N_1/D_1, N_2/D_2, N_3/D_3, . . . are rational fraction approximations to the required ratio, approaching it in an oscillatory manner, such that any two adjacent fractions bracket the required ratio. Usually the sequence is ended when the denominator becomes larger than the maximum allowable number of teeth, if a single gear pair is to be used. For four gears, the sequence might be continued until the denominator is as large as the square of the largest allowable number of teeth on one gear. Unfortunately, the continued fraction process may skip over some rational fractions that are better approximations and have acceptable tooth numbers, so this list may not be optimal. It is, however, a beginning. A computer program for generating continued fraction approximations is given at the end of this discussion. The result is a table that gives, for each approximation, integer values for the numerator and the denominator, and the ratio of the two expressed as a decimal fraction.

A *proper fraction* is the ratio of two integers that have a value between zero and one and do not have any common factors. The entries in the continued fraction list are proper fractions, such that if N_i/D_i and N_{i+1}/D_{i+1} represent two adjacent entries in the list, the following identity will be true:

$$N_i \cdot D_{i+1} - N_{i+1} \cdot D_i = \pm 1$$

Such ratios are called *conjugate fractions.* (Unfortunately, this is a second use of the word conjugate. In this case, "conjugate" has nothing to do with tooth profiles and constant angular velocity ratios.) In choosing gear tooth numbers to approximate a required ratio, conjugate fractions are important because *between any two conjugate fractions, there are no other fractions with a denominator less than the larger of the given pair.* Consider the two fractions 7/19 and 11/30; these are conjugate because

$$7 \cdot 30 - 11 \cdot 19 = +1$$

Applied to this case, this statement says that there are no fractions lying between 11/30 and 7/19 having a denominator less than 30. The proof of this property is given by Rasch.[3] Any two *consecutive* entries in the continued fraction approximation form a conjugate pair. The property of conjugacy is important in the use of the next tool to be discussed, the Brocot Table.

The Brocot Table

If all the proper fractions that have denominators less than a specified value B_o, are arranged in increasing order, the result is a Brocot Table. The number B_o is called the *base*. This table is attributed to Achille Brocot, a French horologist. The construction of a Brocot Table between a pair of conjugate fractions is fairly involved. It is implemented in a computer program provided at the end of this discussion. Alternatively, published Brocot Tables are available, such as the one by Buckingham.[4]

[3] W. H. Rashe, "Gear Train Design," *Virginia Polytech. Inst. Engr. Expt. Sta. Ser. Bull. 14*, 1933.

[4] E. Buckingham, *Gear Ratio Tables for 4- 6- and 8-Gear Combinations*, Industrial Press, 1958.

A Brocot Table consists of integer values for the numerator and the denominator, and a decimal fraction for their ratio. Every pair of adjacent fractions in the Brocot Table are conjugates. However, the Table often contains nonadjacent conjugate pairs as well. If two conjugate fractions have been obtained as consecutive entries in the continued fraction approximation, the Brocot Table between these two values can be constructed to look for better approximations, that lie between the first two conjugate fractions. The value chosen for the base, B_0, should be the largest allowable tooth number if a single gear pair is being considered, or the square of the largest allowable tooth number if two pairs are being considered, and so forth.

Prime Factor Algorithm

Either the continued fraction approximation or the Brocot Table provide integer values for the numerator and denominator that have no common factors; they are said to be *relatively prime.* They are frequently not prime numbers in the absolute sense, and, if not prime, they can be factored. So, to have maximum flexibility to create tooth number groups, it is convenient to first express the numerator and denominator each as a product of prime factors. This can be accomplished using a prime factor table, of which there are several published. Alternatively, a computer program can be used to reduce each integer to prime factors. One such program follows shortly. A word of caution should be given regarding this algorithm; it is not suited to factoring very large numbers, such as 500,000 or more. However, for the size of numbers commonly encountered in gear design calculations, it is quite satisfactory.

5.10.1 Gear Train Design Example

Choose the tooth numbers for a gear train to approximate a train ratio of 0.457. Two, four, or six gears may be used.

To begin, the continued fraction approximation is used to get starting values. The continued fraction approximation to 0.457 gives

Continued Fraction
Approximation to 0.457

N	*D*	*Ratio*
1	2	.5
5	11	.454545455
16	35	.457142857
85	186	.456989247
186	407	.457002457
457	1000	.457000000

If this ratio is to be accomplished with a single pair of gears with tooth numbers less than 120, the next step is a Brocot Table with base 120 constructed between 5/11 and 16/35, as follows:

Brocot Table Between 5/11 and 16/35

N	*D*	*Ratio*
5	11	.454545455
51	112	.455357143
46	101	.455445545
41	90	.455555556
36	79	.455696203
31	68	.455882353
26	57	.456140351
47	103	.456310680
21	46	.456521739
37	81	.456790123
53	116	.456896552
16	35	.457142857

The best choice for a single gear pair will be 53/116, for which the error = $0.457000000 - 0.456896552 = 1.034(10^{-4})$.

If a four-gear train is considered, the two previous tables can be searched for factorable entries. In the continued fraction list, two promising fractions appear: 85/186 and 186/407.

$$\frac{85}{186} = \frac{5 \times 17}{2 \times 3 \times 31} = \frac{5}{6} \times \frac{17}{31} = \frac{17}{30} \times \frac{25}{31} = .45698924$$

Thus, a train having a 17-tooth pinion in mesh with a 30-tooth gear on the same shaft with a 25-tooth pinion meshed with a 31-tooth gear has an error of $1.075(10^{-5})$. (This ignores the questionable use of a 17-tooth pinion.)

$$\frac{186}{407} = \frac{2 \times 3 \times 31}{11 \times 37} = \frac{6}{11} \times \frac{31}{37} = \frac{24}{37} \times \frac{31}{44} = .45700245$$

$$\text{error} = 2.457(10^{-6})$$

This train consists of a 24-tooth pinion meshed with a 37-tooth gear on the same shaft with a 31-tooth pinion in mesh with a 44-tooth gear; the error is as previously indicated. The smaller error is clearly with the second choice, and the tooth numbers are reasonable. Notice that in each of these last two combinations, a factor had to be introduced to adjust the magnitudes of the numbers (5/5 in the first case, 4/4 in the second). These factors were distributed in different gear meshes, thus avoiding the problem of a common factor between the tooth numbers in any single gear mesh.

When six gears are allowed, an even better approximation is to be expected. However, the continued fraction list and the Brocot Table shown before offer no basis for investigating directly a six-gear train. One approach is to use a pair of values from the Brocot Table to assign one pair of gears; any such pair will roughly approximate the required ratio. For this example, let 41/90 be chosen for this purpose. This fraction is then divided into the required gear ratio to produce a new ratio to be approximated such that

$$\text{Required Ratio} = 41/90 \times \text{New Ratio}$$

$$\text{New Ratio} = \frac{.457}{\frac{41}{90}} = 1.0031707 = \frac{1}{.99683929}$$

The design procedures are directed toward obtaining proper fractions, so the final step is required to get a train ratio value between zero and one. With one of the three pairs assumed as 41/90, the search continues for the remaining four-gear train with the ratio 0.99683929. The continued fraction approximation for this ratio is as follows:

Continued Fraction
Approximation to 0.99683929

N	*D*	*Ratio*
1	1	1.0
315	316	.99683543
631	633	.996840442
946	949	.99683877
1577	1582	.99683944
2523	2531	.996839194
4100	4113	.99683929

In this new continued fraction list, several ratios appear promising. Using the ratio 946/949 and the previously assigned pair 41/90, the overall train ratio is

$$\frac{41}{90} \times \frac{949}{946} = \frac{41 \times 13 \times 73}{90 \times 2 \times 11 \times 43} = \frac{26}{43} \times \frac{41}{44} \times \frac{73}{90} = .45700023$$

$$\text{error} = 2.349(10^{-7})$$

In this calculation, a factor 2/2 was introduced and distributed to avoid having a common factor between the tooth numbers in any mesh. There are other possibilities that could be investigated, including the ratio 1577/1582, but those are not pursued here. Where tooth numbers have been assigned according to the Brocot Table fraction 41/90, assign instead 21/46 (also from the same Brocot Table). With these new values, the new ratio for the remaining four gears is

$$\frac{.457}{\frac{21}{46}} = 1.0010476 = \frac{1}{.998953477}$$

This is the basis for another continued fraction approximation, the results of which follow:

Continued Fraction
Approximation to 0.998953477

N	*D*	*Ratio*
1	1	1.0
954	955	.9989529
955	956	.9989540
1909	1911	.9989534
10500	10511	.9989535

Three of these ratios look promising: 954/955, 955/956, and 1909/1911. When factored, 955 and 956 are found to have prime factors greater than 120, so the ratios involving those values are discarded. Using the last ratio, the overall train ratio is

$$\frac{21}{46} \times \frac{1911}{1909} = \frac{3^2 \times 7^3 \times 13}{2 \times 23^2 \times 83} = \frac{21}{23} \times \frac{49}{46} \times \frac{39}{83} = .457000023$$

$$\text{error} = 2.3(10^{-8})$$

This result is a full order of magnitude better than the previous six-gear train.

It should be emphasized that the choices just explored are by no means exhaustive. The number of combinations is immense, even with the limitation that all gears have between 20 and 120 teeth. A hint of the magnitude of the number of possibilities can be seen by just looking at the part of the process where the required train ratio is "traded-off" for a new train ratio by assigning two tooth numbers. Consider then how many possibilities there are for the initial trade-off, each of which opens up anew the entire search process.

A computer program to develop continued fraction approximations to a specified ratio follows.

```
1000 REM  CONTINUED FRACTION APPROXIMATION
1010 DIM N(50),D(50),Q(50),X(50)
1020 HOME
1030 PRINT"ENTER THE REQUIRED RATIO, 0 < RO < 1.0"
1040 INPUT RO
1050 E7 = 0
1060 V1 = RO*10^E7
1070 V2 = INT(V1)
1080 IF ABS(V2 - V1) < 0.1 THEN GOTO 1100
1082 E7=E7+1
1090 GOTO 1060
1100 REM  SET FIRST TWO VALUES IN EACH COLUMN OF
     TABLE
1110 X(0) = INT(10^E7 + 0.5)
```

```
1120 X(1) = RO*X(0)
1130 Q(1) = INT(X(0)/X(1))
1140 X(2) = X(0) - Q(1)*X(1)
1150 N(1) = 1
1160 D(1) = Q(1)
1170 J=1
1180 GOSUB 1500
1190 Q(2) = INT (X(1)/X(2))
1200 X(3)= X(1) - Q(2)*X(2)
1210 N(2) = Q(2)
1220 D(2) = Q(1)*Q(2) + 1
1230 J=2
1240 GOSUB 1500
1250 REM EXTEND THE TABLE
1260 J=J+1
1270 Q(J) = INT(X(J-1)/X(J))
1280 X(J+1) = X(J-1) - Q(J)*X(J)
1290 N(J) = Q(J)*N(J-1) + N(J-2)
1300 D(J) = Q(J)*D(J-1) + D(J-2)
1310 GOSUB 1500
1320 IF X(J+1) = 0 THEN GOTO 1340
1330 GOTO 1260
1340 END
1500 REM  OUTPUT SUBROUTINE
1510 PRINT J;"   ";N(J);"       ";D(J);"
";N(J)/D(J)
1520 RETURN
1530 END
```

This completes the program listing for the continued fraction approximation.

The next computer program performs the construction of the Brocot Table.

```
2000 REM       BROCOT'S TABLE CALCULATION
2010 DIM N(100),D(100)
2020 HOME
2030 PRINT"ENTER FIRST RATIONAL FRACTION"
2040 PRINT"ENTER A FOR FORM A/B"
2050 INPUT A
2060 PRINT"ENTER B FOR FORM A/B"
2070 INPUT B
2080 PRINT
2090 PRINT"ENTER SECOND RATIONAL FRACTION"
2100 PRINT"ENTER  C FOR FORM C/D"
2110 INPUT C
2120 PRINT"ENTER  D FOR FORM C/D"
2130 INPUT D
2140 PRINT
2150 PRINT"ENTER BASE, BO"
```

```
2160 INPUT B0
2170 REM  BEGIN THE BROCOT TABLE
2180 N(1)=A
2190 N(2)=C
2200 D(1)=B
2210 D(2)=D
2220 K=2     : REM K IS LENGTH OF BROCOT TABLE
2230 F=0
2240 FOR I=1 TO K-1
2250 I1=I+1
2260 IF D(I)+D(I1)>B0 THEN GOTO 2380
2270 F=F+1
2280 REM  OPEN A SPACE IN THE TABLE
2290 FOR J=K TO I1 STEP -1
2300 J1=J+1
2310 N(J1)=N(J)
2320 D(J1)=D(J)
2330 NEXT J
2340 K=K+1    : REM  INCREASE THE TABLE LENGTH BY ONE
2350 REM  CALCULATE THE NEW TABLE ENTRIES
2360 N(I1)=N(I)+N(I+2)
2370 D(I1)=D(I)+D(I+2)
2380 NEXT I
2390 REM  QUIT THE TABLE EXPANSION ONLY WHEN ALL
2400 REM  DENOM SUMS EXCEED THE BASE
2410 REM  THIS WILL BE INDICATED BY F=0 AT THE END
2420 REM  OF THE LOOP ON I
2430 IF F>0 THEN GOTO 2230
2440 REM  PRINT THE BROCOT TABLE
2450 DEF FN R5(Q) = INT(Q*10E5 + .5)/10E5
2460 PRINT"           BROCOT TABLE"
2470 PRINT"    N        D                 RATIO*
2480 PRINT
2490 FOR I=1 TO K
2500 PRINT"    ";N(I);"      ";D(I);"           ";FN
     R5(N(I)/D(I))
2510 NEXT I
2520 END
```

This completes the program listing for the Brocot Table construction.

The next program is useful for determining the prime factors of an integer.

```
1000 REM  PRIME FACTOR DECOMPOSITION
1010 HOME
1020 PRINT"    PRIME FACTOR DECOMPOSITION"
1030 PRINT
1040 PRINT"ENTER THE NUMBER TO BE FACTORED"
1050 INPUT N0
1060 N = N0
1070 F = 1
```

```
1080 F = F + 1
1090 IF F > SQR(N)   THEN GOTO 1160
1100 T = N/F
1110 IF INT(T) < > T THEN GOTO 1080
1120 PRINT F
1130 N = T
1140 F = 1
1150 GOTO 1080
1160 S$ = ""
1170 IF N = NO THEN S$ = " IS PRIME"
1180 PRINT N; S$
1190 PRINT"FACTORING IS COMPLETE"
1200 END
```

This completes the program listing for the factoring algorithm.

The preceding three programs are constructed as independent programs. To be used extensively, it is desirable to combine all three operations into a single program under menu control. This combination facilitates moving quickly from one operation to the next.

5.11 CONCLUSION

Familiarity with gearing is essential for mechanical engineers. Gears have been and continue to be one of the most important means of mechanical power transmission, and this situation will no doubt continue in the future. They offer highly efficient, precise, positive power transmission at relatively low cost in large volume.

This chapter has considered only spur gears, but many other types of gears are also used commonly. Knowledge of the properties of the involute curve and of the type of information contained in the AGMA standards is applicable to many different types of gearing, such as helical gears and bevel gears. The concepts of contact ratio, interference, and undercutting also apply to many of these other types of gearing as well.

Consideration of gear trains requires a systems approach, in which the details of any single gear mesh are peripheral to the effect of the entire train. Much has been written about the design of gears with regard to strength, but relatively little is written about the kinematic aspect of design—the matter of determining acceptable tooth number combinations. The discussion presented in this chapter regarding ways to establish acceptable tooth number combinations will help to remedy that situation.

REFERENCES

Buckingham, E., *Analytical Mechanics of Gears.* New York: Dover, 1963.

Mabie, H. H. and Ocvirk, F. W., *Mechanisms and Dynamics of Machinery,* 3rd Ed. New York: John Wiley, 1975.

Merritt, H. E., *Gear Trains.* London: Sir Isaac Pitman & Sons, Ltd., 1947.

Usher, A. P., *A History of Mechanical Inventions.* Boston: Beacon Press, 1959, p. 171.

PROBLEM SET

5.1 Using only a compass, dividers, and straightedge, construct an approximate geometrical layout (full-size) of an involute curve on a base circle of radius 100 mm. Locate points on the involute curve at 13 evenly spaced values of the angle ($A_f + B$) from zero to $\pi/2$ (including both end points). Connect the points to form a smooth curve representing the involute. Measure the final radius of curvature, P, and compare this measured value to the length computed for one-quarter circumference of the base circle.

5.2 Plot an involute curve on a base circle of radius 120 mm. Let R, the radius to a point on the involute curve, increase in steps of 2 mm, and compute the polar angle B for each point. Put this data in a tabular form, and then plot the involute curve at full size. This process should be continued until the flank angle exceeds $\pi/4$.

5.3 In the figure shown, the base circle radius is 1.775 in. and the angle C is 0.82 radians. Make an iterative solution for the flank angle, A_f.

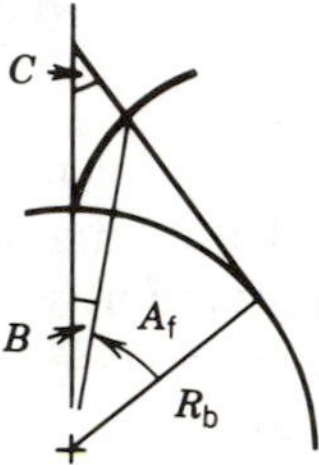

5.4 Consider an involute cam (a single gear tooth) operating with a flat-faced, translating follower (beveled plunger), as shown. When the cam rotation is A, the plunger position is $H(A)$. The base circle radius for the involute is R_b, a known value. If A_1 and A_2 are two distinct cam positions,

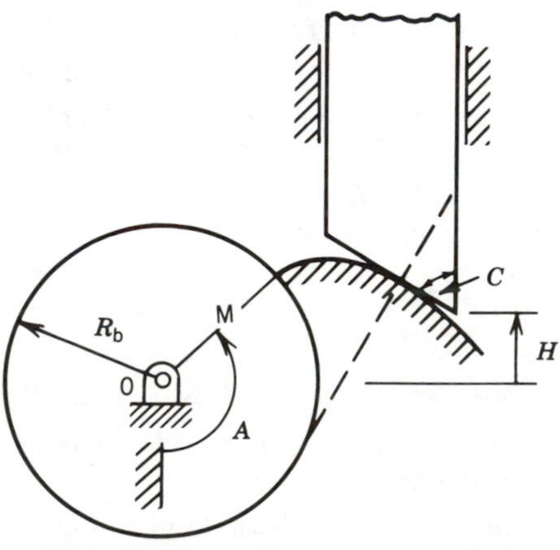

a. Determine the change in follower position $H(A_2) - H(A_1)$;

b. Express $\dot{H}$ in terms of $\dot{A}$.
Hint: Try drawing the system in two nearby positions to see the relation of $H(A_2) - H(A_1)$ to $A_2 - A_1$.

5.5 Consider two intersecting involute curves, as shown in Figure 5.6. The base radius is 5.25 in., and at a radius of 5.87 in. the circular thickness is 1.17 in. At what radius do the two involute curves intersect?

5.6 For a gear made on the AGMA 20-degree full-depth system with 32 teeth and a diametral pitch of 4 T/in., determine the profile radius of curvature

a. At the nominal pitch radius;

b. At the addendum circle.

5.7 Consider a gear with 38 teeth and a pinion with 27 teeth. The teeth are AGMA standard teeth with a nominal 25-degree pressure angle and a diametral pitch of 16 T/in.

a. For each gear, determine the base circle radius and the nominal pitch radius;

b. Determine the nominal center distance for the pair;

c. If the gears are mounted at seven percent over the nominal center distance, determine the actual pressure angle.

5.8 A pinion with 22 teeth drives a 70-tooth gear. The gears are 90 T/in. diametral pitch and cut on the AGMA 20-degree fine-pitch system; they are mounted on their nominal center distance. Determine the contact ratio.

5.9 A 20-tooth pinion drives a 48-tooth gear. The gears are made on the AGMA 20-degree full-depth system, with a diametral pitch of 1 T/in. The speed of the pinion is 500 rpm. Make a plot of sliding velocity versus contact position (displacement from the pitch point) through the zone of contact. Consider the sliding velocity to be positive at the initial point of contact. Be sure to determine the initial and final contact positions, as well as points between.

5.10 A 23-tooth pinion engages a 74-tooth gear. The gears are made to the AGMA 20-degree full-depth system, and the diametral pitch is 4 T/in. The gears are mounted at 3.5 percent more than their nominal center distance. Consideration is being given to remaking the gears with increased addendum radii. What is the maximum addendum radius allowable for each gear without undercutting?

5.11 A gear is made to the AGMA 20-degree full-depth system, with 18 teeth and a diametral pitch of 4 T/in. Is the base circle inside or outside the root circle, and by how much?

5.12 A gear with 27 teeth and diametral pitch of 4 T/in. is made according to the AGMA 25-degree full-depth system. Is the base circle inside or outside the root circle, and by how much?

5.13 The gears shown in the figure turn on axes fixed in space, and the tooth numbers are as indicated. The input shaft is given a rotation of 0.1 radians CW as seen from the right side. What is the rotation of the output shaft (magnitude and sense)?

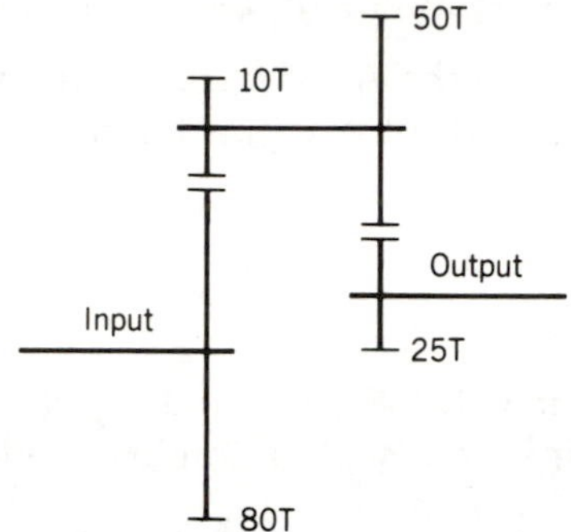

5.14 A compound gear train is shown with tooth numbers as indicated; the gear axes are fixed in space. If the input shaft is given a rotation of 0.1 radians CW as seen from the right side, what is the rotation of the output shaft (magnitude and sense)?

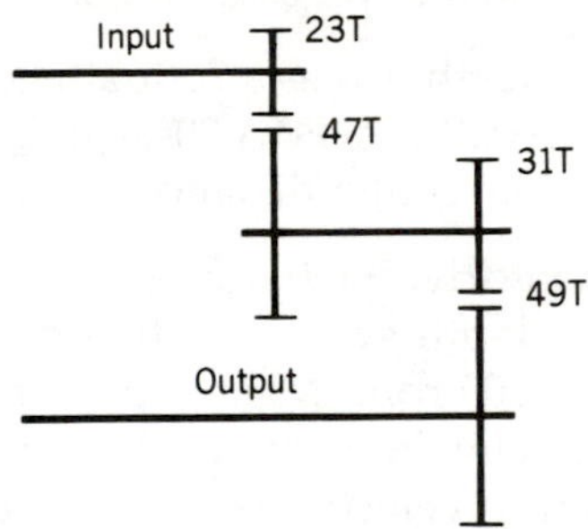

5.15 The bull gear with 131 teeth drives two compressor impellers, one through a 33-tooth pinion and the other through a 37-tooth pinion, as

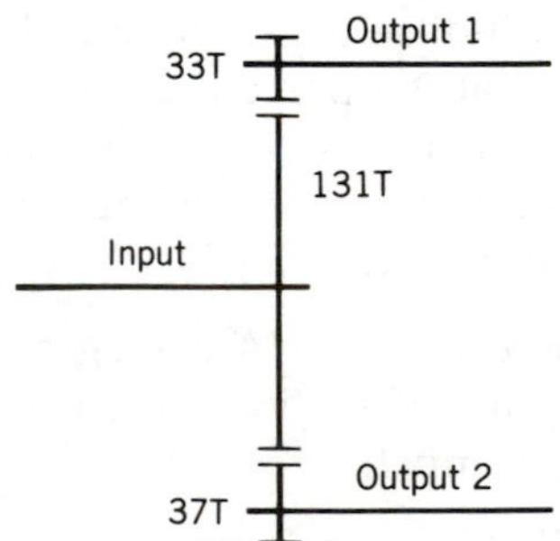

shown; all shaft axes are fixed in space. If the input shaft is given a rotation of 0.15 radians CCW as seen from the right side, what are the rotations of the output shafts (magnitude and sense for each)?

5.16 Determine the train ratio for the gear train considered in

a. Problem 5.13;

b. Problem 5.14.

5.17 A simple planetary train has 43 teeth on the sun, 19 teeth on the planets, and the gears are made with the (nonstandard) diametral pitch of 3.2 T/in.

a. Determine the number of teeth on the ring gear;

b. Determine the planet carrier radius.

5.18 For the simple planetary gear train described in Problem 5.17, the ring gear will be held stationary, the planet carrier is driven, and the output rotation is taken from the sun gear. What is the train ratio?

5.19 For the simple planetary gear train described in Problem 5.17, the sun gear is fixed, the ring gear is driven, and the output is the planet carrier rotation. What is the train ratio?

5.20 A compound planetary gear set is shown, for which the sun gear has 31 teeth, the planet engaging the sun gear has 47 teeth, the planet gear engaging the ring gear has 25 teeth, and the ring gear has 102 teeth. The sun gear is driven while the ring gear is fixed, and the planet carrier rotation is the output. Determine the train ratio.

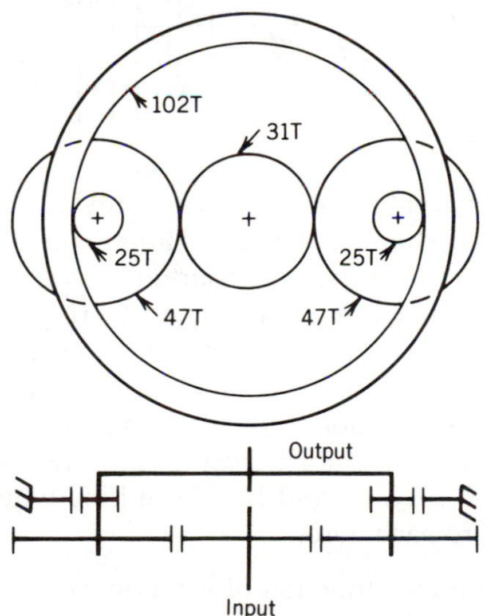

5.21 For the compound planetary gear set shown, the tooth numbers are as indicated. Note that the sun gear on the right is fixed. The planet carrier is driven, and the output is the rotation of the sun gear on the left. Determine the train ratio.

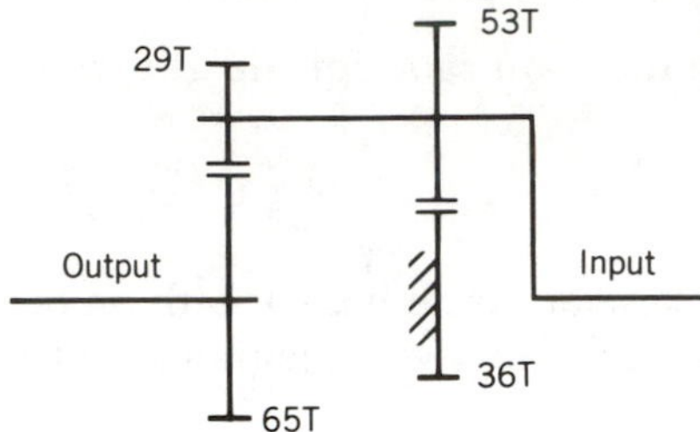

5.22 The planetary gear set shown is called a *spur gear differential*. This is usually considered as a two degree of freedom mechanism, with the rotations A_1 and A_2 as inputs (primary variables). The rotation of the carrier, A_3, is the output (secondary variable). The two gears rotated through A_1 and A_2 each have N_1 teeth and pitch radius R_1. The planets each have pitch radius R_p, and have N_p teeth. Express the rotation A_3 in terms of A_1, A_2, and the various tooth numbers.

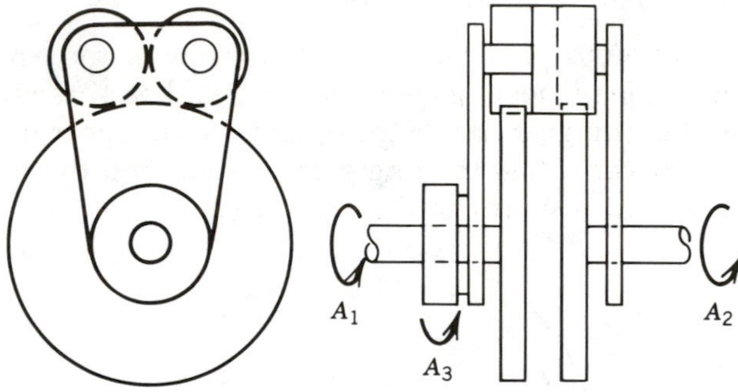

5.23 The pinion rotation q_1 drives the outside edge of a ring gear, causing the rotation G. The ring gear turns about a fixed center, guided by supports not shown in the figure. A second gear, carried on an arm driven with rotation q_2, is meshed with the inside of the ring gear. The rotation of the second gear is B. At a radius C from the center of the second gear, a connecting rod of length L drives a slider moving in a guide; the slider position is X. The geared portion of the system is shown at left in a section view where the ring gear is shown cross ruled. The following values are known: R_1, R_2, R_3, R_4, R_5, C, and L. At the time when $q_1 = q_2 = 0$, the angles A, B, and G are each zero.

a. Express the angle B as a function of rotations q_1 and q_2;
b. With B known from part a, determine the velocity coefficients for the slider with respect to the two inputs, K_{x1} and K_{x2}.

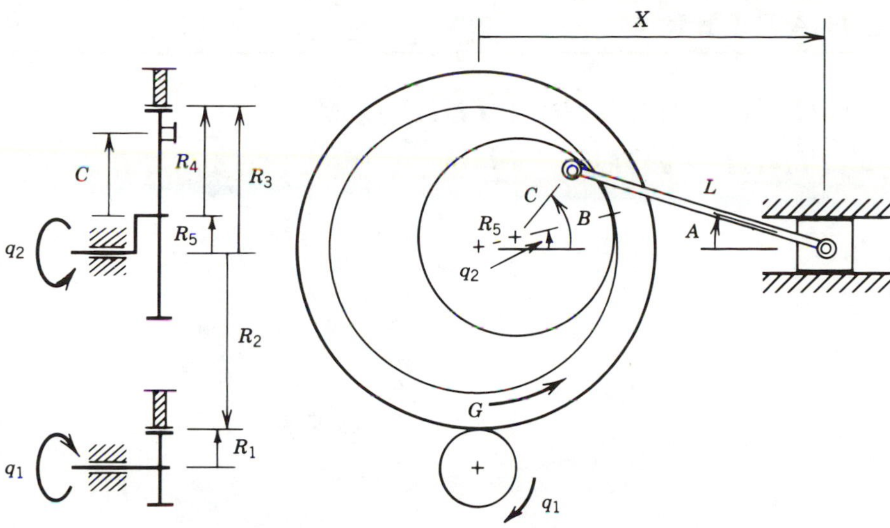

5.24 The train ratio 0.7922 is to be approximated using no more than four gears on stationary axes, with tooth numbers between 20 and 120, inclusively. Determine a satisfactory combination that differs from the desired value by no more than 10^{-6}.

5.25 Using gears with tooth numbers in the range 20 to 140, what is the best approximation to the train ratio 1.554 that can be achieved with a single gear pair?

5.26 The train ratio 3.225 is to be approximated using no more than six gears on stationary axes, with tooth numbers in the range 20 to 100. Determine a satisfactory combination that differs from the desired value by no more than 10^{-5}.

CHAPTER 6

Statics and the Principle of Virtual Work

6.1 GENERAL COMMENTS

The subject of statics is the familiar science of equilibrium. In an introductory mechanics sequence, this is usually one of the first topics discussed, and almost invariably presented in terms of Newton's First Law for a particle and its various extentions to translation and rotation of rigid bodies. Thus, the conditions for equilibrium are usually associated with the vanishing of certain force and moment sums. Although these methods are relatively simple and work well for simple problems, another approach is available that is often superior in the context of machinery problems.

This alternative approach is called the Principle of Virtual Work. In a related form called the Principle of Virtual Velocities, this principle can be traced back to the third century B.C. In western Europe, the Principle of Virtual Work is associated with such famous names of science as Leonardo da Vinci, Descartes, Lagrange, Gauss, Navier, Clapeyron, and Kirchoff. It is also the oldest of what are termed the Energy Principles of Mechanics, and has been suggested to be the fundamental axiom of mechanics. Consequently, the Principle of Virtual Work has played a major role in the theoretical development of mechanics. However, it is often omitted or only lightly included in current engineering curricula. Because it is especially useful in applications to machines, this chapter will attempt to remedy that slight.

Statics problems can be classified in two types, depending on whether the geometry is constant or non-constant. For the first type, many situations exist in which there is no gross motion as the loads are applied, so that the loaded equilibrium geometry is very nearly the same as the unloaded geometry, that is, the geometry is essentially constant. Most civil structures, such as bridges, towers, and dams, are examples of this type. The second type includes those cases in which gross motion may occur as the loads are applied, and the equilibrium position is often not a known part of the problem. For these cases, the geometry is not constant. The second type includes most devices designed to move, particularly machines. Examples include a pair of pliers compressing a spring, the pendulum scale mechanism, and the slider–crank mechanism used in automotive engines. Situations of the second type are often difficult to address by means of Newton's First Law (Equilibrium). The Principle of Virtual Work applies to both types of problems, but it is especially suited to the second type, where application of force and moment sums often leads to difficulty.

Most mechanisms transform mechanical power of one sort to another, such as a gear box that transforms high-speed rotary power to low-speed rotary power. If it is necessary to determine the relation between input force (torque) and output force (torque) by application of the First Law, a sequence of free-body diagrams, including internal forces and reactions, is required with an associated large number of equations. When this problem is approached from the Principle of Virtual Work, the reactions and internal forces are not involved in the formulation, and the number of equations is greatly reduced.

The following sections of this chapter will present the Principle of Virtual Work, a presentation based on Newton's First Law, and then show applications of this principle to machine problems. In the applications, it is useful to consider also the First Law formulation to gain an insight into the difficulties to be expected there and to see the types of problems best suited to each approach.

6.2 PRINCIPLE OF VIRTUAL WORK

The Principle of Virtual Work is a statement of the conditions for equilibrium in terms of work, a scalar quantity, significantly different from the vectorial force and moment sums derived from Newton's First Law as extended to rigid bodies. The concept of work is assumed to be familiar, but the modifier *virtual* requires some explanation. It is necessary to define a *virtual displacement* before the *virtual work* can be defined. Following these definitions, the *Principle of Virtual Work* can then be stated and justified.

Consider a particle located by the position vector $\mathbf{r}$. A virtual displacement of that particle, denoted $\delta\mathbf{r}$, is a postulated, arbitrary differential displacement of the particle, performed in zero-elapsed time, and consistent with any applicable constraints. Note the features of the virtual displacement:

1. It is a differential quantity, an infinitesimal;
2. It is postulated, or proposed for discussion, rather than the displacement associated with an actual motion;

3. It occurs in zero elapsed time, which says that time is constant during the virtual displacement;
4. It is arbitrary, insofar as that is possible consistent with the applicable constraints.

The first three characteristics are reasonably clear in themselves, but the fourth may require an illustration. For a single particle moving without any constraints, the virtual displacement is completely arbitrary in both infinitesimal magnitude and direction. If the particle is then constrained to remain a constant distance from a fixed point, then the particle must move on the surface of a sphere. The virtual displacement is then arbitrary in magnitude and in direction *on the surface of the sphere.* A virtual displacement normal to the surface of the sphere would not be consistent with an applicable constraint. If the particle is part of a rigid body, then the virtual displacement of the particle must be consistent with the virtual displacement of the rigid body as a whole.

The delta notation, using the lowercase Greek delta, denotes an operator similar to the more common differential operator, d. The four characteristics just noted are actually the properties of the delta operator. For comparison, consider the two operators are applied to a function $f(X, Y, Z, t)$:

$$df = \frac{\partial f}{\partial X}\, dX + \frac{\partial f}{\partial Y}\, dY + \frac{\partial f}{\partial Z}\, dZ + \frac{\partial f}{\partial t}\, dt$$

$$\delta f = \frac{\partial f}{\partial X}\, \delta X + \frac{\partial f}{\partial Y}\, \delta X + \frac{\partial f}{\partial Z}\, \delta Z$$

Notice that δf involves the delta operator applied to each of the coordinates, X, Y, and Z. Because of the third characteristic item in the initial list, the delta operator applied to the time variable is zero.

As just used, applying the delta operator to the position vector $\mathbf{r}$ produces $\delta\mathbf{r}$, called a "virtual displacement." In other circumstances, the result of applying the delta operator to the function f is called the "variation of f." This is a carryover from the Calculus of Variations, an advanced mathematical topic in which the delta operator is widely used. However, the calculus of variations will not be required for any work in this book. It is well to be aware that the terms "virtual something" and "variation of something" are, in fact, the same operations, and both denote the application of the delta operator.

If a force $\mathbf{F}$ is applied to a particle undergoing a virtual displacement $\delta\mathbf{r}$, then the force does *virtual work* δW, a scalar expressed in terms of a dot product:

$$\delta W = \mathbf{F} \cdot \delta\mathbf{r}$$

Thus, virtual work is the work associated with a finite force applied through a virtual displacement. The four characteristics, associated with the use of the delta operator, apply to virtual work.

Principle of Virtual Work for a Particle

For a single particle, the principle of virtual work says that the particle is in equilibrium if the virtual work of all forces acting on the particle is zero. Because the determination of equilibrium conditions based on Newton's First Law is a familiar approach, the first objective is to show that the Principle of Virtual Work is fully equivalent to Newton's First Law. This requires two steps: (1) to show that the Principle of Virtual Work is a necessary consequence of the First Law (necessity); and (2) to show that the First Law is a consequence of the Principle of Virtual Work (sufficiency). These will both be given in detail for the present situation. Later cases will show necessity only; the sufficiency arguments are similar to but more involved than the one given here.

To show that the Principle of Virtual Work is a necessary consequence of Newton's First Law for the case of a single particle, consider the conditions for equilibrium according to the First Law:

$$\sum_i \mathbf{F}_i = \mathbf{0}$$

The particle is located by the position vector $\mathbf{r}$, and if given a virtual displacement $\delta\mathbf{r}$, the virtual work is

$$\delta W = \left(\sum_i \mathbf{F}_i\right) \cdot \delta\mathbf{r}$$

Because one factor in the dot product is zero, the product is zero. Thus, if the First Law is satisfied, the virtual work is necessarily zero.

To show that the Principle of Virtual Work is sufficient for equilibrium, the argument will begin with the vanishing of the virtual work and show that the First Law conditions are a consequence. The Principle of Virtual Work says that equilibrium exist provided the virtual work is zero. That is,

$$\delta W = \left(\sum_i \mathbf{F}_i\right) \cdot \delta\mathbf{r} = 0$$

This demonstration is by contradiction, so assume that

$$\sum_i \mathbf{F}_i = |\mathbf{F}|\mathbf{E}_f, \quad |\mathbf{F}| \neq 0$$

that is, assume that the sum of forces is non-zero. The quantity $\mathbf{E}_f$ is a unit vector in the direction of the resultant force, whereas $|\mathbf{F}|$ is the magnitude of the resultant. Now because $\delta\mathbf{r}$ is completely arbitrary, it could be that

$$\delta\mathbf{r} = |\delta\mathbf{r}|\mathbf{E}_f, \quad |\delta\mathbf{r}| \neq 0$$

in which case the virtual work is

$$\delta W = |\mathbf{F}|\ |\delta\mathbf{r}|\mathbf{E}_f \cdot \mathbf{E}_f = |\mathbf{F}|\ |\delta\mathbf{r}|$$

There is no way that this can be zero for arbitrary $|\delta\mathbf{r}|$, except to violate the assumption that $|\mathbf{F}|$ is non-zero. Thus, it must be concluded that the vanishing of the virtual work implies the vanishing of the resultant force on the particle, the condition for equilibrium. This completes the sufficiency argument for the single particle case.

Principle of Virtual Work for a Rigid Body

Next consider the extension of the Principle of Virtual Work to a rigid body. Figure 6.1 shows a rigid body typified by three particles, labeled ①, ②, and ③. For each of the particles there is a position vector, $\mathbf{r}_1$, $\mathbf{r}_2$, and $\mathbf{r}_3$. On particle 1 there are three forces shown: $\mathbf{F}_1$ is an external force applied to the body at the location of this particle, whereas $\mathbf{f}_{12}$ and $\mathbf{f}_{13}$ are internal forces required to enforce the rigid body condition. Similar forces act on the other particles. If the body is in equilibrium, then each particle is in equilibrium and the First Law requires that

$$\mathbf{F}_1 + \mathbf{f}_{12} + \mathbf{f}_{13} = \mathbf{0}$$

$$\mathbf{F}_2 + \mathbf{f}_{21} + \mathbf{f}_{23} = \mathbf{0}$$

$$\mathbf{F}_3 + \mathbf{f}_{31} + \mathbf{f}_{32} = \mathbf{0}$$

Because each force sum is zero, the sum of the virtual work of these force sums is also zero:

$$\delta\mathbf{r}_1 \cdot (\mathbf{F}_1 + \mathbf{f}_{12} + \mathbf{f}_{13}) + \delta\mathbf{r}_2 \cdot (\mathbf{F}_2 + \mathbf{f}_{21} + \mathbf{f}_{23}) + \delta\mathbf{r}_3 \cdot (\mathbf{F}_3 + \mathbf{f}_{31} + \mathbf{f}_{32}) = 0$$

Now look at a pair of the terms in this summation, and recall that the internal forces must be equal and opposite to each other, $\mathbf{f}_{ij} = -\mathbf{f}_{ji}$

$$\begin{aligned}\delta\mathbf{r}_1 \cdot \mathbf{f}_{12} + \delta\mathbf{r}_2 \cdot \mathbf{f}_{21} &= \delta\mathbf{r}_1 \cdot \mathbf{f}_{12} - \delta\mathbf{r}_2 \cdot \mathbf{f}_{12} \\ &= (\delta\mathbf{r}_1 - \delta\mathbf{r}_2) \cdot \mathbf{f}_{12}\end{aligned}$$

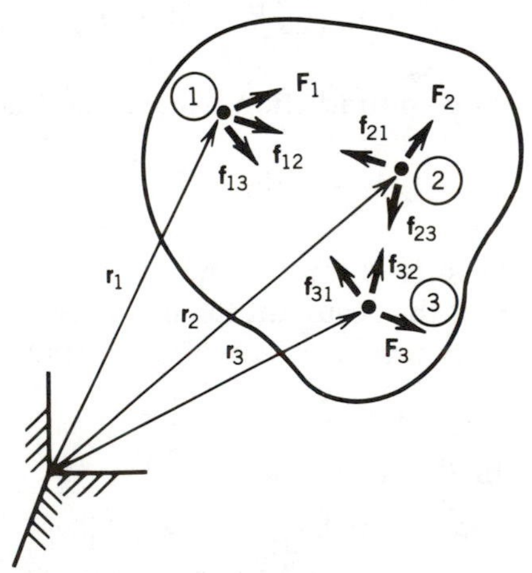

FIGURE 6.1 A Single Rigid Body Represented by Three Typical Particles

Let $\mathbf{E}_{12}$ denote a unit vector directed from particle 1 toward particle 2, and use $\mathbf{E}_{12}$ to express the force $\mathbf{f}_{12}$:

$$\mathbf{E}_{12} = \frac{\mathbf{r}_2 - \mathbf{r}_1}{|\mathbf{r}_2 - \mathbf{r}_1|}$$

$$\mathbf{f}_{12} = |\mathbf{f}_{12}|\mathbf{E}_{12}$$

Then, the difference in virtual displacements appearing in the virtual work of the internal forces is

$$\begin{aligned}\delta\mathbf{r}_1 - \delta\mathbf{r}_2 &= \delta(\mathbf{r}_1 - \mathbf{r}_2)\\ &= \delta(|\mathbf{r}_1 - \mathbf{r}_2|\mathbf{E}_{12})\\ &= \mathbf{E}_{12}\,\delta|\mathbf{r}_1 - \mathbf{r}_2| + |\mathbf{r}_1 - \mathbf{r}_2|\,\delta\mathbf{E}_{12}\end{aligned}$$

Because the body is rigid, $|\mathbf{r}_1 - \mathbf{r}_2|$ is constant and $\delta|\mathbf{r}_1 - \mathbf{r}_2|$ must be zero. Any change in the unit vector $\mathbf{E}_{12}$ must be normal to $\mathbf{E}_{12}$, so that $\mathbf{E}_{12} \cdot \delta\mathbf{E}_{12}$ must also be zero. The two terms of the virtual work considered are

$$\begin{aligned}&\delta\mathbf{r}_1 \cdot \mathbf{f}_{12} + \delta\mathbf{r}_2 \cdot \mathbf{f}_{21}\\ &\quad= (\mathbf{E}_{21}\,\delta|\mathbf{r}_1 - \mathbf{r}_2| + |\mathbf{r}_1 - \mathbf{r}_2|\,\delta\mathbf{E}_{12}) \cdot |\mathbf{f}_{12}|\mathbf{E}_{12}\\ &\quad= 0\end{aligned}$$

In a similar manner, the virtual work of each of the other internal force pairs can be shown to be zero. This reduces the original virtual work sum to

$$\delta\mathbf{r}_1 \cdot \mathbf{F}_1 + \delta\mathbf{r}_2 \cdot \mathbf{F}_2 + \delta\mathbf{r}_3 \cdot \mathbf{F}_3 = 0$$

The conclusion is that equilibrium of the rigid body requires that the virtual work of the *external forces* be zero. This, then, is the Principle of Virtual Work for a Rigid Body. Note that in forming the sum, the virtual displacements are subject to the constraint that the particles must move as parts of a rigid body.

Principle of Virtual Work for a System of Rigid Bodies

Before attempting to extend the Principle of Virtual Work to multiple rigid bodies, it is necessary to point out one of the key features in the preceding discussion for a single rigid body. The rigid body assumption prevents the absorption of any work by the system, either by elastic deformation or by friction with eventual conversion to heat. This avoidance of energy absorption by the system is a key assumption in dealing with systems of rigid bodies. This does not, however, prevent the inclusion of friction forces external to the system, but it does preclude friction at the contact points between the bodies. A system of rigid bodies for which there is no energy absorption at the points of interconnection is called an *ideal system*.

An ideal system of two pin-jointed rigid bodies is shown in Figure 6.2(a); free-body diagrams for the two bodies are shown in Figure 6.2(b).

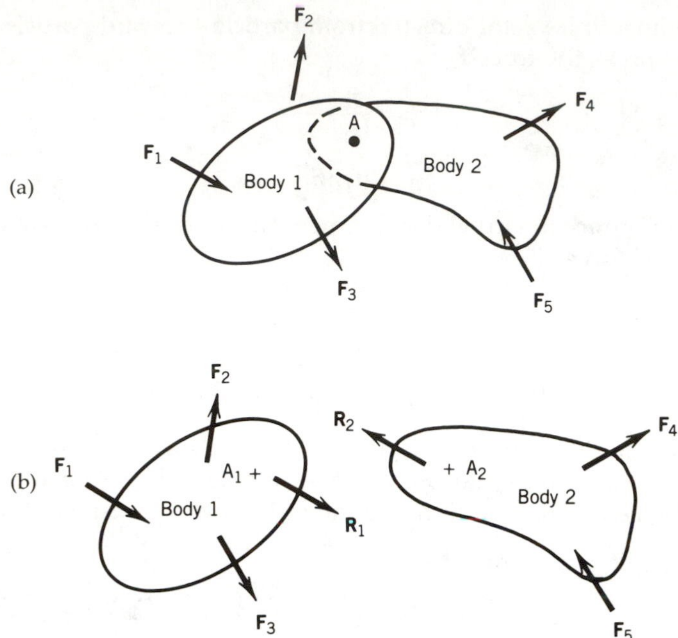

FIGURE 6.2 (a) An Ideal System of Rigid Bodies
(b) Freebody Diagrams for the Two Rigid Bodies

Note that the reactions forces, $\mathbf{R}_1$ and $\mathbf{R}_2$, are unknown in both direction and magnitude. However, it is known that they are equal and opposite, $\mathbf{R}_1 = -\mathbf{R}_2$. The equilibrium condition in terms of virtual work for each rigid body, taken individually, has already been established. It is as follows:

$$\text{Body 1:} \quad \delta\mathbf{r}_1 \cdot \mathbf{F}_1 + \delta\mathbf{r}_2 \cdot \mathbf{F}_2 + \delta\mathbf{r}_3 \cdot \mathbf{F}_3 + \delta\mathbf{r}_{A1} \cdot \mathbf{R}_1 = 0$$

$$\text{Body 2:} \quad \delta\mathbf{r}_4 \cdot \mathbf{F}_4 + \delta\mathbf{r}_5 \cdot \mathbf{F}_5 + \delta\mathbf{r}_{A2} \cdot \mathbf{R}_2 = 0$$

The two bodies are to remain connected at A, so there is a constraint on the virtual displacements $\delta\mathbf{r}_{A1}$ and $\delta\mathbf{r}_{A2}$:

$$\delta\mathbf{r}_{A1} = \delta\mathbf{r}_{A2} = \delta\mathbf{r}_A$$

Adding the two virtual work expressions together and eliminating the internal force terms because of the relations between $\mathbf{R}_1$ and $\mathbf{R}_2$ and between $\delta\mathbf{r}_{A1}$ and $\delta\mathbf{r}_{A2}$ results in

$$\delta\mathbf{r}_1 \cdot \mathbf{F}_1 + \delta\mathbf{r}_2 \cdot \mathbf{F}_2 + \cdot \cdot \cdot + \delta\mathbf{r}_5 \cdot \mathbf{F}_5 = 0$$

The Principle of Virtual Work for an Ideal System of Rigid Bodies says that the system will be in equilibrium provided the virtual work of the *forces external to the system* is zero. Note that the forces $\mathbf{R}_1$ and $\mathbf{R}_2$ are internal to the system and, therefore, are not included in the statement of the Principle of Virtual Work. Similarly, reactions at stationary supports are external forces, but they do no virtual work and, thus, are not required in the application of the Principle of Virtual Work. The fact that these two types of terms are not required is one of the major advantages of the Principle of Virtual Work. This development shows why the statement is limited to ideal systems—if non-zero work could be done at the point of interconnection during a virtual displacement, then other terms would need to be included in the virtual work expression. A simple example will help to clarify the ideas presented above and to demonstrate their application.

6.2.1 Crank–Lever Mechanism

Consider the crank–lever mechanism shown in Figure 6.3. The external loads applied to the system that can do virtual work include the vertical force F and the moment M acting on the crank. The system is in equilibrium under these loads. Determine the relation between F, M, and the angle q in the equilibrium condition.

According to the Principle of Virtual Work, the equilibrium relation is described by

$$\delta W = M\,\delta q - F\,\delta Y = 0$$

Notice that the work of the moment is positive because M and q are in the same sense. Similarly, the work of the force F has a negative sign because F and the coordinate Y are in opposite senses. This is a single degree of freedom mechanism, and it is convenient to associate that degree of freedom with the angle q. There must then be a kinematic relation expressing Y in terms of q.

The position loop equations are

$$X\cos A - C - R\cos q = 0$$

$$X\sin A - R\sin q = 0$$

Eliminating X allows A to be determined from

$$\tan A = \frac{R\sin q}{C + R\cos q}$$

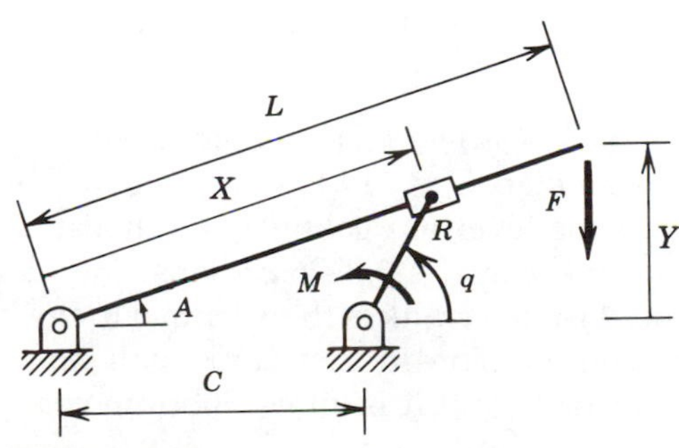

FIGURE 6.3 Crank-Lever Mechanism

The coordinate Y is then related to A (and thus indirectly to q) by

$$Y = L \sin A$$

A virtual change in Y is required for the virtual work expression, and this will necessitate an expression for the virtual change in A expressed in terms of δq. These are obtained by differentiation:

$$\delta A = \frac{dA}{dq}\,\delta q = K_a\,\delta q$$

$$= \frac{\cos^2 A(C\,R\cos q + R^2)}{(C + R\cos q)^2}\,\delta q$$

$$\delta Y = \frac{dY}{dA}\,\delta A = L\cos A\,\delta A$$

$$= \frac{L\cos^3 A(C\,R\cos q + R^2)}{(C + R\cos q)^2}\,\delta q$$

Note that the coefficient of δq is actually the velocity coefficient, $K_y = dY/dq$. Using the previous expression for the virtual displacement δY, the virtual work expression is

$$\delta W = M\,\delta q - F\,\frac{L\cos^3 A(C\,R\cos q + R^2)}{(C + R\cos q)^2}\,\delta q$$

Due to the fact that this expression must be zero for equilibrium, and because δq is arbitrary and not necessarily zero, the coefficient of δq must be zero, with the following result:

$$M - F\,\frac{L\cos^3 A(C\,R\cos q + R^2)}{(C + R\cos q)^2} = 0$$

This last equation involves the variables M, F, A, and q; an earlier equation related A and q. If any two of the three variables M, F, or q are given, the third variable and angle A can be determined from these two equations.

One way to present the relation between M, F, and q in graphical form is to rearrange the equation to read

$$\frac{M}{FL} = f(q)$$

and then plot the ratio $M/(FL)$ versus q. A plot of this type is presented in Figure 6.4 for the parameters $C = 11$ in. and $R = 5$ in. This plot shows that when the crank and the lever are colinear, the maximum moment is required. When the crank angle is 2.0426 radians, the required moment is zero. This occurs because the crank is then normal to the lever and the load F is supported by compression in the crank with no moment required. (This position is not stable, but it is an equilibrium configuration.)

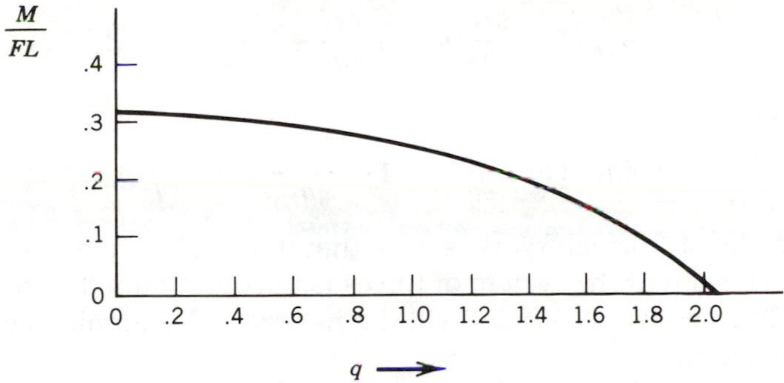

FIGURE 6.4 Moment to Load Ratio for Crank-Lever Mechanism

Principle of Virtual Work with Multiple Degrees of Freedom

In the preceding example, the virtual angular displacement, δq, was eventually displayed as one factor of a product that must be zero. The argument made was that δq is arbitrary, and that there could be no assurance that it is zero. On this basis, it was then concluded that the coefficient of δq must be zero because the relation was to hold for any δq. This simple reasoning was possible because of the fact that the system had only one degree of freedom. However, note that even this argument could not be made until δA and δY were expressed in terms of a virtual change in the single generalized coordinate, q. In the event that the system has multiple degrees of freedom, a more involved argument is required.

Now consider a system with N_1 degrees of freedom associated with the generalized coordinates $q_1, q_2, \ldots q_{N1}$. The position vectors for the load application points can be expressed as

$$\mathbf{r}_1 = \mathbf{r}_1(q_1, q_2, \ldots q_{N1})$$
$$\mathbf{r}_2 = \mathbf{r}_2(q_1, q_2, \ldots q_{N1})$$
$$\vdots \quad \vdots$$

For this system, the virtual work of the external loads will be

$$\begin{aligned}
\delta W &= \mathbf{F}_1 \cdot \delta\mathbf{r}_1 + \mathbf{F}_2 \cdot \delta\mathbf{r}_2 + \cdots \\
&= \mathbf{F}_1 \cdot \left(\frac{\partial \mathbf{r}_1}{\partial q_1}\delta q_1 + \frac{\partial \mathbf{r}_1}{\partial q_2}\delta q_2 + \cdots + \frac{\partial \mathbf{r}_1}{\partial q_{N1}}\delta q_{N1}\right) \\
&\quad + \mathbf{F}_2 \cdot \left(\frac{\partial \mathbf{r}_2}{\partial q_1}\delta q_1 + \frac{\partial \mathbf{r}_2}{\partial q_2}\delta q_2 + \cdots + \frac{\partial \mathbf{r}_2}{\partial q_{N1}}\delta q_{N1}\right) \\
&\quad + \cdots \\
&= \delta q_1 \left(\mathbf{F}_1 \cdot \frac{\partial \mathbf{r}_1}{\partial q_1} + \mathbf{F}_2 \cdot \frac{\partial \mathbf{r}_2}{\partial q_1} + \cdots\right)
\end{aligned}$$

$$+ \delta q_2 \left(\mathbf{F}_1 \cdot \frac{\partial \mathbf{r}_1}{\partial q_2} + \mathbf{F}_2 \cdot \frac{\partial \mathbf{r}_2}{\partial q_2} + \cdots \right)$$
$$+ \cdots$$
$$+ \delta q_{N1} \left(\mathbf{F}_1 \cdot \frac{\partial \mathbf{r}_1}{\partial q_{N1}} + \mathbf{F}_2 \cdot \frac{\partial \mathbf{r}_2}{\partial q_{N1}} + \cdots \right)$$

The Principle of Virtual Work states that the virtual work expression shown will be zero if the system of forces is in equilibrium, but this is only a single, scalar equation. Where will the necessary N_1 simultaneous equations come from?

The answer to this question comes from the definition of a generalized coordinate as discussed in Section 1.2. One of the characteristics of a generalized coordinate discussed there was that generalized coordinates are *independent.* Each generalized coordinate can be varied independently, having no effect on the others. In the current situation, this means that each of the δq_i is independent of all the others. The last form for the virtual work that was just shown is a sum of terms, each involving the virtual change in a generalized coordinate multiplied by a coefficient. Assume first that these coefficients are not all zero. The Principle of Virtual Work states that this sum is zero for arbitrary δq_i. This means that the virtual work sum is zero, *no matter how the* δq_i *are chosen.* A non-zero δq_1, multiplied by a non-zero coefficient, makes a non-zero contribution to the sum. Without much difficulty, it should be possible to choose δq_2 such that the second term will also make a non-zero contribution of the same sign as that of the first term, and so on. The eventual consequence of this reasoning is that the virtual work sum can be made non-zero by some choices of the variations. The only way to assure that the sum must be zero, no matter what choices are made for the δq_i, is for each of the coefficients to be zero. There are N_1 coefficients, so this provides the required N_1 equilibrium equations. In the abstract, this appears quite awkward, but in practice it is quite simple to apply. An example will demonstrate the application.

6.2.2 Summing Linkage

One very simple multidegree of freedom mechanism is the summing linkage. This linkage was the subject of one of the problems at the end of Chapter 3. This example will deal with the force analysis for that same linkage, and will draw on the kinematic results from that problem. Figure 6.5 shows the summing linkage in equilibrium under the applied loads P_1, P_2, and F. Determine the values of P_1 and P_2 in terms of F.

For this situation, the Principle of Virtual Work states

$$\delta W = P_1 \,\delta q_1 + P_2 \,\delta q_2 - F \,\delta X = 0$$

From the earlier problem solution, q_1 and q_2 were chosen as generalized coordinates associated with the two degrees of freedom. The solution to

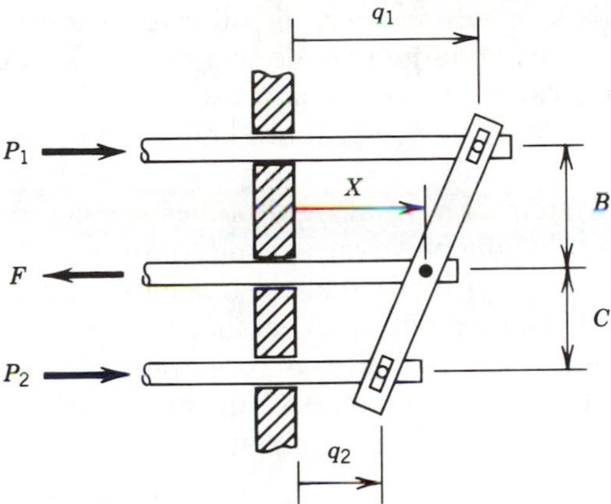

FIGURE 6.5 Summing Linkage in Equilibrium Under Forces P_1, P_2, and F

the loop equations provided the relation

$$X = \frac{C\,q_1 + B\,q_2}{B + C}$$

from which the virtual displacement δX is

$$\delta X = \frac{C}{B + C}\,\delta q_1 + \frac{B}{B + C}\,\delta q_2$$

The virtual work of the external forces, expressed in terms of virtual displacements in the generalized coordinates, is

$$\delta W = \left(P_1 - F\frac{C}{B + C}\right)\delta q_1 + \left(P_2 - F\frac{B}{B + C}\right)\delta q_2$$

According to the preceding argument, the coefficients of δq_1 and δq_2 must each be zero. From this the forces P_1 and P_2 are determined:

$$P_1 = F\frac{C}{B + C} \qquad P_2 = F\frac{B}{B + C}$$

These results are readily verified by using moments about the top and bottom pin joints. As is seen in the verification, using moment sums for this simple problem is probably a quicker way to the result. The purpose here, however, has been to demonstrate the method—a method of great power in complicated problems.

Potential Energy Representation of Conservative Forces

In extending the Principle of Virtual Work to systems of rigid bodies, care was taken to exclude systems that could absorb energy in a virtual displacement. It is now useful to relax that restriction, allowing the inclusion of interconnections in which there is fully recoverable energy storage. Therefore, conservative energy storage will now be included, but nonconservative energy absorption will remain excluded. Two obvious examples of effects to be included are springs (both linear and nonlinear) and gravitational forces.

Consider a system of several rigid bodies subject to external loads F_i (which are inherently nonconservative) and to conservative, working internal loads f_j. Any internal loads that do not move during the virtual displacement are called *nonworking* internal loads. An example of such a nonworking load is the reaction at a fixed support; it cannot move and, thus, does no work. When the virtual work of the forces acting on each body are summed, the virtual work of this force system is

$$\delta W = \mathbf{F}_1 \cdot \delta \mathbf{r}_1 + \mathbf{F}_2 \cdot \delta \mathbf{r}_2 + \cdots + \mathbf{f}_1 \cdot \delta \mathbf{R}_1 + \mathbf{f}_2 \cdot \delta \mathbf{R}_2 + \cdots$$

where $\mathbf{r}_i$ are the position vectors for the points of application of the external loads and $\mathbf{R}_j$ are the position vectors for the points of application of the working, conservative internal forces. For every conservative force there exists a potential function, V, such that

$$f_j = -\nabla V_j$$

where the del operator, denoted by the inverted capital Greek delta, indicates the following operation in rectangular cartesian coordinates:

$$\nabla = \mathbf{i}\frac{\partial}{\partial X} + \mathbf{j}\frac{\partial}{\partial Y} + \mathbf{k}\frac{\partial}{\partial Z}$$

Note that this is just an operator; it requires an operand on which to operate. Note also that the form given for the del operator here is specifically for rectangular cartesian coordinates; in other coordinate systems the del operator takes different forms. In terms of the potential function, the virtual work of one of the conservative, working internal forces is

$$\begin{aligned}\delta W_j &= (-\nabla V_j) \cdot \delta \mathbf{R}_j \\ &= -\left(\mathbf{i}\frac{\partial V_j}{\partial X} + \mathbf{j}\frac{\partial V_j}{\partial Y} + \mathbf{k}\frac{\partial V_j}{\partial Z}\right) \cdot (\mathbf{i}\,\delta X + \mathbf{j}\,\delta Y + \mathbf{k}\,\delta Z) \\ &= -\delta V_j\end{aligned}$$

When the virtual work sum is rewritten, there are two types of terms to be included:

$$\delta W = \sum_i \mathbf{F}_i \cdot \delta \mathbf{r}_i - \sum_j \delta V_j$$

The Principle of Virtual Work, modified to allow for the inclusion of some forces in terms of potential functions, requires that the preceding expression vanish for equilibrium.

The potential functions just described are nothing other than the familiar potential energy functions. Thus, if a gravitational force is to be represented, with Z positive along the upward vertical, the potential function is

$$V = MgZ$$

where M is the mass of the body and g is the gravitational constant. The variation of this potential is

$$\delta V = M\, g\, \delta Z$$

The case of a spring is slightly more complicated. Consider an unstretched spring oriented along the X–axis with free length S_0. The spring constant is denoted K. In the equilibrium configuration, the ends of the spring are at X_1 and X_2, respectively, and there is a tensile force F in the spring:

$$F = K(X_2 - X_1 - S_0)$$

Now consider small displacements of both ends from their equilibrium positions, δX_2 and δX_1. Because these displacements are small, the force in the spring is approximately constant during the displacements. The work done on the rigid bodies during these displacements consists of the sum of the works done at the ends:

$$\begin{aligned} &-F\, \delta X_2 + F\, \delta X_1 \\ &= -K\,(X_2 - X_1 - S_0)\,(\delta X_2 - \delta X_1) \\ &= -\delta[0.5\, K\,(X_2 - X_1 - S_0)^2] \\ &= -\delta[0.5\, K\,(\text{deformation})^2] \end{aligned}$$

It is evident that the negative variation of the potential energy function for the spring provides the correct terms for the virtual work done on the rigid bodies. These ideas are illustrated in the following example problem.

6.2.3 Spring Supported Lever

The spring supported lever shown in Figure 6.6 supports the weight $W = 15$ lb. The spring is anchored at a distance $3C = 42$ in. above the pivot point, and the spring is attached to the lever at a distance $3C$. The spring rate is $K = 50$ lb/in., and the free length of the spring is $3C$. The full length of the lever is $4C = 56$ in., and the lever is weightless. There is an external moment $M = 35$ in.-lb acting to help support W. What is the equilibrium value for the angle A? What is the tension in the spring at equilibrium?

First consider the kinematic aspects of the problem. Let Y_w denote the elevation of the end of the rod:

$$Y_w = 4C \sin A$$

$$\delta Y_w = 4C\, \delta A \cos A$$

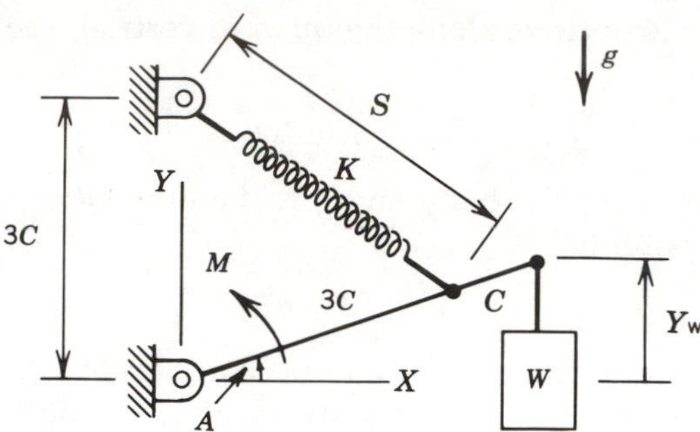

FIGURE 6.6 Spring Supported Lever Holding Weight W

The length of the spring is S, while the free length is denoted S_0.

$$S = \sqrt{(3C - 3C \sin A)^2 + (3C \cos A)^2}$$

$$= 3C\sqrt{2 - 2 \sin A}$$

$$\delta S = \frac{-9C^2 \cos A}{S}\,\delta A$$

With these geometric relations established, the potential energy terms are

$$V_w = W\,Y_w$$

$$\delta V_w = W\,\delta Y_w = 4C\,W\,\delta A \cos A$$

$$V_s = \frac{K}{2}(S - S_0)^2$$

$$\delta V_s = K(S - S_0)\,\delta S$$

$$= \frac{K(S - S_0)}{S}(-9C^2 \cos A)\,\delta A$$

The condition for equilibrium, expressed in terms of virtual work is

$$\delta W = M\,\delta A - \delta V_w - \delta V_s = 0$$

After substitutions and minor rearrangement of terms, the equilibrium equation is

$$M - 4C\,W \cos A + 9C^2\,K \cos A \left[1 - \frac{1}{\sqrt{2 - 2 \sin A}}\right] = 0$$

to be solved for A. Using the parameter values just given, an iterative solution gives $A = 0.51302$ radians. With this known, it is relatively straightforward to determine the equilibrium length of the spring, S, and then the spring tension, $T_s = K(S - S_0) = 19.2178$ lb. Note that both gravitational and spring potential energy are demonstrated in this problem. There will be further examples of the use of potential energy terms in the applications section.

6.3 APPLICATIONS OF THE PRINCIPLE OF VIRTUAL WORK

In this section, two more applications of the Principle of Virtual Work are given. For each case, it is well to consider the alternative approach, specifically the application of force and moment sums to the same problems. The power of the Principle of Virtual Work becomes more evident with the increasing complexity of the application.

6.3.1 Chain Drive Phase-Shifting Device

The mechanism shown in Figure 6.7 is a chain drive with provisions for adjusting the phase relation between the driving motion, q_1, and the output motion, A. The phase adjustment is controlled by the second primary variable, q_2. For this example, consider the system to be in equilibrium under the external force P_2 (holding the control block stationary), the driving moment M_1, and the reaction moment M_a. The problem is to determine the relations between M_a, M_1, and P_2. Note that the chain may be moving or not, but in either case the system speed is considered constant.

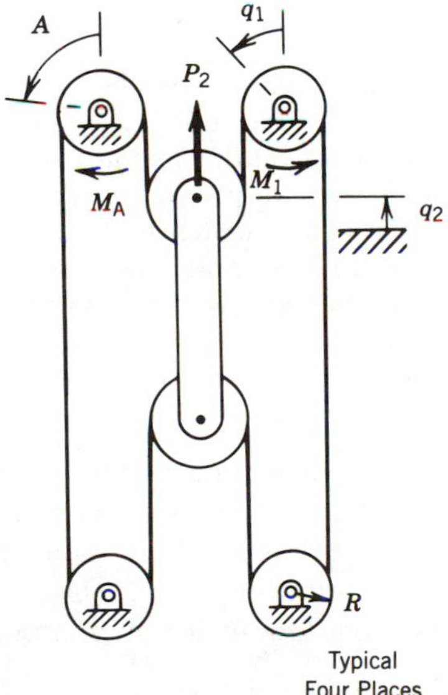

FIGURE 6.7 Chain Drive Phase Shifting Device in Equilibrium Under Moments M_a, q_1, and Force P_2

The kinematic relation between q_1, q_2, and A is

$$A = q_1 + 2q_2/R$$

which gives the following relation among the system virtual displacements:

$$\delta A = \delta q_1 + (2/R)\,\delta q_2$$

where R is the radius of each of the four corner sprockets with fixed centers. The condition for equilibrium is that the virtual work vanish:

$$\begin{aligned}\delta W &= M_1\,\delta q_1 + P_2\,\delta q_2 - M_a\,\delta A\\ &= M_1\,\delta q_1 + P_2\,\delta q_2 - M_a[\delta q_1 + (2/R)\,\delta q_2]\\ &= (M_1 - M_a)\,\delta q_1 + (P_2 - 2M_a/R)\,\delta q_2\\ &= 0\end{aligned}$$

By virtue of the fact that q_1 and q_2 are generalized coordinates, and thus are mutually independent, the coefficients for each virtual displacement must be zero to assure the vanishing of the virtual work for all possible virtual displacements. From those two statements come the final relations:

$$M_1 = M_a$$

$$P_2 = 2M_a/R$$

6.3.2 L-shaped Bracket

Consider a weightless L-shaped bracket with leg lengths B and C as shown in Figure 6.8(a). This bracket is constrained to roll on a stationary cylinder of radius R under the influence of a spring and an external force F (Figure 6.8(b)). When the legs of the bracket are parallel to the X– and Y–axes (shown in broken line), the bracket extends a distance a below the X–axis, and the spring force is zero. The lower end of the spring is anchored at the point (X_0, Y_0), the spring rate is K, and the free length is S_0. As the horizontal force F is increased, the point of contact between the bracket and the stationary cylinder moves through the angle q. Determine the equilibrium value for angle q and the corresponding tension in the spring.

This problem involves a single degree of freedom that is readily associated with the angle q. It is useful to define body coordinates U and V along the legs of the bracket. Any point on the bracket can then be located in terms of the body coordinates for that point, (U_p, V_p). From Figure 6.8(b), it is evident that the base coordinates for such a point are given by

$$\begin{Bmatrix} X_p \\ Y_p \end{Bmatrix} = \begin{bmatrix} \cos q & -\sin q \\ \sin q & \cos q \end{bmatrix} \begin{Bmatrix} R + U_p \\ V_p - a - Rq \end{Bmatrix}$$

$$= \begin{Bmatrix} (R + U_p) \cos q - (V_p - a - Rq) \sin q \\ (R + U_p) \sin q + (V_p - a - Rq) \cos q \end{Bmatrix}$$

$$\begin{Bmatrix} \delta X_p \\ \delta Y_p \end{Bmatrix} = \delta q \begin{Bmatrix} dX_p/dq \\ dY_p/dq \end{Bmatrix} = \delta q \begin{Bmatrix} K_{xp} \\ K_{yp} \end{Bmatrix}$$

$$= \delta q \begin{Bmatrix} -U_p \sin q + (a + Rq - V_p) \cos q \\ U_p \cos q + (a + Rq - V_p) \sin q \end{Bmatrix}$$

Note that the derivatives are, in fact, the velocity coefficients for the particular point with respect to the generalized coordinate q.

Let 1 and 2 denote the points of spring attachment and load application, respectively. The virtual work of the applied force will be simply $F\,\delta X_2 = F\,K_{x2}\,\delta q$, so no further analysis is required for this term. For the potential energy of the spring, it is necessary to first express the length of the spring as a function of q:

$$S(q) = \sqrt{(X_1 - X_0)^2 + (Y_1 - Y_0)^2}$$

$$\delta S = \frac{(X_1 - X_0)\,\delta X_1 + (Y_1 - Y_0)\,\delta Y_1}{S}$$

$$= \frac{(X_1 - X_0)\,K_{x1} + (Y_1 - Y_0)\,K_{y1}}{S}\,\delta q$$

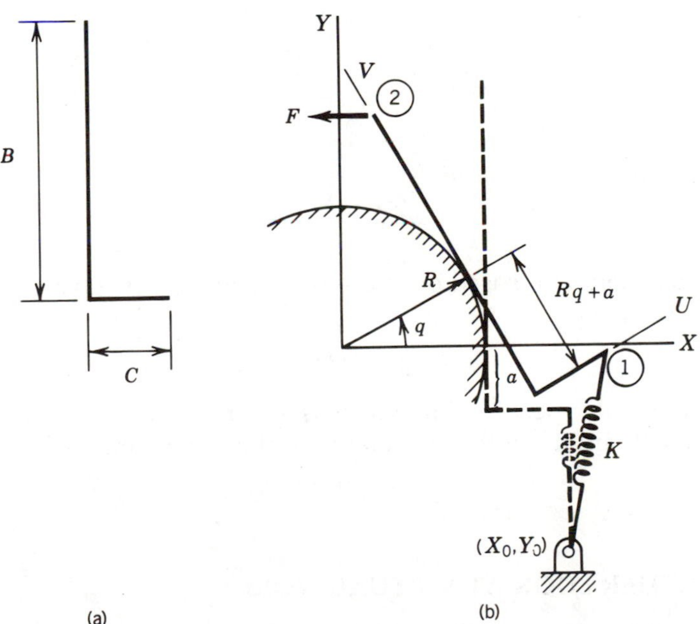

FIGURE 6.8 L-shaped Bracket in Equilibrium Under External Force F

The free length value, S_0, was given as part of the problem statement. The spring potential energy is V_s:

$$V_s = 0.5\,K\,(S - S_0)^2$$

$$\delta V_s = \frac{K(S - S_0)}{S}\,[(X_1 - X_0)\,K_{x1} + (Y_1 - Y_0)\,K_{y1}]\,\delta q$$

The virtual work expression, including the modification to allow for representation of conservative forces through the use of the potential energy, is

$$\delta W = -F\,K_{x2}\,\delta q - \frac{K(S - S_0)}{S}\,[(X_1 - X_0)\,K_{x1} + (Y_1 - Y_0)\,K_{y1}]\,\delta q$$
$$= 0$$

Because the virtual displacement δq is not necessarily zero, the coefficient must be zero and the equilibrium equation is

$$F\,K_{x2} + \frac{K(S - S_0)}{S}\,[(X_1 - X_0)\,K_{x1} + (Y_1 - Y_0)\,K_{y1}] = 0$$

This equation, along with the expressions for S, X_1, Y_1, K_{x1}, K_{y1}, and K_{x2}, must be solved for the equilibrium value of q. Because of the complexity of the equations involved, an iterative solution is indicated.

Numerical Values. For numerical work, take the following values for the system parameters:

$B = 14$ in. $\quad C = 3$ in.

$R = 5$ in. $\quad a = 2$ in.

$X_0 = 6$ in. $\quad Y_0 = -7$ in.

$K = 25$ lb/in. $\quad S_0 = 4$ in.

$F = 10$ lb

An iterative solution, based on these parameter values, gives the equilibrium position

$$q = 0.97233 \text{ radian}$$

With the angular position known, it is a simple matter to evaluate the spring length and from that to determine the spring tension T_s:

$$T_s = 165.176 \text{ lb}$$

6.4 ANOTHER LOOK AT VIRTUAL WORK

Earlier in this chapter, vector notation involving dot products was used extensively, even though the resulting virtual work expressions were scalars. It is useful to reexamine the principle of virtual work using matrix notation, as it offers a new perspective on that principle and on the meaning of the generalized force.

Consider a force $\mathbf{F}_i$ acting at the point $\mathbf{r}_i$ on a rigid body, where $\mathbf{r}_i$ is a function of the generalized coordinates. The virtual work of this force is then

$$\begin{aligned} \delta W &= \mathbf{F}_i \cdot \delta \mathbf{r}_i \\ &= (F_{ix}\,\mathbf{i} + F_{iy}\,\mathbf{j}) \cdot (\delta X_i\,\mathbf{i} + \delta Y_i\,\mathbf{j}) \\ &= F_{ix}\,\delta X_i + F_{iy}\,\delta Y_i \end{aligned}$$

The components of the virtual displacement are

$$\delta X_i = \sum_j \frac{\partial X_i}{\partial q_j}\,\delta q_j$$

$$\delta Y_i = \sum_j \frac{\partial Y_i}{\partial q_j}\,\delta q_j$$

These sums are readily indicated by the matrix products:

$$\begin{Bmatrix} \delta X_i \\ dY_i \end{Bmatrix} = \begin{bmatrix} K_{ix1} & K_{ix2} & K_{ix3} & \cdots \\ K_{iy1} & K_{iy2} & K_{iy3} & \cdots \end{bmatrix} \begin{Bmatrix} \delta q_1 \\ \delta q_2 \\ \vdots \end{Bmatrix}$$

where the Ks are the velocity coefficients for the particular point. For several forces applied at several different points, the notation is readily expanded:

$$\begin{Bmatrix} \delta X_1 \\ \delta Y_1 \\ \delta X_2 \\ \delta Y_2 \\ \vdots \end{Bmatrix} = \begin{bmatrix} K_{1x1} & K_{1x2} & K_{1x3} & \cdots \\ K_{1y1} & K_{1y2} & K_{1y3} & \cdots \\ K_{2x1} & K_{2x2} & K_{2x3} & \cdots \\ K_{2y1} & K_{2y2} & K_{2y3} & \cdots \\ \vdots & \vdots & \vdots & \end{bmatrix} \begin{Bmatrix} \delta q_1 \\ \delta q_2 \\ \delta q_3 \\ \vdots \end{Bmatrix}$$

For the virtual work of these several forces, matrix notation is again useful to represent the necessary summations:

$$\delta W = (F_{1x}, F_{1y}, F_{2x}, F_{2y}, \ldots) \begin{Bmatrix} \delta X_1 \\ \delta Y_1 \\ \delta X_2 \\ \vdots \end{Bmatrix}$$

$$= (F_{1x}, F_{1y}, F_{2x}, \ldots) \begin{bmatrix} K_{1x1} & K_{1x2} & K_{1x3} & \ldots \\ K_{1y1} & K_{1y2} & K_{1y3} & \ldots \\ \vdots & \vdots & \vdots & \end{bmatrix} \begin{Bmatrix} \delta q_1 \\ \delta q_2 \\ \vdots \end{Bmatrix}$$

$$= (Q_1, Q_2, Q_3, \ldots) \begin{Bmatrix} \delta q_1 \\ \delta q_2 \\ \delta q_3 \\ \vdots \end{Bmatrix}$$

where

$$\begin{Bmatrix} Q_1 \\ Q_2 \\ Q_3 \\ \vdots \end{Bmatrix} = [K]^t \begin{Bmatrix} F_{1x} \\ F_{1y} \\ F_{2x} \\ \vdots \end{Bmatrix}$$

This last product defines the generalized forces associated with each of the generalized coordinates. The usual terminology is to speak of the generalized force *conjugate* to a particular generalized coordinate because neither the generalized force nor the generalized coordinate has a direction in the usual sense. With this definition, the virtual work is written as $\{Q\}^t \{\delta q\}$. Because the generalized coordinates are independent and their variations are completely arbitrary, the vanishing of the virtual work requires the vanishing of each of the generalized forces. This is a parallel to Newton's First Law, but is a more general statement. It does, however, help to focus attention on the generalized forces as the cause of equilibrium or, in the dynamic situation, as the cause of nonequilibrium.

Velocity coefficients play a very important role in the definition of the generalized force. For a large, complicated problem, it may be very useful to evaluate the generalized forces using the matrix product as indicated. Alternatively for less complex problems, a less formal approach may be useful, but, in either event, the expression for the generalized force should be a sum of terms, with each term the product of a velocity coefficient with an actual force or moment.

6.5 CONCLUSION

The Principle of Virtual Work is a very powerful tool that has been frequently slighted in recent years. This approach makes possible omitting all internal forces and nonworking external constraint forces in the problem formulation, which is a great advantage in the area of machinery statics.

The generalized coordinate and velocity coefficient concepts were employed again in this chapter, as they will be in Chapters 7 and 8. The velocity coefficient concept facilitates the application of the Principle of Virtual Work because it neatly connects the virtual displacement at the point of application of a force to the virtual changes in the generalized coordinates. The idea of generalized force follows naturally from the expression of virtual work in terms of virtual changes in the generalized coordinates.

REFERENCES

Beer, F. P. and Johnston, E. R., *Vector Mechanics for Engineers, Statics and Dynamics*, 4th ed. New York: McGraw–Hill, 1984, ch. 10.

Meriam, J. L., *Engineering Mechanics—Statics and Dynamics, SI Version*. New York: John Wiley, 1980, Ch. 7.

Timoshenko, S., and Young, D. H., *Engineering Mechanics*, 4th ed. New York: McGraw–Hill, 1956, Ch. 5.

PROBLEM SET

All problems in this set are intended to be worked by the method of virtual work, in order to acquire practice with the technique. Do not consider any gravitational effects unless the gravity vector is shown in the figure.

6.1 The lever shown is in equilibrium under the action of the force F and the weight W. Find the angle q in terms of F, W, and the dimensions.

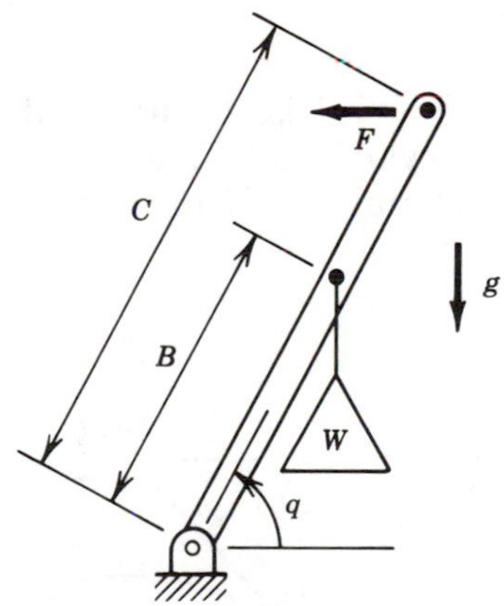

6.2 The force F and the effect of the weight W acting through the cord hold the lever in equilibrium in the position shown. There is no friction between the cord and the edge of the guide hole. Determine the equilibrium position of the lever.

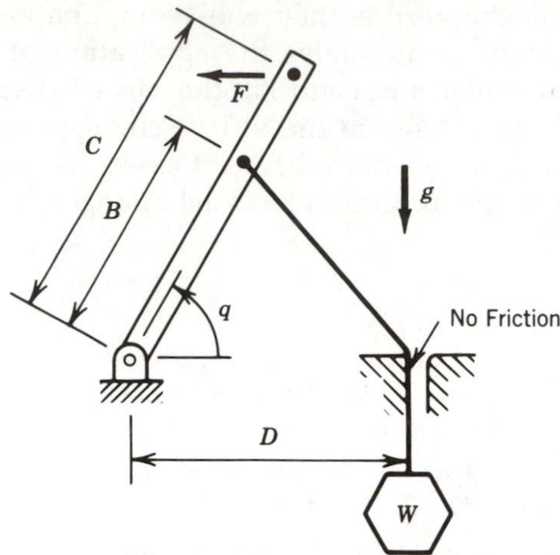

6.3 For the slider–crank mechanism shown, the connecting rod extends beyond the crank a distance H to a point where the horizontal force F_2 acts. The system is in equilibrium under the action of the forces F_1, F_2, and the couple C. Determine F_2 in terms of F_1, C, A, q, and the velocity coefficient K_a.

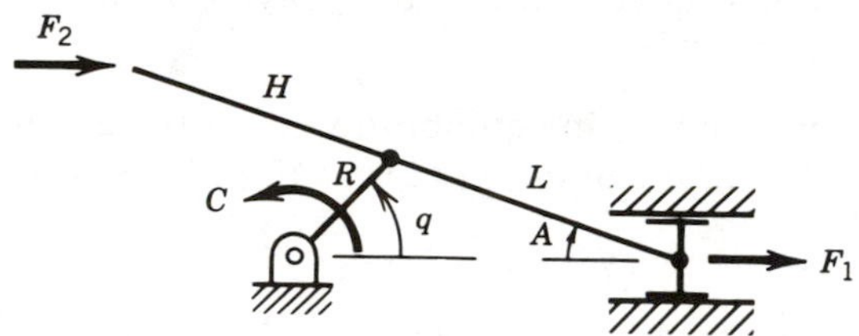

6.4 A slender rod of length L is attached to the collar at B and rests against a smooth, circular cylinder of radius R. The collar slides freely along the vertical guide. Determine the equilibrium value for F_2 in terms of F_1 and q.

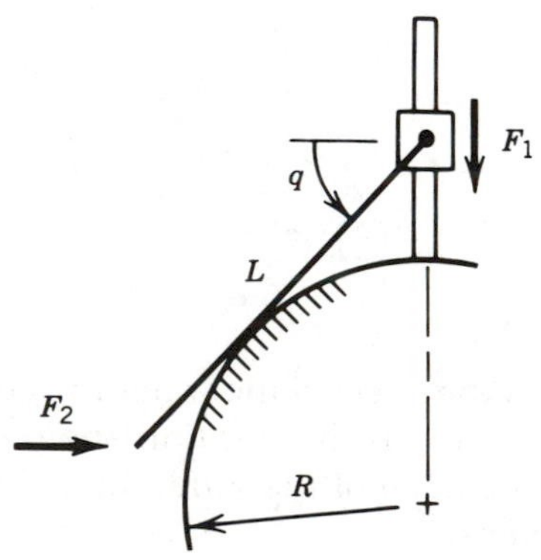

6.5 The two weights W_1 and W_2 move along a smooth, circular guide of radius R; they are separated by a weightless link of length $2B$. Find the equilibrium value for the angle q.

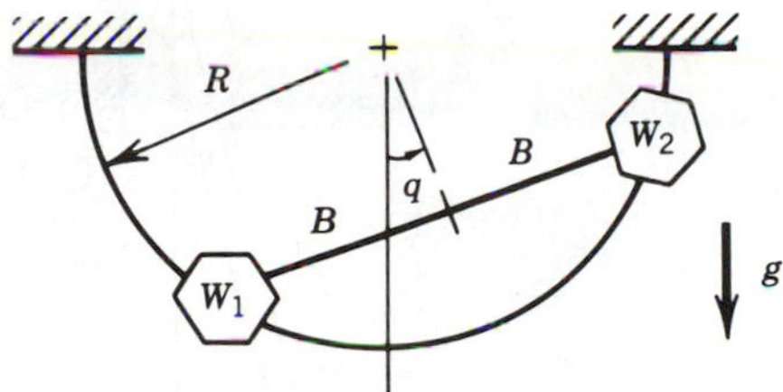

6.6 The articulated slider–crank mechanism shown is in equilibrium under the forces F_x and F_y.

a. Set up (but do not solve) the equations required for a complete position analysis;
b. Assuming that solutions are known for the equations of part a, develop the equations required to determine the velocity coefficients (do not solve);
c. Assuming that complete solutions have been determined for the positions and velocity coefficients, express the equilibrium value for F_x in terms of F_y.

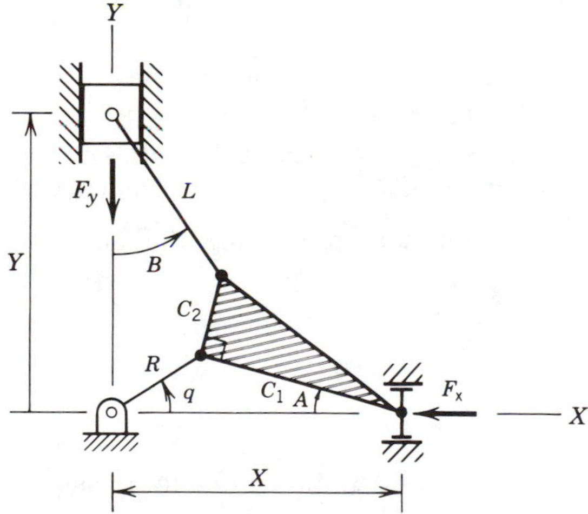

6.7 The mechanism shown is an air-actuated press that moves upward for the working stroke. The crank is defined by the three lengths S_1, S_2, and S_3. Air is supplied to the cylinder at a pressure P_0, and the effective piston area for the cylinder is A_0. The working force reaction on the ram is F_w.

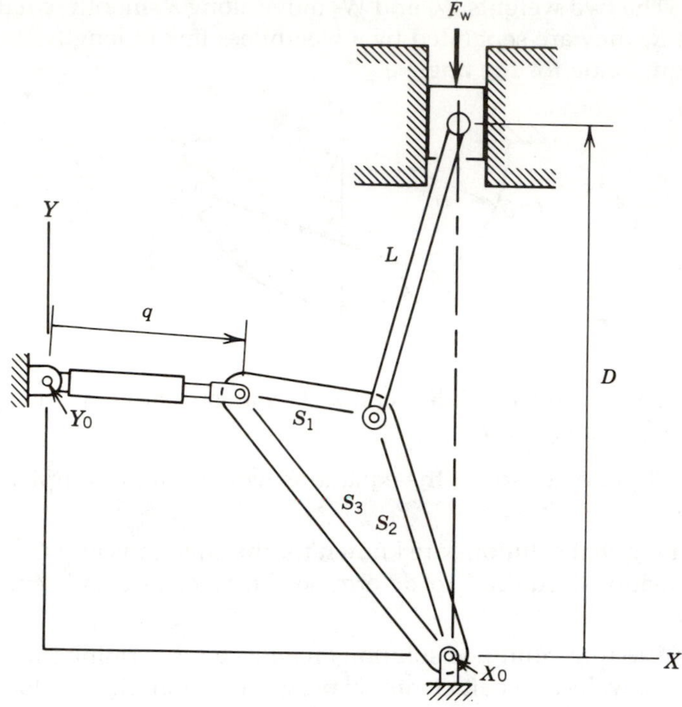

a. Set up the equations required for a complete position analysis (do not solve);

b. Assuming that the position solutions are known, develop the matrix equation required to determine the velocity coefficients (do not solve);

c. With positions and velocity coefficients known, express the static working force, F_w, in terms of P_0, A_0, q, and the dimensions.

6.8 For the mechanism of Problem 6.7, the following numerical data applies:

$S_1 = 10$ in. $\quad X_0 = 30$ in. $\quad P_0 = 80$ psig

$S_2 = 19$ in. $\quad Y_0 = 20$ in. $\quad A_0 = 4.22$ in.^{2}

$S_3 = 25$ in. $\quad L = 32$ in.

For q values ranging from 14.0 in. up to 19.0 in. in steps of 0.5 in.,

a. Obtain numerical solutions for the position equations and tabulate the ram position, D, versus q;

b. Obtain numerical values for the static equilibrium working force, F_w, and tabulate F_w versus q;

c. Plot the equilibrium working force versus ram position.

6.9 The figure shows the mechanism of a pendulum scale. The load to be measured is F, applied at the lower bar and transmitted upward through the loading tapes that wrap around the outer curve of the sector elements. The sector elements and the indicator rack are supported on the support tapes, wrapped around the inner curve of the sector elements, and attached at the top to the scale frame. The inner and outer curves for the sector element are described in polar coordinates by the functions $R_1(q)$ and $R_2(q)$; see the lower detail. When the load is applied, the sector elements rotate and swing the counterweights out. In addition to rotation, the sector elements must also rise, causing the rack to drive the pointer. The rotation axis of the pointer is mounted in stationary bearings. The load tapes and the support tapes are relatively long, and may be assumed to remain vertical at all times.

a. Working in terms of the functions $R_1(q)$, $R_2(q)$, the applied load F, the counterweight values $W/2$ each, and the length L, write the equations that describe the equilibrium position (do not solve);

b. Consider the functions $R_1(q)$ and $R_2(q)$ as constants, so that the sector element curves are concentric circular arcs. With this simplification, develop an expression that describes the position of the rack as a function of the applied load F; consider the rack position zero when q is zero.

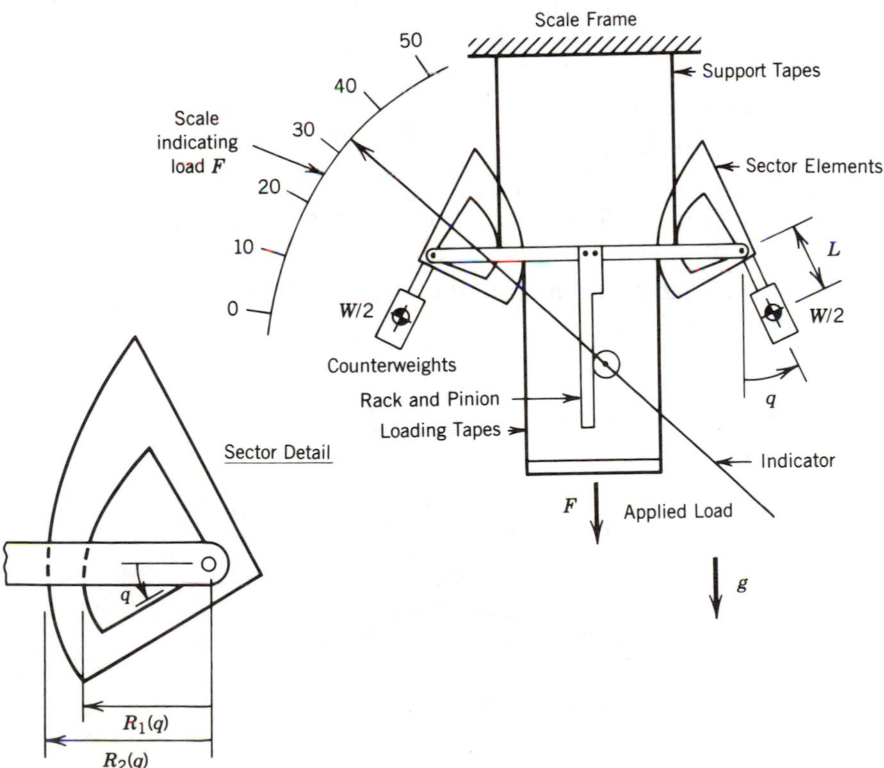

6.10 The spring–lever system is in equilibrium under the applied horizontal force F. The spring rate is K and the free length of the spring is L_0.

a. Determine the equilibrium value for A in terms of F, K, L_0, and the dimensions in closed form;

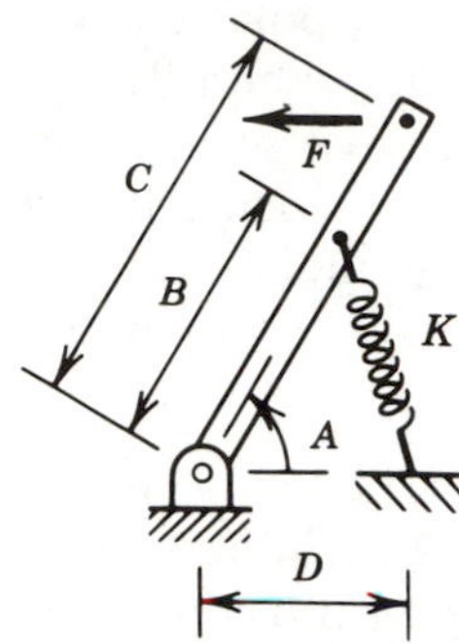

b. Using the data that follows part c, determine equilibrium values for A as F varies from 10 lb to 100 lb in 10-lb steps;

c. Plot A as a function of F, using the data of part b.

$K = 80$ lb/in. $\quad L_0 = 10$ in.

$B = 8.5$ in. $\quad C = 16$ in.

$D = 5.5$ in.

6.11 The spring–lever system shown is in equilibrium under the applied vertical load F. The spring rate is K, and the free length of the spring is L_0.

a. Determine the equilibrium equation, solvable for the equilibrium value of q (do not solve);

b. Obtain a numerical solution to the equation determined in part a, and use the following data.

$F = 22$ lb $\quad C = 15$ in.

$R = 10$ in. $\quad L_0 = 3$ in.

$K = 10$ lb/in.

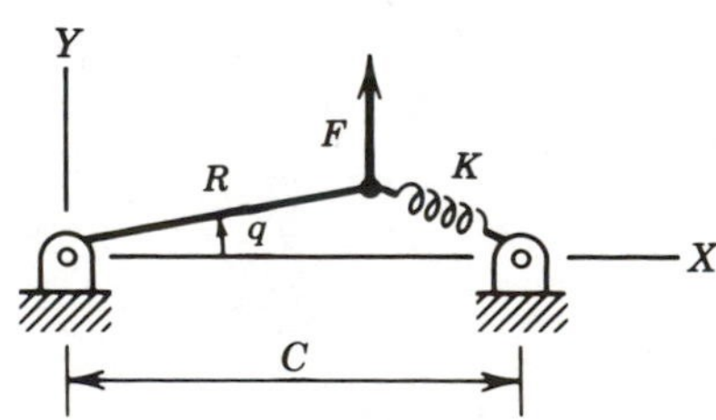

6.12 The slider–crank mechanism is in equilibrium under the action of the applied load F and the spring shown. The free length of the spring is $C/4$, and the spring rate is K.

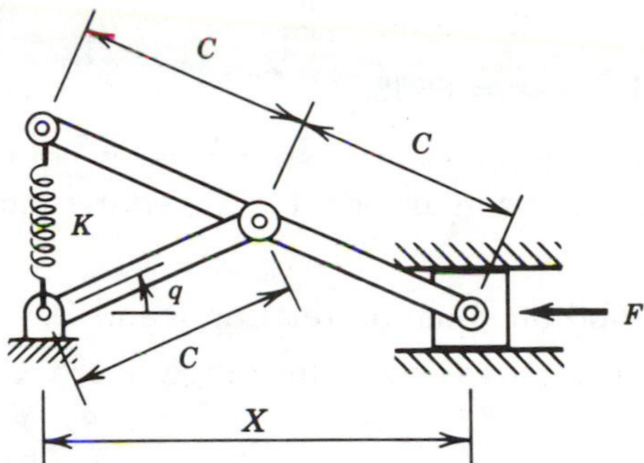

a. Determine the equilibrium equation, solvable for the equilibrium value of q;

b. Obtain two numerical solutions to the equation determined in part a, using the data that follows part c;

c. Explain why there are two solutions found in part b.

$F = 140$ lb $\qquad C = 23$ in. $\qquad K = 45$ lb/in.

6.13 The system shown is in equilibrium under the applied horizontal force F and the torsion spring acting between the two links. The spring rate is K, and the spring is relaxed when the angle between the two links is zero. Establish the equilibrium equation and the kinematic position equations that must be solved simultaneously to determine the equilibrium configuration (do not solve).

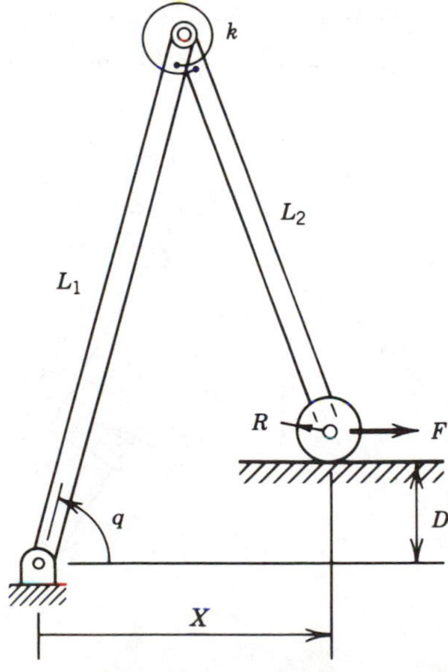

6.14 Using the following data for the system described in Problem 6.13, obtain a numerical solution for the equilibrium values of q and X.

$L_1 = 78$ in. $\quad L_2 = 70$ in.

$D = 26$ in. $\quad R = 3$ in.

$K = 150$ in.-lb/rad $\quad F = 85$ lb

6.15 The spring–lever system shown is in equilibrium under the action of the applied horizontal force F. The spring rate is K and the free length is L_0.

a. Obtain the equilibrium equation relating F and G;

b. Make a plot of F versus G over the range $\pi/4 < G < \pi/2$, using the following data.

$A = 14$ in. $\quad B = 3$ in.

$C = 10$ in. $\quad D = 5$ in.

$L_0 = 6$ in. $\quad K = 32$ lb/in.

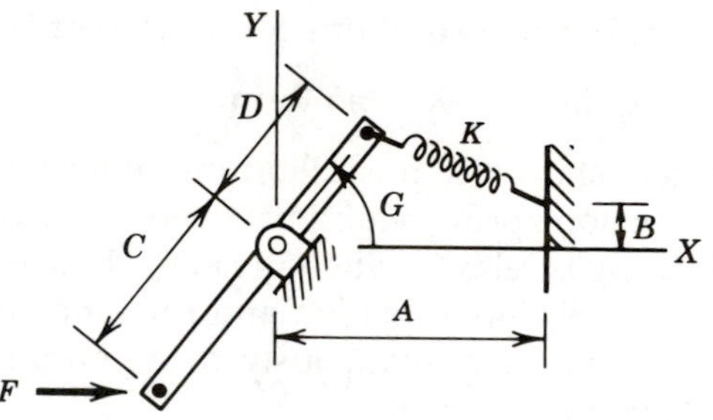

6.16 The right-angle bracket shown is pivoted at the vertex, and the two arms are massless. A weight W_2 is attached to the end of the left arm,

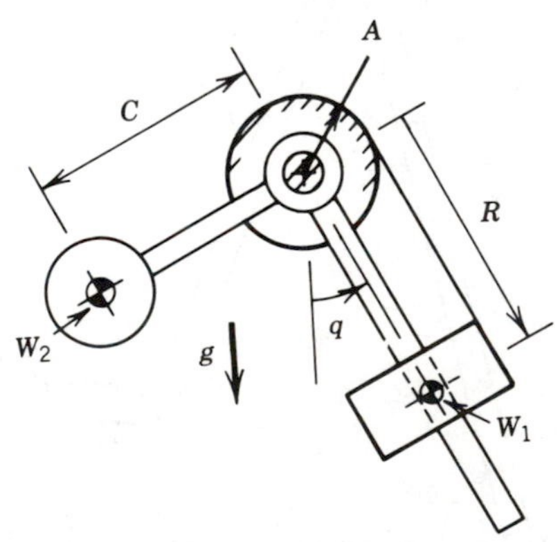

while a sliding weight W_1 moves freely on the right arm. The position of W_1 is controlled by a tape attached to W_1 and wrapped around a stationary circular drum. When $q = 0$, the position of W_1 is $R = C$. Determine the equilibrium equation for this system.

• **6.17** The mechanism shown is in equilibrium under the applied force F and the spring tension. The spring rate is K, and the free length of the spring is L_0.

a. Develop any kinematic relations needed in the equilibrium solution;
b. Determine the equilibrium equation, solvable with the kinematic relations for q;
c. Obtain a numerical solution for q, using the following data.

$R = 0.1$ m	$L = 0.35$ m	$C = 0.22$ m
$L_0 = 0.15$ m	$K = 40$ N/m	$F = 12$ N

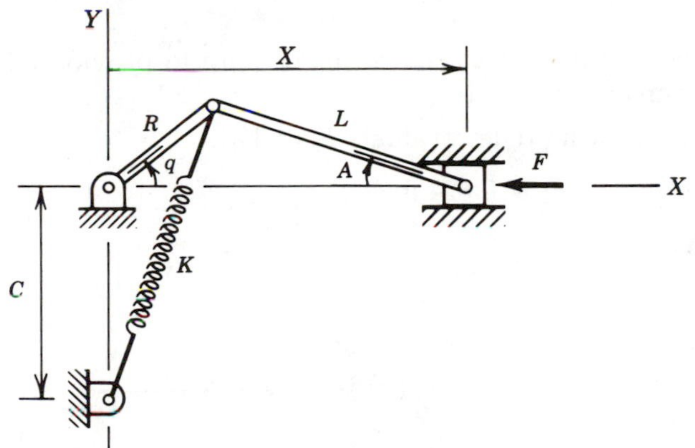

6.18 The figure shows a ball detent, a device that prevents relative horizontal motion between the two parts until a specified minimum force has been exceeded. When the ball is fully seated in the socket, the angular coordinate for the ball center is q_0 and the spring is compressed an amount Y_0. The spring rate is K. Do not consider any friction between the ball and the guide wall.

a. Determine q_0 in terms of R and B;
b. Determine the general relation between the applied force F and the position of the ball (described by the angle q) for $q_0 < q < \pi/2$;
c. Determine the minimum force required to begin motion, F_1;
d. Determine the minimum force required to fully disengage the ball from the socket, F_2.

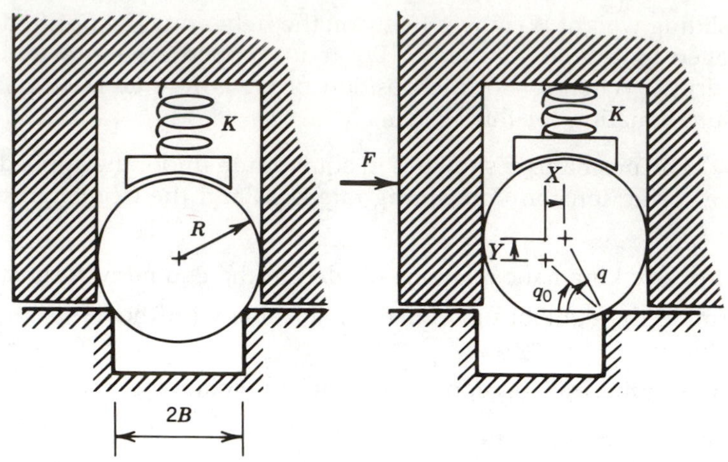

6.19 The mechanism shown is in equilibrium with F_1, F_2, and q_1 known.

a. Determine F_r and q_2 in closed form. Be sure to provide all necessary kinematic analysis;

b. Using the following data, evaluate F_r and q_2.

$F_1 = 0.875$ lb $\qquad F_2 = 0.674$ lb

$q_1 = 14.4$ in.

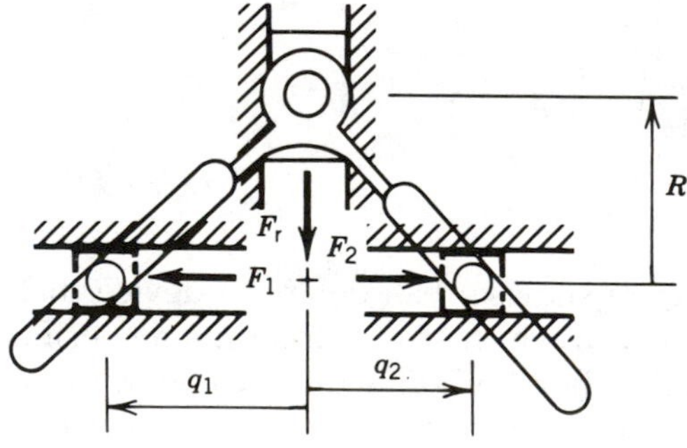

6.20 The mechanism shown is the "floating" slider–crank mechanism, considered previously (Problem 3.1). It is in equilibrium under the forces F_1, F_2, and the couple C. For given values of q_1 and q_2, express the equilibrium values of C and F_1 in terms of F_2 and the velocity coefficients K_a and K_x.

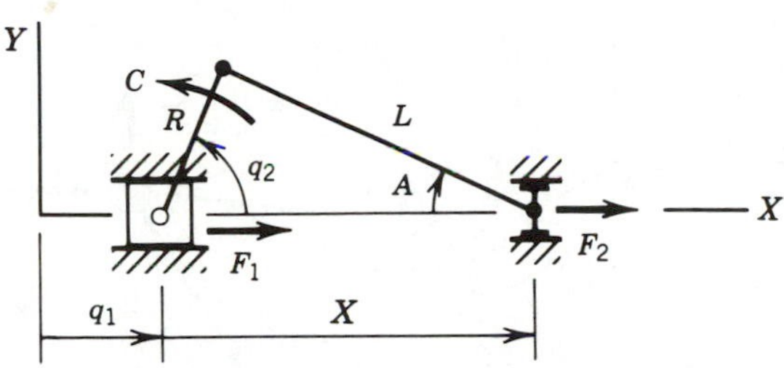

• **6.21** For the mechanism shown, assume that kinematic solutions are known for $A(q_1, q_2)$, $B(q_1, q_2)$, $K_{a1}(q_1, q_2)$, and $K_{a2}(q_1, q_2)$. The system is in equilibrium under the specified forces F_1, F_2, F_x, and F_y. Determine the equations of equilibrium, solvable for q_1 and q_2 if the forces are all specified (do not solve).

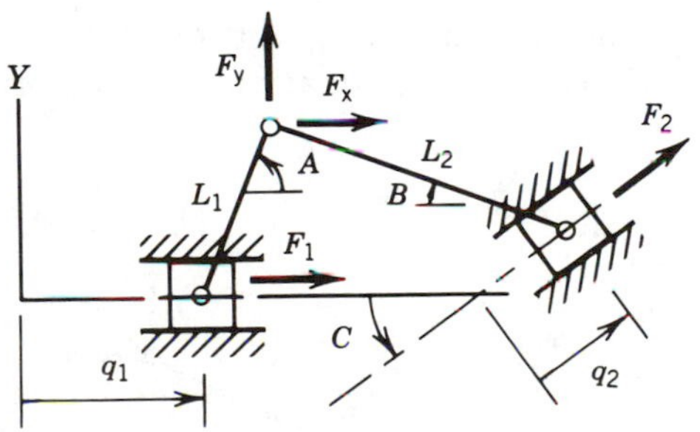

6.22 For the mechanism shown, kinematic solutions are known for $A(q_1, q_2)$, $B(q_1, q_2)$, $K_{a1}(q_1, q_2)$, $K_{a2}(q_1, q_2)$, $K_{b1}(q_1, q_2)$, and $K_{b2}(q_1, q_2)$. The system is in equilibrium under the specified forces F_1, F_2, F_x, and F_y. Determine the equations of equilibrium, solvable for q_1 and q_2 (do not solve).

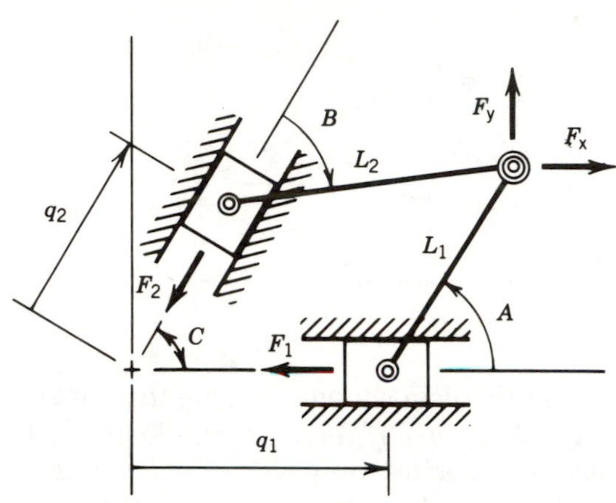

6.23 For the system shown, kinematic solutions have been previously determined for $A(q_1, q_2)$, $B(q_1, q_2)$, $K_{a1}(q_1, q_2)$, $K_{a2}(q_1, q_2)$, $K_{b1}(q_1, q_2)$, and $K_{b2}(q_1, q_2)$. The system is in equilibrium under the specified forces F_1, F_2, F_x, and F_y. Determine the equations of equilibrium solvable for q_1 and q_2 (do not solve).

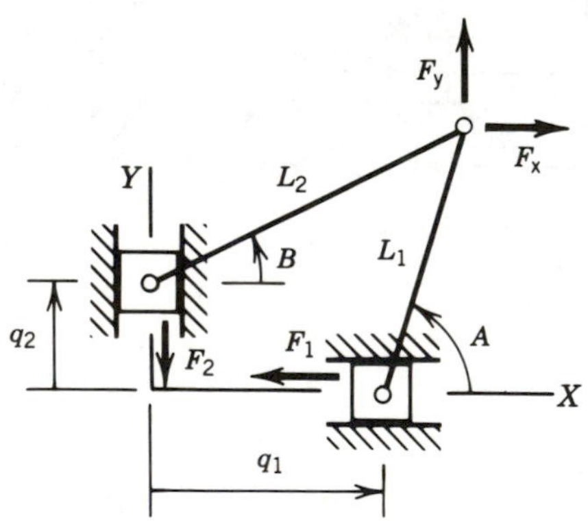

6.24 For the system, assume that kinematic solutions have been previously determined for $A_2(q_1, q_2, q_3)$, $A_3(q_1, q_2, q_3)$, and

$$[K] = \begin{bmatrix} K_{a21} & K_{a22} & K_{a23} \\ K_{a31} & K_{a32} & K_{a33} \end{bmatrix}$$

The system is in equilibrium under the specified forces F_1, F_2, F_3, and F_4. Determine the equilibrium equations solvable for q_1, q_2, and q_3 (do not solve).

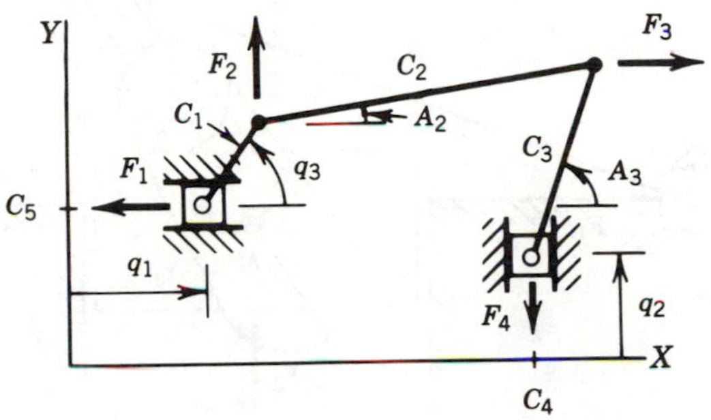

6.25 The slider–crank mechanism is used as a quasi-static compactor, with a working foot moved from place to place by the variable displacement q_2. For given values of q_1, q_2, and M_1, express the reaction force F_r and the required holding force F_2. Provide any required kinematic results.

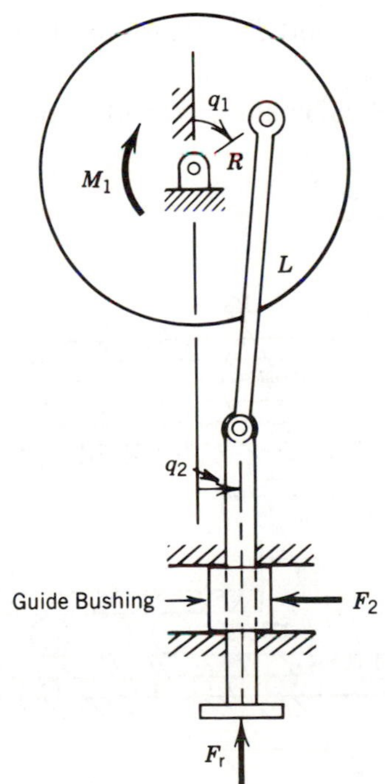

6.26 The circular cylinder of radius R_2 rolls on the surface of the table while a massive arm extends downward beyond the edge of the table. The upper extension of that arm is connected by a massless link to the slider at the far left. Determine the forces F_1 and F_2 required to hold the system in equilibrium for specified values of q_1 and q_2.

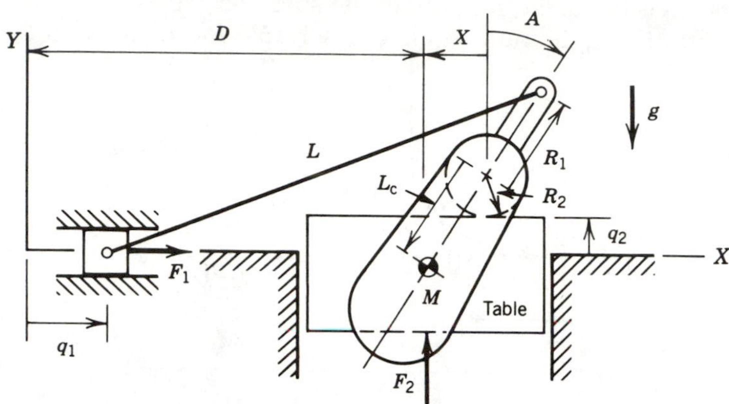

6.27 The two platforms shown are separated by rollers that do not slip on either contact surface. View (a) shows the springs in the system relaxed; view (b) shows the system under load. The larger rollers between the two platforms are of radius R_1, while the smaller rollers have radius R_2.

a. For given values of F_1 and F_2, determine the equilibrium equations, solvable for q_1 and q_2;
b. Using the following data, evaluate q_1 and q_2 for $F_1 = 17$ lb and $F_2 = 12$ lb.

$K_1 = 20$ lb/in. $\quad S_{10} = 4$ in.

$K_2 = 8$ lb/in. $\quad S_{20} = 6$ in.

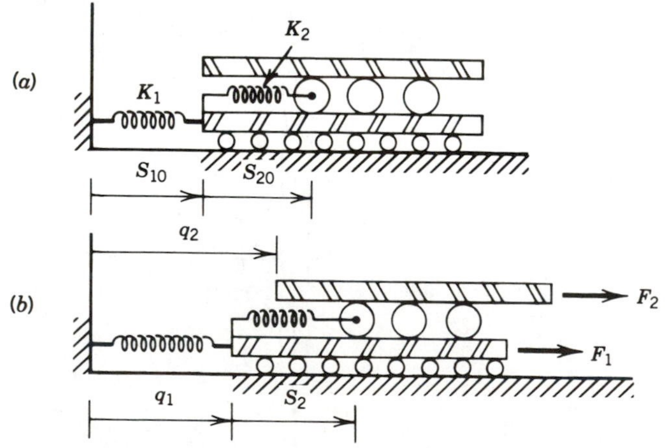

6.28 The system shown is in equilibrium under the force F and the applied torque T. The spring has rate K and free length such that it is relaxed when $q_1 = q_2 = 0$.

a. Set up the equations required for position analysis (do not solve);
b. Develop the equations required to determine the velocity coefficients;
c. Determine the equilibrium equations solvable for q_1 and q_2 for given values of F and T (do not solve).

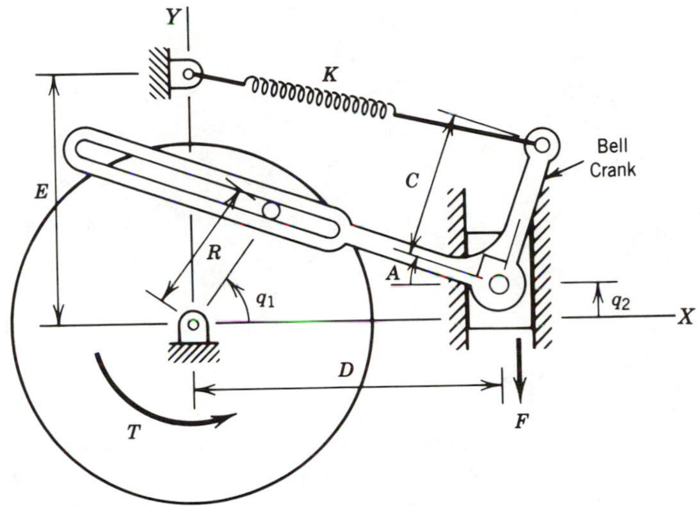

CHAPTER 7

Dynamics of Single Degree of Freedom Machines: Eksergian's Equation of Motion

Dynamics is the science of systems in motion, and it consists of kinematics and kinetics. This chapter will focus on the kinetics of systems having only a single degree of freedom. Although this is a somewhat special class, the large number of useful single degree of freedom mechanisms justify separate consideration. In Chapter 8, the kinetics of multiple degree of freedom systems will be considered. The significance of kinematics in the study of dynamics cannot be overestimated; the material of Chapters 2 and 3 is essential to the study of Chapters 7 and 8.

7.1 KINETIC ENERGY OF A SYSTEM OF RIGID BODIES

The kinetic energy of a single rigid body can be separated into two terms, one that depends on the velocity of the center of mass and the other that depends on the angular velocity of the body. For a review of this development, see Appendix A6. Mechanism problems involve multiple rigid body components, for which the total kinetic energy is simply the sum of the kinetic energies of the individual components. Thus, for such a system the kinetic energy may be written as

$$\begin{aligned} T = {} & 0.5\, M_1\{V_{1cm}\}^t\{V_{1cm}\} + 0.5\,\{\omega_1\}^t[I_{1cm}]\{\omega_1\} \\ & + 0.5\, M_2\{V_{2cm}\}^t\{V_{2cm}\} + 0.5\,\{\omega_2\}^t[I_{2cm}]\{\omega_2\} \\ & + \cdots \end{aligned}$$

where $\{V_{1cm}\}$, $\{V_{2cm}\}$, $\{\omega_1\}$, $\{\omega_2\}$, and so forth are velocity components measured in an inertial coordinate system. Now, let all of the center of mass velocity components for the various mechanism elements be assembled into a column vector. Similarly, assemble all of the angular velocity components for the several parts as a second column vector. The results are

$$\{V_{cm}\} = \text{Col}(V_{1x}, V_{1y}, V_{1z}, V_{2x}, V_{2y}, V_{2z}, V_{3x}, \ldots)$$
$$\{\omega\} = \text{Col}(\omega_{1x}, \omega_{1y}, \omega_{1z}, \omega_{2x}, \omega_{2y}, \omega_{2z}, \omega_{3x}, \ldots)$$

Then, with the appropriate definitions for the mass matrix $[M]$ and the moment of inertia matrix $[I_{cm}]$, the system total kinetic energy can be written as

$$T = 0.5\,\{V_{cm}\}^t[M]\{V_{cm}\} + 0.5\,\{\omega\}^t[I_{cm}]\{\omega\}$$

For a single degree of freedom system, the vectors $\{V_{cm}\}$ and $\{\omega\}$ can be written in terms of the generalized velocity, $\dot{q}$, and a vector of the appropriate velocity coefficients:

$$\{V_{cm}\} = \dot{q}\,\{K_v\}$$
$$\{\omega\} = \dot{q}\,\{K_\omega\}$$

With these definitions, the entire kinetic energy can be written in terms of $\dot{q}$:

$$T = 0.5\,\dot{q}^2\,(\{K_v\}^t[M]\{K_v\} + \{K_\omega\}^t[I_{cm}]\{K_\omega\})$$

By analogy with the expression for the kinetic energy of a single particle, the coefficient of $0.5\,\dot{q}^2$ is called the *generalized inertia,* a scalar quantity. Because the velocity coefficients are usually functions of the generalized coordinate, q, the generalized inertia is also a function of q. There is no universal notation for the generalized inertia; the notation $\mathcal{I}(q)$ or simply, $\mathcal{I}$, will be used here.

A word of caution is required at this point. There are single degree of freedom systems with base displacement excitations, say an $X_0(t)$, for which the velocity required to express the kinetic energy is of the form $(\dot{X}_o + \dot{q}K_v)$. In this case, the definition of the generalized inertia is not useful and the following discussion is not valid. The problem develops because the term $\dot{q}K_v$ describes a *relative* velocity, rather than a velocity measured with respect to an inertial coordinate system. All such systems are excluded from the discussions of this chapter; they can be handled by the methods to be discussed in Chapter 8.

To illustrate the idea of generalized inertia, consider the slider–crank mechanism shown in Figure 7.1 (for which the kinematics were studied in Section 2.2). Let M_2 and M_3 denote the masses of the connecting rod and the slider, respectively, while I_{1cm} and I_{2cm} are the mass moments of inertia for the crank and the connecting rod, each with respect to the component center of mass. The connecting rod center of mass is located by the body

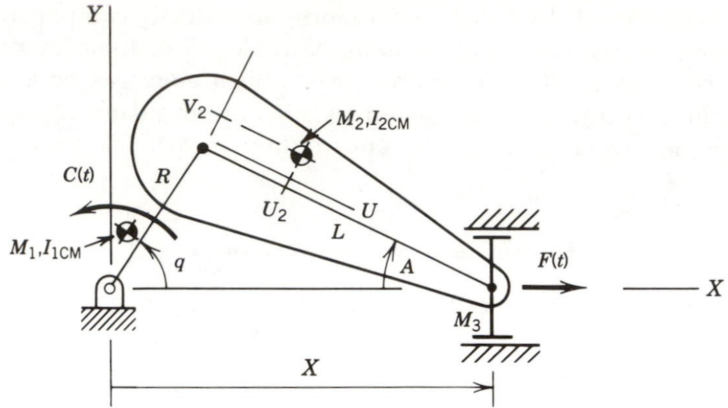

FIGURE 7.1 Slider-Crank Mechanism

coordinates (U_2, V_2) from which the base coordinates (X_2, Y_2) are determined. The kinetic energy of the assembly is

$$\begin{aligned}
T &= T_{\text{crank}} + T_{\text{con rod}} + T_{\text{slider}} \\
&= 0.5\, M_1 V_{1\text{cm}}^2 + 0.5\, I_{1\text{cm}}\, \dot{q}^2 \qquad \text{(Crank)} \\
&\quad + 0.5\, M_2 V_{2\text{cm}}^2 + 0.5\, I_{2\text{cm}}\, \dot{A}^2 \qquad \text{(Con Rod)} \\
&\quad + 0.5\, M_3 V_{3\text{cm}}^2 \qquad \text{(Slider)} \\
&= 0.5\, M_1(\dot{X}_1^2 + \dot{Y}_1^2) + 0.5\, I_{1\text{cm}}\, \dot{q}^2 \\
&\quad + 0.5\, M_2(\dot{X}_2^2 + \dot{Y}_2^2) + 0.5\, I_{2\text{cm}}\, \dot{A}^2 \\
&\quad + 0.5\, M_3 \dot{X}_3^2 \\
&= 0.5\, M_1\, \dot{q}^2(K_{1x}^2 + K_{1y}^2) + 0.5\, I_{1\text{cm}}\, \dot{q}^2 \\
&\quad + 0.5\, M_2\, \dot{q}^2(K_{2x}^2 + K_{2y}^2) + 0.5\, I_{2\text{cm}}\, \dot{q}^2\, K_a^2 \\
&\quad + 0.5\, M_3\, \dot{q}^2 K_x^2 \\
&= 0.5\, \dot{q}^2[M_1(K_{1x}^2 + K_{1y}^2) + I_{1\text{cm}} \\
&\quad + M_2(K_{2x}^2 + K_{2y}^2) + I_{2\text{cm}}\, K_a^2 + M_3\, K_x^2] \\
&= 0.5\, \dot{q}^2[I_{1o} + M_2\, (K_{2x}^2 + K_{2y}^2) + I_{2\text{cm}}\, K_a^2 + M_3\, K_x^2]
\end{aligned}$$

The parallel axis theorem has been used to replace the terms $M_1(K_{1x}^2 + K_{1y}^2) + I_{1\text{cm}}$ by I_{1o}, the mass moment of inertia for the crank with respect to the rotation axis. This is to be expected for a body rotating about a fixed center. No similar simplification is possible for the connecting rod terms; the connecting rod does not rotate about a fixed center. The generalized inertia for the slider–crank mechanism is the coefficient of $0.5\, \dot{q}^2$. The generalized inertia is clearly dependent on q and, in fact, oscillates about a non-zero mean value with two maxima and two minima for each crank rotation.

7.2 GENERALIZED FORCES

All of the forces and couples that work on the system influence its dynamic response. In Section 6.4, the generalized forces associated with the multiple degrees of freedom were developed; the single degree of freedom case is considered briefly here. The objective of this section is to determine a single generalized force that, when acting through a virtual coordinate change δq, will do virtual work $Q\ \delta q$ equal to the sum of the virtual work of the actual forces and moments moving through their associated virtual displacements. Consider external forces $\mathbf{F}_i$ applied at locations defined by the position vectors $\mathbf{r}_i$, and similarly, couples $\mathbf{C}_j$ acting on angles A_j. The virtual work of this force system is

$$\delta W = \sum_i \mathbf{F}_i \cdot \delta \mathbf{r}_i + \sum_j \mathbf{C}_j \cdot \delta \mathbf{A}_j$$

All the positions are functions of the single generalized coordinate, q, so the virtual displacements can be written in terms of the virtual change in q:

$$\delta \mathbf{r}_i = \frac{d\mathbf{r}_i}{dq}\,\delta q$$

$$\delta \mathbf{A}_j = \frac{d\mathbf{A}_j}{dq}\,\delta q$$

When these expressions are applied in the virtual work expression, the result is

$$\delta W = \delta q \left(\sum_i \mathbf{F}_i \cdot \frac{d\mathbf{r}_i}{dq} + \sum_j \mathbf{C}_j \cdot \frac{d\mathbf{A}_j}{dq} \right)$$

The coefficient of δq is the generalized force, Q:

$$Q = \sum_i \mathbf{F}_i \cdot \frac{d\mathbf{r}_i}{dq} + \sum_j \mathbf{C}_j \cdot \frac{d\mathbf{A}_j}{dq}$$

Although this looks formidable, the actual determination of the generalized force is usually quite simple. To continue with the illustration involving the slider–crank mechanism, Figure 7.1 shows a couple C acting on the crank and a force F acting on the slider. These are the only two external forces that do work on the system. Other external forces, such as the bearing reactions or the transverse reaction on the slider, do no work and need not be included. The virtual work of this force system is

$$\begin{aligned} \delta W &= C\,\delta q + F\,\delta X \\ &= C\,\delta q + F\frac{dX}{dq}\,\delta q \\ &= \delta q(C + F\,K_x) \end{aligned}$$

with the result that the generalized force is

$$Q = C + F\,K_x$$

Notice the role of the velocity coefficient, K_x. Why is there no velocity coefficient written with C?

7.3 EKSERGIAN'S EQUATION OF MOTION

One of the theorems established in introductory dynamics courses states that the work done on a mechanical system is equal to the change in kinetic energy of the system. For application here, that statement is considered in differentiated form.

$$\text{Power (into system)} = \frac{d(\text{Kinetic Energy})}{dt}$$

For the single degree of freedom system under consideration, the power into the system is

$$\begin{aligned} \text{Power} &= \sum_{i} (F_{xi}\dot{X}_i + F_{yi}\dot{Y}_i + M_i\dot{A}_i) \\ &= \sum_{i} (F_{xi}K_{xi} + F_{yi}K_{yi} + M_iK_{ai})\dot{q} \\ &= Q\,\dot{q} \end{aligned}$$

while the kinetic energy is

$$T = 0.5\ \mathcal{I}(q)\ \dot{q}^2$$

Differentiating the kinetic energy with respect to time and equating this result to the power expression gives

$$0.5\,\frac{d\mathcal{I}}{dq}\,\dot{q}\,\dot{q}^2 + \mathcal{I}\,\dot{q}\,\ddot{q} = Q\,\dot{q}$$

Eliminating the common factor $\dot{q}$ gives Eksergian's form for the equation of motion of a single degree of freedom system:

$$\mathcal{I}\ddot{q} + 0.5\,\frac{d\mathcal{I}}{dq}\,\dot{q}^2 = Q$$

This, then, is the generalized equation of motion applicable to all single degree of freedom systems. This form for the equation of motion, and also the ideas of velocity coefficients and velocity coefficient derivatives, were used repeatedly by Eksergian in a series of papers on machinery dynamics.[1] If the generalized inertia is constant, the equation of motion reduces to a familiar form: inertia × acceleration = force. For nonconstant

[1] R. Eksergian, "Dynamical Analysis of Machines," in 15 parts, *J. of the Franklin Institute*, Vols. 209, 210, 211, 1930–1931.

Eksergian seems to have been the first to present this equation of motion in an English-language publication, but it had appeared previously in the German literature. His work was far ahead of its time, and it is only now practical because of the general availability of digital computers.

generalized inertia, the second term, known as the *centripetal* term, must also be included. The coefficient in the centripetal term, 0.5 $d\mathcal{I}/dq$, is given the symbol $\mathcal{C}(q)$ and is called the centripetal coefficient. With this notation, Eksergian's form for the equation of motion is

$$\mathcal{I}(q)\ddot{q} + \mathcal{C}(q)\dot{q}^2 = Q$$

To apply Eksergian's equation of motion to the slider–crank mechanism used in earlier illustrations, it is necessary to determine the generalized inertia, $\mathcal{I}(q)$. From the kinetic energy calculation, the generalized inertia is

$$\mathcal{I}(q) = I_{lo} + M_2\,(K_{2x}{}^2 + K_{2y}{}^2) + I_{2cm}\,K_a{}^2 + M_3\,K_x{}^2$$

Differentiation with respect to q gives the centripetal coefficient, $\mathcal{C}(q)$:

$$\begin{aligned}\mathcal{C}(q) &= 0.5\,\frac{d\mathcal{I}(q)}{dq}\\ &= M_2\,(K_{2x}\,L_{2x} + K_{2y}\,L_{2y}) + I_{2cm}\,K_a\,L_a + M_3\,K_x\,L_x\end{aligned}$$

Using these expressions in Eksergian's equation of motion, along with the generalized force previously identified, gives the equation of motion for the slider–crank mechanism:

$$\begin{aligned}&[I_{lo} + M_2\,(K_x{}^2 + K_y{}^2) + I_{2cm}\,K_a{}^2 + M_3\,K_x{}^2]\ddot{q}\\ &\qquad + [M_2\,(K_{2x}\,L_{2x} + K_{2y}\,L_{2y}) + I_{2cm}\,K_a\,L_a + M_3\,K_x\,L_x]\dot{q}^2\\ &= C(t) + F(t)K_x\end{aligned}$$

This is an extremely complex nonlinear differential equation with variable coefficients, and there is no hope of obtaining an analytical solution for it. There is, however, every reason to expect that a numerical solution can be obtained, and that matter is taken up later in this chapter.

7.4 POTENTIAL ENERGY REPRESENTATION OF CONSERVATIVE FORCES

An alternative exists to including conservative forces in the generalized force. A potential energy term can be included in Eksergian's equation of motion to account for them, while the nonconservative forces continue to be included in the generalized force. This modification is developed here; a parallel development was given for statics in Section 6.2.

Let the force acting on the system at point $\mathbf{r}_i$ consist of two parts, one part conservative and the other nonconservative. The conservative force can be written as the negative gradient of its associated potential function.

$$\begin{aligned}\mathbf{F}_i &= \mathbf{F}_i^{c} + \mathbf{F}_i^{nc}\\ &= -\nabla V_i + \mathbf{F}_i^{nc}\end{aligned}$$

If this form is used to determine the generalized force, that result also consists of two terms:

$$Q = \sum_i \mathbf{F}_i \cdot \frac{d\mathbf{r}_i}{dq}$$

$$= \sum_i (-\nabla V_i + \mathbf{F}_i^{nc}) \cdot \frac{d\mathbf{r}_i}{dq}$$

$$= -\sum_i \frac{dV_i}{dq} + \sum_i \mathbf{F}_i^{nc} \cdot \frac{d\mathbf{r}_i}{dq}$$

$$= -\frac{dV}{dq} + Q^{nc}$$

Note that V without a subscript is used for the total potential energy of the system, whereas Q^{nc} is the nonconservative generalized force. With this result applied to Eksergian's equation, and the potential energy term shifted to the left side, the modified form is

$$\mathcal{I}(q)\,\ddot{q} + \mathcal{C}(q)\,\dot{q}^2 + \frac{dV}{dq} = Q^{nc}$$

This modified form is particularly useful in cases that involve springs that change direction as the mechanism moves. For such situations, expressing the potential energy of the spring in a typical position, and then including its effect through the indicated potential energy term is relatively simple. Direct inclusion of the generalized force term is also possible, although usually more difficult.

Continuing with the slider–crank example used in previous sections, let the force on the slider be replaced by a spring and dashpot arrangement as shown in Figure 7.2. The free length of the spring is S_o. The potential energy of the spring is then

$$V = 0.5\,K\,(X_o - X - S_o)^2$$

with the derivative

$$\frac{dV}{dq} = K\,(X_o - X - S_o)\left(-\frac{dX}{dq}\right)$$

$$= -K\,K_x\,(X_o - X - S_o)$$

To include the dashpot and the time-dependent couple, the nonconservative virtual work is determined as follows:

$$\delta W^{nc} = -B\,\dot{X}\,\delta X + C(t)\,\delta q$$

$$= -B\,(\dot{q}\,K_x)(K_x\,\delta q) + C(t)\,\delta q$$

$$= [-B\,K_x^2\,\dot{q} + C(t)]\,\delta q$$

$$Q^{nc} = -B\,K_x^2\,\dot{q} + C(t)$$

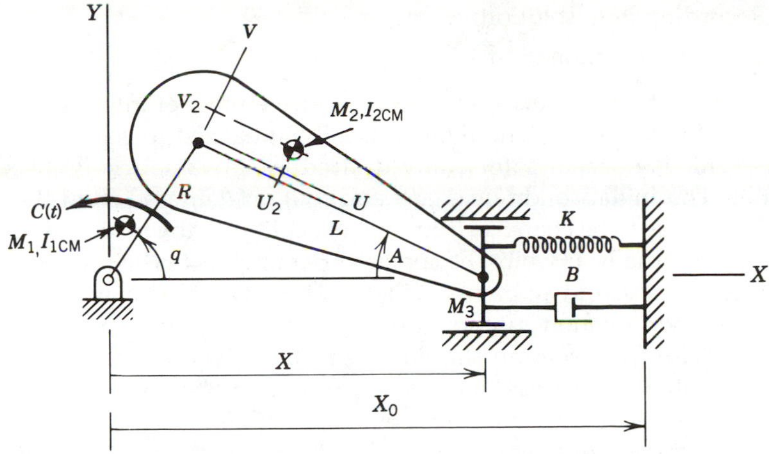

FIGURE 7.2 Slider—Crank Mechanism with Spring/Dashpot Assembly

Note that time-dependent external forces are necessarily nonconservative, and therefore must enter through Q^{nc}. With these modifications, the equation of motion for the system is

$$\mathcal{I}(q)\,\ddot{q} + \mathcal{C}(q)\,\dot{q}^2 - K\,K_x\,(X_o - X - S_o) = -B\,K_x^2\,\dot{q} + C(t)$$

where the expressions for $\mathcal{I}(q)$ and $\mathcal{C}(q)$ were previously determined.

7.5 MECHANISM SIMULATION

The differential equation that describes the motion of a single degree of freedom mechanism is often quite formidable because in most cases it is nonlinear with variable coefficients. Such equations are generally not solvable in closed form, but they are usually amenable to numerical solution. The Runge–Kutta method is one numerical solution technique that usually gives good results for mechanism problems. It is presented in Appendix A3, with the particular form applicable to a single, second-order differential equation given in Appendix A3.2. That appendix should be reviewed for details of the Runge–Kutta method; the present section addresses its implementation in a computer program. The resulting program is often referred to as a "simulation," because the calculations performed in the program "simulate" or behave like the physical system from which the differential equation was derived.

The essential parts of any simulation program are

1. Initialization
2. Iterative advancement of the solution
   a. Evaluating derivatives
   b. Determining new solution values

c. Recording new solution values
d. Testing for termination

In the initialization phase, all initial values are either set internally or entered from the keyboard. The time is usually initialized to zero internally, although this is not mandatory (any desired initial value can be used for the time). The initial conditions, $q(0)$ and $\dot{q}(0)$, are entered from the keyboard if they are to be changed from one execution to the next; if not, they may be set internally as well. The constant parameters of the problem such as lengths, masses, and moments of inertia should also be set in this phase. Some type of termination criterion must be established and any parameters associated with that criterion should be set during the initialization. As a final step, column headings should be printed and the initial conditions recorded as the first entries in the tabulated solution.

Standard practice is to program the evaluation of the second derivative as a subroutine. This is then called with the appropriate argument sets for each of the four derivative evaluations required by the Runge–Kutta algorithm. Because of the nature of BASIC, some care is required in terms of the naming of variables in the subroutine and in the main program. To be specific, consider a simulation program for the system described by the differential equation

$$\ddot{q} = f(t, q, \dot{q})$$

When the second derivative subroutine is written, the variables used could be T, Q, and Q1 for t, q, and $\dot{q}$. Then, in the main program, the variable names might be T0, Q0, and QD for the values of t, q, and $\dot{q}$. Before calling the subroutine to evaluate the derivative, the appropriate values must be defined for T, Q, and Q1. For the first evaluation, this is simply

```
 T = T0
 Q = Q0
Q1 = QD
```

For the second evaluation, the appropriate values are

```
 T = T0 + H/2
 Q = Q0 + H*QD/2 + H*M1/8
Q1 = QD + M1/2
```

where M1 is determined from the first evaluation of the derivative. This will be illustrated in the examples in Section 7.6.

After the four evaluations of the derivative, the new values of the solution must be calculated or, as it is sometimes phrased, the solution must be "updated." In addition to updating the values for Q and QD, it is also necessary to update the time, T0.

For the simplest type of simulation, the results are printed at every time step. Thus, after every update, the new results are printed. If very small time steps are being used, it may be desirable to print less frequently, perhaps every 10 steps. If this is required, the process of advancing the solution will involve two nested loops. The printed results usually include the values for t, q, and $\dot{q}$, but any desired additional quantities such as forces, stresses, relative displacements, or relative velocities can also be evaluated and printed at the same time.

It is necessary to provide some means to end the simulation program. This can be done in several ways, one of the most common being to end after a fixed time interval has been simulated. Other termination criteria may be based on the occurrence of some specified event, such as when the solution is greater than a prescribed value. For any of these criteria, a test is needed to terminate the execution, usually in the form of an IF-statement test following each output sequence. Failure to include an achievable termination criterion causes the simulation to run until the operator intervenes; that is wasteful. The term "achievable termination criterion" refers to an event that actually occurs at some time in the solution, as opposed to a criterion defined by an event that never occurs and hence, would allow the solution to run indefinitely. If the termination criterion is not met, control should be transferred to the beginning of the loop to move forward another time step.

These steps are illustrated in both of the example problems given in the next section. The reader should locate each of these steps in those computer programs and trace the sequence of calculations.

The four derivative evaluations required for the Runge–Kutta integration method pose a difficulty in the simulation of mechanism systems. In the most direct approach, each evaluation of the derivative will require a new solution for the position loop equations, followed by new evaluations of the velocity coefficients and velocity coefficient derivatives. This process is very time consuming if iterative position solutions are involved, and the whole process must be repeated four times for each integration time step! This is unacceptable if the work is done on any other than the fastest mainframe computers, and even there it is very wasteful. An attractive alternative is to first generate the kinematic solution functions—positions, velocity coefficients, and velocity coefficient derivatives—over the expected range of the motion, and to then store these solutions in tabular form in the simulation program. The kinematic solution values required in the derivative evaluations can then be determined as needed from the stored data table by interpolation.

To minimize the size of the tabulated solution required, it is useful to perform interpolation using a polynomial approximation to the stored data points, rather than using the more common linear interpolation. Polynomial approximation assures that the estimate obtained by interpolation lies on a smooth curve that passes through more than the two nearest data points. If it is convenient to store the kinematic solution data at evenly spaced increments, the cubic interpolation scheme described in Appendix A5.2 can be used. For cases in which the data will be stored at unevenly spaced intervals, the quadratic interpolation method given in Appendix A5.1 is useful. The latter case applies when the data cannot be obtained readily at evenly spaced intervals, or when uneven spacing is needed to describe adequately a function that varies rapidly in one region while varying slowly elsewhere.

As a further means to speed up the evaluation of the kinematic solution functions by interpolation, it is usually reasonable to locate the proper interval in the table only once for every time step. This is justified if the movement of the solution in a single time step is small compared to the kinematic solution tabulation interval. Assuming that all of the kinematic solution functions have been tabulated for the same values of the generalized coordinate, the procedure for evaluating the second derivative is as follows:

1. On the first evaluation of the second derivative,
   a. Locate the tabular interval, including the current value of the generalized coordinate;
   b. Evaluate the coefficients for the interpolating polynomial applied to that interval *for each of the kinematic solution functions* such as secondary position variables, velocity coefficients, and velocity coefficient derivatives;
   c. Evaluate the second derivative, using kinematic solution values determined by evaluating the interpolating polynomials.
2. On the second, third, and fourth calls to the derivative subroutine, evaluate the second derivative function using kinematic solution function values determined by evaluating the interpolating polynomials at the appropriate values of the generalized coordinate, using the previously determined polynomial coefficients.

This process is demonstrated in the second example problem of the next section.

7.6 MECHANISM SIMULATION EXAMPLES

This section presents two examples of mechanism simulation using the Runge–Kutta integration algorithm and other ideas discussed in the preceding sections. Each example will begin with a problem statement and formulation, followed by a computer program and results.

7.6.1 Rocker Response

This problem involves the dynamic response of the rocker shown in Figure 7.3 to an impulsive force applied at point ①; this problem is, in some respects, similar to the statics example problem involving an L-shaped bracket given in Section 6.3. For the current problem, the rocker is constrained to roll without slipping on the semicircular support, under the influence of the impulsive

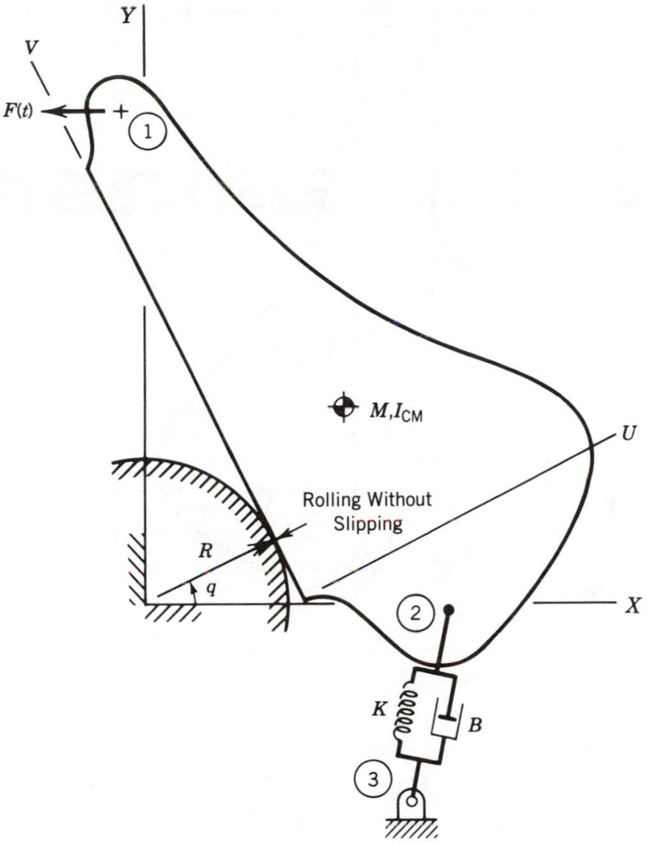

FIGURE 7.3 Rocker System Geometry

applied load, $F(t)$, and a spring and dashpot assembly attached at point ②. The spring and dashpot assembly is anchored at point ③, a point with base coordinates (3, −4).

The rocker outline is shown in Figure 7.4 on a one-inch grid, and the rocker thickness is 2.25 in. Area, centroid location, and moments of inertia for the rocker were determined using the methods of Appendix A7, "Geometric Calculations for Irregular Planar Areas." The rocker material is steel; the rocker weight is 8.7991 lb, which implies a mass of 2.27904 (10^{-2}) lb-s^2/in. The centroid is located by the body coordinates $U_c = 1.27684$ in. and $V_c = 1.58609$ in. The centroidal mass moment of inertia about the axis of rotation is $I_{cm} =$ 8.46351 (10^{-2}) lb-s^2-in.

The rocker support has a radius $R = 1.5$ in. For the restraining spring, the spring rate is $K = 55$ lb/in. and the free length of the assembly is $S_o = 2.5$ in. The dashpot coefficient is $B = 3.5$ lb-s/in. The applied force is a single rectangular pulse acting horizontally to the left; its magnitude is 50 lb lasting for a duration of 0.01 s. The system is initially at rest in an upright position, with the U–axis parallel to the X–axis, the V–axis parallel to the Y–axis, and supported by a stop (not shown) when the impulsive load is first applied.

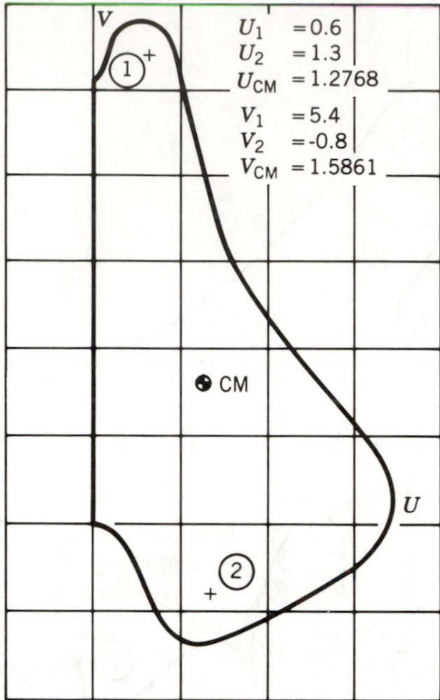

FIGURE 7.4 Rocker Profile Details

The two primary objectives for this simulation are:

1. Determining the maximum excursion of the rocker;
2. Determining the rocker velocity as it returns to the stop.

One significant aspect of this problem is the simplicity of the kinematic relations. With the contact angle taken as the generalized coordinate, q, there are no loop equations to be solved and no secondary variables. Each required base coordinate is readily expressible in terms of the generalized coordinate and the body coordinates for the point in question. Thus, for any point with body coordinates (U_p, V_p), the base coordinates, velocity coefficients, and velocity coefficient derivatives are

$$\begin{Bmatrix} X \\ Y \end{Bmatrix} = \begin{bmatrix} \cos q & -\sin q \\ \sin q & \cos q \end{bmatrix} \begin{Bmatrix} R + U_p \\ V_p - R\, q \end{Bmatrix}$$

$$= \begin{Bmatrix} (R + U_p) \cos q + (R\, q - V_p) \sin q \\ (R + U_p) \sin q + (V_p - R\, q) \cos q \end{Bmatrix}$$

$$\begin{Bmatrix} K_{px} \\ K_{py} \end{Bmatrix} = \begin{Bmatrix} (R\, q - V_p) \cos q - U_p \sin q \\ (R\, q - V_p) \sin q + U_p \cos q \end{Bmatrix}$$

$$\begin{Bmatrix} L_{px} \\ L_{py} \end{Bmatrix} = \begin{Bmatrix} (R - U_p) \cos q - (R\, q - V_p) \sin q \\ (R - U_p) \sin q + (R\, q - V_p) \cos q \end{Bmatrix}$$

To apply Eksergian's form of the equation of motion, the kinetic energy is required. This can be expressed as the sum of terms dependent on translational and rotational velocities:

$$T = 0.5\, M(\dot{X}_{cm}^2 + \dot{Y}_{cm}^2) + 0.5\, I_{cm}\, \dot{q}^2$$
$$= 0.5\, \dot{q}^2\, [M(K_{cmx}^2 + K_{cmy}^2) + I_{cm}]$$

From the kinetic energy, the generalized inertia is readily identified as the coefficient of $0.5\, \dot{q}^2$,

$$\mathcal{I}(q) = M(K_{cmx}^2 + K_{cmy}^2) + I_{cm}$$

Because the generalized inertia is not constant, there is a non-zero centripetal coefficient:

$$\mathcal{C}(q) = M(K_{cmx}\, L_{cmx} + K_{cmy}\, L_{cmy})$$

The effects of the spring and dashpot assembly are dependent on the length of that assembly, S, and the time derivative, $\dot{S}$. The length is

$$S(q) = \sqrt{(X_2 - X_3)^2 + (Y_2 - Y_3)^2}$$

With the required differentiation, the velocity coefficient K_s is determined:

$$K_s = \frac{(X_2 - X_3)}{S(q)}\, K_{2x} + \frac{(Y_2 - Y_3)}{S(q)}\, K_{2y}$$

For this example, the effect of the spring will be included through a potential energy term. The potential energy of the stretched spring is

$$V = 0.5\, K\, [S(q) - S_o]^2$$

with the derivative

$$\frac{dV}{dq} = K\, K_s\, [S(q) - S_o]$$

The final step in preparation for writing the equation of motion is to determine the nonconservative generalized force. The nonconservative virtual work is

$$\delta W^{nc} = -F(t)\, \delta X_1 - B\, \dot{S}\, \delta S$$
$$= -\dot{q}\, [F(t)\, K_{1x} + B\, \dot{q}\, K_s^2]$$

from which the nonconservative generalized force is

$$Q^{nc} = -F(t)\, K_{1x} - B\, \dot{q}\, K_s^2$$

The equation of motion is then obtained by applying Eksergian's form for the equation of motion:

$$\mathcal{I}(q)\, \ddot{q} + \mathcal{C}(q)\, \dot{q}^2 + \frac{dV}{dq} = Q^{nc}$$

where expressions have been obtained previously for $\mathcal{I}(q)$, $\mathcal{C}(q)$, V, and Q^{nc} for this problem. For use in the numerical solution, it is necessary to solve the equation of motion for the second derivative:

$$\ddot{q} = \frac{1}{\mathcal{I}(q)}\left[Q^{nc} - \mathcal{C}(q)\,\dot{q}^2 - \frac{dV}{dq}\right]$$

The foregoing analysis is the basis for the computer program that follows shortly. The first part of the program contains the initialization phase, the Runge-Kutta loop, and provisions for printing the results. Most of the results from the previous analysis can be found in the derivative evaluation subroutine that begins at line 3000. Note that in that subroutine the values of time, q, and $\dot{q}$ are denoted by T, Q, and Q1; in the main program, these same variables are denoted by T0, Q0, and QD.

The input statement at line 2070, "INPUT V$," is included only to force a pause to examine the results appearing on the screen. To continue the calculation, key "Return." If the results are being recorded on a printer, line 2070 can be changed to a REM statement to allow the program to run continuously.

The termination criterion is based on the need to continue the simulation until the rocker returns to its original position against the stop. Thus, as long as the position is greater than zero there is no need to test the direction of motion; the simulation clearly must continue. When the position is negative, suggesting that the rocker has hit the stop, then the second test is used to assure that this condition really does imply the return rather than the initial departure of the rocker.

The program listing follows.

```
1000 REM  SIMULATION OF ROCKER RESPONSE
1010 REM  USING RUNGE-KUTTA INTEGRATION
1100 REM   X1,Y1:  BASE COORD FOR PT1
1110 REM   X2,Y2:  BASE COORD FOR PT2
1120 REM   X3,Y3:  BASE COORD FOR PT3
1130 REM   R:      RADIUS OF SUPPORT
1140 REM   F:      APPLIED FORCE
1150 REM   K:      SPRING RATE
1160 REM   B:      DASHPOT COEFF
1170 REM   S:      LENGTH OF SPRING/DASHPOT
1180 REM   S0:     FREE LENGTH OF SPRING
1190 REM   KX,KY:  VEL COEF FOR CENTER OF MASS
1200 REM   LX,LY:  VEL COEF DERIV FOR CM
1210 REM   K1,K2:  K1X,K1Y VEL COEFF FOR PT1
1220 REM   K3,K4:  K2X,K2Y VEL COEFF FOR PT2
1230 REM   KS:     VEL COEF FOR LENGTH S
1240 REM   UC,VC:  BODY COORD FOR CENTER OF MASS
1250 REM   U1,V1:  BODY COORD FOR PT1
1260 REM   U2,V2:  BODY COORD FOR PT2
1270 REM   M:      BODY MASS
```

```
1280 REM   IC:     CENTROIDAL MASS MOMENT OF INERTIA
1300 REM   *****      PROBLEM DATA     *****
1310 X3=3
1320 Y3=-4
1330 R=1.5
1340 F0=50
1350 K=55
1360 B=3.5
1370 S0=2.5
1380 UC=1.27684237
1390 VC=1.58609453
1400 U1=0.6
1410 V1=5.4
1420 U2=1.3
1430 V2=-0.8
1440 M=2.27904E-2
1450 IC=8.46351E-2
1460 REM   *****      END OF PROBLEM DATA      *****
1500 REM  ENTER INITIAL VALUES
1510 HOME
1520 PRINT"ENTER INITIAL ANGLE,Q"
1530 INPUT Q0
1540 PRINT"ENTER INITIAL ANGULAR VEL, QDOT"
1550 INPUT QD
1560 REM INITIAL TIME IS ZERO
1570 T0=0
1580 PRINT"ENTER TIME STEP, H"
1590 INPUT H
1600 REM  PRINT INITIAL VALUES
1610 PRINT"ROCKER RESPONSE SIMULATION"
1620 PRINT"USING RUNGE-KUTTA INTEGRATION"
1630 PRINT"      INTEGRATION TIME STEP = ";H
1640 PRINT
1650 PRINT"TIME = ";T0
1660 PRINT"  Q   = ";Q0
1670 PRINT"  QD = ";QD
1690 FOR II=1 TO 100
1700 REM  FIRST EVAL OF DERIV
1710 Q=Q0
1720 Q1=QD
1730 T=T0
1740 GOSUB 3000
1750 M1=H*Q2
1760 REM SECOND EVAL OF DERIV
1770 Q=Q0+H*QD/2+H*M1/8
1780 Q1=QD+M1/2
1790 T=T0+H/2
1800 GOSUB 3000
1810 M2=H*Q2
```

```
1820 REM THIRD EVAL OF DERIV
1830 Q=Q0+H*QD/2+H*M2/8
1840 Q1=QD+M2/2
1850 T=T0+H/2
1860 GOSUB 3000
1890 M3=H*Q2
1900 REM FOURTH EVAL OF DERIV
1910 Q=Q0+H*QD+H*M3/2
1920 Q1=QD+M3
1930 T=T0+H
1940 GOSUB 3000
1950 M4=H*Q2
1960 REM  UPDATE ALL VALUES
1970 Q0=Q0+H*(QD+(M1+M2+M3)/6)
1980 QD=QD+(M1+2*M2+2*M3+M4)/6
1990 T0=T0+H
2000 REM  PRINT UPDATED VALUES
2010 PRINT"TIME = ";T0
2020 PRINT"  Q  = ";Q0
2030 PRINT"  QD = ";QD
2040 REM TEST FOR TERMINATION
2050 IF Q>0 THEN GOTO 2070
2060 IF QD <0 THEN GOTO 2100
2070 INPUT V$
2080 NEXT II
2100 END
3000 REM EVALUATION OF SECOND DERIVATIVE
3010 REM GENL COORD & VEL, Q & Q1 ARE INPUT
3020 X2=(R+U2)*COS(Q)+(R*Q-V2)*SIN(Q)
3030 Y2=(R+U2)*SIN(Q)+(V2-R*Q)*COS(Q)
3040 S=SQR((X2-X3)^2+(Y2-Y3)^2)
3050 K1=(R*Q-V1)*COS(Q)-U1*SIN(Q)
3060 K2=(R*Q-V1)*SIN(Q)+U1*COS (Q)
3070 K3=(R*Q-V2)*COS(Q)-U2*SIN(Q)
3080 K4=(R*Q-V2)*SIN(Q)+U2*COS(Q)
3090 KX=(R*Q-VC)*COS(Q)-UC*SIN(Q)
3100 KY=(R*Q-VC)*SIN(Q)+UC*COS(Q)
3110 LX=(R-UC)*COS(Q)+(V2-R*Q)*SIN(Q)
3120 LY=(R-U2)*SIN(Q)+(R*Q-V2)*COS(Q)
3130 KS=((X2-X3)*K3+(Y2-Y3)*K4)/S
3140 IN=M*(KX^2+KY^2)+IC
3150 C=M*(KX*LX*KY*LY)
3160 F=0
3170 IF T<0.01 THEN F=F0
3180 QN=-K1*F-B*Q1*KS^2
3190 DV=K*KS*(S-S0)
3200 Q2=(QN-C*Q1^2-DV)/IN
3210 REM  THIS IS THE SECOND DERIV VALUE
3220 RETURN
3230 END
```

The results of this simulation follow.

```
RUN
ENTER INITIAL ANGLE, Q
?0
ENTER INITIAL ANGULAR VEL, QDOT
?0
ENTER TIME STEP, H
?.001
ROCKER RESPONSE SIMULATION
USING RUNGE-KUTTA INTEGRATION
     INTEGRATION TIME STEP = 1E-03
TIME = 0
  Q  = 0
  QD = 0
TIME = 1E-03
  Q  = 6.12152931E-04
  QD = 1.21808098
TIME = 2E-03
  Q  = 2.42378211E-03
  QD = 2.39900877
       .
       .
       .
TIME = 9E-03
  Q  = .045552262
  QD = 9.60712793
TIME = .01
  Q  = .055597102
  QD = 10.2183079
TIME = .011
  Q  = .0654734321
  QD = 9.53551372
       .
       .
       .
TIME = .027
  Q  = .139672653
  QD = .411665997
TIME = .028
  Q  = .139869676
  QD = -.0146754554
       .
       .
       .
TIME = .0600000002
  Q  = 3.28155275E-03
  QD = -6.7004407
TIME = .0610000002
  Q  = -3.45405218E-03
  QD = -6.76996064
```

Notice that the maximum velocity coincides with the end of the impulse as it should.

Maximum excursion occurs between 27 and 28 milliseconds with a value of approximately $q = 0.1398$ radians.

The rocker returns to the stop between 60 and 61 milliseconds with a velocity of approximately $\dot{q} = -6.73$ rad/s.

This completes the partial listing of the computer results. Note that the major objectives regarding maximum excursion and velocity on return to the stop have been answered by selecting the appropriate results from the complete computer output. The observation that maximum velocity coincides with the end of the impulse serves as a check on the validity of the calculations.

7.6.2 Four-Bar Mechanism

This example considers the response of a four-bar mechanism under the influence of an impulsive torque applied to the input crank and the retarding effect of a dashpot acting on the pin joining the coupler and the second crank. In this system, the kinematic solution functions are essential; their approximation by polynomial interpolation is demonstrated.

The overall mechanism is shown in Figure 7.5(a), and the details of the coupler link geometry are given in Figure 7.5(b). As mentioned before, a time-dependent torque of brief duration is applied to the mechanism, which is initially at rest in the position $q = 0.2$ radians. The dashpot that retards the motion is anchored at the point (X_b, Y_b). The necessary numerical data for the system are:

$C_1 = 5.0$	Length of input crank, in.
$C_2 = 9.0$	Length of coupler link, in.
$C_3 = 7.0$	Length of second crank, in.
$C_4 = 10.0$	Length of stationary link, in.
$J_{1o} = 0.075$	Mass moment of inertia for input crank with respect to the fixed pivot, lb-s^2-in.
$J_{2c} = 0.213444$	Mass moment of inertia of the coupler with respect to its center of mass, lb-s^2-in.
$M_2 = 0.0181416$	Mass of coupler link, lb-s^2/in.
$J_{3o} = 0.15$	Mass moment of inertia for the second crank with respect to the fixed pivot, lb-s^2-in.
$U_{cm} = 5.11111$	Body coordinate for coupler center of mass, in.
$V_{cm} = 1.33333$	Body coordinate for coupler center of mass, in.
$B = 5.5$	Dashpot coefficient, lb-s/in.
$X_b = 13.0$	Dashpot anchor coordinate, in.
$Y_b = 1.2$	Dashpot anchor coordinate, in.

To begin the motion, a counterclockwise torque pulse of magnitude 25 in./lb and duration 0.075 s is applied to the input crank. After the end of the torque pulse, the system slows down under the influence of the dashpot. The objectives for this simulation will be:

1. To determine the maximum angular velocity of the input crank;
2. To estimate the time and position of the system when the angular velocity drops below 0.05 rad/s following the end of the torque pulse.

For all of the data given, inch-pound-second units (IPS units) are indicated. As long as consistent units are used, the system of units used makes no difference on the results. Consistency is the key. Thus, the following simulation applies equally well to a much larger mechanism for which $C_1 =$

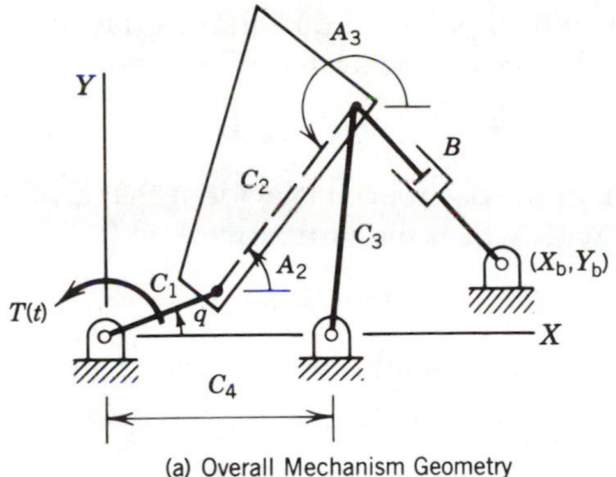

(a) Overall Mechanism Geometry

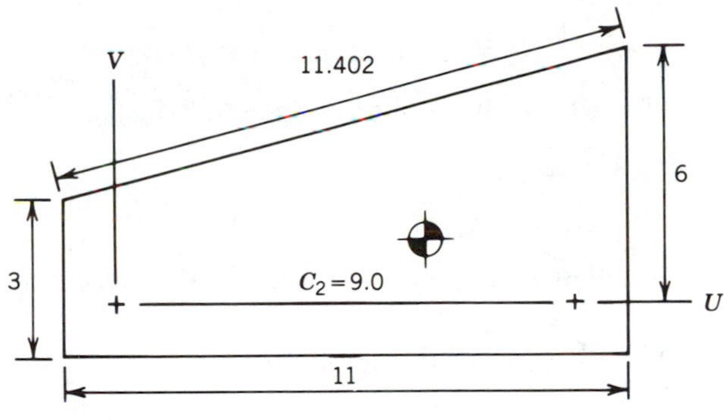

(b) Coupler Link Details

FIGURE 7.5 Four Bar Mechanism

5.0 m, $C_2 = 9.0$ m, . . . $J_{1o} = 0.075$ kg-m^2, $J_{2c} = 0.213444$ kg-m^2, . . . $B = 5.5$ N-s/m, and so forth, and that is subject to a brief torque pulse of 25 N-m. Regardless of whether the system is described in IPS or SI units, the computed response values for q and $\dot{q}$ will remain in units of radians and radians per second, respectively.

The input crank angle is chosen as the generalized coordinate, and the equations of motion must be determined. As indicated in Figure 7.5, the secondary variables are A_2 and A_3, measured in the same manner as the four-bar mechanism considered in Section 2.3. As a first step, the kinetic energy of the system is

$$T = 0.5\,\dot{q}^2\,[J_{1o} + J_{3o}\,K_{a3}^2 + J_{2cm}\,K_{a2}^2 + M_2\,(K_{2x}^2 + K_{2y}^2)]$$

From the kinetic energy, the generalized inertia is identified as

$$\mathcal{I}(q) = J_{1o} + J_{3o}\,K_{a3}^2 + J_{2cm}\,K_{a2}^2 + M_2(K_{2x}^2 + K_{2y}^2)$$

and the centripetal coefficient is

$$\mathcal{C}(q) = J_{3o}\,K_{a3}\,L_{a3} + J_{2cm}\,K_{a2}\,L_{a2} + M_2\,(K_{2x}\,L_{2x} + K_{2y}\,L_{2y})$$

The virtual work of the applied torque and the dashpot force is used to define the generalized force on the system:

$$\delta W = T_a(t)\,\delta q - B\,\dot{D}\,\delta D$$

where D is the length of the dashpot. It is evident that a velocity coefficient K_d will be required, with which the virtual work is

$$\delta W = [T_a(t) - B\,\dot{q}\,K_d^2]\,\delta q$$

The generalized force is identified from the virtual work expression as

$$Q = T_a(t) - B\,\dot{q}\,K_d^2$$

Let point ② be the joint that connects the coupler link and the second crank. This is the same joint to which the dashpot is attached. Base coordinates for point ② are

$$X_2 = C_1 \cos q + C_2 \cos A_2$$

$$Y_2 = C_1 \sin q + C_2 \sin A_2$$

The dashpot length is $D(q)$:

$$D(q) = \sqrt{(X_b - X_2)^2 + (Y_b - Y_2)^2}$$

The velocity coefficient K_d is then determined by differentiation:

$$K_d = \frac{(X_b - X_2)}{D(q)}(C_1 \sin q + C_2 K_{a2} \sin A_2) - \frac{(Y_b - Y_2)}{D(q)}(C_1 \cos q + C_2 K_{a2} \cos A_2)$$

From the terms just described, the equation of motion can be assembled to evaluate the second derivative, as will be required for the numerical integration. A review of the terms shows that the required kinematic solutions are $A_2(q)$, $A_3(q)$, $K_{a2}(q)$, $K_{a3}(q)$, $L_{a2}(q)$, and $L_{a3}(q)$. All of the other required information, such as the base coordinates for point ②, the velocity coefficients, and velocity coefficient derivatives for the center of mass, and so on, can be determined directly from these six functions. The necessary function values can then be determined using the example program in Section 2.3. The values are then stored in a data array within the program, with the required solution values at intermediate points determined by cubic polynomial interpolation.

An alternative to be considered is to tabulate and interpolate values for only $A_2(q)$ and $A_3(q)$. For once these two values are known, the other four function values can be determined as the solutions of two systems of linear algebraic equations. This requires less preparation because making tables for $K_{a2}(q)$, $K_{a3}(q)$, $L_{a2}(q)$, and $L_{a3}(q)$ becomes unnecessary. However, there are two problems with this approach as regards accuracy and speed. Any error in the interpolated values for $A_2(q)$ and $A_3(q)$ is propagated into the values of the other four functions, and results in less accurate values for the velocity coefficients and velocity coefficient derivatives in most cases. In addition, evaluation by interpolation is actually faster than solving the system of linear simultaneous equations. To compare these two approachs, the computer program for this problem has been written both ways, and the running times compared. The program using the linear equations to evaluate the velocity coefficients and velocity coefficient derivatives required approximately 19 percent more time than the version that evaluated all six functions by interpolation. In view of these problems, the use of interpolation is recommended for all of the secondary variables and their derivatives with respect to q.

Returning to the mainstream of the problem, the equation of motion is given by Eksergian's form as

$$\mathcal{I}(q)\,\ddot{q} + \mathcal{C}(q)\,\dot{q}^2 = Q(t)$$

with the initial conditions

$$q(0) = 0.2 \qquad \dot{q}(0) = 0.0$$

All of the required coefficients have been expressed already. Solving this for the second derivative, as will be required for the Runge–Kutta solution, gives:

$$\ddot{q} = \frac{1}{\mathcal{I}(q)}\,[Q(t) - \mathcal{C}(q)\,\dot{q}^2]$$

To construct the computer program, it is convenient to begin with the part that deals with the kinematic function evaluation by polynomial interpolation. The necessary data is generated in increments $\Delta q = 0.174533$ radians over the range $0 < q < 1.745329$ radians. This data is entered using the DATA statements that begin at line 4000 near the end of the program. The first action of the program, lines 1210 to 1320, reads this data into the storage array DD. The interpolation scheme is then written in lines 3800 through 3995. Notice that the interval location algorithm is written as a subroutine separate from that used to evaluate the polynomial coefficients. As the four polynomial coefficients for each function are evaluated (G0, G1, G2, and G3), they are stored in the polynomial coefficient array, PC. From there they are called, as needed, to evaluate the approximations. A short dummy main program is useful to test this part of the program.

The second part of the program that should be constructed is the subroutine that evaluates the second derivative, $\ddot{q}$, in lines 3000 through 3520. Two entry points to this routine are identified, one used for the first call for each time step, and a second used for all other calls. Entry at the first location, at line 3100, causes two actions—the interval search routine is called to locate the appropriate tabular interval (line 3120) and the polynomial coefficients are evaluated for that interval (lines 3130 through 3190). For the second, third, and fourth derivative evaluations, these steps are bypassed by entry at line 3200. All derivative evaluations, first through fourth passes, must pass through lines 3200 through 3520. In this section, the polynomial approximations are evaluated for the six required kinematic solution values, the related kinematic expressions for the center of mass and the dashpot length are also evaluated, and the generalized acceleration, Q2, is evaluated and returned at line 3520.

Lastly, the program initialization and the Runge–Kutta loop must be written. These are very similar to the corresponding parts of the first example and require only one further comment. The program is written to terminate when the angular velocity drops below 0.05 rad/s. This is based on the second objective for the simulation—to determine the time and position at which the angular velocity drops below 0.05 rad/s.

The program listing follows.

```
1000 REM  FOUR BAR MECHANISM SIMULATION
1010 REM  USING INTERPOLATED KINEMATIC SOLUTIONS
1020 REM  WITH CUBIC INTERPOLATION FOR EVENLY SPACED DATA
1100 DIM DD (11,7),PC(3,7)
1190 HOME
1200 REM  INITIALIZATION
1210 REM READ KINEMATIC DATA TABLE
1220 READ SS
1230 FOR I=1 TO 11
1240 FOR J=1 TO 3
1250 READ DD(I,J)
1260 NEXT J
1270 NEXT I
1280 FOR I=1 TO 11
1290 FOR J=4 TO 7
1300 READ DD(I,J)
1310 NEXT J
1320 NEXT I
1400 REM    **** PROBLEM DATA ****
1410 C1=5            : REM LENGTH OF LINK 1
1420 C2=9            : REM LENGTH OF LINK 2
1430 C3=7            : REM LENGTH OF LINK 3
1440 C4=10           : REM LENGTH OF STATIONARY LINK 4
1450 J1=.075         : REM MASS MOI OF LINK 1
WRT FIXED PT
1460 J2=0.213444     : REM MASS MOI OF LINK 2 WRT CM
1470 MM=1.81416E-2   : REM MASS OF LINK 2
1480 J3=.15          : REM MASS MOI OF LINK 3
WRT FIXED PT
1490 UC=5.11111112   : REM BODY COORD FOR CM
1500 VC=1.33333333   : REM BODY COORD FOR CM
1510 BB=5.5          :REM DASHPOT COEF
1520 XB=13           : REM DASHPOT ANCHOR COORD
1530 YB=1.2          : REM DASHPOT ANCHOR COORD
1540 REM    **** END OF PROBLEM DATA ****
1600 REM INITIAL VALUES
```

```
1610 Q0 = 0.2          : REM INITIAL VALUE OF Q
1620 QD = 0.0          : REM INITIAL VALUE OF Q DOT
1630 T0=0              : REM INITIAL TIME
1640 H=.005            : REM INTEGRATION TIME STEP
1900 REM  PRINT HEADER AND BEGIN TABULAR SOLN
1910 PRINT"FOUR BAR MECHANISM SIMULATION"
1920 PRINT"USING INTERPOLATED KINEMATIC SOLUTIONS"
1930 PRINT"TIME STEP = ";H
1940 PRINT
1950 PRINT"T, Q, QDOT ...."
1960 PRINT
1970 PRINT T0,Q0,QD
1980 REM
2000 REM RUNGE-KUTTA INTEGRATION OF EQUATION OF MOTION
2010 REM 1ST EVAL OF SECOND DERIV
2020 Q=Q0
2030 Q1=QD
2040 T=T0
2050 GOSUB 3100
2060 M1=H*Q2
2070 REM 2ND EVAL OF SECOND DERIV
2080 Q=Q0+H*QD/2+H*M1/8
2090 Q1=QD+M1/2
2100 T=T0+H/2
2110 GOSUB 3200
2120 M2=H*Q2
2130 REM 3RD EVAL OF SECOND DERIV
2140 Q=Q0+H*QD/2+H*M2/8
2150 Q1=QD+M2/2
2160 T=T0+H/2
2170 GOSUB 3200
2180 M3=H*Q2
2190 REM 4TH EVAL OF SECOND DERIV
2200 Q=Q0+H*QD+H*M3/2
2210 Q1=QD+M3
2220 T=T0+H
2230 GOSUB 3200
2240 M4=H*Q2
2300 REM  UPDATE SOLUTION VALUES
2310 Q0=Q0+H*(QD+(M1+M2+M3)/6)
2320 QD=QD+(M1+2*M2+2*M3+M4)/6
2330 T0=T0+H
2400 REM  PRINT UPDATED VALUES
2410 PRINT T0, Q0, QD
2500 REM  TEST FOR TERMINATION
2510 IF QD<0.05 THEN GOTO 2900
2520 REM OTHERWISE, RETURN TO BEGINNING OF R-K CYCLE
2530 GOTO 2020
2900 END
3000 REM  EVALUATION OF THE SECOND DERIVATIVE
3002 REM  Q:   Q, INPUT
```

```
3004 REM  Q1:  Q DOT, INPUT
3006 REM  Q2:  Q DOT DOT, OUTPUT
3008 REM  IN:  GENERALIZED INERTIA
3010 REM  CC:  CENTRIPETAL COEF
3012 REM  K2:  KA2, VEL COEF
3014 REM  K3:  KA3, VEL COEF
3016 REM  KX:  KCM2X, CM VEL COEF
3018 REM  KY:  KCM2Y, CM VEL COEF
3019 REM  DB:  DASHPOT LENGTH
3020 REM  KD:  KD, DASHPOT LENGTH VEL COEF
3021 REM  L2:  LA2, VEL COEF DERIV
3022 REM  L3:  LA3, VEL COEF DERIV
3024 REM  LX:  LCM2X, CM VEL COEF DERIV
3026 REM  LY:  LCM2Y, CM VEL COEF DERIV
3028 REM  UC:  BODY COORDINATE FOR CM
3030 REM  VC:  BODY COORDINATE FOR CM
3100 REM FIRST CALL ENTRY POINT
3110 X=Q
3120 GOSUB 3850
3130 FOR JJ=2 TO 7
3140 GOSUB 3920
3150 PC(0,JJ)=G0
3160 PC(1,JJ)=G1
3170 PC(2,JJ)=G2
3180 PC(3,JJ)=G3
3190 NEXT JJ
3200 REM ENTRY POINT FOR SECOND AND LATER CALLS
3210 D=(Q-DD(I1,1))/SS
3220 A2=PC(0,2)+D*(PC(1,2)+D*(PC(2,2)+PC(3,2)*D))
3230 A3=PC(0,3)+D*(PC(1,3)+D*(PC(2,3)+PC(3,3)*D))
3240 K2=PC(0,4)+D*(PC(1,4)+D*(PC(2,4)+PC(3,4)*D))
3250 K3=PC(0,5)+D*(PC(1,5)+D*(PC(2,5)+PC(3,5)*D))
3260 L2=PC(0,6)+D*(PC(1,6)+D*(PC(2,6)+PC(3,6)*D))
3270 L3=PC(0,7)+D*(PC(1,7)+D*(PC(2,7)+PC(3,7)*D))
3280 KX=-(C1*SIN(Q)+K2*(UC*SIN(A2)+VC*COS(A2)))
3290 KY=C1*COS(Q)+K2*(UC*COS(A2)-VC*SIN(A2))
3300 LX=-C1*COS(Q)-L2*(UC*SIN(A2)+VC*COS(A2))
-K2^2*(UC*COS(A2)-VC*SIN(A2))
3310 LY=-C1*SIN(Q)+L2*(UC*COS(A2)-VC*SIN(A2))
-K2^2*(UC*SIN(A2)+VC*COS(A2))
3320 X2=C1*COS(Q)+C2*COS(A2)
3330 Y2=C1*SIN(Q)+C2*SIN(A2)
3340 DB=SQR((XB-X2)^2+(YB-Y2)^2)
3350 KD=(C1*SIN(Q)+C2*K2*SIN(A2))*(XB-X2)/DB
-(C1*COS(Q)+C2*K2*COS(A2))*(YB-Y2)/DB
3400 REM  EVALUATION OF GENL INERTIA 7 CENTRIP COEF
3410 IN=J1+J2*K2^2+MM*(KX^2+KY^2)+J3*K3^2
3420 CC=J2*K2*L2+MM*(KX*LX+KY*LY)+J3*K3*L3
3430 REM  EVALUATION OF GENL FORCE
3440 TA=0
```

```
3450 IF T<.075 THEN TA=25
3460 QQ=TA-BB*Q1*KD^2
3500 REM  EVALUATION OF SECOND DERIVATIVE
3510 Q2=(QQ-CC*Q1^2)/IN
3520 RETURN
3800 REM  INTERPOLATION ROUTINE
3810 REM  USING CUBIC APPROXIMATION
3820 REM  JJ DENOTES THE DATA COLUMN INDEX
3830 REM  X IS INDEPENDENT VARIABLE VALUE
3840 REM LOCATE THE INTERVAL
3850 FOR II=1 TO 11
3860 IF X<DD(II,1) THEN GOTO 3880
3870 NEXT II
3880 I1=II-2
3890 I2=II-1
3900 I3=II
3910 I4=II+1
3912 RETURN
3918 REM EXTRACT DATA VALUES, DETERMINE COEF
3920 V1=DD(I1,JJ)
3930 V2=DD(I2,JJ)
3940 V3=DD(I3,JJ)
3950 V4=DD(I4,JJ)
3960 G0=V1
3970 G1=-(11/6)*V1+3*V2-(3/2)*V3+V4/3
3980 G2=V1-(5/2)*V2+2*V3-V4/2
3990 G3=-V1/6+V2/2-V3/2+V4/6
3995 RETURN
4000 REM  TABULATED FOUR-BAR LINKAGE SOLUTION DATA
4002 REM  TABULATION INTERVAL, H = SS
4004 DATA . 174533
4010 REM  Q, A2, A3, ...
4020 DATA 0.0, .884943,4.612222
4030 DATA .174533,.718139,4.466885
4040 DATA .349066,.578997,4.387930
4050 DATA .523599,.472529,4.370945
4060 DATA .698132,.393422,4.401303
4070 DATA .872665,.334725,4.464951
4080 DATA 1.047198,.290661,4.551333
4090 DATA 1.221730,.257188,4.653189
4100 DATA 1.396263,.231627,4.765557
4110 DATA 1.570796,.212260,4.884936
4120 DATA 1.745329,.198053,5.008710
4130 REM  KA2, KA3, LA2, LA3 ...
4140 DATA -1,-1,.201008,1.636776
4150 DATA -.889107,-.647550,.963387,2.249333
4160 DATA -.701629,-.263037,1.097732,2.058619
4170 DATA -.524802,.053105,.904271,1.551687
4180 DATA -.388285,.281256.,664561,1.081449
4190 DATA -.289718,.438151,.475192,.737204
```

```
4200 DATA -.218995,.544950,.343894,.501803
4210 DATA -.167084,.617648,.257113,.341125
4220 DATA -.127434,.666764,.201519,.227781
4230 DATA -.095460,.698806,.167988,.142919
4240 DATA -.067820,.717550,.151259,.073582
```

Executing the program produces the following results.

```
RUN
FOUR BAR MECHANISM SIMULATION
USING INTERPOLATED KINEMATIC SOLUTIONS
TIME STEP = 5E-03

T, Q, QDOT ....

0              .2             0
5E-03          .200572559     .222145051
.01            .202160926     .407832711
.              .              .
.              .              .
.              .              .
.07            .26811508      1.69854426
.075           .276904461     1.76462805
.08            .285264068     1.58402791
.              .              .
.              .              .
.              .              .
.515           .41040956      .049642676
```

The maximum angular velocity for the input crank is 1.7646 rad/s, occurring at $t = 0.075$ s. This coincides with the end of the impulse, as expected. The program terminated as intended when the angular velocity dropped below 0.05 rad/s for the first time at $t = 0.515$ s with the crank position $q = 0.4104$ radians. A brief look at the data preceding termination shows the expected very slowly decaying motion because the process is controlled by viscous friction only. As in the previous example, the objectives for the simulation are met by study of the computed time history of the motion.

7.7 CONCLUSION

Eksergian's equation provides a simple, direct way to obtain the equation of motion for single degree of freedom mechanisms. The alternative approach, based on the application of Newton's Second Law to a group of free bodies, is much longer and more prone to errors for systems of even modest complexity. Furthermore, the alternative approach produces a large system of equations, from which the force terms must be eliminated to obtain the equation of motion. With application of Eksergian's equation, the equation of motion follows in just a few steps after the system kinetic energy is written.

The generalized inertia shows the role of each mass within the system as it effects the complete system inertia. The velocity coefficients tie this all together because they relate the individual component velocities to the generalized velocity. However, perhaps the most useful aspect of Eksergian's equation is the manner in which the effect of varying generalized inertia is taken into account through the centripetal coefficient. The use of velocity coefficient derivatives to express the centripetal coefficient shows the varying effect of each individual mass on the system as a whole.

The generalized force, previously introduced in Chapter 6 as a term that must vanish for equilibrium, is here seen to drive the motion. In some cases, it is useful to split the generalized force into a nonconservative term and a conservative term, the latter expressed as the gradient of a potential function. The velocity coefficients are again useful in expressing these terms.

REFERENCES

Myklebust, A., "Dynamic Response of an Electric Motor–Linkage System During Startup," *J. Mechanical Design, Trans. ASME*, Vol. 104, Jan. 1982.

Myklebust, A., Fernandez, E. F., and Choy, T. S., "Dynamic Response of Slider—Crank Machines During Startup," *J. Mechanisms, Transmissions, and Automation in Design, Trans. ASME*, Vol. 106, Dec. 1984.

Paul, B., "Analytical Dynamics of Mechanisms—A Computer Oriented Overview," *Mechanism and Machine Theory*, Vol. 10, 1975.

Paul, B. *Kinematics and Dynamics of Planar Machinery*. Englewood Cliffs, N.J.: Prentice–Hall, 1979.

PROBLEM SET

For all of the problems in this set, the objective is to formulate the equation of motion and the appropriate initial conditions. For that purpose, the following five parts are required for each problem:

a. Kinematic analysis as required;
b. Determination of kinetic energy, generalized inertia, and centripetal coefficient;
c. Determination of the complete generalized force, or the nonconservative generalized force and the appropriate potential function;
d. Determination of the equation of motion;
e. Determination of the appropriate initial conditions.

With regard to item a, in those cases where an analytical position solution appears practical, carry it out. In cases where no analytical position solution is evident, set up the position equations and note that a numerical solution is required. Then, develop the velocity coefficients, velocity coefficient derivatives, base coordinates, and so forth as will be required for the later parts of the problem. For the later parts, do not substitute the expressions for secondary variables, velocity coefficients, and velocity coefficient derivatives, but simply refer to them by the standard notations. Be careful to use consistent subscripting throughout.

Many of the data items are indicated on the figures, such as M_1 or I_{4c}. These are understood to be known values, with the subscripts identifying the body with which they are associated. A subscript c on a mass moment of inertia is understood to indicate the center of mass as the reference point for that value. Do not consider any gravitational effects unless the gravity vector is shown in the figure.

In most cases the initial values can be determined in closed form, but there may be some cases where a numerical process is required. If so, set up the equations to be solved and identify the numerical procedure to be used, but do not attempt to complete the solution.

In those cases where a generalized coordinate has been denoted by q, the equation of motion should be written in terms of that variable. If no generalized coordinate is indicated, then a suitable choice must be made.

• **7.1** The two sliders with masses M_1 and M_2 are separated by a constant distance L. The center of mass for the connecting link is located by the body polar coordinates (R_c, A_c). Forces F_x and F_y act at the point located by the body polar coordinates (R_f, A_f). Note the dashpot acting on the vertical slider, developing a force equal to the dashpot coefficient B multiplied by the velocity of the slider. The slider M_1 is initially moving to the right with speed V_{xo}, and slider pivot is at position X_o.

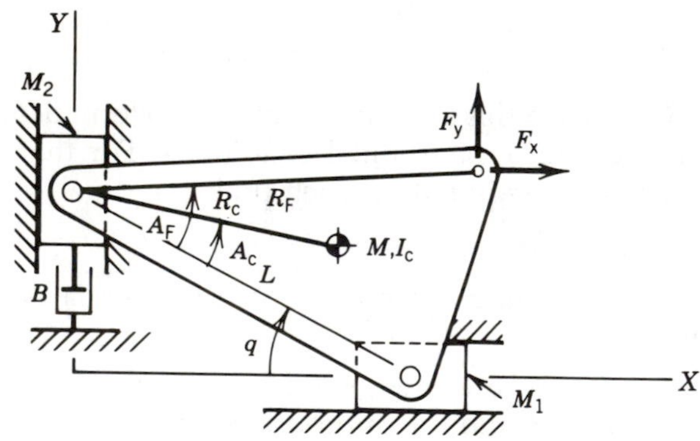

7.2 The distance between the two slider pivots is L and the link has mass M_3 and mass moment of inertia I_{3c}; the center of mass for the link is located by the body coordinates (U_c, V_c). The motion of the system is driven by the force $F(t)$ and is also subject to the spring force. The spring rate is K and the free length is such that the spring is relaxed for $q = q_o$.

The system is released from rest with the link at an angle $A = A_o$ above the horizontal.

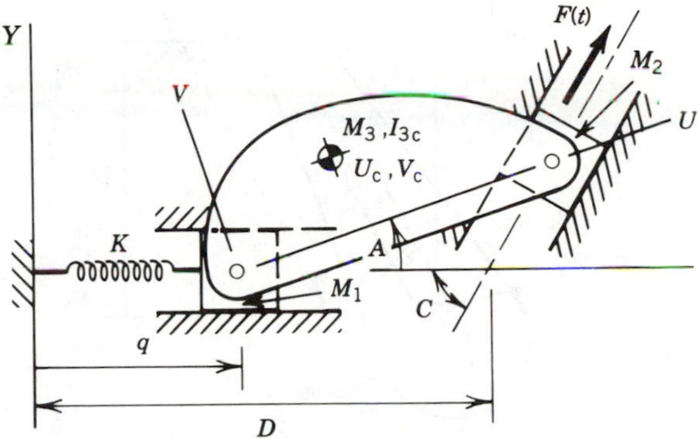

7.3 Each of the two links has uniform mass distribution, m = mass/length, and the sliders have masses M_1, M_2, and M_3. The motion of the system is driven by the force F and opposed by the dashpot action. The system is initially located such that $X_2 = X_{20}$, while the velocity of M_1 is V_{10}.

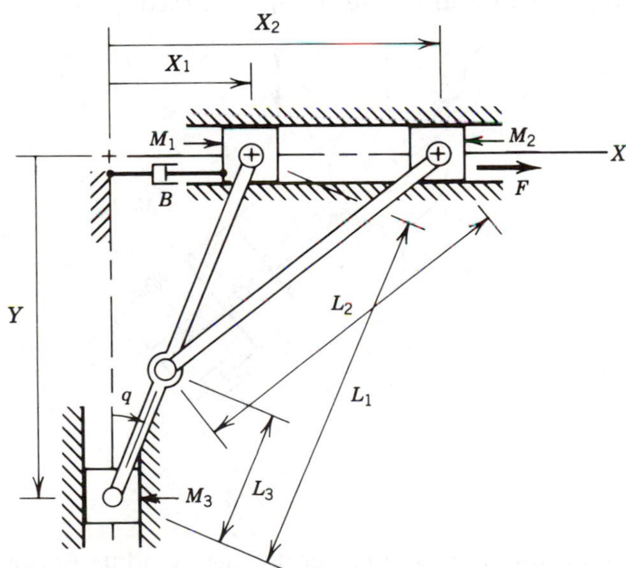

7.4 The roller rolls without slipping on the horizontal surface, while the lever pivots about the stationary pin in the slot. Gravity acts vertically downward. The system is initially at $q = 0$ with the roller moving to the right at the speed V_{xo}.

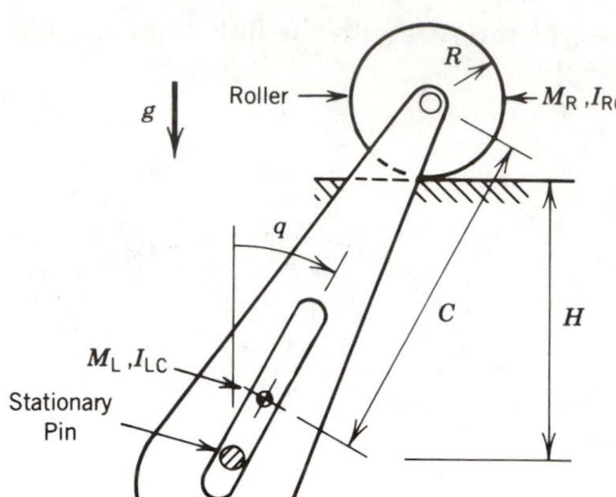

7.5 A slider of mass M_s and centroidal mass moment of inertia I_{sc} moves along the pendulum shaft under the control of an inextensible tape. The other end of the tape is wrapped around a stationary circular drum. When the pendulum is vertical, the slider position is $R = R_o$. There is also a spring–dashpot assembly acting between the pendulum and the slider. The pendulum center of mass is at radius A and the pendulum has mass M_p and mass moment of inertia I_{pc}. The system is released from rest with $R = L/2$ and, at that position, the spring is relaxed.

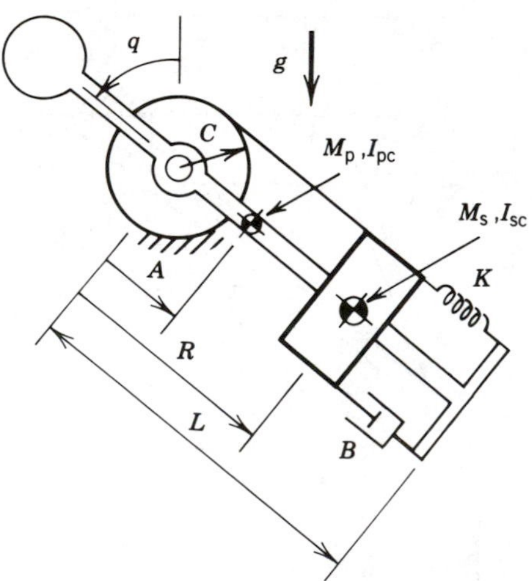

7.6 The pendulum has its center of mass at radius A; the mass is M_p and the centroidal mass moment of inertia is I_{pc}. The slider position is controlled by an inextensible tape that wraps around the stationary circular drum. When the pendulum is vertical, the slider position is $R = R_o$. A spring with rate K and free length S_o is attached between the upper end of the pendulum and a stationary support. The system is initially vertical, and the slider is moving up the pendulum shaft with speed V_{so}.

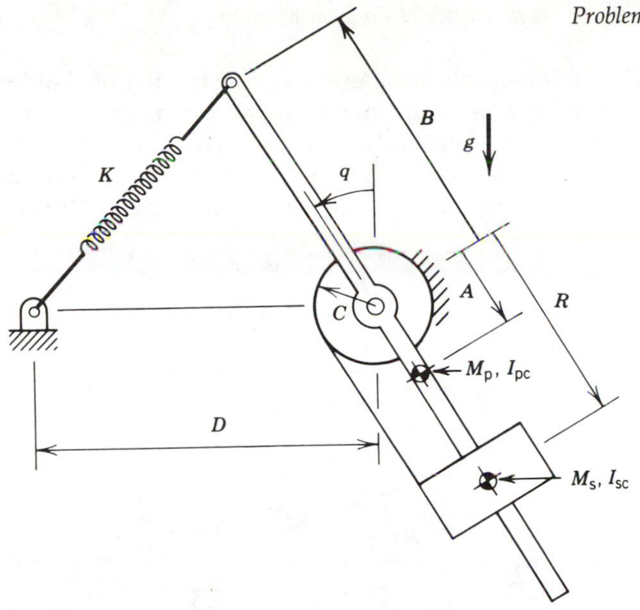

• **7.7** The mass M_3 (with mass moment of inertia I_{3c}) slides without friction on both the ramp and the pendulum shaft. When the pendulum shaft is vertical, the slider center of mass is a distance B below the pivot. The pendulum shaft is of length L and mass M_1, while the bob is a thin, uniform disk of mass M_2 and diameter D. The associated mass moments of inertia should be determined. The system is released with $q = 0$, and the slider is moving with speed V_{30} relative to the ramp, measured parallel to the ramp.

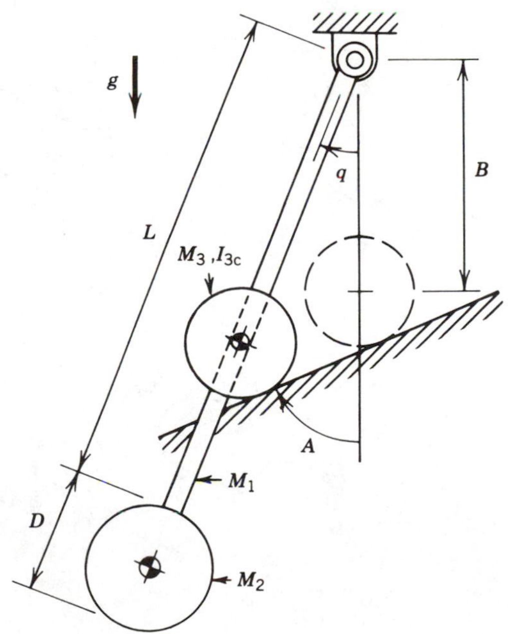

7.8 The slider–crank mechanism is constructed of slender bars, with masses and mass moments of inertia as indicated; the slider mass is M_s. The system motion is controlled by the applied force F, the applied couple C, and the viscous force in the dashpot. At time $t = 0$, the connecting rod obliquity is A_o, and the slider is moving to the right with velocity V_o.

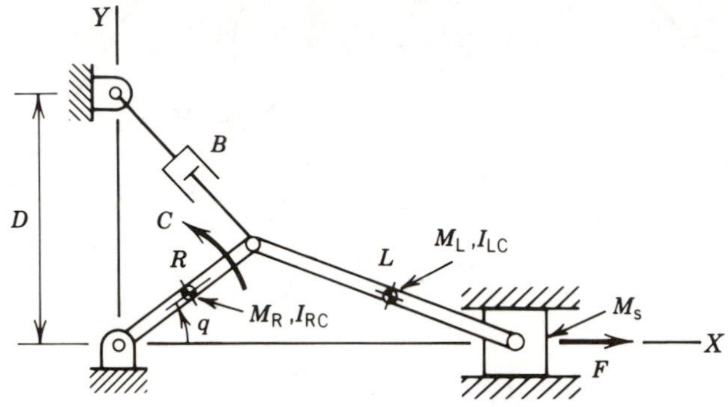

• **7.9** The slider–crank mechanism shown falls under the influence of gravity. There is viscous friction between the slider and the guide, described by viscous coefficient B. The crank mass is M_1, the connecting rod mass is M_2, and the mass of the slider is M_3. The associated centroidal mass moments of inertia are I_{1c}, I_{2c}, and I_{3c}. The system is released with the crank and the connecting rod colinear, and with the connecting rod rotating clockwise at ω_o rad/s, $\omega_o > 0$.

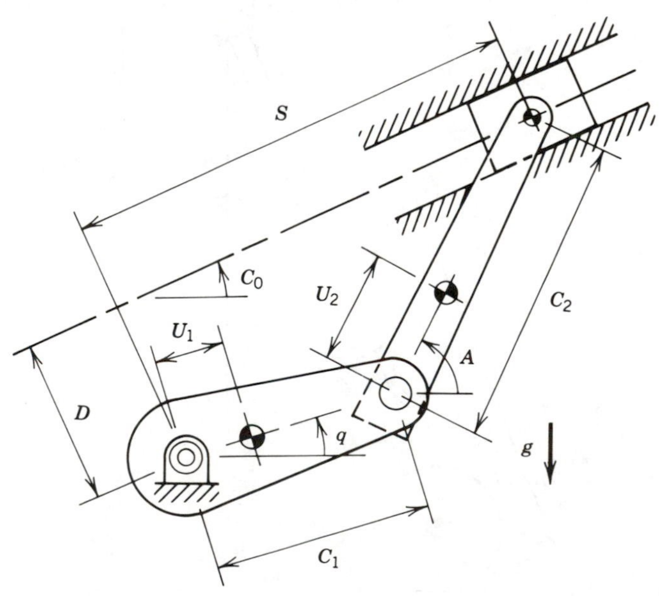

7.10 The figure shows a double slider–crank mechanism with an articulated connecting rod. The crank has mass moment of inertia I_o with respect to the fixed pivot point, and the sliders have masses M_x and M_y, respectively. The master rod is a uniform right triangle with legs C_1 and C_2, mass M_2, and centroidal mass moment of inertia I_{2c}. The slave rod is a slender uniform link with mass M_3 and centroidal mass moment of inertia I_{3c}. Forces F_x and F_y act on the mechanism as shown, and the system is released with M_y at its highest point and M_x traveling to the left at speed V_{xo}.

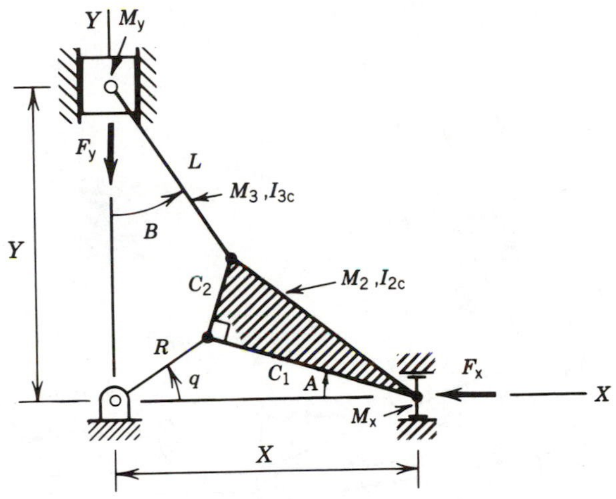

7.11 The trammel crank drive involves a drive disk with mass moment of inertia I_{wc}, two sliders with mass M_s and centroidal mass moment of inertia I_{sc}, a third slider with mass M_x, and a uniform connecting rod with mass distribution m = mass/length. Viscous friction, described by the coefficient B, acts at three places as indicated. The system is driven by the

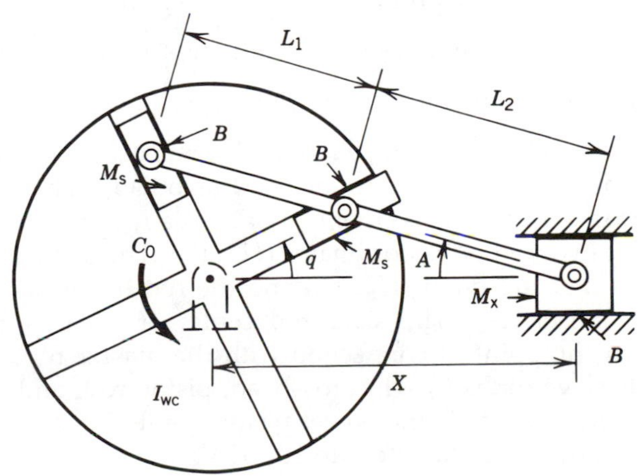

couple C_o as indicated. Initially, the slider M_x is at the midpoint of its stroke and its speed is V_{xo} to the right.

• **7.12** The figure shows one crank throw of an *integral engine–compressor,* a type of machine commonly used in the gas pipeline industry. The double-acting compressor cylinder is horizontal, whereas the power cylinder is inclined to the horizontal by the angle C. An articulated connecting rod mechanism is used, with the master rod attached to the compressor crosshead. The slave rod goes to the power piston.

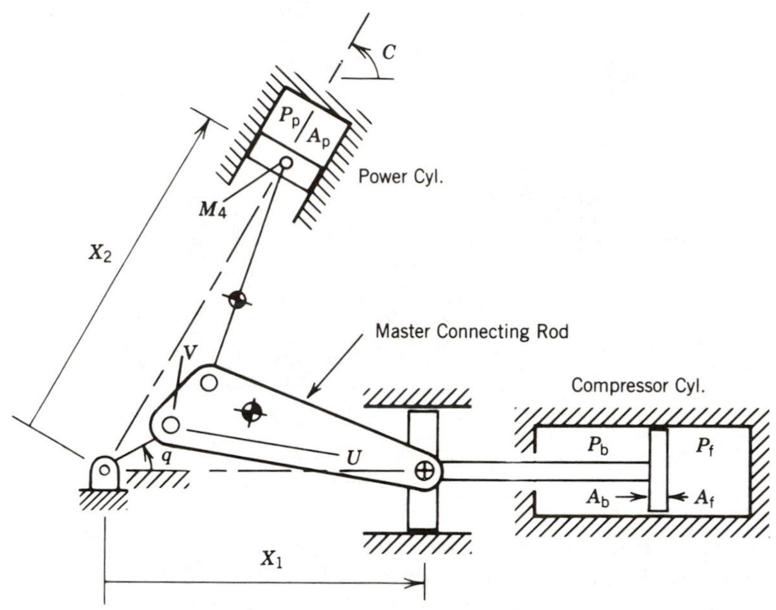

Consider the cylinder pressures to be the following known functions:

$P_f = P_f(X_1, \dot{X}_1)$ compressor front-side pressure

$P_b = P_b(X_1, \dot{X}_1)$ compressor back-side pressure

$P_p = P_p(X_2, \dot{X}_2)$ power cylinder pressure

These pressures act on areas A_f, A_b, and A_p, respectively. The crank mass moment of inertia with respect to the fixed pivot is I_o. The master rod has length L_1, mass M_1, and mass moment of inertia I_{1c}; the master rod center of mass is located by body coordinates (U_c, V_c). The slave rod attaches to the master rod at the point (U_p, V_p) and the rod has mass M_2 and mass moment of inertia I_{2c}. The slave rod center of mass is a distance U_2 upward from the point of connection with the master rod, and the total length of the slave rod is L_2. The crosshead, piston rod, and piston assembly has mass M_3. At $t = 0$, the power piston is at top dead center, and the crosshead is moving to the left with speed V_{xo}.

7.13 Each link has uniform mass distribution, with m = mass/length, and the slider mass is M_s. The system is subject to the applied force F and the viscous force on the lower face of the slider. At $t = 0$, the system is at rest with the length L_2 perpendicular to L_3.

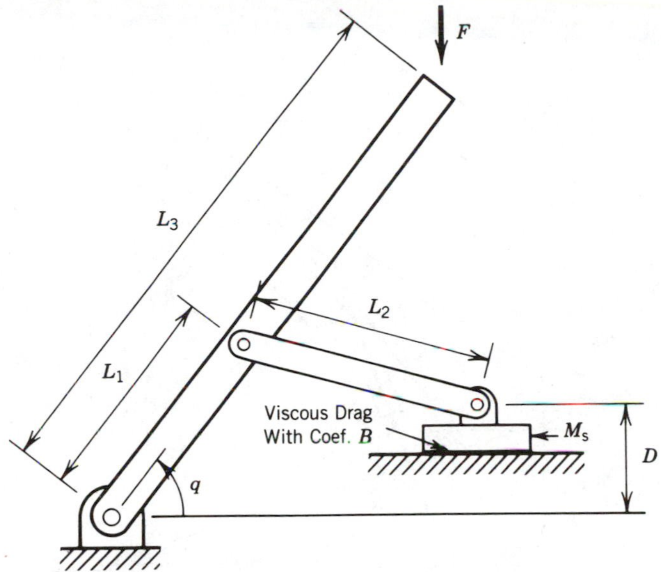

7.14 The figure shows a four-bar linkage with a massive coupler link; the other links are massless. The center of mass of the coupler link is located by (U_c, V_c), as indicated. The system is driven by applied forces F_x and F_y, and subject to a retarding force developed in the dashpot. When the motion begins, the link C_1 is along the Y-axis, and the coupler center of mass has the horizontal velocity V_{xo}.

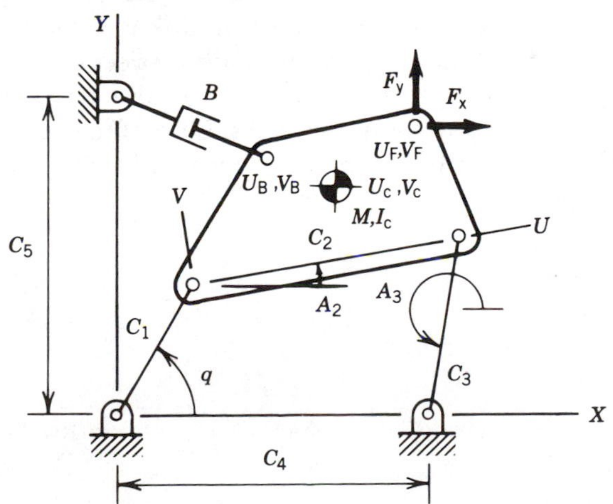

• **7.15** The system shown is a safety steam shutoff valve, designed to function by gravity in the event of an emergency. Essentially, it is a four-bar linkage, with the valve plug as the only significant mass. Note that the closing motion is retarded by a dashpot to reduce the impact on the valve seat. The system is released from rest with the lengths L_2 and L_3 colinear.

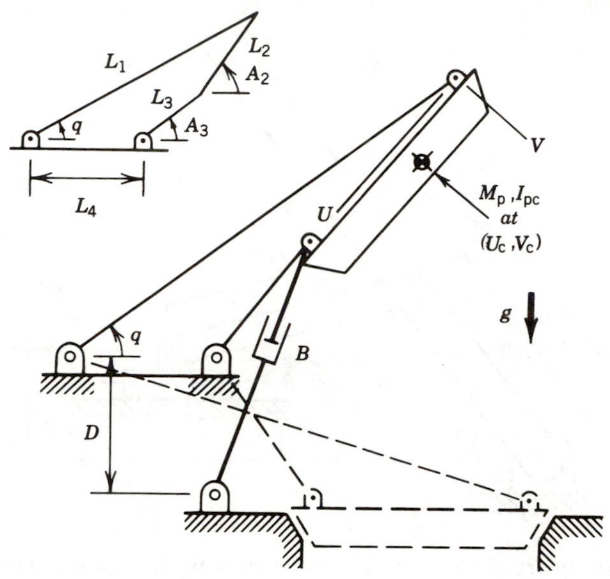

7.16 The quick return mechanism consists of two sliders of mass M_1 and M_2, respectively, and uniform bars for the connecting link and the crank. The length of the connecting link is L and the crank radius is R. The slider where the crank drives the connecting link is considered massless. The mass moment of inertia of the crank with respect to the fixed pivot is I_r. The system is driven by the moment M_o on the crank and F_x on the horizontal slider. Viscous friction, described by coefficients B_1 and B_2, acts on the two sliders. At $t = 0$, the crank is perpendicular to the connecting link, and rotating clockwise at 20 rad/s.

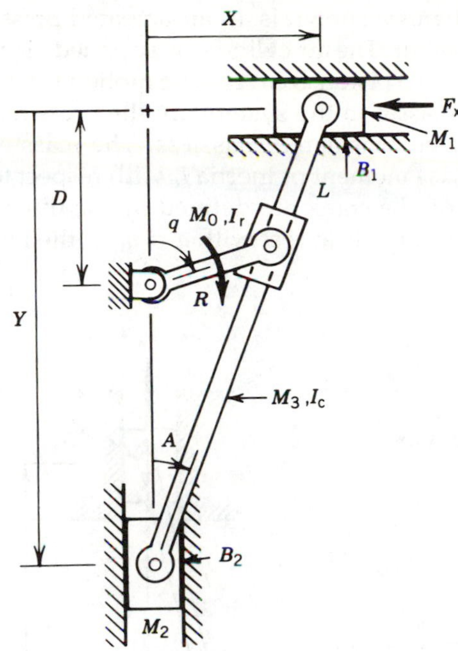

• **7.17** The figure shows another variation on the quick return mechanism. The lever has mass moment of inertia with respect to the fixed point I_o. The system is driven by the moment M_o and the force F acting on the horizontal slider. Viscous friction, described by B_s, acts on the slider. Viscous friction, described by B_p, also acts on *both* pins sliding in the lever slots. The system is initially at $q = 0$ with the slider moving to the right with speed V_{xo}.

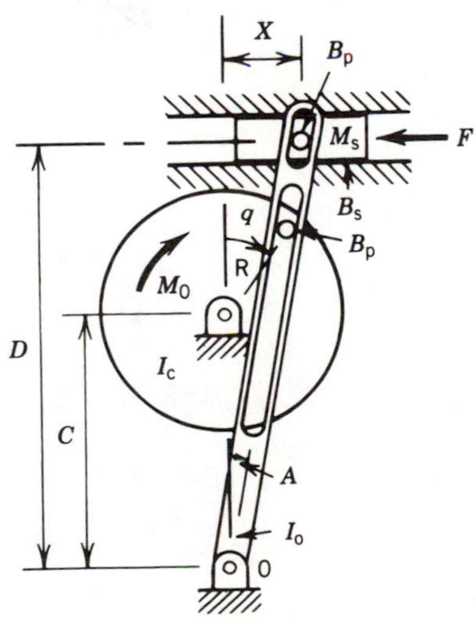

7.18 The mechanism shown is an air-actuated press that is used in a manufacturing operation. The air cylinder is supplied air at gauge pressure P_a, acting on an effective piston area A_a. The motion is resisted by the force F_w. The significant masses in the system are the press ram and the crank; the other parts may be considered massless. The mass of the ram is M_r, and the crank has mass moment of inertia I_o with respect to the fixed point. The pivot locations on the crank are defined by the three distances S_1, S_2, and S_3. The system is initially at rest with $q = q_{min}$, the minimum length of the air cylinder.

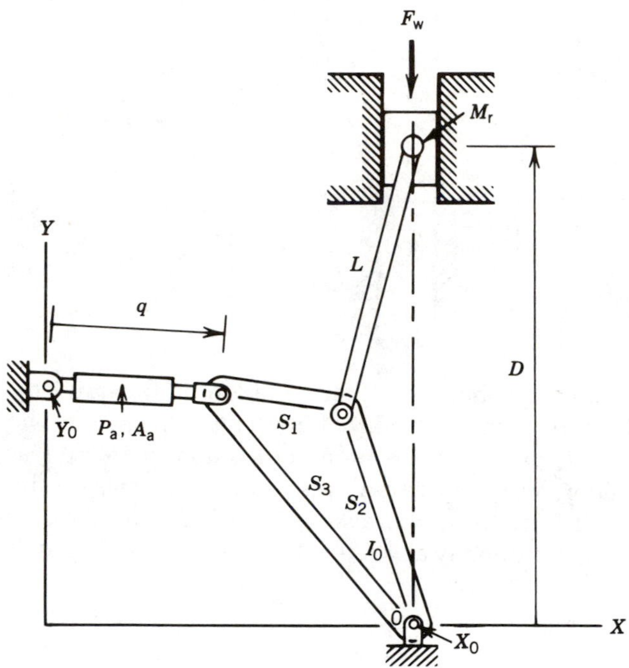

CHAPTER 8

Dynamics of Multidegree of Freedom Machines: The Lagrange Equation of Motion

Chapter Seven dealt with the dynamics of single degree of freedom machines, and obtained the equation of motion for such systems from work and energy considerations by using Eksergian's form for the equation of motion. This chapter complements that chapter by presenting a parallel development for the dynamics of multidegree of freedom machines, using the Lagrange form for the equation of motion. As in the previous case, the kinetic energy plays a central role, and the use of potential energy is again optional. The Lagrange form can also be applied to single degree of freedom systems, although it is slightly less convenient than Eksergian's equation. The two approaches reduce to exactly the same describing equation for a single degree of freedom problem.

8.1 KINETIC ENERGY FOR A MULTIDEGREE OF FREEDOM MACHINE

The kinetic energy for each rigid body in the system can again be written as two terms, one that represents the energy associated with the motion of the center of mass, and a second that reflects the rotational kinetic energy of the body (see Appendix A6 for a review of this development). Consequently, the kinetic energy of the system can be written as

$$T = 0.5\, M_1\{V_{1c}\}^t\{V_{1c}\} + 0.5\,\{\omega_1\}^t[I_{1cm}]\{\omega_1\} + 0.5\, M_2\{V_{2c}\}^t\{V_{2c}\} + 0.5\,\{\omega_2\}^t[I_{2cm}]\{\omega_2\} + \cdot\cdot\cdot\cdot$$

where the $\{V_{ic}\}$ is the velocity vector for the center of mass of the ith body, and $\{\omega_i\}$ is the angular velocity vector for the same body. From Chapter 3, recall that the velocity components for any center of mass or other point of interest can be written as a linear combination of the generalized coordinate velocities. For multiple degrees of freedom, the velocity coefficients in general form a rectangular matrix, so that:

$$\begin{Bmatrix} V_{icx} \\ V_{icy} \\ V_{icz} \\ \omega_{ix} \\ \omega_{iy} \\ \omega_{iz} \end{Bmatrix} = \begin{bmatrix} K_{icx1} & K_{icx2} & K_{icx3} & \cdots \\ K_{icy1} & K_{icy2} & K_{icy3} & \cdots \\ K_{icz1} & K_{icz2} & K_{icz3} & \cdots \\ K_{i\omega x1} & K_{i\omega x2} & & \cdots \\ K_{i\omega y1} & \cdot & \cdot & \\ K_{i\omega z1} & \cdot & \cdot & \end{bmatrix} \begin{Bmatrix} \dot{q}_1 \\ \dot{q}_2 \\ \vdots \\ \dot{q}_n \end{Bmatrix}$$

A complete velocity coefficient matrix may be assembled, giving all center of mass velocities and rotation rates in terms of the generalized coordinate velocities, so that all of the velocities can be written as

$$\begin{Bmatrix} \dot{X}_{c1} \\ \dot{X}_{c2} \\ \vdots \\ \dot{Y}_{c1} \\ \dot{Y}_{c2} \\ \vdots \\ \omega_{1x} \\ \omega_{2x} \\ \vdots \end{Bmatrix} = \begin{bmatrix} K_{11} & K_{12} & K_{13} & \cdots \\ K_{21} & K_{22} & K_{23} & \cdots \\ \cdot & \cdot & \cdot & \cdots \\ \cdot & \cdot & \cdot & \cdots \\ \cdot & \cdot & \cdot & \cdots \\ \cdot & \cdot & \cdot & \cdots \\ \cdot & \cdot & \cdot & \cdots \\ \cdot & \cdot & \cdot & \cdots \\ \cdot & \cdot & \cdot & \cdots \end{bmatrix} \begin{Bmatrix} \dot{q}_1 \\ \dot{q}_2 \\ \vdots \\ \dot{q}_n \end{Bmatrix}$$

or,

$$\{V\} = [K_c]\,\{\dot{q}\}$$

The kinetic energy can then be written as

$$T = 0.5\,\{\dot{q}\}^t[K_c]^t[M][K_c]\{\dot{q}\}$$

where the inertia matrix is

$$[M] = \begin{bmatrix} M_1 & & & & & \\ & M_2 & & & & \\ & & \ddots & & & \\ & & & [I_{c1}] & & \\ & & & & [I_{c2}] & \\ & & & & & \ddots \end{bmatrix}$$

The M_i values in the inertia matrix are the masses of the various components, and the $[I_{ci}]$ are their respective moment of inertia matrices. In the event that the motion is planar, three scalar components replace each moment of inertia matrix, and the entire inertia matrix becomes diagonal.

A note of caution is needed at this point similar to one raised in Chapter 7. The preceding kinetic energy expression depends on being able to express the required velocities as the product of a velocity coefficient matrix with the vector of generalized velocities. If there is a base displacement type of excitation, say an $X_o(t)$, one of the velocity components required to develop the kinetic energy may be of the form $(\dot{X}_o + \dot{q}\,K_v)$. In that case, the scalar forms given below for the Lagrange equation can still be applied, but the matrix forms that follow no longer apply.

The use of matrix products to express the kinetic energy in terms of generalized velocities and velocity coefficient matrices is a very powerful tool, as will be illustrated later in Section 8.4.2. It is not always useful, however, to rely on the matrix operations formalism. At times, it is more convenient simply to write out the appropriate sum of products for the kinetic energy; this is especially true when the center of mass velocity coefficient matrix contains many zeroes. For large, complicated problems, however, it is nice to know that the matrix approach is available.

8.2 FIRST GENERAL FORM FOR THE LAGRANGE EQUATION

In the following discussion, the Lagrange equation is presented without a derivation (for a derivation, see Appendix A9). The first general form for the Lagrange equation expresses the equation of motion in terms of only the kinetic energy and the generalized force. The generalized forces for a multidegree of freedom system were developed in Section 6.4. As shown there, if $[K_x]$ is a velocity coefficient matrix that relates the virtual displacements at the points of application of the external forces to the virtual changes in the generalized coordinates, then

$$\begin{Bmatrix} \delta X_1 \\ \delta X_2 \\ \delta X_3 \\ \vdots \end{Bmatrix} = [K_x] \begin{Bmatrix} \delta q_1 \\ \delta q_2 \\ \delta q_3 \\ \vdots \end{Bmatrix}$$

The generalized forces are then determined from the actual forces by

$$\begin{Bmatrix} Q_1 \\ Q_2 \\ Q_3 \\ \vdots \end{Bmatrix} = [K_x]^t \begin{Bmatrix} F_1 \\ F_2 \\ F_3 \\ \vdots \end{Bmatrix}$$

The first general form for the Lagrange equation is

$$\frac{d}{dt}\frac{\partial T}{\partial \dot{q}_j} - \frac{\partial T}{\partial q_j} = Q_j$$

For each generalized coordinate, q_j, there is an equation of motion of this form. The application of this result is rather automatic. First, the system kinetic energy is written and the required derivatives determined. Second, the generalized forces are determined. Finally, the equations of motion are assembled. Having said that, it should also be added that these steps often involve quite a lot of effort, but the process remains basically quite straightforward.

For problems involving many degrees of freedom, it may be useful to express the Lagrange equation in matrix form, so that all of the equations of motion are developed at once:

$$\frac{d}{dt}\frac{\partial T}{\partial \{\dot{q}\}} - \frac{\partial T}{\partial \{q\}} = \{Q\}$$

The right side of this equation is simply a column vector composed of the generalized forces; the terms on the left side will be developed in the discussion that follows. For that purpose, consider the kinetic energy in the form

$$T = 0.5\,\{\dot{q}\}^t[K_c]^t[M][K_c]\{\dot{q}\}$$

as in the previous section. The second term on the left side of the Lagrange equation is the derivative of the kinetic energy with respect to a particular generalized coordinate, q_j:

$$\frac{\partial T}{\partial q_j} = 0.5\,\{\dot{q}\}^t([L_j]^t[M][K_c] + [K_c]^t[M][L_j])\{\dot{q}\}$$

where the $[L_j]$ are velocity coefficient partial derivative matrices, as developed in Chapter 3. Because this pattern will appear repeatedly in this section, it is useful to define a matrix $[N_j]$:

$$[N_j] = 0.5\,([L_j]^t[M][K_c] + [K_c]^t[M][L_j])$$

The second matrix product in the definition of $[N_j]$ is the transpose of the first, so only one product of three matrices must be computed. The matrix $[N_j]$ is said to be the *symmetric part* of the matrix $[L_j]^t[M][K_c]$. For the Lagrange equation in matrix form, the corresponding term is the partial derivative of the kinetic energy with respect to the column vector of generalized coordinates:

$$\frac{\partial T}{\partial \{q\}} = \begin{Bmatrix} \underline{\{\dot{q}\}^t[N_1]\{\dot{q}\}} \\ \underline{\{\dot{q}\}^t[N_2]\{\dot{q}\}} \\ \underline{\{\dot{q}\}^t[N_3]\{\dot{q}\}} \\ \underline{\{\dot{q}\}^t[N_4]\{\dot{q}\}} \\ \vdots \quad \vdots \quad \vdots \end{Bmatrix}$$

$$= \left(\begin{bmatrix} \{\dot{q}\}^t \\ \hline (-0-) \\ \hline (-0-) \\ \hline \vdots \end{bmatrix} [N_1] + \begin{bmatrix} (-0-) \\ \hline \{\dot{q}\}^t \\ \hline (-0-) \\ \hline \vdots \end{bmatrix} [N_2] + \begin{bmatrix} (-0-) \\ \hline (-0-) \\ \hline \{\dot{q}\}^t \\ \hline \vdots \end{bmatrix} [N_3] + \cdots \right) \{\dot{q}\}$$

This last form facilitates the eventual combination of similar terms in the Lagrange equation.

The derivative of the kinetic energy with respect to the column vector of generalized velocities is also required for the Lagrange equation in matrix form. This is

$$\frac{\partial T}{\partial \{\dot{q}\}} = [K_c]^t[M][K_c]\{\dot{q}\}$$

A time differentiation of this quantity is then needed to complete the first term of the Lagrange equation:

$$\begin{aligned} \frac{d}{dt}\frac{\partial T}{\partial \{\dot{q}\}} &= [K_c]^t[M][K_c]\{\ddot{q}\} + \dot{q}_1([L_1]^t[M][K_c] + [K_c]^t[M][L_1])\{\dot{q}\} \\ &\quad + \dot{q}_2([L_2]^t[M][K_c] + [K_c]^t[M][L_2])\{\dot{q}\} + \cdots \\ &= [K_c]^t[M][K_c]\{\ddot{q}\} + (2\dot{q}_1[I][N_1] + 2\dot{q}_2[I][N_2] + \cdots)\{\dot{q}\} \end{aligned}$$

In this expression $[I]$ is an identity matrix of the appropriate order. At this point, the left side of the Lagrange equation can be assembled in matrix form:

$$\begin{aligned} \frac{d}{dt}\frac{\partial T}{\partial \{\dot{q}\}} - \frac{\partial T}{\partial \{q\}} &= [K_c]^t[M][K_c]\{\ddot{q}\} \\ &\quad + (2\dot{q}_1[I][N_1] + 2\dot{q}_2[I][N_2] + \cdots)\{\dot{q}\} \\ &\quad - \left(\begin{bmatrix} \{\dot{q}\}^t \\ \hline (-0-) \\ \hline (-0-) \\ \hline \vdots \end{bmatrix} [N_1] + \begin{bmatrix} (-0-) \\ \hline \{\dot{q}\}^t \\ \hline (-0-) \\ \hline \vdots \end{bmatrix} [N_2] + \cdots \right) \{\dot{q}\} \end{aligned}$$

For a more compact notation, $[W_j]$ is defined such that:

$$[W_j] = \left(2\dot{q}_j[I] - \begin{bmatrix} (-0-) \\ \hline \vdots \\ \hline \{\dot{q}\}^t \\ \hline (-0-) \\ \hline \vdots \end{bmatrix} \right)$$

Here only the jth row of the second matrix is non-zero. With this notation, the matrix equation of motion is finally reduced to the form

$$[K_c]^t[M][K_c]\{\ddot{q}\} + \left(\sum_j [W_j][N_j]\right)\{\dot{q}\} = \{Q\}$$

8.3 SECOND GENERAL FORM FOR THE LAGRANGE EQUATION

The second general form results when the potential energy is used to represent the conservative parts of the generalized forces. This type of modification was made in Section 6.2 for virtual work applied to statics, and again in Section 7.4 for single degree of freedom dynamics. With this separation of the generalized force into conservative and nonconservative parts, the Lagrange equation reads as follows:

$$\frac{d}{dt}\frac{\partial T}{\partial \dot{q}_j} - \frac{\partial T}{\partial q_j} + \frac{\partial V}{\partial q_j} = Q_j^{nc}$$

where the potential energy term has been shifted to the left side of the equation. This modification to include the use of a potential energy function can also be made in the matrix form for the equation of motion. In general, there is little that can be done with the potential energy term other than to include a column vector of partial derivatives of the potential energy with respect to the coordinates. This happens because the details depend on the form for the potential function.

It is a common practice to define a new function, called the *Lagrangian function*, that is the difference between the kinetic and potential energy functions:

$$L = T - V$$

The Lagrange equation of motion can then be written as

$$\frac{d}{dt}\frac{\partial L}{\partial \dot{q}_j} - \frac{\partial L}{\partial q_j} = Q_j^{nc}$$

The additional term introduced into the equation in this last form, $\partial V/\partial \dot{q}_j$, is zero because the potential energy is a function of position only. The usefulness of the Lagrangian function itself is questionable, but it is a widely used notation.

The Lagrange form for the equation of motion is a very powerful tool for multidegree of freedom systems. Its application will be demonstrated in the next section of this chapter.

8.4 APPLICATIONS OF THE LAGRANGE EQUATION

This section presents two examples, each involving two degrees of freedom, in which the Lagrange form is used to develop the equations of motion. The first example is relatively simple in terms of the kinematics involved, and is formulated without the matrix formalism. For the second example, the matrix formulation is useful because of the greater complexity of the mechanism.

8.4.1 Centrifugal Governor Mechanism Response

A type of centrifugal governor mechanism is shown in Figure 8.1, where the sensing mass moves radially on a spoke of the flywheel. The sensing element has mass M and centroidal mass moment of inertia, I_c. The mass moment of inertia for the flywheel, with respect to the axis of rotation, is I_o. The motion of the sensing mass is resisted by a spring of stiffness K. When the coordinate X is zero, the spring is relaxed. The external torque $T_0(A, X, t)$ acts on the flywheel. The equations of motion are to be determined.

Let the two degrees of freedom be associated with the angle A and the displacement X. The position and velocity of the center of mass of the sensing mass are

$$\mathbf{R}_c = (R_o + X)\,\mathbf{E}_r$$
$$\dot{\mathbf{R}}_c = \dot{X}\,\mathbf{E}_r + (R_o + X)\,\dot{A}\,\mathbf{E}_a$$

The kinetic energy of the system is written by inspection:

$$T = 0.5\,I_o\,\dot{A}^2 + 0.5\,M[\dot{X}^2 + (R_o + X)^2\dot{A}^2] + 0.5\,I_c\,\dot{A}^2$$
$$= 0.5\,\dot{A}^2\,[I_o + I_c + M(R_o + X)^2] + 0.5\,M\,\dot{X}^2$$

The potential energy of the system consists of the gravitational potential energy of the sensing mass and the strain energy of the spring:

$$V = M\,g(R_o + X)\sin A + 0.5\,K\,X^2$$

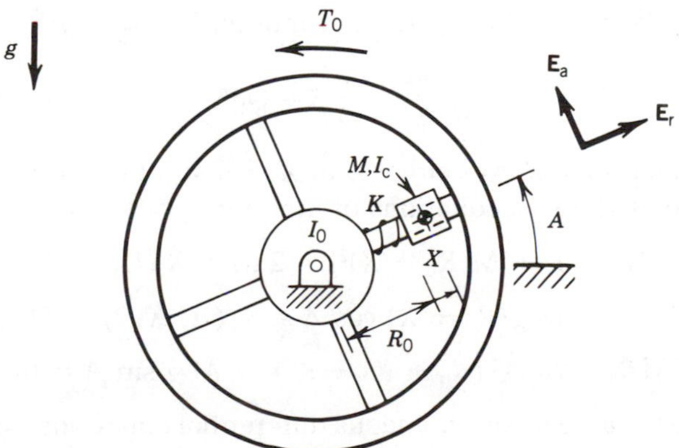

FIGURE 8.1 Centrifugal Governor Mechanism

The kinetic and potential energies can then be used to express the Lagrangian function:

$$\begin{aligned} L &= T - V \\ &= 0.5\,\dot{A}^2\,[I_o + I_c + M(R_o + X)^2] + 0.5\,M\,\dot{X}^2 \\ &\quad - M\,g(R_o + X)\sin A - 0.5\,K\,X^2 \end{aligned}$$

The required derivatives of the Lagrangian function are

$$\frac{\partial L}{\partial A} = -M\,g(R_o + X)\cos A$$

$$\frac{\partial L}{\partial \dot{A}} = \dot{A}\,[I_o + I_c + M(R_o + X)^2]$$

$$\frac{d}{dt}\frac{\partial L}{\partial \dot{A}} = \ddot{A}\,[I_o + I_c + M(R_o + X)^2] + 2\,M\,\dot{A}\,\dot{X}\,(R_o + X)$$

$$\frac{\partial L}{\partial X} = M\,\dot{A}^2\,(R_o + X) - K\,X - M\,g\sin A$$

$$\frac{\partial L}{\partial \dot{X}} = M\,\dot{X}$$

$$\frac{d}{dt}\frac{\partial L}{\partial \dot{X}} = M\,\ddot{X}$$

The last step is to define the nonconservative generalized forces. To that end, first the nonconservative virtual work is determined:

$$\delta W^{nc} = T_o(A, X, t)\,\delta A$$

By inspection, the nonconservative generalized forces are

$$Q_a^{nc} = T_o(A, X, t)$$

$$Q_x^{nc} = 0$$

The Lagrange form for the equation of motion is

$$\frac{d}{dt}\frac{\partial L}{\partial \dot{q}_j} - \frac{\partial L}{\partial q_j} = Q_j^{nc}$$

where, for this problem, q_j is either A or X. Substitution of the information developed gives the two equations of motion:

$$\begin{aligned} \ddot{A}[I_o + I_c + M(R_o + X)^2] + 2\,M\,\dot{A}\,\dot{X}\,(R_o + X) \\ + M\,g(R_o + X)\cos A = T_o(A, X, t) \end{aligned}$$

$$M\,\ddot{X} - M\,\dot{A}^2\,(R_o + X) + K\,X + M\,g\sin A = 0$$

These are the nonlinear, simultaneous differential equations for the centrifugal governor, for which a closed-form solution is probably not available. Consequently, a numerical solution must be performed.

8.4.2 Four-Bar Mechanism with Translating Crank Pivot

The mechanism for this example is shown in Figure 8.2 and Figure 8.3. It is not actually a four-bar mechanism because there is no actual fourth bar, but the name is descriptive. Note that the input crank pivot translates along the X–axis; this translation and the input crank rotation will be used to describe the two degrees of freedom. The centers of mass for the three links are located by known body coordinates (U_{c1}, V_{c1}), (U_{c2}, V_{c2}), and (U_{c3}, V_{c3}) in their respective body coordinate systems. The motion is driven by the force F, acting parallel to the X–axis as shown. Determine the equations of motion for this system.

This mechanism was considered previously in Section 3.2, in which the loop equations were written, the velocity coefficient matrix relating $\dot{A}_2$ and $\dot{A}_3$ to $\dot{q}_0$ and $\dot{q}_1$ was developed, and the velocity coefficient matrix for the center of mass was determined. From Section 3.2, that center of mass velocity relation was

$$\left\{\begin{matrix} \dot{X}_{c1} \\ \dot{Y}_{c1} \\ \dot{X}_{c2} \\ \dot{Y}_{c2} \\ \dot{X}_{c3} \\ \dot{Y}_{c3} \\ \dot{q}_1 \\ \dot{A}_2 \\ \dot{A}_3 \end{matrix}\right\} = \begin{bmatrix} & [K_1] & \\ & [K_2] & \\ & [K_3] & \\ 0 & & 1 \\ K_{a20} & & K_{a21} \\ K_{a30} & & K_{a31} \end{bmatrix} \begin{Bmatrix} \dot{q}_0 \\ \dot{q}_1 \end{Bmatrix} = [K_c] \begin{Bmatrix} \dot{q}_0 \\ \dot{q}_1 \end{Bmatrix}$$

$$(9 \times 1) \qquad (9 \times 2) \quad (2 \times 1)$$

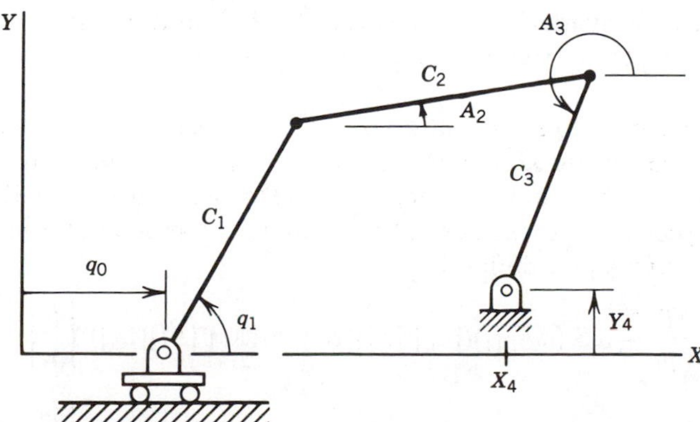

FIGURE 8.2 Kinematic Skeleton for Four-Bar Mechanism with Translating Crank Pivot

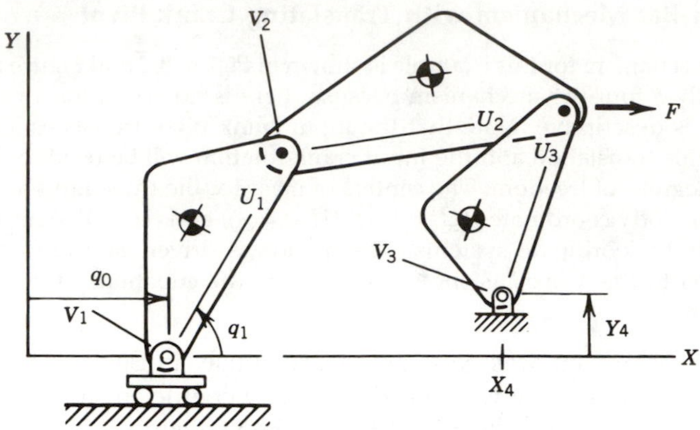

FIGURE 8.3 Pictorial Sketch for Four-Bar Mechanism with Translating Crank Pivot

where $[K_c]$ is a (9×2) matrix. The kinetic energy for the system is

$$T = 0.5\,(\dot{q}_0, \dot{q}_1)[K_c]^t[M][K_c]\begin{Bmatrix}\dot{q}_0\\ \dot{q}_1\end{Bmatrix}$$

where the required inertia matrix is

$$[M] = \text{Diag}(M_1, M_1, M_2, M_2, M_3, M_3, I_{c1}, I_{c2}, I_{c3})$$

Now, define the two velocity coefficient derivative matrices, $[L_0]$ and $[L_1]$ as the partial derivatives of the velocity coefficient matrix $[K_c]$ with respect to q_0 and q_1:

$$[L_0] = \frac{\partial [K_c]}{\partial q_0}$$

$$[L_1] = \frac{\partial [K_c]}{\partial q_1}$$

Each of these derivative matrices will be (9×2) and will play a significant role in forming the derivatives required for use in the Lagrange equation of motion.

To prepare for the application of the Lagrange equation, consider first the derivatives of the kinetic energy with respect to the coordinates. The dependence of the kinetic energy on the coordinates is in the velocity coefficient matrix, so these derivatives will be expressed in terms of the $[L_0]$ and $[L_1]$ matrices just defined:

$$\frac{\partial T}{\partial q_0} = 0.5\,(\dot{q}_0, \dot{q}_1)([L_0]^t[M][K_c] + [K_c]^t[M][L_0])\begin{Bmatrix}\dot{q}_0\\ \dot{q}_1\end{Bmatrix}$$

$$\frac{\partial T}{\partial q_1} = 0.5\,(\dot{q}_0, \dot{q}_1)([L_1]^t[M][K_c] + [K_c]^t[M][L_1])\begin{Bmatrix}\dot{q}_0\\ \dot{q}_1\end{Bmatrix}$$

As in Section 8.2, it is convenient to define the matrices $[N_0]$ and $[N_1]$ such that

$$[N_0] = 0.5\,([L_0]^t[M][K_c] + [K_c]^t[M][L_0])$$
$$[N_1] = 0.5\,([L_1]^t[M][K_c] + [K_c]^t[M][L_1])$$

The two derivatives of the kinetic energy with respect to the generalized coordinates can then be written as

$$\frac{\partial T}{\partial q_0} = \{\dot{q}\}^t[N_0]\{\dot{q}\}$$

$$\frac{\partial T}{\partial q_1} = \{\dot{q}\}^t[N_1]\{\dot{q}\}$$

The second term of the Lagrange equation is then written in matrix form as

$$\frac{\partial T}{\partial \begin{Bmatrix} q_0 \\ q_1 \end{Bmatrix}} = \left(\begin{bmatrix} \dot{q}_0 & \dot{q}_1 \\ 0 & 0 \end{bmatrix} [N_0] + \begin{bmatrix} 0 & 0 \\ \dot{q}_0 & \dot{q}_1 \end{bmatrix} [N_1] \right) \{\dot{q}\}$$

The partial derivative of the kinetic energy with respect to the generalized velocities is expressed as the partial derivative with respect to the velocity vector:

$$\frac{\partial T}{\partial \begin{Bmatrix} \dot{q}_0 \\ \dot{q}_1 \end{Bmatrix}} = [K_c]^t[M][K_c] \begin{Bmatrix} \dot{q}_0 \\ \dot{q}_1 \end{Bmatrix}$$

For the time differentiation, the chain rule is necessary to account for the change in the velocity coefficient matrix arising from the change in the coordinates:

$$\begin{aligned} \frac{d}{dt} \frac{\partial T}{\partial \begin{Bmatrix} \dot{q}_0 \\ \dot{q}_1 \end{Bmatrix}} &= [K_c]^t[M][K_c] \begin{Bmatrix} \ddot{q}_0 \\ \ddot{q}_1 \end{Bmatrix} \\ &\quad + \dot{q}_0([L_0]^t[M][K_c] + [K_c]^t[M][L_0]) \begin{Bmatrix} \dot{q}_0 \\ \dot{q}_1 \end{Bmatrix} \\ &\quad + \dot{q}_1([L_1]^t[M][K_c] + [K_c]^t[M][L_1]) \begin{Bmatrix} \dot{q}_0 \\ \dot{q}_1 \end{Bmatrix} \\ &= [K_c]^t[M][K_c]\{\ddot{q}\} + (2\,\dot{q}_0[N_0] + 2\,\dot{q}_1[N_1]) \begin{Bmatrix} \dot{q}_0 \\ \dot{q}_1 \end{Bmatrix} \end{aligned}$$

The generalized forces are then evaluated from the virtual work of the applied force. The force F is parallel to the X–axis, so only that component of displacement will be required for the virtual work expression. The X–coordinate for the point of application of F is

$$X = X_4 - C_3 \cos A_3$$

for which the virtual displacement is

$$\begin{aligned} \delta X &= C_3 \sin A_3 \, \delta A_3 \\ &= C_3 \sin A_3 \, (K_{a30} \, \delta q_0 + K_{a31} \, \delta q_1) \end{aligned}$$

The virtual work is

$$\delta W = F \, \delta X = F \, C_3 \sin A_3 \, (K_{a30} \, \delta q_0 + K_{a31} \, \delta q_1)$$

so the generalized forces are

$$Q_0 = F \, C_3 \sin A_3 \, K_{a30}$$

$$Q_1 = F \, C_3 \sin A_3 \, K_{a31}$$

At this point, all of the components have been prepared, and the Lagrange equation of motion can be applied.

Remember that in matrix notation, the first form of the Lagrange equation of motion was

$$\frac{d}{dt} \frac{\partial T}{\partial \begin{Bmatrix} \dot{q}_0 \\ \dot{q}_1 \end{Bmatrix}} - \frac{\partial T}{\partial \begin{Bmatrix} q_0 \\ q_1 \end{Bmatrix}} = \begin{Bmatrix} Q_0 \\ Q_1 \end{Bmatrix}$$

Substituting the various terms previously prepared yields the equations of motion:

$$[K_c]^t[M][K_c] \begin{Bmatrix} \ddot{q}_0 \\ \ddot{q}_1 \end{Bmatrix} + \left(\begin{bmatrix} \dot{q}_0 & -\dot{q}_1 \\ 0 & 2\dot{q}_0 \end{bmatrix} [N_0] + \begin{bmatrix} 2\dot{q}_1 & 0 \\ -\dot{q}_0 & \dot{q}_1 \end{bmatrix} [N_1] \right) \begin{Bmatrix} \dot{q}_0 \\ \dot{q}_1 \end{Bmatrix}$$
$$= \begin{Bmatrix} F \, C_3 \, K_{a30} \sin A_3 \\ F \, C_3 \, K_{a31} \sin A_3 \end{Bmatrix}$$

These, then, are the equations of motion for the four-bar mechanism with a translating crank pivot. They appear rather involved, and a numerical solution is surely indicated. For a numerical solution, it is first necessary to solve for the highest derivatives, $\ddot{q}_0$ and $\ddot{q}_1$. The matrix form facilitates this operation. Although this approach appears rather heavy handed for a system with only two degrees of freedom, the same general approach can be used for systems with any number of degrees of freedom.

8.5 MULTIDEGREE OF FREEDOM SIMULATION EXAMPLES

The following examples demonstrate the modeling and mathematical formulation of multidegree of freedom machine systems and the numerical solution of the resulting equations of motion. The Lagrange equation of motion is used to develop the describing equations, and the solutions are obtained by a computer simulation using the Runge–Kutta algorithm.

8.5.1 Induction Motor Starting a Blower

Consider a system consisting of an induction motor that drives a blower through a friction clutch and a gear box. Such a system is shown in Figure 8.4. The motor runs initially at no-load speed and the clutch is disengaged. When the clutch begins to engage, the motor speed drops due to the load imposed, and the blower train accelerates. During initial engagement, the clutch slips. At some time the relative motion across the clutch ceases, and thereafter the entire assembly rotates together. While the clutch is slipping, the system has two degrees of freedom; with the clutch locked, the system is reduced to one degree of freedom. The description and modeling of each component follows. The program listing for the computer simulation is given at the end of the example, and follows the discussion of the program and the computer results. Questions of interest include:

How long does it take to reach final operating speed?

What is the final operating speed of the motor?

What is the minimum motor speed?

What is the maximum torque through the clutch?

How much energy is dissipated in the clutch?

Motor. The motor specified for this system has the following characteristics:

3 Phase Induction Motor

460 volts, 59.5 amps, p.f. 0.86, 60 Hz

Power	50 hp
Synchronous speed	1800 rpm
No-load speed	1790 rpm
Rated speed	1700 rpm
Rotor WR^2	506 lb-in.^{2}

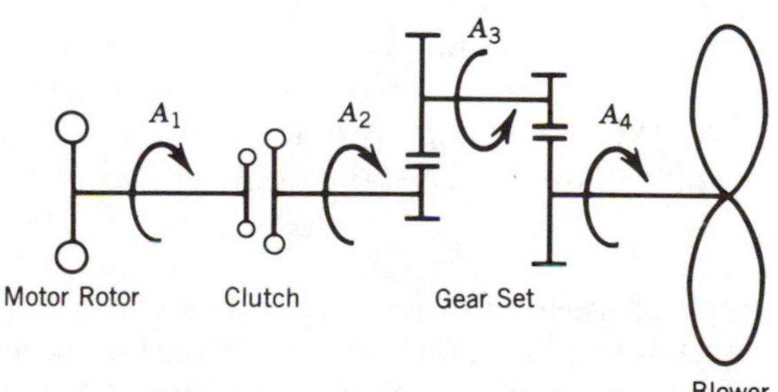

FIGURE 8.4 Blower Driven by an Induction Motor

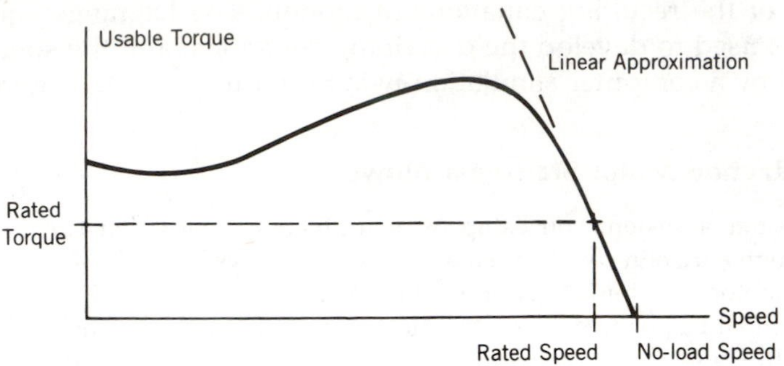

FIGURE 8.5 Induction Motor Torque—Speed Curve

Induction motors are characterized by an almost linear torque–speed relationship near the synchronous speed. Therefore, the motor torque can be modeled as a linear function of motor speed, as shown in Figure 8.5. The no-load speed is the speed at which there is no usable torque developed by the motor; at no-load speed, the motor develops only enough torque to overcome its internal mechanical losses. The rated torque is the torque developed at rated speed, and together the rated torque and rated speed provide the rated power. Thus, from the data provided, two points on the linear portion of the torque–speed curve are known. A linear function fitted to those two points gives the linear approximation to the torque–speed curve to be used in the simulation:

$$T_m = C_0 - C_1 \dot{A}_1$$

where

$$C_0 = 367867.8 \text{ in.-lb}$$

$$C_1 = 196.682 \text{ in.-lb-s/rad}$$

The other aspect of the motor that enters into the system dynamics is the mass polar moment of inertia of the rotor. It is common industrial practice not to give the moment of inertia directly, but rather to give what is called "WR^2," or sometimes "WK^2." The quantity WR^2 is the product of the acceleration of gravity with the mass polar moment of inertia, and it has the units force − length^2. Note that WR^2 for the motor rotor is given, so the moment of inertia may be determined readily. Because moments of inertia are additive, WR^2 values are also additive. This analysis will continue using WR^2 values until the final moments of inertia are prepared for the simulation.

Clutch. The clutch proposed for this system is an air-actuated, multiple-disk, dry friction clutch. The manufacturer's data for the clutch indicate that the disk contact forces build as shown in Figure 8.6, requiring a finite time to reach the rated slip torque for the clutch. The particular model selected for this application has a rated slip torque (M5) of 2050 in.-lb, and a

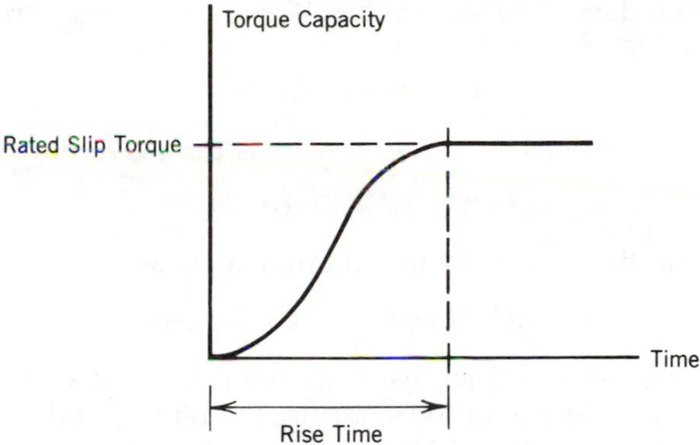

FIGURE 8.6 Clutch Torque Capacity as a Function of Time

rise time (T5) of 0.45 s. (The notations M5 and T5 in parentheses are used to indicate the names used for these quantities in the computer program.) The torque capacity of the clutch is a function of time, approximately

$$T_c = 0.5\ \text{M5}\ [1 - \cos(\pi t/\text{T5})] \qquad 0 \le t < \text{T5}$$
$$= \text{M5} \qquad t \le \text{T5}$$

The torque capacity is the torque carried by the slipping clutch. If the clutch is not slipping, the actual torque transmitted may have any value less than or equal to the torque capacity.

The inertia data for the clutch has two parts—one value for the driving side (the components connected to the motor) and a second value for the driven side (those components connected to the driven train). For this clutch, the manufacturer's values are as follows:

$$WR^2 \text{ (driving side)} = 12 \text{ lb-in.}^2$$
$$WR^2 \text{ (driven side)} = 40 \text{ lb-in.}^2$$

Gear Set. The necessary speed reduction is accomplished in a two-stage gear reduction. Both stages are identical, and each involves a 29-tooth pinion driving a gear with 114 teeth. The gear ratio of a single gear pair is

$$N = 29/114$$

The inertias for each pinion and gear are

$$WR^2 \text{ (pinion)} = 7.9 \text{ lb-in.}^2$$
$$WR^2 \text{ (gear)} = 1890 \text{ lb-in.}^2$$

Blower. This train is to drive an industrial blower, rated at 43 Hp when driven at 110 rpm. Lacking other data about the power required for various speeds, the blower is assumed to follow the usual cubic power law:

$$\text{Power Required} = C\ (\text{Rotational Speed})^3$$

From the given data, the constant C (C4 in the computer program) can be determined:

$$C = 185.67 \text{ in.-lb-s}^2$$

The blower load torque at various speeds is then given by the equation

$$T_b = C \dot{A}_4^2$$

The blower inertia is given by the manufacturer as

$$WR^2 \text{ (blower)} = 106149 \text{ lb-in.}^2$$

For the present example, losses in the gears and elsewhere, are neglected, as are the inertias of the shafting, couplings, and so forth. In practice, all of these items should be taken into account.

The first step to determine the system equations of motion through the Lagrange equation is to write the kinetic energy. Referring to Figure 8.4, the kinetic energy of the system is

$$\begin{aligned} T &= 0.5\, \dot{A}_1^2 (J_m + J_{c1}) + 0.5\, \dot{A}_2^2 (J_{c2} + J_p) \\ &\quad + 0.5\, \dot{A}_3^2 (J_g + J_p) + 0.5\, \dot{A}_4^2 (J_g + J_b) \\ &= 0.5\, \dot{A}_1^2 (J_m + J_{c1}) + 0.5\, \dot{A}_2^2 [J_{c2} + J_p + N^2 (J_g + J_p) + N^4 (J_g + J_b)] \\ &= 0.5\, \dot{A}_1^2 J_1 + 0.5\, \dot{A}_2^2 J_2 \end{aligned}$$

where

J_m = motor mass moment of inertia
J_{c1} = clutch driving side mass moment of inertia
J_{c2} = clutch driven side mass moment of inertia
J_g = gear mass moment of inertia
J_p = pinion mass moment of inertia
J_b = blower mass moment of inertia

In the final expression, J_1 and J_2 are the effective inertias associated with the coordinates A_1 and A_2, which are chosen as generalized coordinates. The inertia values are determined by summing the WR^2 values with the appropriate multipliers, and then dividing the result by the acceleration of gravity, as summarized in the table that follows.

Item	*MOI*	*WR*^{2}	Multiplier
Motor Rotor	J_m	506	1.0
Clutch Half	J_{c1}	12	1.0
Clutch Half	J_{c2}	40	1.0
Pinion	J_p	7.9	1.0
Gear	J_g	1890	N^2
Pinion	J_p	7.9	N^2
Gear	J_g	1890	N^4
Blower	J_b	106149	N^4

First Shaft: $WR^2 = 518$ lb-in.^{2}
$J_1 = 1.341656$ lb-s^2-in.
Second Shaft: $WR^2 = 623.1491$ lb-in.^{2}
$J_2 = 1.613999$ lb-s^2-in.

With the kinetic energy determined, and because potential energy does not have to be considered, look next at the nonconservative virtual work. There is work done on the system by the motor torque and by the blower load torque. Work is also done by the clutch torques, a negative work by the torque acting back on the motor shaft, and a positive work by the torque on the clutch output member. Thus, the virtual work is described as

$$\begin{aligned} \delta W^{nc} &= T_m \,\delta A_1 - T_c \,\delta A_1 + T_c \,\delta A_2 - T_b \,\delta A_4 \\ &= (T_m - T_c)\,\delta A_1 + (T_c - N^2 T_b)\,\delta A_2 \end{aligned}$$

The blower load torque is proportional to the square of the blower speed

$$T_b = C\,\dot{A}_4^2$$

but, $\dot{A}_4 = N^2\,\dot{A}_2$, so the two generalized forces are

$$\begin{aligned} Q_1 &= T_m - T_c \\ Q_2 &= T_c - N^2 T_b \\ &= T_c - N^6\,C\,\dot{A}_2^2 \end{aligned}$$

Taking the necessary derivatives, the equations of motion are

$$\begin{aligned} J_1 \ddot{A}_1 &= T_m(\dot{A}_1) - T_c(t) \\ J_2 \ddot{A}_2 &= T_c(t) - N^6\,C\,\dot{A}_2^2 \end{aligned}$$

The expressions for T_m and T_c do not depend on either A_1 or A_2, so these are, in fact, a pair of first-order differential equations in the angular velocities. With the angular velocities denoted V_1 and V_2, the equations of motion are

$$\begin{aligned} J_1 \dot{V}_1 &= T_m(V_1) - T_c(t) \\ J_2 \dot{V}_2 &= T_c(t) - C\,N^6\,V_2^2 \end{aligned}$$

The initial conditions for these equations are

$V_1(0) = 187.45$ rad/s No-load speed
$V_2(0) = 0.0$ rad/s Initially at rest

At some later time in the simulation, the clutch will cease to slip, and the system will have only one degree of freedom from that time forward. When that occurs, the equation of motion for the single degree of freedom is

$$(J_1 + J_2)\dot{V} = T_m(V) - C\,N^6\,V^2$$

where V represents the single degree of freedom. The initial condition for this equation is simply the value of $V_1 = V_2$ at the time when the clutch ceases to slip. This is the system to be solved numerically in the computer simulation.

As with the example of Section 7.6.1, this problem does not involve complex kinematic relations that require tabulated solutions for the loop equations and velocity coefficients. This condition simplifies that aspect of the programming significantly.

As with the earlier examples, it is convenient to begin with the subroutine that will perform the derivative evaluations. This subroutine nominally begins at line 3000 with a sequence of REMarks to document the notation; actual calculation begins at line 3100. The time, the two angular velocities, and the two angular accelerations are denoted T, V1, V2, V3, and V4 in the subroutine. Each call to the subroutine begins with an evaluation of three items—the motor torque, the torque capacity of the clutch, and the blower load torque as seen at the clutch output shaft. This analysis is implemented in lines 3100 through 3140. It is then necessary to know if the system has one or two degrees of freedom, to determine which equations of motion to use. This information is in the flag F7, received from the main program. If F7 is 1, the system is understood to have only one degree of freedom; if F7 is 2, the system has two degrees of freedom. The assignment of the proper value to F7 is done in the main program, and will be described later.

For F7 = 2, $\dot{V}_1$ and $\dot{V}_2$ are evaluated using the two equations of motion in lines 3210 through 3250. The clutch must be continuing to slip, because the system has two degrees of freedom, so the torque transmitted through the clutch is assigned the current value of the clutch torque capacity. Control is then returned to the main program.

If F7 = 1, the value V is assigned to both V_1 and V_2; this is done in lines 3160 through 3200. For this case, the torque transmitted through the clutch may be less than the torque capacity of the clutch. As long as the system continues to accelerate, the transmitted torque is that required to support the blower load plus the torque required to accelerate the inertia of that portion of the train following the clutch. When the motor reaches steady operating speed, the transmitted torque will be reduced to only the torque required to support the blower load. The transmitted torque is evaluated in line 3190, and then control is returned to the main program.

In the main program, the value of time and the two angular velocities at the end of any integration step are denoted by T0, A1, and A2. The structure of the main program is similar to that used in previous examples. It consists of three general parts: (1) statement of the system parameters, (2) evaluation of other parameters and setting of initial values, and (3) a loop implementing the Runge–Kutta algorithm. The Runge–Kutta process is described in Appendix A3.3, "Systems of Differential Equations," which applies to a system of first-order differential equations such as those that describe the present case.

The question of continuing the simulation as two degrees of freedom or going to one degree of freedom is evaluated at the end of each integration step (line 2130). The flag F7, which indicates the number of degrees of freedom, was set to 2 in line 1640 during the initialization phase, and remains at that value until the calculated angular velocity for the clutch output exceeds that of the motor. When that happens, both angular velocities are set to the motor speed and the flag is set to 1. The simulation continues until the angular acceleration is less than 0.01 rad/s^2, based on the final derivative evaluation from the previous step.

One of the questions to be addressed by the program is "How much energy is dissipated in the clutch?" This energy will be manifest as heat in the clutch components at the end of the engagement interval and will have to be carried away by convection and conduction. Consider the process of energy transfer through the clutch, imagining for the moment that the "clutch" is a physical medium between the clutch plates. The energy into the clutch is the work done by friction from the clutch plate driven by the motor,

$$\begin{aligned} E_{\text{in}} &= \int T_{\text{f}}\, dA_1 \\ &= \int T_{\text{f}}\, \dot{A}_1\, dt \end{aligned}$$

where T_{f} is the friction torque, and the integration extends over the duration of the clutch slippage. The energy transferred through the clutch then leaves the clutch as work done on the output clutch plate:

$$\begin{aligned} E_{\text{out}} &= \int T_{\text{f}}\, dA_2 \\ &= \int T_{\text{f}}\, \dot{A}_2\, dt \end{aligned}$$

The difference between these two is the energy dissipated in the clutch:

$$\begin{aligned} E_{\text{lost}} &= E_{\text{in}} - E_{\text{out}} \\ &= \int T_{\text{f}}\, (\dot{A}_1 - \dot{A}_2)\, dt \end{aligned}$$

This integral is evaluated using the trapezoidal rule in lines 2310 through 2400. The data was saved during the simulation in lines 2170 through 2210 where the angular velocities, A1 and A2, and the transmitted torque, S0, were stored as A5(i), A6(i), and S6(i). This result provides an answer for one of the original questions.

Executing the simulation yields the following results:

```
CALCULATED SYSTEM PARAMETERS
MOTOR COEFFICIENTS
C0   =   36867.7744
C1   =   196.682297
BLOWER COEFFICIENT
C4   =  185.672883
```

```
                      ENGAGEMENT HISTORY
DOF       TO          A1          A2          SO
          SEC         R/S         R/S         IN-LB
2         0           187.45      0           0
2         .01         187.44      .01         2
2         .02         187.42      .04         10
2         .03         187.37      .14         22
2         .04         187.3       .33         40
2         .05         187.21      .64         62
2         .06         187.09      1.1         89
.         .           .           .           .
.         .           .           .           .
.         .           .           .           .
2         .2          183.38      36.18       847
2         .21         183.03      41.18       918
2         .22         182.67      46.49       989
.         .           .           .           .
.         .           .           .           .
.         .           .           .           .
2         .43         177.12      168.62      2040
2         .44         177.07      172.23      2048
2         .45         177.04      175.5       2050
1         .46         177.03      177.03      2050
1         .47         178.16      178.16      1713
1         .48         178.71      178.71      1663
.         .           .           .           .
.         .           .           .           .
.         .           .           .           .
1         .6          179.23      179.23      1616
CLUTCH ENERGY LOSS = 33677.8041
```

It is often desirable to see simulation results in graphic form with all variables plotted as functions of time. Instead of ENDing at line 2420 following the energy loss calculation, control can be passed to a plotting subroutine for graphic output. The data saved for the energy loss calculation, as well as that for a fourth item, the blower load torque as reflected at the clutch output shaft, are plotted in Figure 8.7. The motor speed is seen to drop slowly while the clutch output speed rises rapidly until the two meet at 177.03 rad/s when $t = 0.46$ s. Thereafter, the entire assembly accelerates slightly to a final speed of 179.23 rad/s. The specific questions raised in the beginning can now be answered, based on the graphical and tabular output.

Time to reach steady speed	0.6 s
Steady operating speed	179.23 rad/s
Minimum motor speed	177.03 rad/s
Maximum torque through clutch	2050 in.-lb
Energy dissipated in clutch	33678 in.-lb

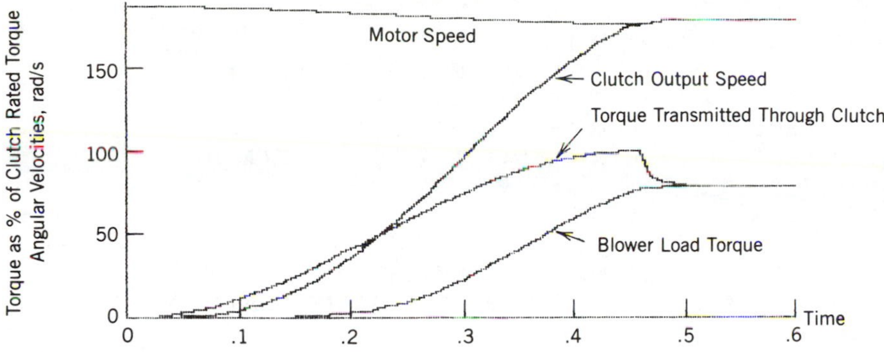

FIGURE 8.7 Engagement History Curves

The original questions have all been answered, but other questions may be addressed as well. Study the program listing, and think about the modifications that would be required to answer the following questions.

- What is the effect of increasing the clutch's rated capacity?
- What is the effect of a shorter clutch rise time?
- What would be the effect of substituting a two-stage belt drive in place of the gear box?
- What are the maximum gear tooth forces in the present system?
- What is the effect of losses in the gear box?

The computer program listing follows.

```
1000 REM      INDUCTION MOTOR STARTING A BLOWER
1010 REM      COUPLED THROUGH A CLUTCH AND A GEAR BOX
1030 DIM A5(100), A6(100),S6(100),B6(100)
1040 PI = 3.1415926535
1050 DEF FN RO(Q) = INT (Q + .5)
1060 DEF FN R2(Q) = INT(100*Q + .5)/100
1100 REM      SYSTEM PARAMETERS
1110 J1 = 1.341656       : REM  MOTOR SHAFT INERTIA
1120 J2 = 1.61399958     : REM  CLUTCH OUTPUT SHAFT
INERTIA
1130 REM
1140 REM     MOTOR PARAMETERS
1150 UQ = 1790 * 2*PI/60   : REM   MOTOR NO LOAD SPEED
1160 U1 = 1700 * 2*PI/60   : REM   MOTOR RATED SPEED
1170 P1 = 50 * 550*12      : REM   MOTOR RATED POWER
1180 M1 = P1 / U1          : REM   MOTOR RATED TORQUE
1190 CO = M1 * UO / (UO - U1)   : REM TORQUE - SPEED COEF
1192 C1 = M1 / (UO - U1)        : REM TORQUE - SPEED COEF
1200 REM  MOTOR TORQUE = CO - C1 * V1
1210 REM  LINEAR APPROX MOTOR TORQUE - SPEED RELATION
1220 REM  WHERE V1 IS MOTOR SHAFT SPEED
```

```
1230 REM
1240 REM     CLUTCH PARAMETERS
1250 M5 = 2050 : REM SLIP TORQUE, IN-LB
1260 T5 = .45  : REM RISE TIME, SEC
1270 REM  CTC = CLUTCH TORQUE CAPACITY
1270 REM      = 0.5 * M5 ( 1 - COS(PI*T/T5) )  0 < T <
T5 < T
1280 REM      =M5
1290 REM
1300 REM      GEAR BOX PARAMETERS
1310 N = 29 /114  : REM  SINGLE STATE GEAR RATIO
1320 REM
1330 REM      BLOWER PARAMETERS
1340 P4 = 43 * 550*12    : REM RATED POWER
1350 U4 = 110 * 2*PI/60 : REM RATED SPEED
1360 C4 = P4 / U4^3      : REM BLOWER COEF
1370 C5 = C4 * N^6       : REM EFFECTIVE BLOWER COEF
1380 REM
1400 HOME
1410 PRINT"      SIMULATION OF AN INDUCTION MOTOR"
1420 PRINT"      STARTING A BLOWER COUPLED THROUGH"
1422 PRINT"      A CLUTCH AND A GEAR BOX"
1430 PRINT
1440 PRINT"DO YOU WANT HARD COPY? Y/N"
1450 INPUT A$
1460 IF A$="N"  THEN GOTO 1470
1462 PRINT CHR$(4);" PR#1"
1464 PRINT CHR$(9);"80N"
1466 PRINT CHR$(9);CHR$(1)
1468 PRINT CHR$(27)  "D" CHR$(10) CHR$(15) CHR$(25)
CHR$(35) CHR$(45) CHR$(0)
1470 PRINT
1480 PRINT"      CALCULATED SYSTEM PARAMETERS"
1490 PRINT"      MOTOR COEFFICIENTS
1500 PRINT"            C0 = ";C0
1510 PRINT"            C1 = ";C1
1520 PRINT"      BLOWER COEFFICIENT"
1530 PRINT"            C4 = ";C4
1540 PRINT
1600 H = 0.01  : REM  INTEGRATION TIME STEP
1610 REM  F7 IS THE FLAG FOR ONE VS TWO DEGREES OF
          FREEDOM
1620 REM  F7 = 1  ---> ONE DOF
1630 REM  F7 = 2  ---> TWO DOF
1640 F7 = 2    : REM  INITIAL CONDITION
1650 T0 = 0    : REM  INITIAL CONDITION
1660 A1 = U0   : REM  INITIAL CONDITION
1670 A2 = 0    : REM INITIAL CONDITION
1680 S0 = 0    : REM INITIAL CLUTCH TORQUE
```

```
1690 IF A$ = "N" THEN GOTO 1700
1691 REM SET-UP FOR PRINTER OUTPUT
1692 PRINT"                    ENGAGEMENT HISTORY"
1693 PRINT
1694 PRINT"          DOF   TO         A1         A2
SO"
1695 PRINT"                SEC        R/S        R/S
IN-LB"
1696 PRINT
1697 PRINT CHR$(9);F7;CHR$(9);FN R2(TO);CHR$(9);FN
R2(A1);CHR$(9);FN R2(A2);CHR$(9);FN RO(SO)
1698 GOTO 1720
1700 REM  SET-UP FOR SCREEN OUTPUT
1701 PRINT"         ENGAGEMENT HISTORY"
1702 PRINT
1703 PRINT"DOF  TO    A1        A2        SO"
1704 PRINT"     SEC   R/S       R/S       IN-LB"
1705 PRINT
1706 PRINT F7;TAB(6);FN R2(TO);TAB(11);FN
R2(A1);TAB(21);FN R2(A2);TAB(31);FN RO(SO)
1720 A5(0)  = A1  :  REM  SAVE  INITIAL  VALUES
1730 A6(0)  = A2
1740 S6(0)  = SO
1750 B6(0)  = 0   :  REM  INITIAL BLOWER LOAD TORQUE
1760 REM
1770 REM       RUNGE - KUTTA INTEGRATION
1780 I = 0
1790 I  = I + 1
1800 REM  FIRST DERIVATIVE EVALUATION
1810 T = TO
1820 V1 = A1
1830 V2 = A2
1840 GOSUB 3100
1850 K1 = H * V3
1860 K2 = H * V4
1870 REM  SECOND DERIVATIVE EVALUATION
1880 T = TO + H/2
1890 V1 = A1 + K1/2
1900 V2 = A2 + K2/2
1910 GOSUB 3100
1920 K3 = H * V3
1930 K4 = H * V4
1940 REM  THIRD DERIVATIVE EVALUATION
1950 T = TO + H/2
1960 V1 = A1 + K3/2
1970 V2 = A2 + K4/2
1980 GOSUB 3100
1990 K5 = H * V3
2000 K6 = H * V4
2010 REM FOURTH DERIVATIVE EVALUATION
```

```
2020 T = TO + H
2030 V1 = A1 + K5
2040 V2 = A2 + K6
2050 GOSUB 3100
2060 K7 = H * V3
2070 K8 = H * V4
2080 REM  UPDATE THE SOLUTION
2090 TO = TO + H
2100 A1 = A1 + (K1 + 2 * (K3 + K5) + K7)/6
2110 A2 = A2 + (K2 + 2 * (K4 + K6) + K8)/6
2120 REM TEST TO CONTINUE AS 2DOF OR NOT
2130 IF A2 < A1 THEN GOTO 2170
2140 REM CHANGE TO SDOF
2150 F7 = 1
2160 A2 = A1
2170 REM STORE SOLUTION VALUES
2180 A5(I) = A1
2190 A6(I) = A2
2200 S6(I) = SO
2210 B6(I) = S3
2220 REM  PRINT NEW SOLUTION VALUES
2222 IF A$ = "N" THEN GOTO 2230
2224 PRINT CHR$(9);F7;CHR$(9);FN R2(TO);CHR$(9);FN
R2(A1);CHR$(9);FN R2(A2);CHR$(9);FN RO(SO)
2226 GOTO 2240
2230 PRINT F7; TAB(6);FN R2(TO);TAB(11);FN
R2(A1);TAB(21);FN R2(A2);TAB(31);FN RO(SO)
2240 REM TEST FOR TERMINATION
2250 IF V4 < 0.01 THEN GOTO 2300
2260 GOTO 1790
2300 I9 = I     : REM  FINAL INTEGRATION STEP INDEX
2310 REM  CALCULATE THE CLUTCH ENERGY LOSS
2320 G6 = 0
2330 FOR I = 1 TO I9
2340 G1 = S6(I-1) * ( A5(I-1) - A6(I-1) )
2350 G2 = S6(I) * ( A5(I) - A6(I) )
2360 G6 = G6 + (G1 + G2) /2
2370 NEXT I
2380 G6 = H * G6    : REM  ENERGY LOSS IN THE CLUTCH
2390 PRINT
2400 PRINT"CLUTCH ENERGY LOSS = ";G6
2410 PRINT CHR$(4);"PR#0"
2420 END
3000 REM       EVALUATION OF DERIVATIVES
3010 REM  INPUTS TO THIS ROUTINE ARE T, V1, V2
3020 REM  OUTPUTS FROM THE ROUTINE ARE V3, V4
3030 REM              V3 = V1 DOT
3040 REM              V4 = V2 DOT
3050 REM  S1 = MOTOR TORQUE VALUE
3060 REM  S2 = CLUTCH TORQUE CAPACITY
```

```
3070 REM   S3 = EFFECTIVE BLOWER LOAD TORQUE
3100 S1  = CO  - C1 * V1
3110 S2  = M5
3120 IF T  T5 THEN GOTO 3140
3130 S2  = 0.5 * M5 * (1 - COS(PI*T/T5) )
3140 S3  = C5 * V2^2
3150 IF F7 = 2 THEN GOTO 3210
3160 REM   F7 = 1   ---> SDOF
3170 V3  = (S1  = S3) / (J1  + J2)
3180 V4  = V3
3190 SO  = J2 * V4 + S3   : REM TORQUE THROUGH THE CLUTCH
3200 RETURN
3210 REM   F7 = 2 ---> 2 DOF
3220 V3  =   (S1  - S2) / J1
3230 V4  = (S2  - S3)   / J2
3240 SO  = S2   : REM TORQUE THROUGH THE CLUTCH
3250 RETURN
3260 END
```

8.5.2 Centrifugal Governor Mechanism Response

Consider again the centrifugal governor mechanism introduced in Section 8.4.1 and shown in Figure 8.1. The system parameters chosen for this simulation are

$I_o = 5.5$ lb-s²-in.	Wheel mass MOI
$I_c = 0.5$ lb-s²-in.	Slider mass MOI
$Mg = 4.0$ lb	Slider weight
$R_o = 11.0$ in.	Slider position with the spring relaxed
$K = 150$ lb/in.	Spring rate

The external torque applied to the wheel is taken as a steady load torque and a driving torque proportional to the displacement of the sensing mass from a set point:

$$T_o = C_1 (X_6 - X) - C_2$$

where

C_1 = proportionality constant, 8000 in.-lb/in.
C_2 = steady load torque, 800 in.-lb;
X_6 = set point position, 0.937 in.

Determine the motion that results from the initial conditions

$A(0) = 0.0$ rad	$X(0) = 0.837$ in.
$\dot{A}(0) = 32.0$ rad/s	$\dot{X}(0) = 0.0$ in./s

For these values, the second equation is approximately satisfied, while the first equation is satisfied on the average in the sense that the non-zero term of the equation has zero average value. This might suggest that the solution will remain close to the initial values. However, this is not true, as is described in the following paragraphs.

From Section 8.4.1, the equations of motion are

$$\ddot{A}[I_o + I_c + M(R_o + X)^2] + 2\,M\,\dot{A}\,\dot{X}\,(R_o + X)$$
$$+ Mg(R_o + X)\cos A - T_o(X) = 0$$
$$M\,\ddot{X} - M\,\dot{A}^2\,(R_o + X) + K\,X + M\,g\sin A = 0$$

To apply the Runge–Kutta algorithm, as described in Appendix A3.3, this system of second-order equations must be converted into a system of first-order differential equations. Following the process described there, define new variables X_1, X_2, V_1, and V_2 such that

$$X_1 = A \qquad V_1 = \dot{A}$$
$$X_2 = X \qquad V_2 = \dot{X}$$

In terms of the new variables, the required system of first-order differential equations is

$$\dot{X}_1 = V_1$$
$$\dot{X}_2 = V_2$$
$$\dot{V}_1 = \frac{[T_o(X_2) - 2\,M\,V_1\,V_2(R_o + X_2) - M\,g(R_o + X_2)\cos X_1]}{I_o + I_c + M(R_o + X_2)^2}$$
$$\dot{V}_2 = V_1^2\,(R_o + X_2) - (K/M)X_2 - g\sin X_1$$

A simulation for this system of differential equations has been developed using the Runge–Kutta algorithm for the solution. Using an integration step of 0.001 s, the following are typical solution values:

t	$A(t)$	$\dot{A}(t)$	$X(t)$	$\dot{X}(t)$
0.0	0.0000	32.0000	0.8370	0.0000
0.005	0.1599	31.9691	0.8367	−0.1945
0.010	0.3197	31.9470	0.8345	−0.7223
0.015	0.4794	31.9492	0.8293	−1.3397
0.020	0.6393	31.9919	0.8215	−1.7058
0.025	0.7994	32.0828	0.8132	−1.5090
⋮	⋮	⋮	⋮	⋮
0.170	5.4297	27.9373	0.7083	−62.6213
0.175	5.5732	29.7324	0.4063	−53.7600
0.180	5.7293	32.8543	0.2145	−18.9686
0.185	5.9022	36.2300	0.2473	33.8690
⋮	⋮	⋮	⋮	⋮
0.300	9.6300	42.8328	8.0752	984.5965
0.305	9.7479	4.3687	11.5265	324.4102
0.310	9.6732	−34.6183	11.0733	−478.3519
0.315	9.3968	−76.0769	7.4949	−841.0964
⋮	⋮	⋮	⋮	⋮

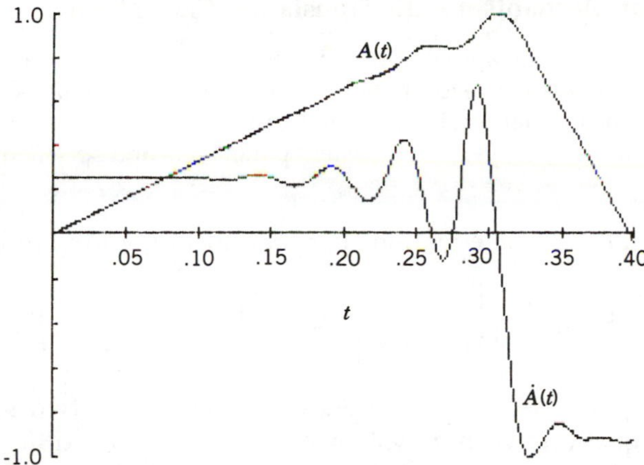

FIGURE 8.8 Angular Position and Velocity of the Flywheel as Functions of Time

Normalized plots of the solution for a little more than one revolution of the wheel are shown in Figures 8.8 and 8.9. The latter part of the simulation is probably unrealistic because the sensing mass will probably strike a stop before the displacement becomes 11 inches. Further, the available driving torque is surely limited, so that the driving torque associated with this large displacement is surely not realistic. These constraints have not been included in the simulation; the solution just tabulated and shown in Figures 8.8 and 8.9 must, therefore, be understood as the result without regard to these constraints.

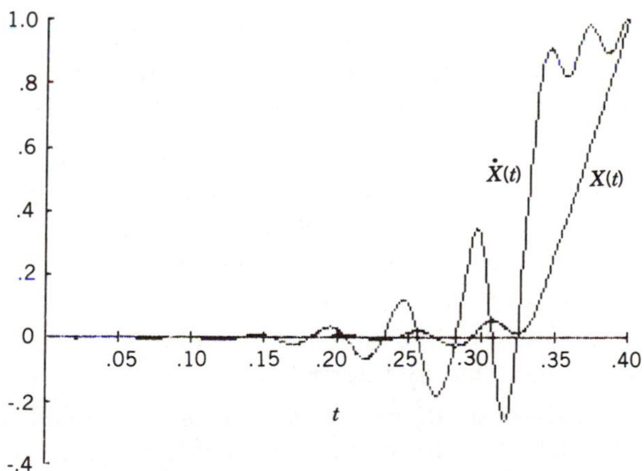

FIGURE 8.9 Sensing Mass Position and Velocity as Functions of Time

8.5.3 Four-Bar Mechanism with Translating Crank Pivot

Equations of motion were developed in Section 8.4.2 for a mechanism similar to a four-bar linkage, except that one of the crank pivots was allowed to translate; this mechanism is shown in Figure 8.2. The system is to be prepared for simulation for a given set of system parameters and specified initial conditions.

In Section 8.4.2, the equations of motion were written in matrix form:

$$[K_c]^t[M][K_c]\{\ddot{q}\} + \left(\begin{bmatrix} \dot{q}_0 & -\dot{q}_1 \\ 0 & 2\,\dot{q}_0 \end{bmatrix}[N_0] + \begin{bmatrix} 2\,\dot{q}_1 & 0 \\ -\dot{q}_0 & \dot{q}_1 \end{bmatrix}[N_1]\right)\{\dot{q}\} = \{Q\}$$

It is necessary at this point to replace this system of two second-order differential equations with a system of four first-order differential equations, as was done in the previous example. For that purpose, define the new variables:

$$\{U\} = \mathrm{Col}(q_0, q_1) \qquad \{V\} = \mathrm{Col}(\dot{q}_0, \dot{q}_1)$$

With this change of variables, the required system of first-order differential equations is

$$\{\dot{U}\} = \{V\}$$

$$\{\dot{V}\} = ([K_c]^t[M][K_c])^{-1}(\{Q\} - ([W_0][N_0] + [W_1][N_1])\{\dot{q}\})$$

where

$$[W_0] = \begin{bmatrix} \dot{q}_0 & -\dot{q}_1 \\ 0 & 2\dot{q}_0 \end{bmatrix} \qquad [W_1] = \begin{bmatrix} 2\dot{q}_1 & 0 \\ -\dot{q}_0 & \dot{q}_1 \end{bmatrix}$$

These four differential equations, with the associated four initial conditions, define the motion of the system in a form suitable for a numerical solution.

One of the major difficulties in carrying out this simulation comes in evaluating the kinematic functions—the positions, the velocity coefficients, and the velocity coefficient derivatives. For this purpose, three options may be considered:

a. Include the complete direct kinematic solution procedure in the derivative evaluation subroutine. This will include the iterative position solution and linear system solutions for the velocity coefficients and velocity coefficient derivatives;

b. Tabulate all of the required kinematic solutions as functions of two variables to be evaluated by interpolation;

c. Tabulate the position solutions as functions of two variables to be evaluated by interpolation, with the velocity coefficients and velocity coefficient derivatives to be determined as linear system solutions.

Any of these three options entail a substantial amount of effort, but the last option will probably give the best results for the least effort. This appears to contradict a related statement in Chapter 7 regarding the simulation of single degree of freedom systems. However, the difference is that complete dependence on tabulation and interpolation in this case requires much more effort because of the two independent variables. The first option, to embed the complete numerical solution process in the derivative evaluation subroutine, will extend the computing time to a degree usually considered unacceptable. Either the second or the third option dictate some amount of interpolation on functions of two variables. Methods to interpolate functions of two variables are presented in the Appendix A5.3, but the storage requirements for a detailed representation of such functions make it unattractive to tabulate very many of them. All of these considerations point to the third option as the way to proceed.

8.6 AN ALTERNATE APPROACH TO FIRST-ORDER EQUATIONS OF MOTION FOR MULTIDEGREE OF FREEDOM SYSTEMS

The numerical solution methods used for multidegree of freedom systems require that the description be a system of first-order differential equations. In the examples in the previous section, the Lagrange equation was used to develop a system of second-order differential equations of motion, and these were subsequently converted into an equivalent set of first-order differential equations with appropriate definitions for new variables. The procedure described in the following paragraphs is attractive in that it uses the Lagrange equation to develop directly the required system of first-order differential equations. Furthermore, the formulation often requires significantly less effort.

In advanced dynamics, a generalized momentum is defined as the derivative of the kinetic energy with respect to a generalized velocity. When applied in matrix form to a system with N degrees of freedom, this definition gives the vector of generalized momenta:

$$\{P\} = \frac{\partial T}{\partial\{\dot{q}\}}$$

where the kinetic energy is understood for the following development to be of the form

$$T = 0.5\,\{\dot{q}\}^t[K_c]^t[M][K_c]\{\dot{q}\}$$

When this definition is substituted into the Lagrange equation, the result is a set of N first-order differential equations, solvable for the derivatives of the generalized momenta:

$$\{\dot{P}\} = \{Q\} - \frac{\partial V}{\partial \{q\}} + \frac{\partial T}{\partial \{q\}}$$

$$= \{Q\} - \frac{\partial V}{\partial \{q\}}$$

$$+ \left(\begin{bmatrix} \{\dot{q}\}^t \\ \hline (-0-) \\ \hline (-0-) \\ \hline \vdots \end{bmatrix} [N_1] + \begin{bmatrix} (-0-) \\ \hline \{\dot{q}\}^t \\ \hline (-0-) \\ \hline \vdots \end{bmatrix} [N_2] + \cdots \right) \{\dot{q}\}$$

When the definition of the generalized momenta is applied to the kinetic energy, the result is a system of linear equations, each of which express a generalized momentum component as a linear combination of the generalized velocities:

$$\{P\} = [K_c]^t[M][K_c]\{\dot{q}\}$$

This expression can be solved for the generalized velocities, which are also the time derivatives of the generalized coordinates. Consequently, these are the additional N first-order differential equations required to complete the system description:

$$\{\dot{q}\} = ([K_c]^t[M][K_c])^{-1}\{P\}$$

The first set can be integrated to give the generalized momenta, $\{P(t)\}$. Integrating the second set yields the generalized coordinates, $\{q(t)\}$.

By introducing the generalized momenta, the often laborious time differentiation step associated with the more common application of the Lagrange equation is rendered trivial. Some might object, saying that the velocity solution, $\{\dot{q}(t)\}$, should be part of the final result. This is not a problem because the velocities must be determined in every derivative evaluation, and they can certainly be recorded if desired. The value of this approach will be seen in the two following example problems.

8.6.1 Centrifugal Governor Mechanism Response

The centrifugal governor was introduced and the equations of motion were determined in Section 8.4.1. Determine a set of four first-order equations of motion for this system using generalized coordinates and generalized momenta as variables.

The kinetic energy takes the form

$$T = 0.5\,\dot{A}^2\,[I_o + I_c + M(R_o + X)^2] + 0.5\,M\,\dot{X}^2$$

so that the generalized momenta are

$$P_a = \frac{\partial T}{\partial \dot{A}} = [I_o + I_c + M(R_o + X)^2]\,\dot{A}$$

$$P_x = \frac{\partial T}{\partial \dot{X}} = M\,\dot{X}$$

The nonconservative generalized forces are

$$Q_a^{nc} = T_o(A, X, t)$$

$$Q_x^{nc} = 0$$

where $T_o(A, X, t)$ is the net torque acting on the flywheel. The potential energy for this system is of the form

$$V = M\,g\,(R_o + X)\sin A + 0.5\,K\,X^2$$

Consequently, the terms involving the derivatives of the potential energy with respect to the coordinates are

$$\frac{\partial V}{\partial A} = M\,g\,(R_o + X)\cos A$$

$$\frac{\partial V}{\partial X} = M\,g\sin A + K\,X$$

The final terms required are the derivatives of the kinetic energy with respect to the generalized coordinates:

$$\frac{\partial T}{\partial A} = 0$$

$$\frac{\partial T}{\partial X} = M\,\dot{A}^2\,(R_o + X)$$

With all of the parts at hand, the momentum derivatives can now be assembled:

$$\dot{P}_a = T_o - M\,g\,(R_o + X)\cos A$$

$$\dot{P}_x = -K\,X - M\,g\sin A + M\,\dot{A}^2\,(R_o + X)$$

while the coordinate derivatives are obtained from the momentum expressions:

$$\dot{A} = P_a / [I_o + I_c + M\,(R_o + X)^2]$$

$$\dot{X} = P_x / M$$

This is the required system of four, first-order equations that describe the motion of the system.

8.6.2 Four-Bar Mechanism with Translating Crank Pivot

This mechanism was introduced in Section 8.4.2, where equations of motion were developed through the usual application of the Lagrange equation. Determine a system of first-order equations of motion for this system using the generalized coordinates and generalized momenta as variables.

The kinetic energy, developed in the earlier presentation, is

$$T = 0.5\,\{\dot{q}\}^t[K_c]^t[M][K_c]\{\dot{q}\}$$

where the inertia matrix is a (9 × 9) diagonal matrix and the velocity coefficient matrix is (9 × 2). The vector of generalized momenta is

$$\{P\} = \frac{\partial T}{\partial\{\dot{q}\}} = [K_c]^t[M][K_c]\{\dot{q}\}$$

When solved for the generalized velocities, this provides a set of first-order differential equations for the time derivatives of the generalized coordinates:

$$\{\dot{q}\} = ([K_c]^t[M][K_c])^{-1}\{P\}$$

In the development given in Section 8.4.2, the generalized force components were determined. The generalized force vector is

$$\{Q\} = \begin{Bmatrix} F\,C_3\,K_{a30}\sin A_3 \\ F\,C_3\,K_{a31}\sin A_3 \end{Bmatrix}$$

The other necessary term is the vector of the kinetic energy derivatives with respect to the generalized coordinates, which was also determined previously:

$$\frac{\partial T}{\partial\begin{Bmatrix} q_0 \\ q_1 \end{Bmatrix}} = \left(\begin{bmatrix} \dot{q}_0 & \dot{q}_1 \\ 0 & 0 \end{bmatrix}[N_0] + \begin{bmatrix} 0 & 0 \\ \dot{q}_0 & \dot{q}_1 \end{bmatrix}[N_1]\right)\begin{Bmatrix} \dot{q}_0 \\ \dot{q}_1 \end{Bmatrix}$$

$$\begin{aligned}\{\dot{P}\} &= \{Q\} + \frac{\partial T}{\partial\{q\}} \\ &= \begin{Bmatrix} Q_0 \\ Q_1 \end{Bmatrix} + \left(\begin{bmatrix} \dot{q}_0 & \dot{q}_1 \\ 0 & 0 \end{bmatrix}[N_0] + \begin{bmatrix} 0 & 0 \\ \dot{q}_0 & \dot{q}_1 \end{bmatrix}[N_1]\right)\begin{Bmatrix} \dot{q}_0 \\ \dot{q}_1 \end{Bmatrix}\end{aligned}$$

These are the second set of N first-order differential equations of motion for the four-bar mechanism with translating crank pivot. The difficulties previously noted regarding the need to evaluate the kinematic solutions as functions of two primary variables remain. However, two-variable interpolation for the position solutions is still the most attractive option.

8.7 CONCLUSION

The Lagrange equation provides the most direct approach to obtain the equations of motion for multidegree of freedom machines. Conversely, applying Newton's Second Law to a system of free bodies, as a means to the equations of motion, is prohibitively difficult for most multidegree of freedom mechanisms. In these same situations, application of the Lagrange equation is relatively straightforward (but sometimes tedious!).

The application of velocity coefficients and velocity coefficient partial derivatives is seen in writing the equations of motion, as in the previous chapter. The multidegree of freedom aspect of the problem complicates their use, yet makes them all the more significant. Velocity coefficients and velocity coefficient derivatives in matrix form are most useful when attacking complicated problems.

The general process of analysis has been developed in this chapter and for any particular problem, the results should have the form shown here. The objective is often to obtain equations for use in a computer simulation, so the alternate approach presented in Section 8.6 will be of use in many cases, especially because formulating the equations of motion requires less effort this way.

REFERENCES

Burr, A. H., *Mechanical Analysis and Design,* Elsevier, 1981.

Doughty, S., "First Order Equations of Motion from Lagrangian Function," *Serie de Mecanique Applique–Revue Romaine des Sciences Techniques,* No. 4, 1974.

PROBLEM SET

For all the problems in this set, follow two steps. First, perform all of the kinematic analysis required for the equations of motion. Carry the analysis in closed form as far as is reasonable; if a numerical solution is necessary, determine the type of numerical solution required and prepare the problem for that type of solution. Second, determine the equations of motion for the system.

8.1 The two rollers are joined by a pair of links with lengths L_1 and L_2. The system is released in a configuration similar to that shown and falls

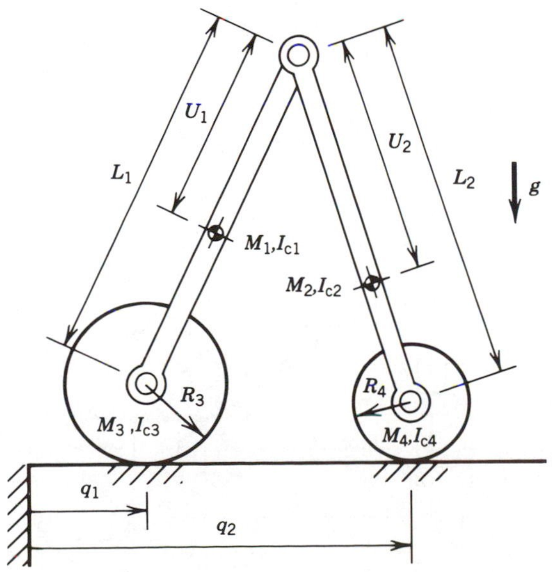

under the influence of gravity. The two links have masses M_1 and M_2, and mass moments of inertia I_{c1} and I_{c2} with respect to their respective centers of mass. The roller masses are M_3 and M_4 with associated mass moments of inertia I_{c3} and I_{c4}.

8.2 The mass M_1 moves horizontally under the influence of the force $F(t)$ and the restraint of the spring and dashpot. The spring is relaxed when the position of the mass is q_{10}. A pendulum having mass M_2 and mass moment of inertia I_{2c} with respect to the center of mass is suspended from the translating mass. The dimension D, which locates the pendulum center of mass from the support, is known.

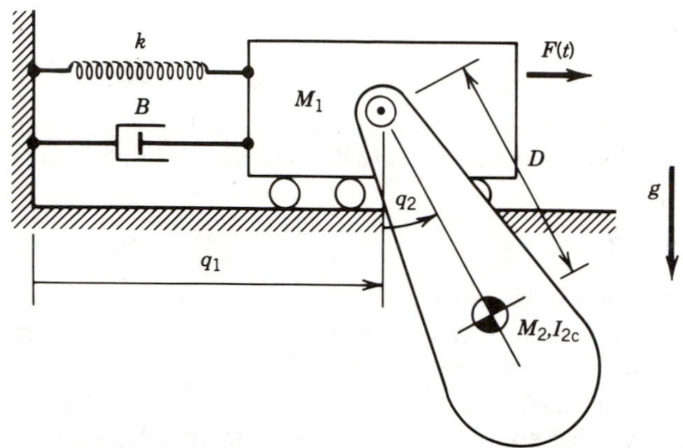

8.3 A system consists of two masses, M_x and M_y, joined by a spring of negligible mass. The spring constant is K, and the spring's free length is S_o. Forces $F_x(t)$ and $F_y(t)$ act on the two masses, and the vertical motion is also influenced by a dashpot with coefficient B.

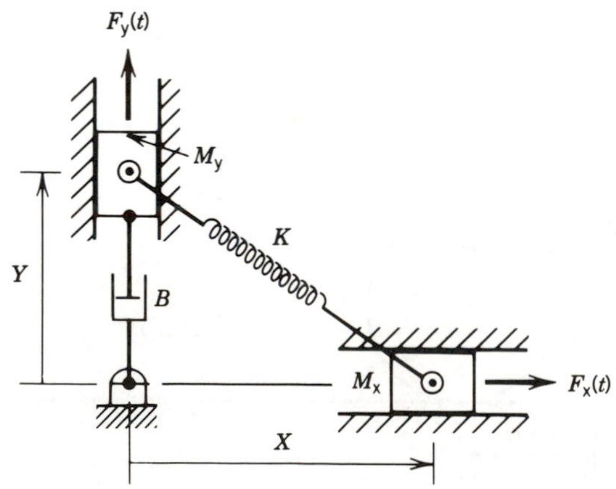

8.4 The masses M_1 and M_2 are coupled by a spring with coefficient K and free length S_o. The masses M_2 moves in a guide inclined at angle C with respect to the guide for mass M_1. These two guides intersect at a known distance D from the reference line for q_1. A force $F(t)$ acts on M_2 while M_1 is subject to the action of a dashpot.

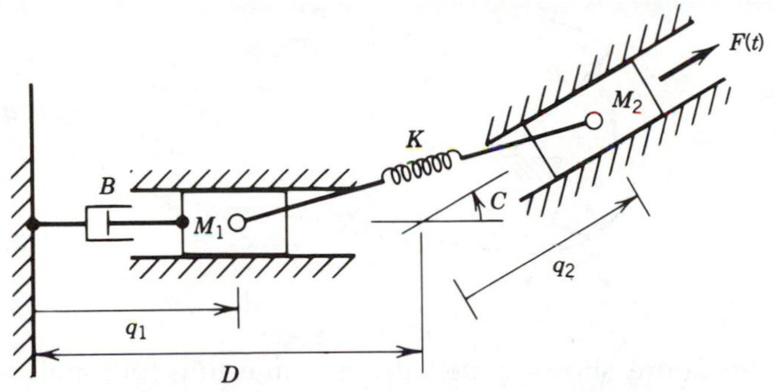

8.5 The two masses M_1 and M_2 are connected by a slotted link. The motion of the system is driven by an external force $F_1(t)$. There is also a viscous friction force described by the coefficient B acting between the pin on the vertical slider and the side of the slot in the link.

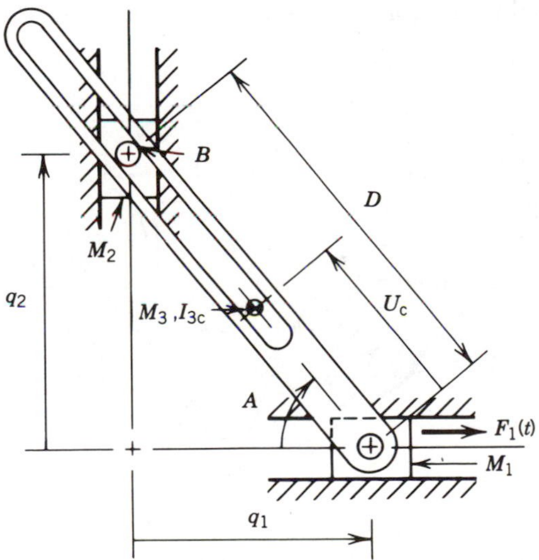

8.6 The figure shows two rotating inertias, I_1 and I_2, turning on fixed axes a distance C apart. The two disks are coupled with a belt drive, shown as a pair of springs to emphasize the elastic nature of the belt material. Although the belt is assumed to stretch in the two unsupported spans, it will be assumed to neither stretch nor creep on the pulleys. The smaller pulley is driven by an external torque, $T_1(t)$, and the power transferred is absorbed by the torque $T_2(\dot{q}_2)$. Note that the assumption of no creep is made for simplicity's sake, but it is not very realistic. For an interesting discussion of belt creep on pulleys, see the text by Burr that is listed in the references for this chapter.

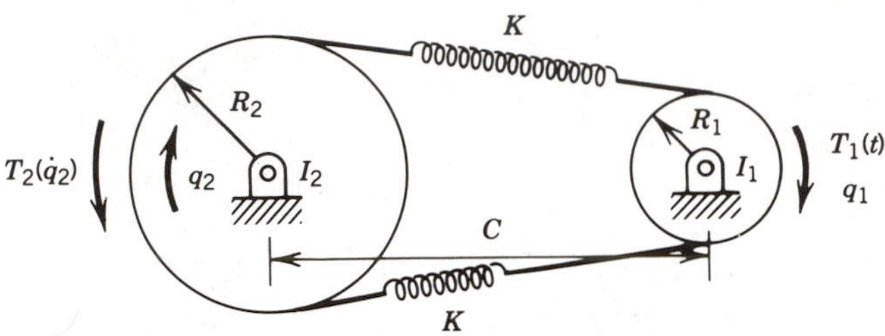

• **8.7** The figure shows a belt drive connecting four pulleys, where segments of the belt have been indicated as springs to emphasize the elasticity of the belt. The large pulley is the driver, with the applied torque $T_1(t)$. The three remaining smaller pulleys are each driven by the belt, supporting load torques $T_2(q_2, \dot{q}_2)$, $T_3(\dot{q}_3)$, and $T_4(\dot{q}_4)$. The short span from pulley 4 to pulley 1 is assumed to have negligible elasticity; the other spans are described by spring constants K_{12}, K_{23}, and K_{34}. Assume no creep on the pulleys. See the previous problem for a note regarding belt creep.

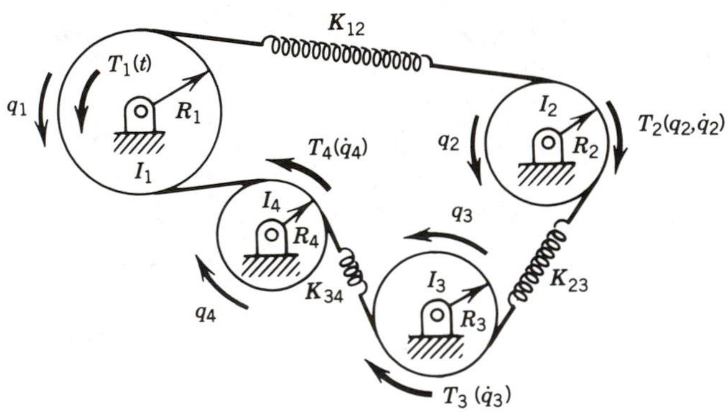

8.8 A belt drive using a spring-loaded idler to maintain tension is shown in the figure. The belt spans are shown as springs, to emphasize the elastic nature of the belt material, with spring rates K_1, K_2, and K_3. With no power transfer, the idler position is a known value, q_{30}. In use, the driving torque, $T_1(t)$, is applied to the larger pulley and the transmitted power is absorbed by the load torque $T_2(\dot{q}_2)$.

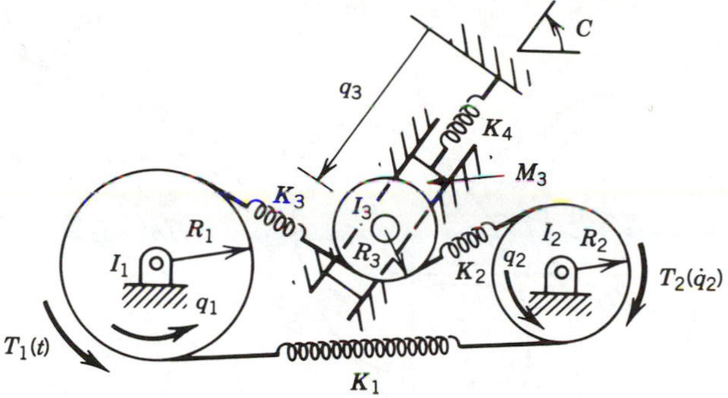

• **8.9** The sliders M_1 and M_2 move in circular guides in a horizontal plane. The two sliders are connected by a linear spring with constant K and free length S_o. The mass M_2 is driven by a tangential force $F(t)$, and viscous friction, described by the coefficient B, acts on both inner and outer curved sides of both sliders. The dimensions R_1, R_2, C_1, and C_2 are known, as are M_1, M_2, I_{1c}, and I_{2c}.

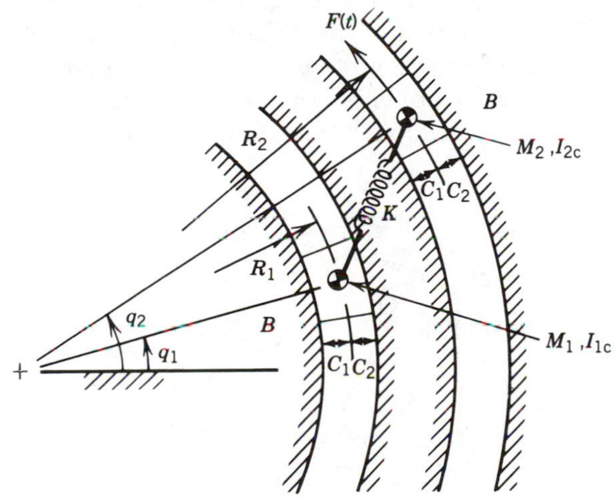

8.10 The figure shows a perspective view of a motor (disk 1) driving a slider–crank mechanism (disk 2 and mass M_3). The motor torque $T_1(t)$ acts on the rotor and is transmitted to the crank through a shaft with torsional stiffness K_1; the shaft is relaxed when $q_1 = q_2$. The connecting rod of the slider–crank mechanism has negligible mass but does have significant compliance, so it is shown as a spring of stiffness K_2. The motion of the slider is influenced by the action of a spring and dashpot assembly acting on the head end as shown. Both springs (K_2 and K_3) are relaxed when $q_2 = \pi$ and $q_3 = C - R$, where C is the free length of the connecting rod.

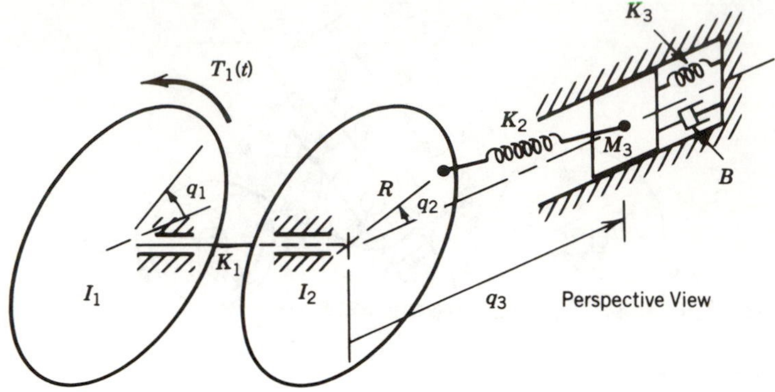

 **8.11** The figure shows a large church bell being rung by a small person swinging on the end of the bell rope. The bell and rope wheel assembly have mass M_b and moment of inertia I_b with respect to the axis of rotation; the bell center of mass is a distance C_1 below the axis of rotation. The clapper is suspended from a point a distance C_2 above the axis of rotation. The clapper has mass M_c and mass moment of inertia I_c with respect to its own center of mass. The clapper center of mass is located a distance C_3 from the pivot point. The person ringing the bell has mass M_p. Consider only those times when the bell ringer (the person) is completely clear of the floor and the clapper is not in contact with the side of the bell. The impact events require special consideration, but they are not part of this problem.

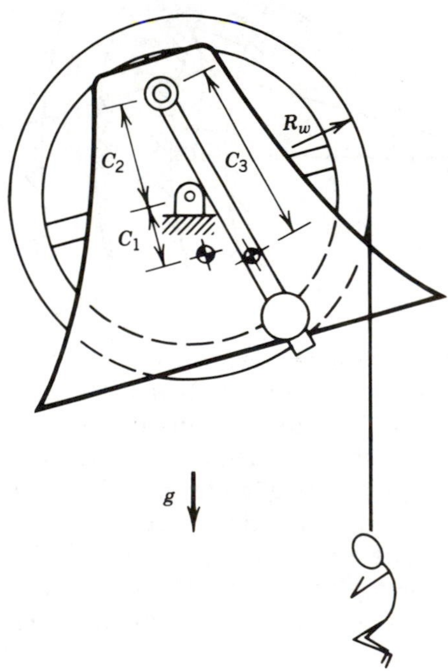

• **8.12** Torsional vibration is frequently excited by a nonsteady prime mover such as an internal combustion engine, often making it necessary to attempt to torsionally isolate the prime mover from the rest of the machine train. There are many different couplings on the market for this purpose, incorporating both elastic and damping elements. One way to construct such a coupling is shown in the figure for this problem where two linear springs are mounted in chordal positions between a pair of disks. Both springs attach to the disks at a radius R. Rotation of the first disk by an amount q_2 stretches the springs, applying a torque to the second disk. Because of the geometry of the system, the torque–displacement relation is nonlinear. The two springs each have spring rate K_c and free length S_{co}. The shaft between the prime mover and the coupling has torsional spring rate K_{12}, and the shaft from the coupling to the load has torsional spring rate K_{34}. The driving torque acts on the first station, $T_1(t)$, whereas the load torque $T_4(q_4, \dot{q}_4)$ opposes the motion of the last station. The mass moments of inertia, I_1, I_2, I_3, and I_4, are known.

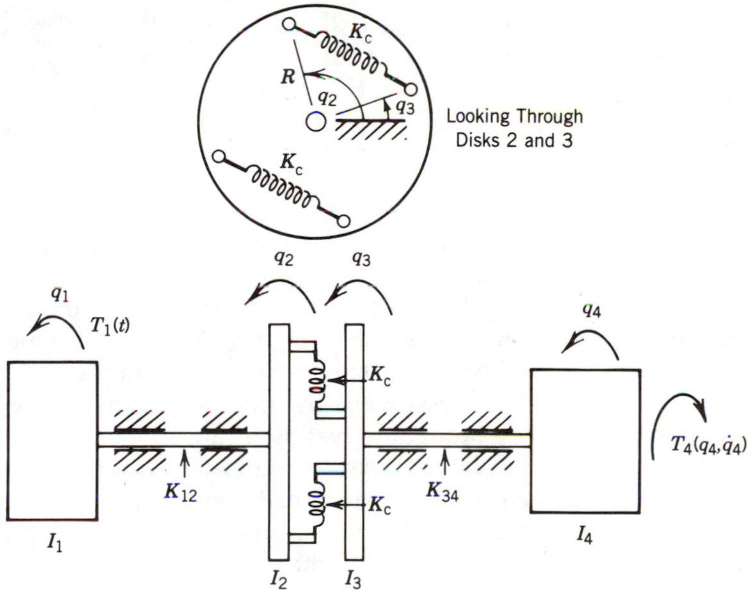

8.13 The process of starting a large electric motor under load requires an extended period of high current and often damages the motor and other components. The alternative is to provide some type of clutch so that the motor can start unloaded, and to have the load subsequently brought up to speed by engaging the clutch. If the clutch continues to slip during operation, this will reduce the efficiency of the system due to the energy converted to heat.

The system shown is a novel alternative to the typical clutch; systems of this general type are available commercially. The central component of

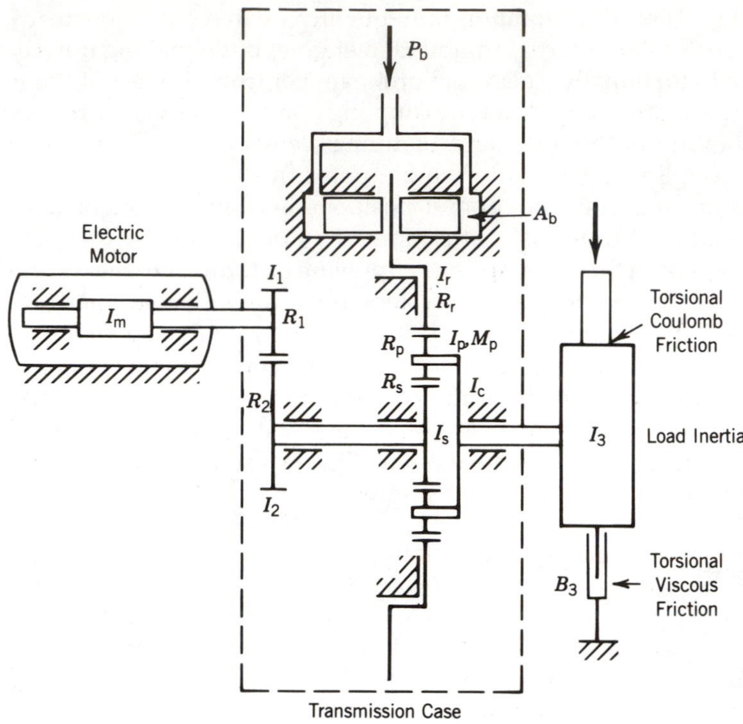

the transmission is a simple planetary gear train, with the sun gear driven by the motor through a single stage of reduction gearing. The output rotation, which must drive the load, is the planet carrier rotation. The clutch action is achieved by braking the ring gear. To unload the motor for starting, the ring gear is allowed to rotate freely; friction in the load causes the planet carrier to remain at rest while the ring gear spins. To drive the load, the ring gear must be brought to rest by means of a brake. A caliper type brake is shown, which acts on a flange attached to the ring gear and is actuated by the fluid pressure P_b. Under steady operating conditions, the ring gear is held motionless by the brake, and all of the motor power, except for losses in the gearing, is sent to the load where it is consumed by the two types of friction shown. The time period of interest is during the deceleration of the ring gear to rest. The following data should be considered as known:

R_1 Pitch radius for the motor pinion
R_2 Pitch radius for the input gear
R_s Pitch radius for the sun gear
R_p Pitch radius for the planets
R_r Pitch radius for the ring gear
R_c Planet carrier radius

A_b Piston area on brake pads
P_b Brake hydraulic pressure, a function of time
μ_b Coulomb friction coefficient for brake
R_b Effective radius for brake friction force, measured from the sun gear axis of rotation
T_m Motor torque
I_m Motor polar mass moment of inertia
I_1 Input pinion polar mass moment of inertia
I_2 Input gear polar mass moment of inertia
I_s Sun gear polar mass moment of inertia
I_p Planet polar mass moment of inertia about its own center; (there are three planets)
M_p Mass of a single planet
I_c Carrier and output shaft polar mass moment of inertia
I_r Ring gear polar mass moment of inertia
I_3 Load polar mass moment of inertia
B_3 Load viscous friction coefficient
T_c Constant Coulomb friction torque due to the load

8.14 The system shown is a trammel mechanism that has a slider moving along the coupling link. The sliders M_1 and M_2 move along the X-

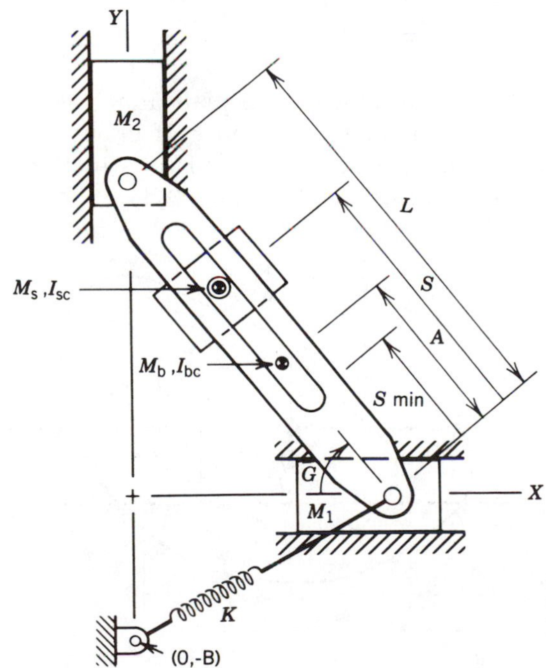

and Y-axes, respectively, and are connected by a link of length L. The center of mass of the coupling link is a distance A away from the lower connection point, and the link has mass M_b and mass moment of inertia I_{bc} with respect to its own center of mass. The slider M_s moves along the coupling link, guided by the slot along the link. The spring acting on the horizontal slider has free length B, and is anchored a distance B below the origin.

• **8.15** A force $F(t)$ is applied to the pin joining the two links as shown. The entire system is restrained horizontally by the spring at the left, and the motion is also influenced by the dashpot acting between the two sliders.

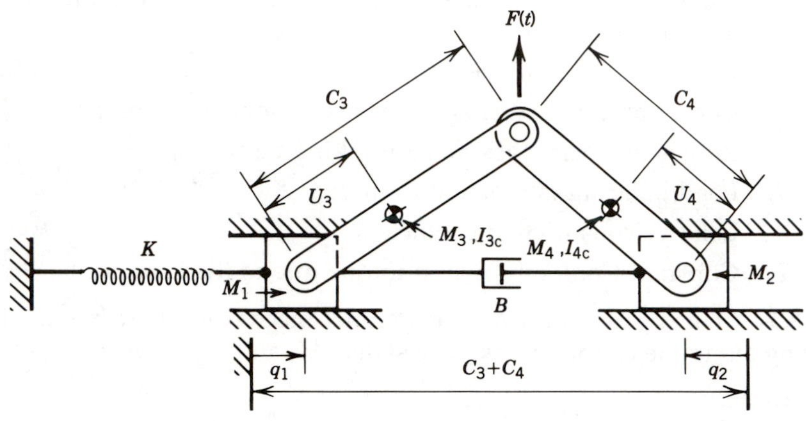

8.16 Two sliders are joined by a plate that has a circular slot as shown. The plate falls under the influence of gravity, pushing the two sliders apart. This motion is resisted by viscous friction on the sliders, described by the viscous coefficients B_1 and B_2.

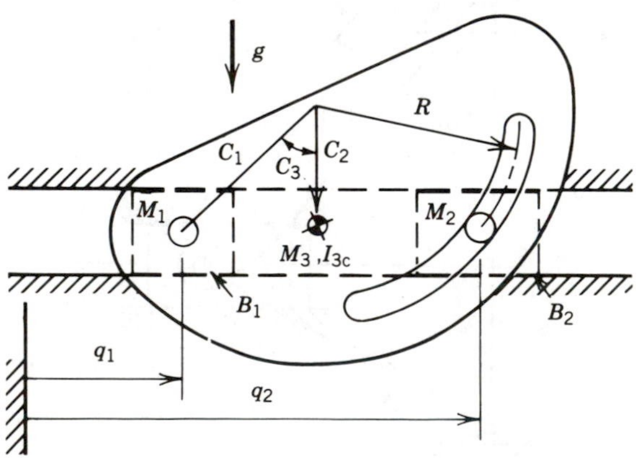

8.17 A harpoon is to be fired horizontally with initial velocity V_o. As the harpoon travels, an attached cord must unwind from a drum as indicated in the figure where the harpoon is modeled as a point mass. The cord is considered massless, but the drum has mass MOI I_c. Angle A describes the point where the cord is tangent to the drum, and angle G describes the rotation of the drum. The length of the unsupported cord is B with the initial value of B_o. The position of the point mass representing the harpoon is described by the two coordinates q_1 and q_2 with initial values R and B_o.

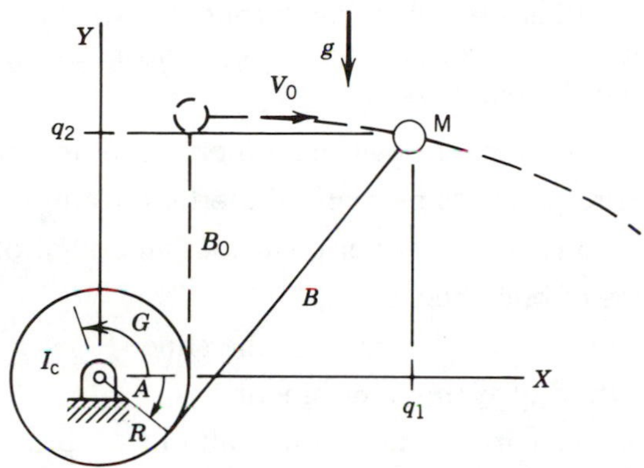

8.18 A bell crank pivots on the slider that moves in a vertical guide. The system is driven by the applied torque, T, and is also influenced by the dashpot with viscous coefficient B. As the disk rotates, the pin engaging the bell crank causes it to rotate or slide. The mass of the bell crank is M_2, and its center of mass is located by the body coordinates (U_c, V_c). The centroidal mass MOI for the bell crank is I_{2c}.

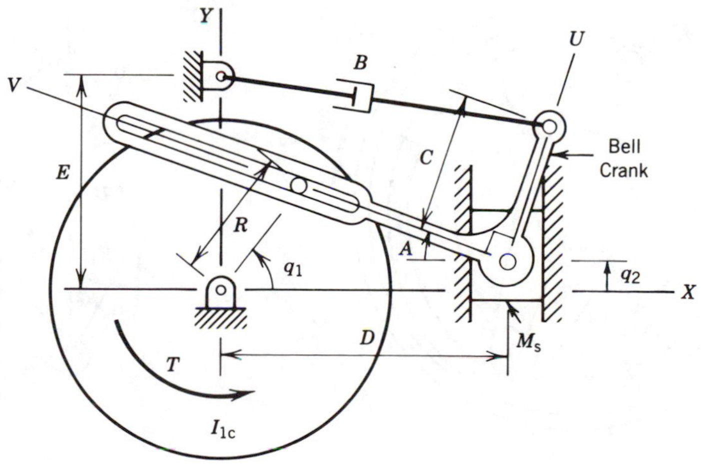

• **8.19** The figure shows a centrifugal clutch mechanism of the type often used with small gasoline engines. When the center member, attached to the crank shaft, rotates slowly, the springs hold the shoes in, preventing contact with the outer, driven member; in this state the clutch is disengaged. When the center member rotates at high speed, the shoes are thrown out against the driven member, and friction causes the entire assembly to rotate as a unit. Between these two conditions are two conditions of interest: (1) the period while the shoes move outward toward the driven member, and (2) the period of sliding contact between the shoes and the driven member. Both are transient conditions that depend on the varying shaft speed. Generalized coordinates q_1, q_2, and q_3 are indicated in the figure. The known data follows.

I_1	Driving member polar mass moment of inertia
I_2	Centroidal mass moment of inertia for one shoe
I_3	Driven (outer) member polar mass moment of inertia
M_2	Mass of one shoe
U_c, V_c	Center of mass location for the shoe
U_s	Location of spring attachment
K	Effective spring rate for one-half of the spring
D_{so}	Free length for one half of the spring
R_1	Radius to shoe pivot point
R_2	Radius to inside of driven member; also outside radius for the shoe
$T_1(t)$	External torque on the driving member
$T_3(\dot{q}_3)$	External torque on the driven member
μ	Coefficient of friction between shoe and driven member

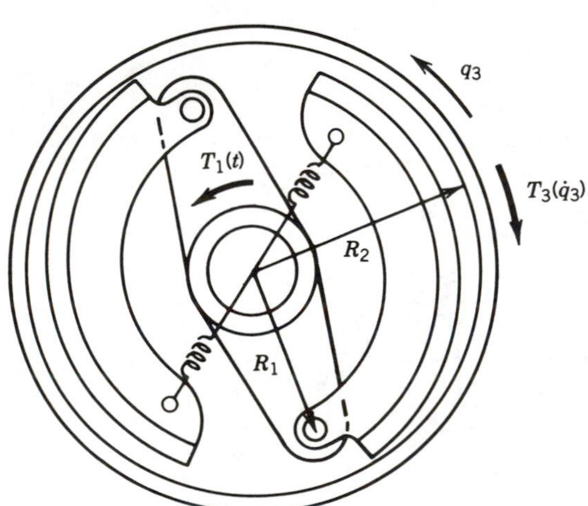

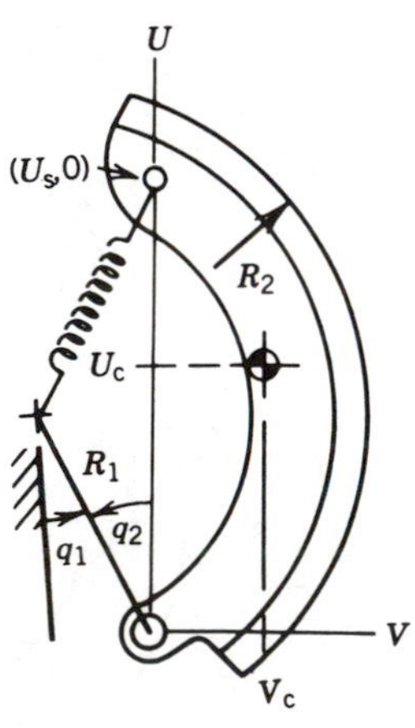

8.20 This problem is similar to problem 8.12, except that the coupling used to isolate the prime mover from the load is different. Here the coupling is constructed between two disks, but here connected by two linkages, each of which consists of two links. At low speed the two links become almost straight, and torque is transmitted. In this configuration no isolation is provided. At high speed, the links move outward to the configuration shown due to centrifugal effects. The four links are identical, and each has mass M_c and centroidal mass moment of inertia I_c. The center of mass is located as shown in the drawing for a typical link. This type of coupling has had only limited commercial exploitation.

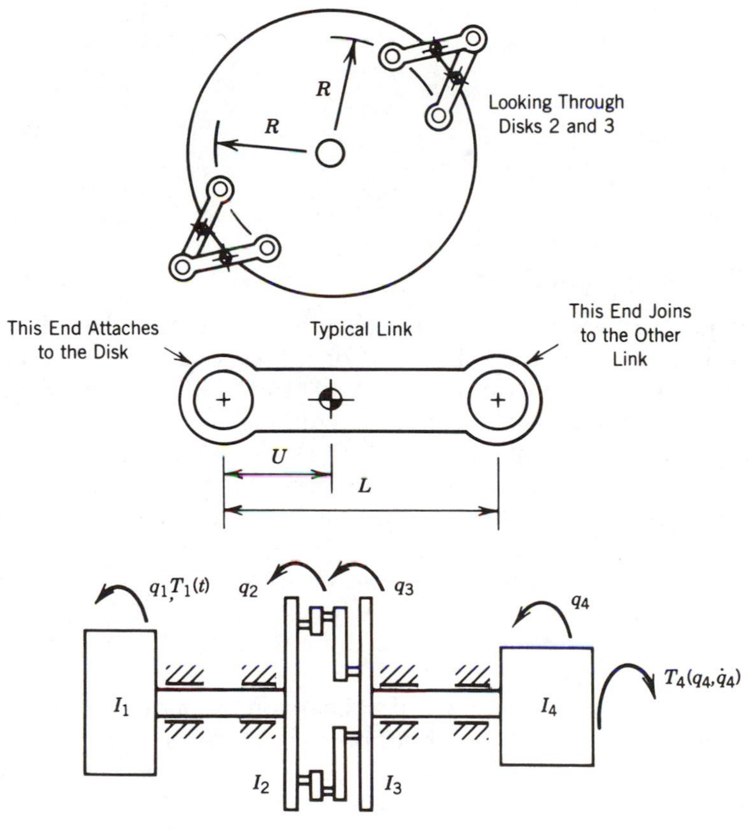

CHAPTER 9

Reactions and Internal Forces

9.1 GENERAL COMMENTS

In previous chapters, the forces that *cause* motion to occur have been considered, but forces that are the *result* of motions have been avoided. In Chapter 6, the forces that caused a body to be in equilibrium (a special state of motion) were analyzed, but the nonworking constraint forces did not enter the analysis. In Chapters 7 and 8, the system motions were described in terms of the external forces that do work on the system, but nonworking external forces and internal forces did not enter the equations of motion. The ability to determine the equations of motion without the need to consider internal forces and nonworking external forces is a great advantage. If, however, the ultimate objective includes determining one or more of these forces, something more must be done. Those additional steps are the subject of this chapter.

An external force or moment that bears on the moving system, often through the foundation or other support structure, is called a *reaction*. A knowledge of the reactions is essential to the design of proper supports for a machine. The time-varying part of the reaction is often called a *shaking force* or *shaking moment*, which are rather descriptive terms because these reactions tend to shake the supporting structure. A machine designer is often asked to provide information about the reactions for use by the person charged with the design of the supports. Alternatively, the designer of the supports may be required to determine those reactions.

Forces of interaction between two moving members of a system are called *internal forces*, a term intended to convey the fact that they are internal to the system. Internal forces are of great concern to a machine designer they affect the choice of sizes for the components. Stress, deflection, bearing loads, wear, and energy consumption are governed by the magnitudes of the internal forces, so no machine design can be considered complete until the internal forces have been determined and found acceptable for the final design.

9.2 THREE APPROACHES TO THE PROBLEM

There are three standard engineering approaches to the problem of determining reactions and internal forces, each of which is appropriate to a particular type of problem. They are the methods of *Statics, Kinetostatics,* and *Dynamics,* and their descriptions follow.

Statics. As the name implies, this approach assumes that all components are in equilibrium, and the forces are determined on this basis. This method is valid for a mechanism that is motionless under load, and is often appropriate for slowly moving (quasistatic) machines. The simplest of the three approaches, this method has been used widely for many years. Unfortunately, it has been applied to many situations where it really was not appropriate.

Kinetostatics. Many mechanisms move without noticeable changes in speed or with an apparently simple speed variation. For either case, it may be possible to assume an approximate description of the motion, often the relatively simple $\dot{q} = constant$. The kinetostatic method attempts to account for the effects of motion by assuming that the approximate motion is the true motion, and allowing for the inertial effects thus induced. For a single degree of freedom system, a description of the type $\dot{q} = constant$ could account for all of the centripetal terms ($\dot{q}^2 L_j$), but would neglect the $\ddot{q}$ terms ($\ddot{q} K_j$).

Dynamics. When speed variation is clearly present, there is often no adequate, simple, approximate description of the motion. In that case, it is necessary first to determine the motion by solving the equation of motion, and then to use that solution in the determination of the internal forces and reactions. For most systems, the dynamic analysis is the most difficult and time consuming of the three approaches. At the same time, however, it is the most correct and reliable.

For a further look at these three alternatives, consider a rigid body moving in two dimensions, for which there are three equations of motion:

$$\sum_i F_{xi} = M \ddot{X}_c$$

$$\sum_i F_{yi} = M \ddot{Y}_c$$

$$\sum_i M_i = I_c \ddot{A}$$

where X_c, Y_c are the center of mass coordinates and A is an angular coordinate describing the orientation of the body. Three equations of this sort can be written for every rigid body in the system. The objective is to evaluate the force and moment terms, but that raises the question of what to do with the inertial terms, $M\ddot{X}_c$, $M\ddot{Y}_c$, and $I_c\ddot{A}$. The three approaches just described propose three different responses to that question:

Static analysis assumes that all of the inertial terms are zero, leaving only sums of force and moment terms.

Kinetostatic analysis recognizes that the inertial terms are not necessarily zero, but asserts that a satisfactory analysis can be performed based on an *assumed* motion.

Dynamic analysis says that the actual motion must be determined by solving the equation of motion. Then, the inertial terms can be correctly evaluated.

In retrospect, it is all a question of just how to handle the inertial terms. The static analysis is strictly applicable only to systems in equilibrium, which often—but not always—means "motionless." It may be a very reasonable approximation in many other cases involving slow, apparently non-accelerated motion. The dynamic analysis is the most difficult and time consuming of the three approaches. The choice of an appropriate method of analysis is an engineering judgment. The three methods of analysis will be illustrated in the next section.

9.3 EXAMPLE OF SLIDER–CRANK FORCE ANALYSES

In this section, the three approaches to the analysis of reactions and internal forces are demonstrated, each as applied to a slider–crank machine. The system considered is a single-cylinder, reciprocating air compressor driven in steady state by an induction motor. After a description of the system and a presentation of that part of the analysis common to all three approaches, the application of each analysis approach is outlined and the results are presented. At the end of the section, key features of all three analyses are compared.

System Description. The system is a reciprocating air compressor driven by an induction motor. The compressor uses poppet valves that open or close in response to differential pressure; there are no springs or lifting mechanisms. Figure 9.1 shows a plot of cylinder pressure versus crosshead position; this is equivalent to a pressure–volume diagram for the machine. The suction manifold is at atmospheric pressure, 14.7 psia, and the discharge is constant at 89.7 psia (75 psig); the crankcase is vented, thus maintaining atmospheric pressure on the underside of the piston. Referring to the pressure–position diagram, compression begins at point ①, with the pressure rising adiabatically to point ②, where the discharge

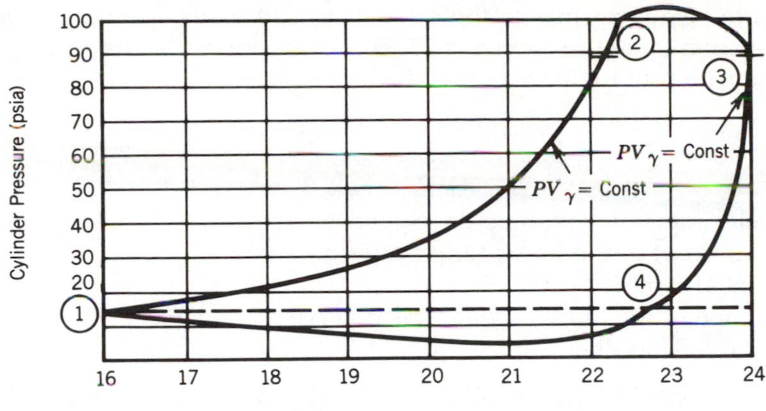

FIGURE 9.1 Cylinder Pressure versus Crosshead Position

valves open. Although the discharge valves are open, the cylinder pressure rises above discharge manifold pressure as the charge is forced out of the cylinder. At point ③, the discharge valves close, trapping some gas in the cylinder.

As the piston moves past top dead center, the trapped gas expands, doing work on the piston; this is also an adiabatic process. At point ④, the intake valves open, and the cylinder pressure falls below atmospheric as the charge is pulled into the cylinder. At point ① the intake valves close, and the cycle begins again. The "corner points" on the pressure–position diagram are:

$P_1 = 14.7$ psia $\quad X_1 = 16.000$ in. $\quad$ BDC

$P_2 = 89.7$ psia $\quad X_2 = 22.166$ in.

$P_3 = 89.7$ psia $\quad X_3 = 24.000$ in. $\quad$ TDC

$P_4 = 14.7$ psia $\quad X_4 = 22.674$ in.

The period of excess pressure while the discharge valves are open is called the "cap," whereas the period of subatmospheric pressure with the suction valves open is called the "sole." Pressure values during these intervals are determined by the flow resistance through the valves and must be read directly from the diagram. During the compression and expansion intervals, the usual adiabatic pressure relation applies:

$$P\,V^{\gamma} = Constant$$

where the polytrophic exponent, γ, is taken as 1.40 for air.

The induction motor driving this compressor is 4.0 Hp, 3 phase, and 440 volts. The synchronous speed is 225 rpm, and rated speed is 208 rpm (there is no information available on the no-load speed, so it will be assumed to be approximately the same as synchronous speed). The motor rotor mass moment of inertia is 23.93 lb-s²-in. From this data, the linear approximation to the torque–speed curve is

$$M_o = C_0 + C_1 \dot{q}$$

where

M_o = electromagnetic torque developed, in.-lb
C_0 = intercept of the linear torque–speed curve
= 16041.5 in.-lb
C_1 = slope of the linear torque–speed curve
= −680.82 in.-lb-s

At rated speed (21.7817 rad/s), this approximation gives the rated torque, 1212.0 in.-lb. (A similar linear torque–speed relation for an induction motor was developed in more detail in Section 8.5.)

Various other mechanical data on the compressor is summarized in the following list. Some of the dimensional information is identified in Figure 9.2.

$D = 5.5$ in.	Bore diameter
$R = 4.0$ in.	Crank radius
$L = 20.0$ in.	Connecting rod length
$V_1 = 202$ in.^{3}	Max cylinder volume
$P_1 = 14.7$ psia	Suction pressure
$P_2 = 89.7$ psia	Discharge pressure
$= 75.0$ psig	
$U_1 = -0.09373$ in.	Position of the crank CM
$W_1 = 399.6$ lb	Crank weight

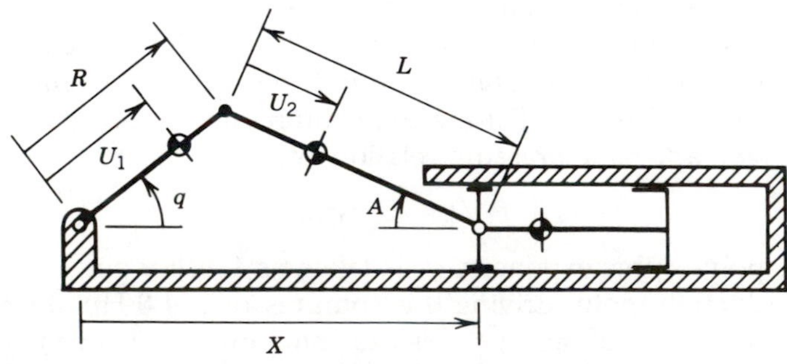

FIGURE 9.2 Schematic Diagram of Air Compressor

$I_1 = 82.36$ lb-s²-in.	Crank mass moment of inertia with respect to the fixed axis of rotation
$W_2 = 34.2$ lb	Connecting rod weight
$I_2 = 2.9543$ lb-s²-in.	Connecting rod mass moment of inertia with respect to the CM
$U_2 = 5.22$ in.	Position of the connecting rod CM
$W_3 = 48.2$ lb	Weight of the crosshead
$W_4 = 13.0$ lb	Weight of the connecting rod
$W_5 = 18.3$ lb	Weight of the piston

One of the necessary steps in the analysis (by any of the three approaches) is to relate the crank position to the cylinder volume and thus to the cylinder pressure. This is required to determine the cylinder pressure during either the compression or expansion phases of the cycle. The top end of the cylinder usually has a relatively complicated shape, but it can be imagined as being a right, circular cylinder. When the piston is at top dead center, the remaining volume is called the *clearance volume*, and the height of an equivalent right circular cylinder is called the *clearance distance*. The crosshead position, X, is readily expressed in terms of the crank angle, q, so the cylinder volume, $V(X)$ is

$$\begin{aligned} V(X) &= V_c + A_c(X_3 - X) \\ &= A_c(D_c + R + L - X) \end{aligned}$$

where

V_c = clearance volume
A_c = cylinder cross-sectional area
X_3 = maximum crosshead excursion
D_c = clearance distance

The maximum cylinder volume, V_1, occurs when X is a minimum, equal to $L - R$, so that

$$V_1 = A_c(D_c + 2R)$$

The clearance distance can be evaluated from this relation. Once the clearance distance is known, the volume can be determined for any crank position by using the general relation.

Component Equations of Motion. Figure 9.3 shows four free-body diagrams, identifying the forces of interest. The system of equations of motion that follow is the result of applying the translational and rotational forms of Newton's Second Law to each of the four free bodies. Note that M_o is the electromagnetic torque of the motor and A_c is the piston area.

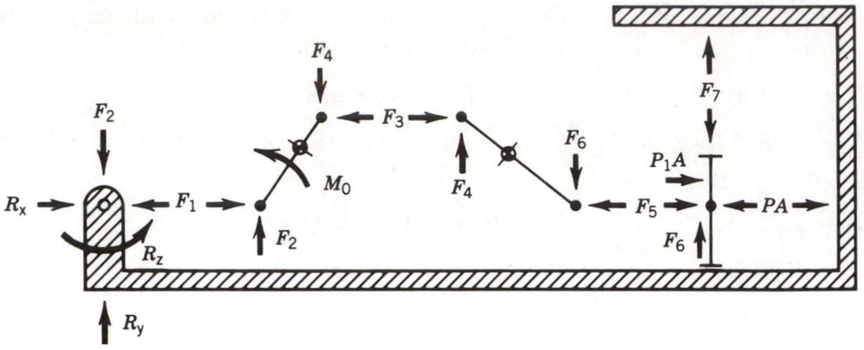

FIGURE 9.3 Free-body Diagrams for Force Analysis

Crank

$$F_1 - F_3 = M_1 \ddot{X}_{c1} = M_1 (-\ddot{q}\, U_1 \sin q - \dot{q}^2\, U_1 \cos q)$$
$$F_2 - F_4 = M_1 \ddot{Y}_{c1} = M_1(\ddot{q}\, U_1 \cos q - \dot{q}^2\, U_1 \sin q)$$
$$M_o + F_3\, R \sin q - F_4\, R \cos q = I_1\, \ddot{q}$$

Connecting Rod

$$F_3 - F_5 = M_2 \ddot{X}_{c2} = M_2 (\ddot{q}\, K_{x2} + \dot{q}^2\, L_{x2})$$
$$F_4 - F_6 = M_2 \ddot{Y}_{c2} = M_2 (\ddot{q}\, K_{y2} + \dot{q}^2\, L_{y2})$$
$$F_3\, U_2 \sin A + F_4\, U_2 \cos A + F_5\, (L - U_2) \sin A$$
$$+ F_6\, (L - U_2) \cos A = I_2\, \ddot{A} = I_2\, (\ddot{q}\, K_a + \dot{q}^2\, L_a)$$

Crosshead and Piston

$$F_5 - (P - P_1)A_c = M_3 \ddot{X}_{c3} = M_3\, (\ddot{q}\, K_{x3} + \dot{q}^2\, L_{x3})$$
$$F_6 - F_7 = 0$$

Frame (Equilibrium)

$$R_x - F_1 + (P - P_1)\, A_c = 0$$
$$R_y - F_2 + F_6 = 0$$
$$R_z + X_3\, F_6 = 0$$

There are eight equations of motion and seven internal forces to be determined. The additional equation may be considered as the system equation of motion for the determination of $q(t)$. Therefore, one of the equations of motion may be omitted from the solution process. It will, however, still be satisfied. The sum of moments on the connecting rod is the most complicated of the equations of motion, and is an obvious choice for the equation to be omitted. This procedure eventually leads to the evaluation of F_4 using the rotational equation for the crank involving a division by cos q. There are two angular positions in the cycle where this results in division by zero. This difficulty can be avoided if the crank rotational equation is the equation omitted. The preceding equations can then be solved for the forces as follows:

$$F_5 = M_3(\ddot{q}\, K_{x3} + \dot{q}^2\, L_{x3}) + (P - P_1)A_c$$

$$F_3 = F_5 + M_2(\ddot{q}\, K_{x2} + \dot{q}^2\, L_{x2})$$

$$F_1 = F_3 - M_1(\ddot{q}\, U_1 \sin q + \dot{q}^2\, U_1 \cos q)$$

$$F_6 = \frac{[-F_3\, U_2 \sin A - F_5(L - U_2) \sin A + I_2(\ddot{q}\, K_a + \dot{q}^2\, L_a) - M_2(\ddot{q}\, K_{y2} + \dot{q}^2\, L_{y2})]}{L \cos A}$$

$$F_4 = F_6 + M_2(\ddot{q}\, K_{y2} + \dot{q}^2\, L_{y2})$$

$$F_2 = F_4 + M_1(\ddot{q}\, U_1 \cos q - \dot{q}^2\, U_1 \sin q)$$

$$F_7 = F_6$$

This completes the solution in closed form and avoids any possibility for division by zero.

With the internal reactions known, the external reactions can be determined easily from the equations of equilibrium for the frame:

$$R_x = F_1 - (P - P_1)A_c$$

$$R_y = F_2 - F_6$$

$$R_z = -X_3\, F_6$$

At this point, the force calculations are complete, provided there is an answer to the question, "What is to be done with the inertia terms?"

Static Analysis. As indicated earlier, the static approach to the problem assumes that all the components are in equilibrium, and the inertia terms are therefore zero. For this analysis, the driving torque is determined exclusively by the load, and it is convenient to supplement the preceding equations with the equilibrium relation between the driving torque and the force of the compressed air in the cylinder:

$$\begin{aligned} 0 = \delta W &= M_o\, \delta q - (P - P_1)A_c\, \delta X \\ &= M_o\, \delta q - (P - P_1)A_c\, K_{x3}\, \delta q \\ &= \delta q\, [M_o - (P - P_1)A_c\, K_{x3}] \end{aligned}$$

from which the required motor torque is

$$M_o = (P - P_1)A_c\, K_{x3}$$

With all the inertial terms assumed to be zero, the equations above are evaluated easily:

$$F_5 = (P - P_1)A_c$$

$$F_3 = F_5$$

$$F_1 = F_3 = F_5$$

$$F_6 = -F_5 \text{ Tan } A$$

$$F_4 = F_6$$

$$F_2 = F_6$$

$$F_7 = F_5$$

The external reactions are

$$R_x = F_1 - (P - P_1)A_c = 0$$

$$R_y = F_2 - F_6 = 0$$

$$R_z = X_3F_6 = -M_o$$

These results can be plotted over the range $0 < q < 2\pi$ to see how the various forces vary according to the static analysis. Other more correct methods of analysis indicate that the static analysis is quite far from accurate for this problem. The translating components in the machine have continually changing linear momentum in the X– and Y–directions, which should require an external force in each direction. The static analysis shows zero reaction forces. The calculated moment reaction, R_z, shows the correct trends, but the values are not correct. With regard to the internal forces, the static analysis values are not correct. (Extreme values from the static analysis are included in the summary table at the end of this example.) It is clear that although the static analysis is simple to perform, the results are poor. This is to be expected when the assumptions of the analysis grossly violate the problem conditions. It cannot even be claimed that the static analysis is conservative, because it underestimates some values and overestimates others.

Kinetostatic Analysis. For the kinetostatic analysis, the machine is assumed to operate at constant speed and electromagnetic torque, both at their rated values. This means that the $\ddot{q}$ terms will all be zero in the equations of motion, but the centripetal terms will have reasonable, nonzero values. The computed reactions are plotted for a little more than one crank revolution in Figure 9.4, and three of the internal forces are plotted in Figure 9.5. It is quite evident that the external reactions R_x and R_y are not zero, as was indicated in the previous static analysis. The reaction moment, R_z, and the internal force, F_1, have the same general trends as in the static analysis, but the values are quite different.

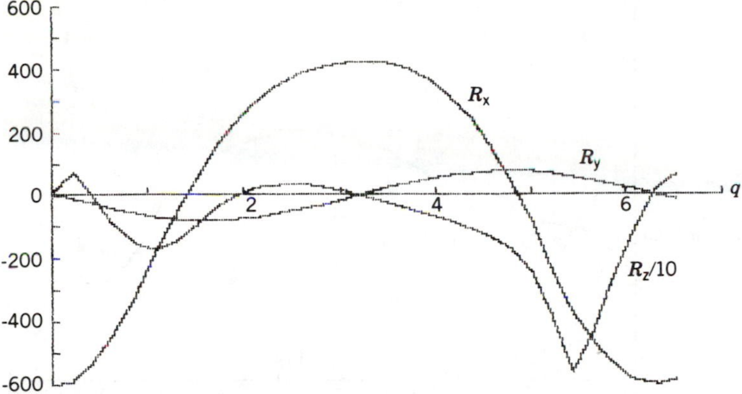

FIGURE 9.4 External Reactions According to the Kinetostatic Analysis

Dynamic Analysis. To carry out the dynamic force analysis, first it is necessary to obtain a solution of the equation of motion. Eksergian's form for the equation of motion is applicable here, so the equation of motion is

$$\mathcal{I}(q)\,\ddot{q} + \mathcal{C}(q)\,\dot{q}^2 = Q$$

where the generalized inertia, the centripetal coefficient, and the generalized force are, respectively,

$$\mathcal{I}(q) = I_1 + M_2(K_{x2}{}^2 + K_{y2}{}^2) + I_2\,K_a{}^2 + M_3\,K_{x3}{}^2$$

$$\mathcal{C}(q) = M_2(K_{x2}\,L_{x2} + K_{y2}\,L_{y2}) + I_2\,K_a\,L_a + M_3\,K_{x3}\,L_{x3}$$

$$Q = M_o - (P - P_1)\,A_c\,K_{x3}$$

The steady-state motion is required but the appropriate initial conditions are not provided. The beginning time, $t = 0$, can be defined as the

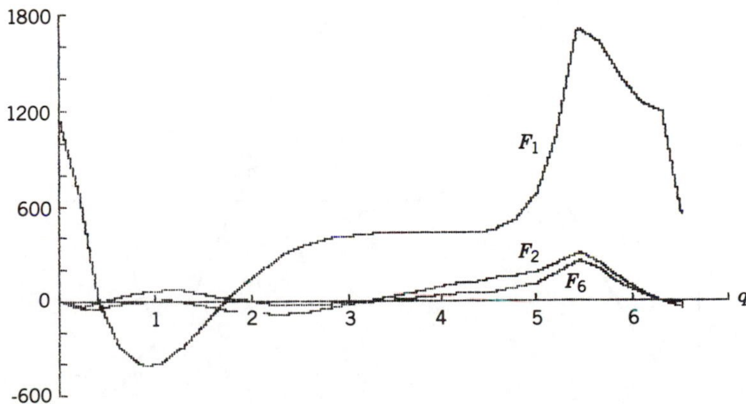

FIGURE 9.5 Internal Forces According to the Kinetostatic Analysis

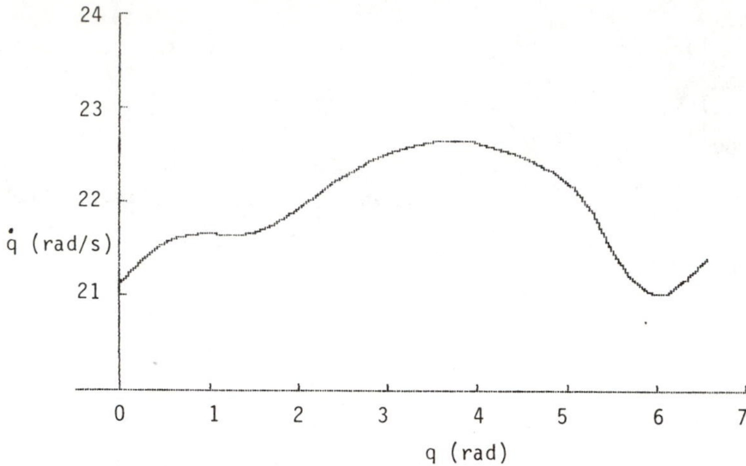

FIGURE 9.6 Crank Speed as a Function of Crank Position as Determined by the Steady State Simulation

time when the crank rotation is zero, establishing one of the initial conditions. In steady state, the crank velocity must be a periodic function of the crank position, so the appropriate initial velocity can be determined by trial and error using the simulation. On this basis, the solution is determined to be periodic when the initial crank angular velocity is 21.13 rad/s. For subsequent simulation runs, the initial conditions

$$q(0) = 0 \text{ rad}$$

$$\dot{q}(0) = 21.13 \text{ rad/s}$$

generate the steady-state motion immediately, so that only a single cycle is required. The manner in which crank speed varies with crank position, as determined from the simulation, is shown in Figure 9.6. The assumption of

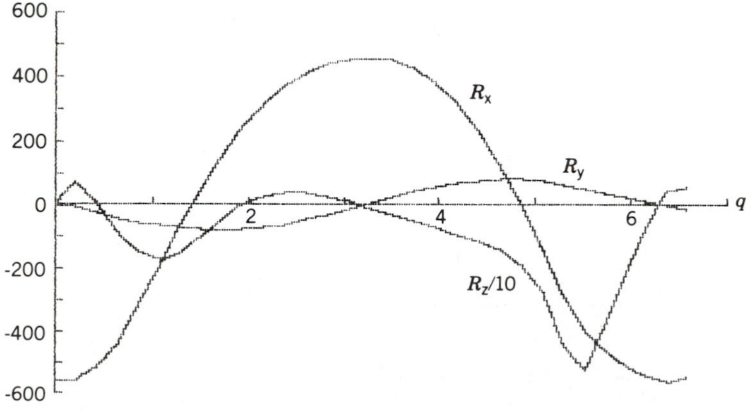

FIGURE 9.7 External Reactions According to the Dynamic Analysis

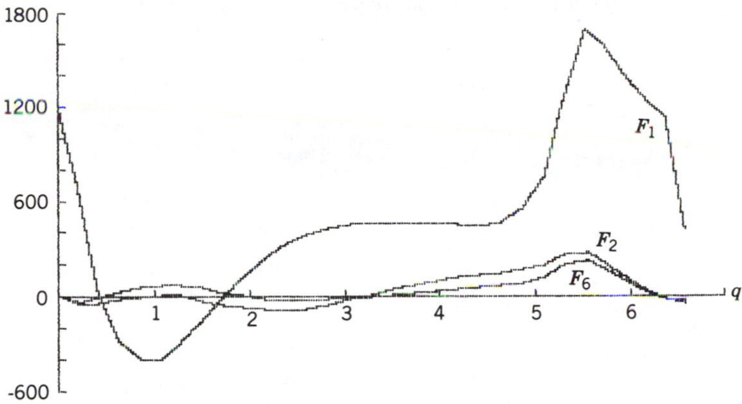

FIGURE 9.8 Internal Forces According to the Dynamic Analysis

constant crank speed, as used in the kinetostatic analysis, is clearly in error.

Using the simulation to provide values for q, $\dot{q}$, and $\ddot{q}$ at each integration time step, the forces are determined according to the equations of motion formulated previously. The results are shown in Figure 9.7 and Figure 9.8. These plots appear to be almost identical to the results obtained from the kinetostatic analysis, but the actual numbers show a few distinct differences. The most notable of these is the approximately 300 in.-lb difference between the computed minimum values for R_z.

One additional plot of interest is shown in Figure 9.9, where the two external reaction forces, R_y and R_x, are plotted against each other (note the distorted scales). The distance from the origin to a point on this curve gives the magnitude of the crank-bearing force at any point in the cycle. The results shown are from the dynamic simulation, and the curve is not quite

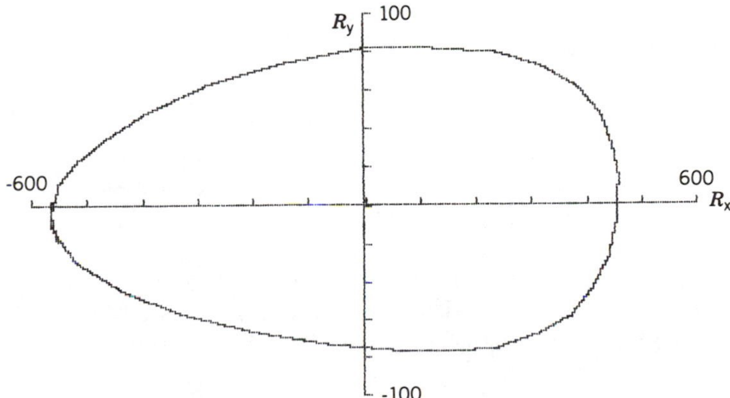

FIGURE 9.9 External Reaction Forces Plotted Against Each Other, as Determined from the Dynamic Analysis

TABLE 9.1 Comparison of Results

	Static Analysis	*Kinetostatic Analysis*	*Dynamic Analysis*
R_x (max)	0.0	425.6	457.3
R_x (min)	0.0	−599.9	−566.9
R_y (max)	0.0	77.8	82.7
R_y (min)	0.0	−78.1	−77.4
R_z (max)	1278	696.3	726.5
R_z (min)	−6982	−5632	−5328
F_1 (max)	2083	1695	1701
F_1 (min)	−227.4	−406.3	−398.6
F_2 (max)	310.9	304.1	282.8
F_2 (min)	−53.5	−82.1	−84.3
F_6 (max)	310.9	250.8	234.8
F_6 (min)	−53.5	−29.2	−30.4

symmetrical about the R_x–axis. The minimum value of R_x occurs for a negative value of R_y, whereas the maximum value of R_x occurs when R_y is positive. A similar plot, using data from the kinetostatic analysis, yields a curve that is exactly symmetric about the horizontal axis.

Summary. Table 9.1 summarizes these results and gives some numerical values for comparison. For each item, the most extreme value has been underlined.

9.4 CALCULATION OF REACTIONS ALONE

There will be times when one or more of the support reactions will be required without need for the internal forces. The previous example may suggest that the internal forces must be determined as a part of evaluating the reactions, but that is not the case. Newton's Second Law, as extended to systems of rigid bodies, provides a convenient alternative when only the reactions are needed.

Let X_{ci} and Y_{ci} denote the coordinates of the center of mass of the ith element of a mechanism, and let A_i be the angular orientation. The total linear momenta of the system in the X– and Y–directions are

$$P_x = \sum_i M_i \dot{X}_{ci}$$

$$P_y = \sum_i M_i \dot{Y}_{ci}$$

and the angular momentum with respect to the origin is

$$H_o = \sum_i [M_i(X_{ci}\,\dot{Y}_{ci} - Y_{ci}\,\dot{X}_{ci}) + I_i\,\dot{A}_i]$$

From Newton's Second Law, the total external forces in the X– and Y–directions are

$$F_x = \dot{P}_x = \frac{d}{dt}\left(\sum_i M_i\,\dot{X}_{ci}\right)$$

$$F_y = \dot{P}_y = \frac{d}{dt}\left(\sum_i M_i\,\dot{Y}_{ci}\right)$$

The total external moment with respect to the origin is

$$M_o = \dot{H}_o = \frac{d}{dt}\sum_i [M_i(X_{ci}\,\dot{Y}_{ci} - Y_{ci}\,\dot{X}_{ci}) + I_i\,\dot{A}_i]$$

The total external force in the X–direction consists of any driving force applied in the X–direction plus the reaction in the X–direction, R_x, and similarly for the Y-direction. The total external moment with respect to the origin consists of any driving moment applied to the system, the moments of any driving forces with respect to the origin, and the reaction moment, R_z. The utility of this approach is shown in the following discussion where it is applied to the slider–crank mechanism of the previous example.

For the air compressor considered in the previous section, the linear momentum in the X–direction is

$$\begin{aligned} P_x &= M_1\,\dot{X}_1 + M_2\,\dot{X}_2 + M_3\,\dot{X}_3 \\ &= \dot{q}\,(-M_1U_1 \sin q + M_2\,K_{x2} + M_3\,K_{x3}) \end{aligned}$$

The only external force in this direction is the reaction, R_x, determined from the derivative of the linear momentum:

$$\begin{aligned} R_x &= \ddot{q}\,(-M_1U_1 \sin q + M_2\,K_{x2} + M_3\,K_{x3}) \\ &\quad + \dot{q}^2\,(-M_1U_1 \cos q + M_2\,L_{x2} + M_3\,L_{x3}) \end{aligned}$$

In the Y-direction, the linear momentum is

$$\begin{aligned} P_y &= M_1\dot{Y}_1 + M_2\dot{Y}_2 \\ &= \dot{q}\,(M_1U_1 \cos q + M_2\,K_{y2}) \end{aligned}$$

and the reaction R_y is the only external force in that direction

$$\begin{aligned} R_y &= \ddot{q}\,(M_1U_1 \cos q + M_2\,K_{y2}) \\ &\quad + \dot{q}^2\,(-M_1U_1 \sin q + M_2\,L_{y2}) \end{aligned}$$

The angular momentum of the internal parts with respect to the origin is

$$\begin{aligned} H_o &= I_1\,\dot{q} - I_2\,\dot{A} + M_2\,(X_2\,\dot{Y}_2 - Y_2\,\dot{X}_2) \\ &= \dot{q}\,[I_1 - I_2\,K_a + M_2\,(X_2\,K_{y2} - Y_2\,K_{x2})] \end{aligned}$$

where there is no angular momentum due to the piston because it is not offset. The total external moment consists of the driving moment, M_o, and the reaction moment, R_z—both positive in the counterclockwise sense:

$$R_z + M_o = \dot{H}_o$$

$$= \frac{d}{dt}\{\dot{q}\,[I_1 - I_2\,K_a + M_2\,(X_2\,K_{y2} - Y_2\,K_{x2})]\}$$

The reaction moment is then

$$R_z = -M_o + \ddot{q}\,[I_1 - I_2\,K_a + M_2\,(X_2\,K_{y2} - Y_2\,K_{x2})] + \dot{q}^2\,[-I_2\,L_a + M_2\,(K_{x2}\,K_{y2} + X_2\,L_{y2} - K_{y2}\,K_{x2} - Y_2\,L_{x2})]$$

$$= -M_o + \ddot{q}\,[I_1 - I_2\,K_a + M_2\,(X_2\,K_{y2} - Y_2\,K_{x2})] + \dot{q}^2\,[-I_2\,L_a + M_2\,(X_2\,L_{y2} - Y_2\,L_{x2})]$$

One of the significant points to be made here is that *each* of these equations can be used *alone*. If the only reaction of interest is R_y, there is no need to calculate R_x, R_z, or any of the internal forces. This can significantly reduce the labor required in some cases. It should also be noted that these equations can be used for either kinetostatic or dynamic analysis, depending on whether the motion is to be assumed or is determined as the solution of the equation of motion.

9.5 CONCLUSION

From the standpoint of engineering design, the analysis of internal loads is often a major concern. Clearly, all components must be strong enough to withstand the various combinations of loads imposed on them, and the complete loading cannot be known until the component masses and motions are known.

The internal forces are determined through Newton's Second Law, with the accelerations specified in one of three ways—statics, kinetostatics, or dynamics. The resulting system of equations is linear in the unknown forces, and can often be solved in closed form. Because the internal forces usually vary as the machine moves, it is often not clear just when a particular force will be a maximum or a minimum. Therefore, it is necessary to evaluate the forces for a variety of positions. Once the forces are known, stress and fatigue analyses are usually the next steps.

REFERENCES

Beer, F. P. and Johnston, E. R., *Vector Mechanics for Engineers*, 4th ed. New York: McGraw–Hill, 1984.

Erdman, A. G., and Sandor, G. N., *Mechanism Design: Analysis and Synthesis*, Vol. 1. Englewood Cliffs, N.J.: Prentice–Hall, 1984.

Hirschhorn, J., *Dynamics of Machinery*. New York: Barnes & Noble, 1968.

Paul, B., *Kinematics and Dynamics of Planar Machinery*. Englewood Cliffs, N.J.: Prentice–Hall, 1979.

PROBLEM SET

For the following problems, assume that

a. All indicated dimensional and inertia data are known;

b. The generalized position, velocity, and acceleration—$q(t)$, $\dot{q}(t)$, and $\ddot{q}(t)$—are known;

c. There is no friction unless stated;

d. When required, the gravitational constant is assumed to be $g = 386$ in./s^2 or $g = 9.81$ m/s^2.

9.1 For the trammel mechanism shown, the center of mass is at the middle of the link. Carry out the following operations:

a. Perform all kinematic analysis required for the later steps;

b. Determine the applied force F required if $\dot{q} = V_o = constant$;

c. Determine the guide reaction forces and the horizontal and vertical forces at the pins;

d. Evaluate the guide reactions, pin forces, and required driving force at an instant when $q = 155$ mm and $\dot{q} = 85$ mm/s, for these system parameters:

$D = 388$ mm $\qquad M_x = 0.225$ kg
$M_d = 1.15$ kg $\qquad M_y = 0.272$ kg
$I_{dc} = 0.0152$ kg-m^2

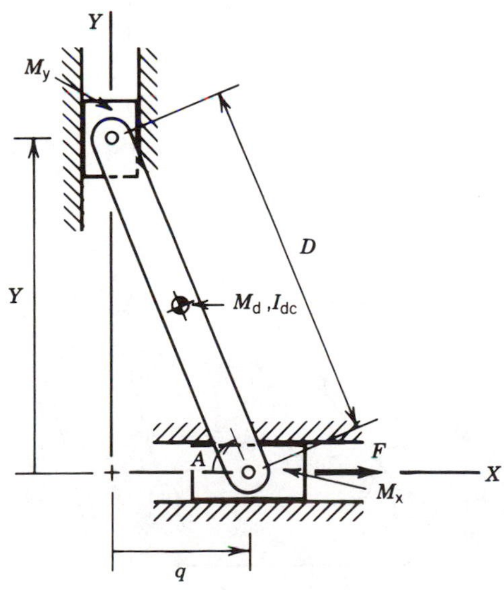

9.2 For the trammel mechanism shown the length between the pins is L. The spring free length is such that the spring is relaxed when $q = q_o$. The motion is driven by the applied force F.

a. Perform all kinematic analysis required for the later steps;
b. Determine the equation of motion;
c. Determine the force in the spring;
d. Determine the horizontal and vertical components of the pin reactions;
e. Determine the guide reactions;
f. Evaluate the spring force, pin reaction components, and guide reactions when $q = 13.55$ inches, $\dot{q} = 210.0$ in./s, and $F = 34.2$ lb, using the following system parameters:

$M_1 g = 0.344$ lb	$C = 0.833$ rad	$D = 26.4$ in.
$M_2 g = 0.771$ lb	$L = 20.29$ in.	
$M_3 g = 6.208$ lb	$K = 38.33$ lb/in.	
$I_{3c} = 0.683$ lb-s²-in.	$q_o = 5.77$ in.	
$U_c = 7.883$ in.	$V_c = 4.966$ in.	

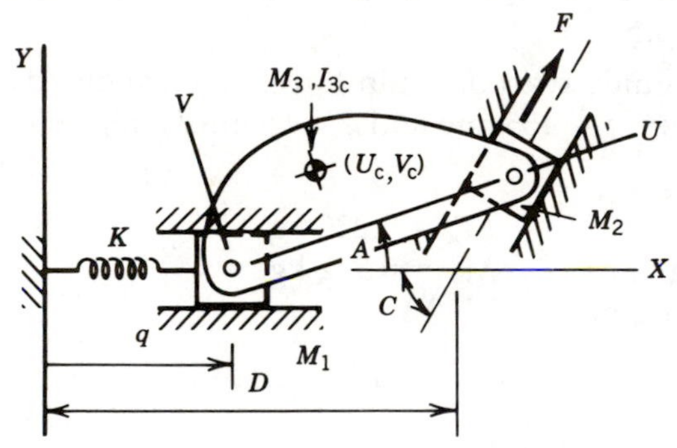

9.3 For the four-bar mechanism shown, the link centers of mass are on the line from pin to pin for each link.

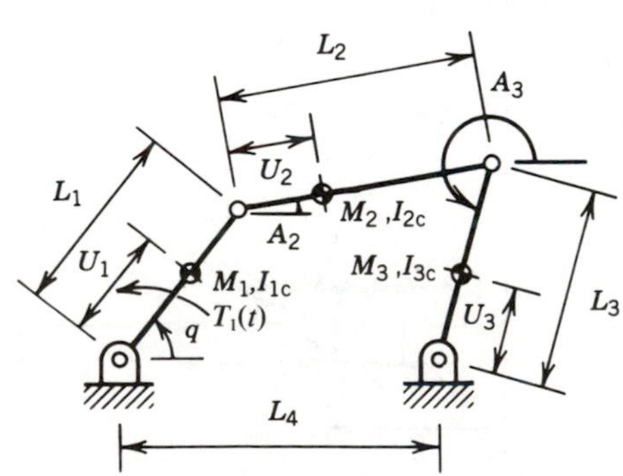

a. Formulate all of the kinematic analysis required for the later steps, but do not solve;

b. Determine the equation of motion;

c. In terms of the driving torque, $T_1(t)$, and the values $q(t)$, $\dot{q}(t)$, and $\ddot{q}(t)$, determine a sequence of closed-form expressions for evaluating the horizontal and vertical forces at all four pin joints.

9.4 The four-bar mechanism shown is a gravity-operated emergency steam shut-off valve. The motion is driven by gravity only.

a. Formulate all the kinematic analysis required for the later steps, but do not solve;

b. Formulate the equation of motion for the system;

c. In terms of the values $q(t)$, $\dot{q}(t)$, and $\ddot{q}(t)$, determine the forces at each pin during the time the plug is falling, prior to impact on the valve seat.

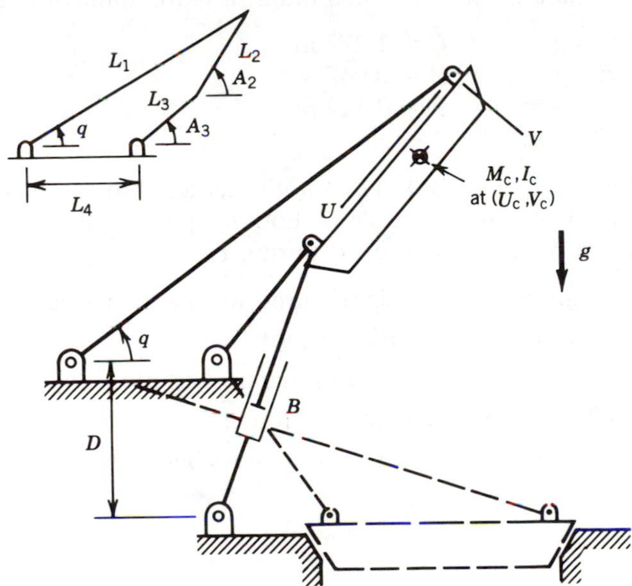

9.5 The pendulum pivots about a stationary pin as the upper end moves horizontally with the roller. There is no driving force other than gravity.

a. Perform all kinematic analysis required for the later steps;

b. Determine the system equation of motion;

c. Determine the force components on the stationary pin;

d. Determine the force components on the roller axle;

e. Determine the force components at the contact between the roller and the table;

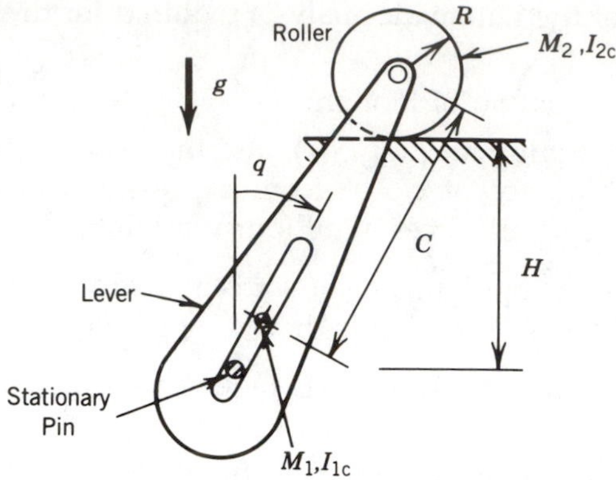

f. Evaluate the force components on the stationary pin, the roller axle, and the shear and normal forces between the roller and the table when $q = 0.0237$ radians and $\dot{q} = 23.68$ rad/s. Use the following data:

$M_1 = 12.86$ kg	$C = 1.277$ m
$M_2 = 18.79$ kg	$H = 0.947$ m
$I_{1c} = 3.854$ kg-m^2	$R = 0.149$ m
$I_{2c} = 0.222$ kg-m^2	

9.6 The length of the link is L, and the center of the mass is a distance U_c from the left end. The disk rolls without slipping on the circular guide. The motion is driven by the applied torque T_o.

a. Perform all kinematic analysis required for the later steps;
b. Determine the system equation of motion;
c. Determine the tension and shear force in the link;
d. Determine the slider-guide reaction force;
e. Determine the normal and tangential forces between the wheel and the guide.

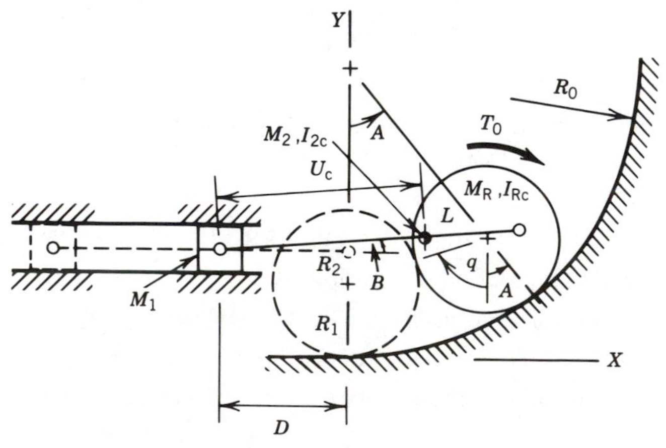

9.7 For the quick-return mechanism shown, the motion is driven by the applied torque M_o and opposed by the working force F.

a. Perform all kinematic analysis required for the later parts;
b. Determine the equation of motion;
c. Determine the slider-guide reaction;
d. Determine the reactions on lever at the bottom pin;
e. Determine the bearing reactions on the flywheel shaft;
f. Determine the contact force between the crank and lever;
g. Evaluate the guide reaction, bottom pin reactions, flywheel bearing reactions, and the crank-to-lever contact force for a situation in which $q = 0.445$ radians, $\dot{q} = 23.33$ rad/s, $M_o = 245$ in.-lb, and $F = 132$ lb, using the following parameters:

$R = 4.55$ in. $\quad I_{1c} = 2.45$ lb-s^2-in.
$C = 19.33$ in. $\quad M_2g = 23.67$ lb
$D = 28.04$ in. $\quad I_{2c} = 3.33$ lb-s^2-in.
$U_c = 34.42$ in. $\quad M_3g = 14.22$ lb

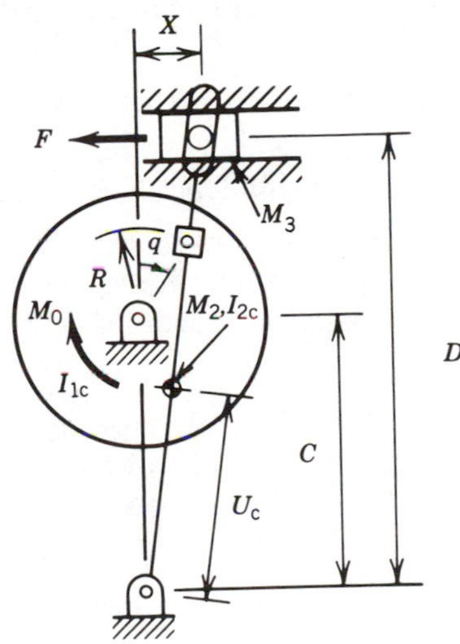

9.8 The cam shown is an eccentric circle with eccentricity R_c and radius of curvature P. The cam shaft rotation rate, $\dot{A}$, is constant. The spring tension is F_{so} when the follower angle B is at its minimum. The indicated moments of inertia are with respect to the axis of rotation in each case. The shafting is considered massless. For each shaft, the center of mass is on the axis of rotation.

a. Perform all of the kinematic analysis required for later steps;
b. Obtain the equation of motion, using A as the generalized coordinate;
c. Determine the required driving torque, T_o;
d. Determine the contact force between the cam and follower;

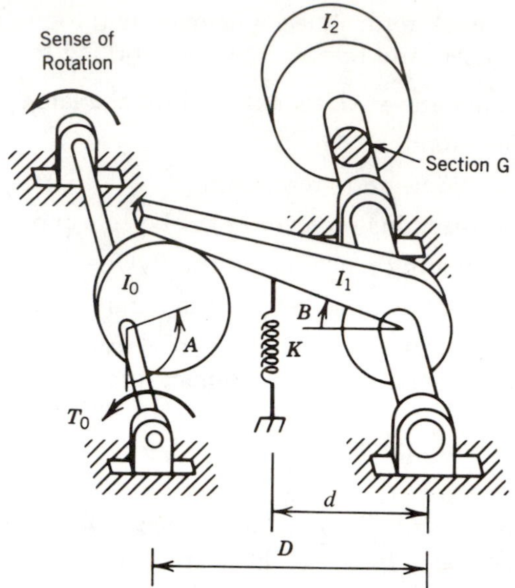

e. Determine the net horizontal and vertical cam shaft bearing reaction force components;

f. Determine the net horizontal and vertical bearing reactions on the follower shaft;

g. Determine the torque in the follower shaft at section G;

h. For the situation described by $A = 1.335$ radians and $\dot{A} = 35.0$ rad/s, evaluate the driving torque, contact force, net cam shaft bearing reactions, net follower shaft bearing reactions, and shaft torque at section G. Use the following data:

$R_c = 3.66$ mm	$I_0 = 1.519\ (10^{-5})$ kg-m^2	$d = 88$ mm
$P = 17.3$ mm	$I_1 = 6.228\ (10^{-5})$ kg-m^2	
$D = 133$ mm	$I_2 = 3.282\ (10^{-3})$ kg-m^2	
$K = 405$ N/mm	$F_{so} = 115$ N	

9.9 The planet carrier motion is the input, q, driven by the torque, T_o, while the planet rotation is the motion of the crank, A. The load rotation is R and the sun gear is stationary. The drive mechanism may be considered massless, but the mass MOI of the load with respect to the axis of rotation is I_c. The input angular velocity, $\dot{q}$, is to be held constant. The sun gear pitch radius is R_s, whereas the pitch radius of the planet gear is R_p.

a. Perform all kinematic analysis required for later parts;

b. Determine the torque required as a function of q;

c. Determine the tangential force between the sun gear and the planet;

d. Determine the contact force between the crank and the slot wall;

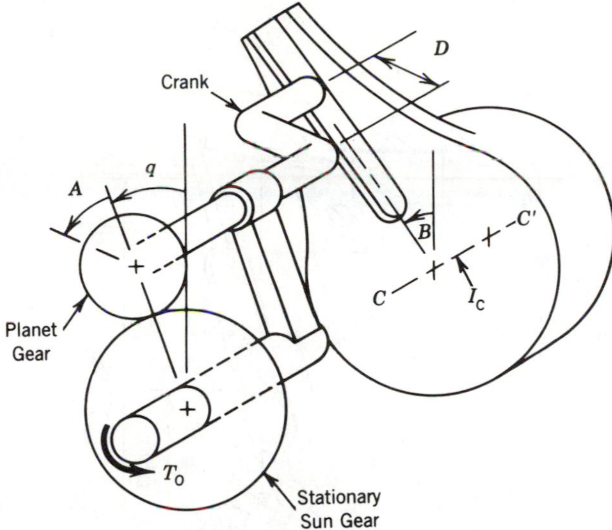

e. Evaluate the drive torque, tangential force between the gear and planet, and the contact force between the crank and slot wall for the situation where $q = 1.122$ radians and $\dot{q} = 15.55$ rad/s. Use the following system data:

$R_s = 15.17$ in. $\quad D = 1.38$ in.
$R_p = 4.83$ in. $\quad I_c = 2.88$ lb-s^2-in.

9.10 The rotation of the crank imparts a reciprocating motion to the slider through the action of the linkage. The center of mass for the final link, M_4, is at the center of the link.

a. Perform all kinematic analysis required for later steps;
b. Determine the equation of motion;
c. Determine the force on the large, stationary pivot pin;
d. Determine the forces on all remaining connecting pins;
e. Determine the force between the slider and the guide;
f. Evaluate the forces on the pivot pin, the connecting pins, and between the slider and guide for the case $q = 0.885$ radians, $\dot{q} = 14.25$ rad/s, based on the system data that follows:

$R = 35$ mm	$M_1 = 0.052$ kg	$T_o = 0.72$ N-m
$D_1 = 92$ mm	$I_{1c} = 5.31\ (10^{-6})$ kg-m^2	
$D_2 = 195$ mm	$M_2 = 0.118$ kg	
$D_3 = 198$ mm	$I_{2c} = 3.84\ (10^{-4})$ kg-m^2	
$D_4 = 77$ mm	$M_4 = 0.078$ kg	
$U_1 = 23$ mm	$I_{4c} = 3.85\ (10^{-5})$ kg-m^2	
$U_2 = 84$ mm	$M_s = 1.42$ kg	

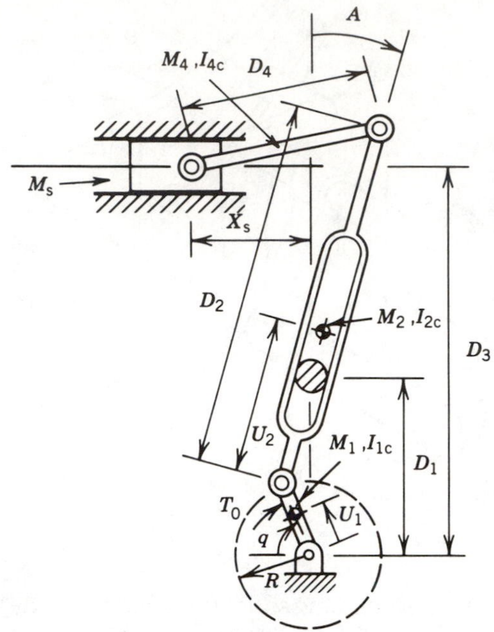

9.11 The two sliders move along the coordinate axes, whereas the link spans the separation between them. The slider motions are driven by the external couple, M_c, acting on the connecting link.

a. Perform all kinematic analysis required for later steps;
b. Determine the equations of motion for the system;
c. Determine the forces on the two pins;
d. Determine the forces between the sliders and the guides;

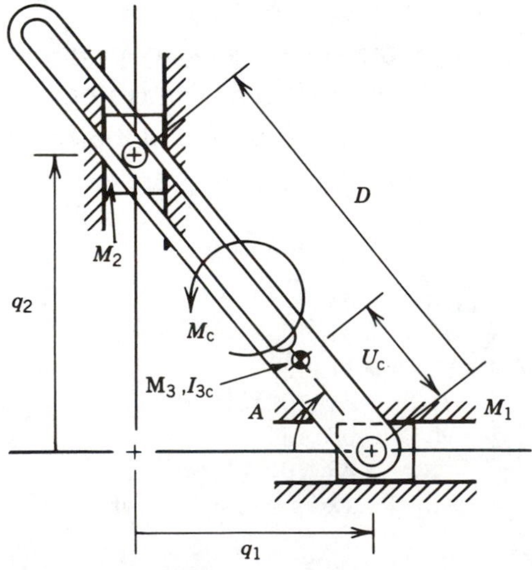

e. For the situation $q_1 = 22.0$ in., $q_2 = 31.4$ in., $\dot{q}_1 = 14.6$ in./s, $\dot{q}_2 = -73.8$ in./s, and $M_c = 188$ in.-lb, evaluate the pin forces and guide forces on both sliders using the following system data:

$M_1 g = 3.32$ lb $M_3 g = 5.35$ lb
$M_2 g = 2.88$ lb $I_{3c} = 2.65$ lb-s^2-in.
$U_c = 17.22$ in.

9.12 The motion of the system of two links and two sliders is driven by the force components F_x and F_y that act at the junction of the two links, subject also to the forces developed in the spring and the dashpot. The forces F_x and F_y should be considered to act directly on the joint pin. In the following steps, carry the analysis as far as is reasonable in closed form. Indicate a numerical solution if necessary.

a. Perform all kinematic analysis required for later steps;
b. Determine the system equations of motion;
c. Determine the forces transferred to each link at the pin where the external forces are applied;
d. Determine the forces applied to each slider by the links;
e. Determine the guide forces acting on each slider.

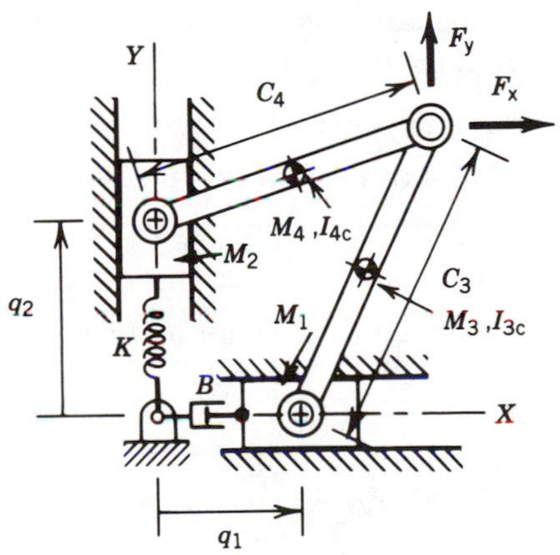

9.13 The pendulum rolls and translates on the table as it lifts and the slider moves to the right, driven by the forces F_1 and F_2. The translation, X, and rotation, A, are zero when $q_1 = q_2 = 0$. The long, slender, connecting link, L, is considered massless, whereas the other components have masses and moments of inertia as indicated. Carry the analysis as far as reasonable in closed form; indicate a numerical solution if necessary.

a. Perform all kinematic analysis required for later steps;
b. Determine the equations of motion for the system;
c. Determine the shear and normal forces acting on the pendulum from the table;

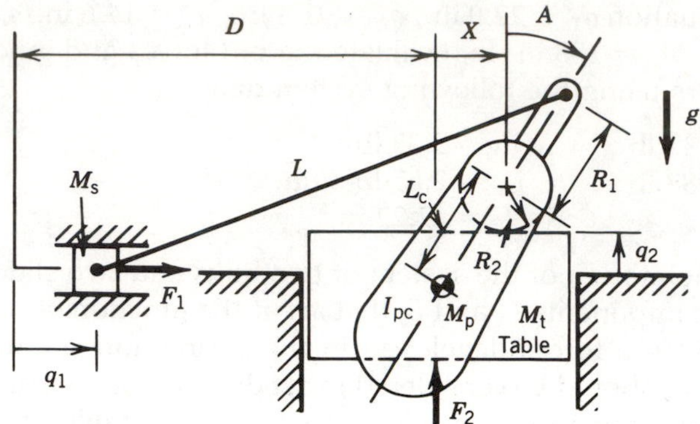

d. Determine the compressive force in the connecting link;

e. Determine the force between the slider and the guide.

9.14 The bell crank pivots on the slider that moves in a vertical guide. The system is driven by the applied torque, T, and the slider force, F, and is subject also to the spring force. As the disk rotates, the pin engaging the bell crank causes it to rotate or slide. The mass of the bell crank is M_2, and its center of mass is located by the body coordinates (U_c, V_c). The centroidal mass MOI for the bell crank is I_{2c}. When $q_1 = -\pi/2$ and $q_2 = 0$, the spring is relaxed. Carry the analysis as far as is reasonable in closed form; indicate a numerical solution if required.

a. Perform all kinematic analysis required for the later steps;

b. Determine the equations of motion for the system;

c. Determine the force components on the bell crank pivot pin;

d. Determine the force between the disk and the bell crank;

e. Determine the force between the slider and the guide.

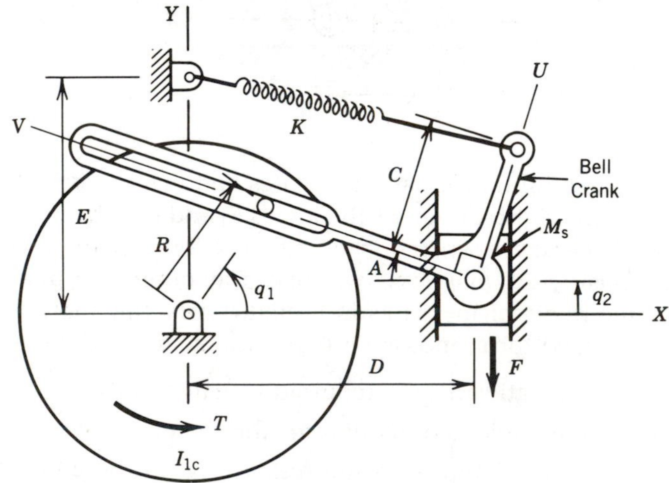

APPENDIX A1

Matrices and Linear Algebraic Equations

A1.1 MATRIX NOTATION

A matrix is a rectangular array of numbers. A typical matrix with r rows and c columns may be denoted $[A]$, and represents the following array:

$$[A] = \begin{bmatrix} A_{11} & A_{12} & A_{13} & \cdots & A_{1c} \\ A_{21} & A_{22} & A_{23} & \cdots & A_{2c} \\ A_{31} & A_{32} & A_{33} & \cdots & A_{3c} \\ \vdots & \vdots & \vdots & & \vdots \\ A_{r1} & A_{r2} & A_{r3} & \cdots & A_{rc} \end{bmatrix}$$

Note that the subscripts refer to the location of the element within the array. In general, A_{ij} denotes the element in row i and column j. The range of the row index (l to r) is the *row dimension* of the matrix. Similarly, the range of the column index (l to c) is the *column dimension* of the matrix.

Two types of matrices are of particular interest: (1) square matrices, and (2) column (or row) matrices. A matrix is square, provided that the row and column dimensions are the same, that is, r = c. A column matrix consists of a single column of n numbers, such as

$$\{V\} = \begin{Bmatrix} V_1 \\ V_2 \\ V_3 \\ \vdots \\ V_n \end{Bmatrix}$$

This may be thought of as the first column of the general rectangular matrix form just given. Because there is only one column, there is no need for a column index and the single index given serves to denote the row in which the element is located. The dimensions of the preceding column matrix are (n × 1), which keeps the column matrix in context as a special case of the rectangular matrix. A column matrix is often termed a *vector* and, conversely, a "vector" is usually understood to mean a column matrix. If there are only two or three elements, this may be considered to represent a physical vector in two- or three-dimensional space. At other times, there will be more than three elements, and the matrix may be considered as a vector in a space of higher dimensions. To conserve space, it is common practice to write the elements on one line, with the prefix "Col," so that the preceding vector would be written as $\{V\} = \text{Col}(V_1, V_2, V_3, \ldots V_n)$. The hybrid term "column vector" is used when it is necessary to distinguish it from the "row vector" to be discussed shortly.

One of the common matrix operations used is the *transpose*. Applied to a rectangular matrix, this operation consists of exchanging rows for columns. The result is that those elements that were originally in the jth row of the original matrix are now found in the jth column of the transposed matrix. The transpose of a matrix is denoted by the same name as the original matrix with the addition of a superscript t. Thus, for the preceding rectangular matrix, the transpose would be

$$[A]^t = \begin{bmatrix} A_{11} & A_{21} & A_{31} & \ldots & A_{r1} \\ A_{12} & A_{22} & A_{32} & \ldots & A_{r2} \\ A_{13} & A_{23} & A_{33} & \ldots & A_{r3} \\ \vdots & \vdots & \vdots & & \vdots \\ A_{1c} & A_{2c} & A_{3c} & \ldots & A_{rc} \end{bmatrix}$$

Note that the subscripts as used here are those of the original array. Thus, A_{23} from the original $[A]$ matrix is found in the transpose in the (3, 2) position. If the transpose operation is applied to a column matrix, the result is to move the single column into a single row. This is sometimes called a "row vector." The dimensions for the row vector are (1 × n). For the preceding column matrix $\{V\}$, the transpose is

$$\{V\}^t = (V_1, V_2, V_3, \ldots V_n)$$

A1.2 MATRIX ADDITION AND SUBTRACTION

Addition and subtraction of two matrices are each defined to produce a new matrix, the elements of which are the sums, or differences, of corresponding elements in the two original matrices. However, addition and subtraction can be defined only when the two matrices have the same dimensions. If $[A]$ and $[B]$ are two rectangular matrices to be added,

$$[A] = \begin{bmatrix} A_{11} & A_{12} & A_{13} & \cdots \\ A_{21} & A_{22} & A_{23} & \cdots \\ A_{31} & A_{32} & A_{33} & \cdots \\ \vdots & \vdots & \vdots & \end{bmatrix} \qquad [B] = \begin{bmatrix} B_{11} & B_{12} & B_{13} & \cdots \\ B_{21} & B_{22} & B_{23} & \cdots \\ B_{31} & B_{32} & B_{33} & \cdots \\ \vdots & \vdots & \vdots & \end{bmatrix}$$

then the definition above gives the result [C]:

$$[C] = [A] + [B] = \begin{bmatrix} A_{11} + B_{11} & A_{12} + B_{12} & A_{13} + B_{13} & \cdots \\ A_{21} + B_{21} & A_{22} + B_{22} & A_{23} + B_{23} & \cdots \\ A_{31} + B_{31} & A_{32} + B_{32} & A_{33} + B_{33} & \cdots \\ \vdots & \vdots & \vdots & \end{bmatrix}$$

Subtraction can be illustrated in a similar manner simply by changing the sign on the second term of each element in [C]. If [A] and [B] are of dimension (r × c), then the sum or difference, [C], is also of dimension (r × c).

A1.3 MATRIX MULTIPLICATION

The product of two matrices is another matrix, the elements of which are sums of products of the elements of the two original matrices. Formally, if the matrices to be multiplied are $[A] = [A_{ij}]$, $(n_1 \times n_2)$, and $[B] = [B_{jk}]$ $(n_2 \times n_3)$, the matrix product

$$\underset{(n_1 \times n_3)}{[C]} = \underset{(n_1 \times n_2)}{[A]} \; \underset{(n_2 \times n_3)}{[B]}$$

has, as the typical element of [C],

$$C_{ik} = \sum_{j=1}^{j=n_2} (A_{ij}B_{jk})$$

Notice that the dimensions for each matrix are indicated in parentheses below the matrix name in the displayed matrix product. The results matrix, [C], will be $(n_1 \times n_3)$, as indicated there. A very important condition for the existence of the matrix product is also indicated—that the column dimension of the first factor must be the same value as the row dimension of the second factor. In the preceding illustration, these dimensions are each n_2. The need for this condition is seen in the expression for the typical term; if this condition is not satisfied, there will not be equal numbers of A_{ij} and B_{jk} factors to be multiplied together. Furthermore, the matrix product [A][B] is, in general, not equal to the matrix product [B][A]. The order of the matrix factors is significant and *must* be maintained.

The typical term expression is the basis for a computer algorithm or the product of matrices. Such an algorithm is given in the Matrix Operations Package listed in Appendix A1.6. For manual evaluation of matrix products, another approach is useful. A specific example serves to present the manual method. Consider the product of two matrices $[A]$ and $[B]$:

$$[A] = \begin{bmatrix} A_{11} & A_{12} & A_{13} \\ A_{21} & A_{22} & A_{23} \\ A_{31} & A_{32} & A_{33} \end{bmatrix} \qquad [B] = \begin{bmatrix} B_{11} & B_{12} \\ B_{21} & B_{22} \\ B_{31} & B_{32} \end{bmatrix}$$

$$[C] = [A][B] = \begin{bmatrix} A_{11} & A_{12} & A_{13} \\ A_{21} & A_{22} & A_{23} \\ A_{31} & A_{32} & A_{33} \end{bmatrix} \begin{bmatrix} B_{11} & B_{12} \\ B_{21} & B_{22} \\ B_{31} & B_{32} \end{bmatrix}$$

To form the (i, j)th element of the product, the steps are as follows:

1. Locate the ith row of the first factor and the jth column of the second factor. These are indicated in the example in broken lines for the case i = 2, j = 1;
2. Consider the ith row to be rotated 90 degrees clockwise and placed adjacent to the elements of the jth column, forming three indicated products;
3. Sum the three indicated products and store the result in the (i, j)th position of the result.

For the case i = 2, j = 1 just identified, this produces

$$\begin{aligned} C_{21} = \; & A_{21}\,B_{11} \\ & + A_{22}\,B_{21} \\ & + A_{23}\,B_{31} \end{aligned}$$

At times, the transpose operation may be applied to an indicated matrix product. The transpose of a product is the reversed product of the individual factors transposed:

$$([A][B])^t = [B]^t[A]^t$$

For the proof of this statement, consult any standard text on matrix theory.

A1.4 MATRIX INVERSION AND SOLUTION OF LINEAR ALGEBRAIC EQUATIONS

Matrix Inversion. Addition, subtraction, and multiplication are each defined for both real numbers and matrices. At this point the similarity ends because there is no division operation defined for matrices. Instead, there is an operation known as *inversion* that produces a matrix called the *inverse* of the original matrix. Inversion is only distantly related to the idea of division, and the two must not be considered in any sense the same.

Before beginning the explanation of the matrix inverse, it is convenient to define the terms *diagonal matrix* and *identity matrix*. For a square matrix, the elements on the diagonal from the upper left corner to the lower right corner make up what is called the *main diagonal*. Some special matrices have all elements zero except for the elements on the main diagonal. Such matrices are called *diagonal matrices* and are sometimes written as

$$[D] = \text{Diag}(D_{11}, D_{22}, D_{33}, \ldots)$$

where only the elements on the main diagonal are written out. One very important diagonal matrix is the *identity matrix*, for which the elements on the main diagonal are each 1.0. A typical identity matrix of dimensions (3×3) is

$$[I] = \text{Diag}(1, 1, 1) = \begin{bmatrix} 1 & 0 & 0 \\ 0 & 1 & 0 \\ 0 & 0 & 1 \end{bmatrix}$$

The symbol $[I]$ is often used for the identity matrix, although it also has other meanings such as the moment of inertia matrix. Care should be taken to make clear what is meant in each case.

One important characteristic of the identity matrix is seen when it is multiplied with a second matrix. The product is simply the second matrix; multiplication by the identity matrix changes nothing at all. This is indicated by the following statement:

$$[I][A] = [A][I] = [A]$$

This is comparable to multiplying a scalar by the value 1.0.

If $[A]$ is a square matrix for which $[B]$ is the inverse, this relation is expressed as

$$[B] = [A]^{-1}$$

where the superscript -1 is reminiscent of the exponent -1 that is used to denote the inverse of a scalar. Because $[B]$ is the inverse of $[A]$, this means that

$$[A][B] = [B][A] = [I]$$

Notice that the matrix product taken in either order, $[A][B]$ or $[B][A]$, results in the identity matrix. This defines the inverse, $[B]$, in terms of the result when $[B]$ is multiplied with $[A]$; it does not tell how to determine $[B]$. There are many methods to determine the inverse of a matrix, and for these a suitable text on matrix algebra or numerical methods should be consulted. The Matrix Operations Package (see Appendix A1.6) includes a numerical method for determining an inverse.

One of the major concerns in dealing with the inverse of a particular matrix is whether or not an inverse actually exists for that matrix. To investigate, let $[A]$ be a square matrix for which $[B]$ is postulated to be the inverse. The matrix $[B]$ exists if, and only if, the determinant of $[A]$ is nonzero. Testing the value of the determinant of $[A]$ becomes a necessary part of assuring that subsequent analysis using $[B]$ will in fact be valid.

There are occasions when the inverse of a matrix product is indicated. For further analysis, this inverse can be replaced by the reversed product of the individual factor inverses:

$$([A][B])^{-1} = [B]^{-1}[A]^{-1}$$

The proof of this relation is available in many texts on matrices.

Solution of Linear Algebraic Equations. One application of the matrix inverse is the formal solution of a system of linear simultaneous algebraic equations. The system of N equations in N variables, which can be represented as

$$\begin{array}{ccccccc} C_{11} X_1 & + & C_{12} X_2 & + & C_{13} X_3 + \cdots & = & B_1 \\ C_{21} X_1 & + & C_{22} X_2 & + & C_{23} X_3 + \cdots & = & B_2 \\ C_{31} X_1 & + & C_{32} X_2 & + & C_{33} X_3 + \cdots & = & B_3 \\ \vdots & & \vdots & & \vdots & & \vdots \end{array}$$

can be written very compactly in matrix form as

$$[C]\{X\} = \{B\}$$

where $[C]$ is $(N \times N)$ while $\{X\}$ and $\{B\}$ are each $(N \times 1)$. The formal solution for this system of equations is obtained by premultiplying by the inverse of $[C]$ (this presumes that such an inverse exists):

$$[C]^{-1}[C]\{X\} = [C]^{-1}\{B\}$$

or

$$\{X\} = [C]^{-1}\{B\}$$

This shows that the solution, $\{X\}$, is obtained from the product of the inverse of the coefficient matrix $[C]$ with the column matrix of constants, $\{B\}$. This is entirely correct and, from an analytic point of view, is the manner in which the solution is usually indicated—a matrix inversion followed by a matrix product. From the numerical standpoint, this is not the best way to compute the solution, $\{X\}$. It involves an unnecessary multiplication that slows the process and an unnecessary matrix inversion with associated numerical round-off errors. In writing computer programs to solve a system of linear equations, the direct solution should always be used; the inverse of the coefficient matrix should be computed only when the inverse itself is required.

The Matrix Operation Package subroutine for solving linear systems is an implementation of the well known Gaussian elimination method with pivoting. The routine also provides a means to compute an inverse. The input to this subroutine is in terms of the *augmented matrix of coefficients*, a vertically partitioned matrix containing the matrix of coefficients, $[C]$, in the left partition. In solving a single system of linear equations, as previously discussed, the right partition consists of the vector $\{B\}$. Consequently, for this case the dimensions of the augmented matrix are (N × (N + 1)), and should be so dimensioned in the program. For inversion, the augmented matrix contains the matrix to be inverted in the left partition; internally, the subroutine will fill the right side with an identity matrix. For this case, the augmented matrix must be dimensioned (N × 2N).

There are times when the system to be solved involves rectangular arrays for both the unknowns and the right side:

$$\underset{(N \times N)}{[C]} \quad \underset{(N \times M)}{[X]} = \underset{(N \times M)}{[B]}$$

Consider both $[X]$ and $[B]$ to be partitioned into columns:

$$[C][X_1 \mid X_2 \mid X_3 \mid \ldots] = [B_1 \mid B_2 \mid B_3 \mid \ldots]$$

Then, the vectors $\{X_1\}$, $\{X_2\}$, and so forth, are the solutions associated with the various right-side vectors $\{B_1\}$, $\{B_2\}$, and so on, so that

$$[C]\{X_i\} = \{B_i\}$$

There will be M such solution vectors, each defining a column in the solution matrix $[X]$. The entire process can be handled in the Matrix Operations Package by setting Z2 equal to M, the number of columns in $[B]$, and storing $[B]$ in the right partition of the augmented coefficient matrix. For this case, the augmented matrix must be dimensioned (N × (N + M)).

A1.5 ANALYTICAL INVERSE FOR A (2 × 2) MATRIX AND THE SOLUTIONS OF TWO SIMULTANEOUS LINEAR EQUATIONS

In many of the examples of the analysis of machines, a (2 × 2) matrix must be inverted or a system of two simultaneous linear equations must be solved. These special solutions are tabulated here for use when needed. They apply only for the inversion of a (2 × 2) matrix or the solution of two simultaneous linear equations.

Analytical Inverse for (2 × 2) Matrix. Consider the (2 × 2) matrix $[C]$:

$$[C] = \begin{bmatrix} C_1 & C_2 \\ C_3 & C_4 \end{bmatrix}$$

The determinant of $[C]$ is Det(C),

$$\text{Det}(C) = C_1C_4 - C_2C_3$$

and the inverse of [C] is

$$[C]^{-1} = \frac{1}{\text{Det}(C)}\begin{bmatrix} C_4 & -C_2 \\ -C_3 & C_1 \end{bmatrix}$$

Solution of Two Simultaneous Linear Equations. For the system of two linear equations,

$$\begin{bmatrix} C_1 & C_2 \\ C_3 & C_4 \end{bmatrix}\begin{Bmatrix} X_1 \\ X_2 \end{Bmatrix} = \begin{Bmatrix} B_1 \\ B_2 \end{Bmatrix}$$

the solution is

$$\begin{Bmatrix} X_1 \\ X_2 \end{Bmatrix} = \frac{1}{\text{Det}(C)}\begin{Bmatrix} C_4 B_1 - C_2 B_2 \\ -C_3 B_1 + C_1 B_2 \end{Bmatrix}$$

A1.6 MATRIX OPERATIONS PACKAGE AND VERIFICATION PROGRAM

When BASIC is implemented completely, one of its attractive aspects is the availability of several very powerful matrix operations statements. Because the versions of BASIC implemented in many microcomputers do not include those matrix operations statements, a package of matrix operations subroutines is provided here. If there are no built-in matrix operations available, it is recommended that this package be used as the starting point for all program development in which matrices will be used. If a copy of these subroutines is put into memory as a first step, then the program for any particular problem can be developed in lines 1 through 4999. If needed, lines 5500 and above may also be used with appropriate transfer of control.

BASIC employs global variable definitions that cause a given variable name to have the same value wherever it is referenced—either in the main routine or in a subroutine. This is different from languages such as FORTRAN where variable names apply only in the routine where they appear. The result is that FORTRAN provides a "masking" feature that is not available in BASIC. To avoid difficulties resulting from inadvertent variable value changes, all variables used in the matrix Operations Package use names beginning with the letters O and Z, such as O1 and Z7. Names beginning with O or Z should not be used in the main program except to communicate data to and from the Matrix Operations Package. Notice that the O-arrays must be appropriately DIMensioned in the main routine.

To make most effective use of the Matrix Operations Package, the code that follows should be entered into the computer with great care and stored on diskette. Once verified, that file should be locked and never modified. To verify the integrity and correctness of the Matrix Operations Package a Verification Program is also provided here. The Verification

Program also demonstrates the manner in which the Matrix Operations Package is used; it provides calls to each of the four subroutines after first setting up all of the input. To use the Verification Program, the following steps are required:

1. Clear the computer memory;
2. Load the Verification Program with the Matrix Operations Package at the end;
3. RUN the Verification Program.

The details of accomplishing these steps will vary from one machine to the next; do whatever is necessary to carry them out on your machine. It may be well to permanently store the Verification Program as well as the Matrix Operations Program.

The Matrix Operations Package follows.

```
5000 REM *****  MATRIX OPERATIONS PACKAGE *****
5010 REM
5020 REM
5100 REM ***** GENERAL MATRIX PRODUCT *****
5101 REM        MAT O4 = MAT O2  *  MAT O3
5102 REM        (Z1,Z3)  (Z1,Z2)   (Z2,Z3)
5103 REM
5110 FOR Z5=1 TO Z1
5111 FOR Z6=1 TO Z3
5112 Z9=0
5113 FOR Z7=1 TO Z2
5114 Z9=Z9+O2(Z5,Z7)*O3(Z7,Z6)
5115 NEXT Z7
5116 O4(Z5,Z6)=Z9
5117 NEXT Z6
5118 NEXT Z5
5119 RETURN
5120 REM
5200 REM ***** PRODUCT OF MATRIX WITH A VECTOR *****
5201 REM        MAT O4 = MAT O2 * VECTOR O1
5202 REM        (Z1,1)   (Z1,Z2)   (Z2,1)
5203 REM
5210 FOR Z5=1 TO Z1
5211 Z9=0
5212 FOR Z6=1 TO Z2
5213 Z9=Z9+O2(Z5,Z6)*O1(Z6)
5214 NEXT Z6
5215 O4(Z5,1)=Z9
5216 NEXT Z5
5217 RETURN
5218 REM
5300 REM ***** MATRIX INVERSION & LINEAR SYSTEM
     SOLUTION *****
5301 REM  FOR THE LINEAR SYSTEM   C       X    =    B
```

```
5302 REM                           (Z1XZ1) (Z1XZ2) (Z1XZ2)
5303 REM  INPUT:
5304 REM  O2 = LOCATION OF THE AUGMENTED MATRIX C:B
5305 REM  Z1 = ORDER OF THE  C  MATRIX
5306 REM  Z3 = 0  FOR DET(C) ONLY,  C = (Z1XZ1) ARRAY
5307 REM     > 0  FOR INV(C)  AND DET(C), O2=(Z1 X
                  2*Z1)
5308 REM    < 0  FOR SOLUTION OF THE SYSTEM  C X = B
5309 REM          WHERE O2 = (Z1 X (Z1+Z2)) ARRAY
5310 REM  Z2 = NUMBER OF SYSTEMS TO BE SOLVED
5311 REM  OUTPUT:
5312 REM  ZZ = DETERMINANT OF COEF MATRIX
5313 REM  Z3 = 0, THERE IS NO OTHER OUTPUT
5314 REM  Z3 > 0, INV(C) IS IN RIGHTMOST Z1 COLS OF O2
5315 REM  Z3 < 0, SOLUTIONS ARE IN RIGHTMOST Z2 COLS
5316 REM          OF O2
5317 REM  NOTE THAT O1 IS USED AS A SCRATCH VECTOR
5400 O5=1
5401 Z0=0
5402 Z4=Z1-1
5403 Z5=2*Z1
5404 Z6=Z1+Z2
5405 IF Z3<=0 THEN GOTO 5413
5406 FOR Z7=1 TO Z1
5407 FOR Z8=Z1+1 TO Z5
5408 O2(Z7,Z8)=0
5409 NEXT Z8
5410 O2(Z7,Z1+Z7)=1
5411 NEXT Z7
5412 Z6=Z5
5413 FOR Z7=1 TO Z4
5414 O7=Z7
5415 O8=ABS(O2(Z7,Z7))
5416 FOR Z9=Z7+1 TO Z1
5417 IF ABS(O2(Z9,Z7))<=O8 THEN GOTO 5420
5418 O7=Z9
5419 O8=ABS(O2(Z9,Z7))
5420 NEXT Z9
5421 IF O7=Z7 THEN GOTO 5428
5422 FOR O9=Z7 TO Z6
5423 O6=O2(Z7,O9)
5424 O2(Z7,O9)=O2(O7,O9)
5425 O2(O7,O9)=O6
5426 NEXT O9
5427 O5=-O5
5428 FOR Z8=Z7+1 TO Z1
5429 IF O2(Z8,Z7)=0 THEN GOTO 5434
5430 O0=-O2(Z8,Z7)/O2(Z7,Z7)
5431 FOR O9=Z7 TO Z6
5432 O2(Z8,O9)=O2(Z8,O9)+O0*O2(Z7,O9)
```

```
5433 NEXT O9
5434 NEXT Z8
5435 NEXT Z7
5436 O6=1
5437 FOR Z7=1 TO Z1
5438 IF O2(Z7,Z7)=0 THEN GOTO 5443
5439 O6=O2(Z7,Z7)*O6
5440 NEXT Z7
5441 ZZ=O5*O6
5442 GOTO 5446
5443 Z0=1
5444 IF Z3=0 THEN GOTO 5461
5445 IF Z3<>1 THEN GOTO 5448
5446 IF Z0<>1 THEN GOTO 5448
5447 GOTO 5461
5448 FOR Z7=Z1+1 TO Z6
5449 FOR Z9=Z1 TO 1 STEP -1
5450 O1(Z9)=O2(Z9,Z7)
5451 IF Z9=Z1 THEN GOTO 5455
5452 FOR Z8=Z9+1 TO Z1
5453 O1(Z9)=O1(Z9)-O1(Z8)*O2(Z9,Z8)
5454 NEXT Z8
5455 O1(Z9)=O1(Z9)/O2(Z9,Z9)
5456 NEXT Z9
5457 FOR O9=1 TO Z1
5458 O2(O9,Z7)=O1(O9)
5459 NEXT O9
5460 NEXT Z7
5461 REM
5462 IF Z0<>0 THEN PRINT"*****  MATRIX IS SINGULAR
     *****"
5463 RETURN
5464 END
```

This completes the Matrix Operations Package listing.

The next listing is the Verification Program for the Matrix Operations Package.

```
1000 REM  *****  VERIFICATION PROGRAM  *****
1002 REM  TEST OF THE MATRIX OPERATIONS PACKAGE
1010 DIM A(3,3),B(3,2),C(3,3),D(3),E(3)
1020 DIM O2(3,6),O3(3,3),O4(3,3),O1(3)
1030 REM
1040 REM  DATA FOR MAT A
1050 DATA -1,2,3
1060 DATA 2,3,4
1070 DATA 3,4,5
1080 REM  DATA FOR MAT B
1090 DATA 1,2
1100 DATA 2,5
```

```
1110 DATA 4,3
1120 REM  DATA FOR VECTOR D
1130 DATA 1,2,3
1140 REM  FILL MAT A AND MAT B WITH DATA
1150 FOR I=1 TO 3
1160 FOR J=1 TO 3
1170 READ A(I,J)
1180 NEXT J
1190 NEXT I
1200 FOR I=1 TO 3
1210 FOR J=1 TO 2
1220 READ B(I,J)
1230 NEXT J
1240 NEXT I
1250 REM  FILL VECTOR D WITH DATA
1260 FOR I= 1 TO 3
1270 READ D(I)
1280 NEXT I
1290 HOME
1300 REM  TEST GENERAL MATRIX PRODUCT
1310 PRINT
1320 PRINT"GENERAL MATRIX PRODUCT TEST"
1330 PRINT
1340 PRINT"CORRECT RESULT IS"
1350 PRINT"-1 2 3      1 2      15 17"
1360 PRINT" 2 3 4   X  2 5  =   24 31"
1370 PRINT" 3 4 5      4 3      31 41"
1380 PRINT
1390 PRINT"COMPUTED RESULT"
1400 PRINT
1410 Z1=3
1420 Z2=3
1430 Z3=2
1440 FOR I=1 TO 3
1450 FOR J=1 TO 3
1460 O2(I,J)=A(I,J)
1470 NEXT J
1480 NEXT I
1490 FOR I=1 TO 3
1500 FOR J=1 TO 2
1510 O3(I,J)=B(I,J)
1520 NEXT J
1530 NEXT I
1540 GOSUB 5100
1550 FOR I=1 TO 3
1560 PRINT O4(I,1);"  ";O4(I,2)
1570 NEXT I
1580 PRINT
1590 PRINT"KEY RETURN TO CONTINUE"
1600 INPUT V$
```

```
1610 HOME
1620 REM TEST PRODUCT OF RECTANGULAR MATRIX WITH VECTOR
1630 PRINT"MATRIX - VECTOR PRODUCT TEST"
1640 PRINT
1650 PRINT"CORRECT RESULT IS"
1660 PRINT"-1 2 3      1      12"
1670 PRINT" 2 3 4  X   2   =  20"
1680 PRINT" 3 4 5      3      26"
1690 PRINT
1700 PRINT"COMPUTED RESULT"
1710 Z1=3
1720 Z2=3
1730 FOR I=1 TO 3
1740 FOR J=1 TO 3
1750 O2(I,J)=A(I,J)
1760 NEXT J
1770 NEXT I
1780 FOR I=1 TO 3
1790 O1(I)=D(I)
1800 NEXT I
1810 GOSUB 5200
1820 FOR I=1 TO 3
1830 PRINT O4(I,1)
1840 NEXT I
1850 PRINT
1860 PRINT"KEY RETURN TO CONTINUE"
1870 INPUT V$
1890 HOME
1900 REM TEST MATRIX INVERSION
1910 PRINT"MATRIX INVERSION TEST"
1920 PRINT
1930 PRINT"CORRECT RESULT IS"
1940 PRINT"      -1 2 3      -0.5  1.0 -0.5"
1950 PRINT"INV    2 3 4  =    1.0 -7.0  5.0"
1960 PRINT"       3 4 5      -0.5  5.0 -3.5"
1970 PRINT
1980 PRINT"COMPUTED RESULT IS"
1990 FOR I=1 TO 3
2000 FOR J=1 TO 3
2010 O2(I,J)=A(I,J)
2020 NEXT J
2030 NEXT I
2040 Z1=3
2042 Z3=1
2050 GOSUB 5400
2060 FOR I=1 TO 3
2070 PRINT O2(I,4);"  ";O2(I,5);"  ";O2(I,6)
2080 NEXT I
2090 PRINT
2100 PRINT"KEY RETURN TO CONTINUE"
```

```
2110 INPUT V$
2120 HOME
2130 REM  TEST LINEAR SYSTEMS SOLUTION
2140 PRINT"LINEAR SYSTEMS SOLUTION"
2150 PRINT
2160 PRINT"CORRECT RESULT IS"
2170 PRINT"-1  2  3     X1     12"
2180 PRINT" 2  3  4  X  X2  =  20"
2190 PRINT" 3  4  5     X3     26"
2200 PRINT
2210 PRINT"GIVES  X1,X2,X3 = 1, 2, 3"
2220 PRINT
2230 PRINT"COMPUTED RESULT IS"
2240 FOR I=1 TO 3
2250 FOR J=1 TO 3
2260 O2(I,J)=A(I,J)
2270 NEXT J
2280 NEXT I
2290 O2(1,4)=12
2300 O2(2,4)=20
2310 O2(3,4)=26
2320 Z1=3
2322 Z2=1
2324 Z3=-1
2330 GOSUB 5400
2340 FOR I=1 TO 3
2350 PRINT O2(I,4)
2360 NEXT I
2370 END
```

This completes the Matrix Operations Package Verification Program.

REFERENCES

Hildebrand, F. B., *Introduction to Numerical Analysis.* New York: McGraw–Hill, 1956.

Hohn, F. E., *Elementary Matrix Algebra.* New York: Macmillan, 1964.

APPENDIX A2

Newton–Raphson Solution for Simultaneous Nonlinear Equations

The Newton–Raphson solution method is a technique for solving systems of simultaneous, nonlinear equations. Such systems of equations are frequently encountered in kinematic position analysis. Suppose the simultaneous, nonlinear equations to be solved are

$$
\begin{aligned}
f_1(q_1, q_2, \ldots q_{n1}, s_1, s_2, \ldots s_{n2}) &= 0 \\
f_2(q_1, q_2, \ldots q_{n1}, s_1, s_2, \ldots s_{n2}) &= 0 \\
\vdots \qquad\qquad\qquad & \quad \vdots \\
f_{n2}(q_1, q_2, \ldots q_{n1}, s_1, s_2, \ldots s_{n2}) &= 0
\end{aligned}
$$

The values $q_1, q_2 \ldots$ are considered known for this problem (primary variables), whereas $s_1, s_2, \ldots$ are the unknown values to be determined (secondary variables). It is convenient to think of the s values and the f values each as column vectors,

$$
\begin{aligned}
\{S\} &= \text{Col}(s_1, s_2, \ldots s_{n2}) \\
\{F\} &= \text{Col}(f_1, f_2, \ldots f_{n2})
\end{aligned}
$$

The solution process begins with an initial estimate for the unknowns, denoted $\{S\langle 1\rangle\}$ where the $\langle 1\rangle$ indicates the first iteration (the notation $\langle \text{j}\rangle$ is used here to denote the jth iteration). This initial estimate may be determined by a scale drawing, by an approximate calculation that neglects some terms, or by just a good guess. The better the initial estimate, the faster the process will converge to a satisfactory result. On the other hand, if the initial estimate is too far removed from the actual solution, the process may not converge at all. For most kinematics problems, there is little difficulty making a reasonable initial estimate by one of these approaches.

If the vector function $\{F\}$ is evaluated at $\{S\langle 1\rangle\}$, in general it will not be zero; if it were zero, then the initial estimate would, in fact, be a solution. Assuming that the components of $\{F\}$ are not zero (these values are often called the *residual,* indicating those values that remain non-zero), the problem becomes one of adjusting the estimates for $\{S\}$ to make $\{F\}$ more nearly zero. To accomplish this, consider a Taylor series expansion of the vector function $\{F(\{S\})\}$ about the initial estimate $\{S\langle 1\rangle\}$:

$$\{F(\{S\langle 2\rangle\})\} = \{F(\{S\langle 1\rangle\})\} + \left[\frac{\partial f}{\partial s}\right]\Bigg|_{\{S\langle 1\rangle\}}(\{S\langle 2\rangle - S\langle 1\rangle\}) + \cdots$$

The objective is to find $\{S\}$ that makes $\{F\}$ zero, so suppose that $\{S\langle 2\rangle\}$ is the choice of argument for which $\{F\}$ is identically zero. Assign zero to $\{F(\{S\langle 2\rangle\})\}$, truncate the series after the first-order difference term, and solve for the *adjustment,* $\{S\langle 2\rangle - S\langle 1\rangle\}$. Note that the coefficient of the adjustment is the Jacobian Matrix.

$$\{S\langle 2\rangle - S\langle 1\rangle\} = -\left[\frac{\partial f}{\partial s}\right]^{-1}\Bigg|_{\{S\langle 1\rangle\}}\{F(\{S\langle 1\rangle\})\}$$

If this adjustment is added to $\{S\langle 1\rangle\}$, the result will be $\{S\langle 2\rangle\}$, and $\{F(\{S\langle 2\rangle\})\}$ was intended to be zero. It will not, in fact, be exactly zero because of the higher order terms in the Taylor expansion that were neglected. If $\{S\langle 1\rangle\}$ was sufficiently close to the actual solution, then $\{S\langle 2\rangle\}$ will be closer. This process can be repeated as many times as necessary to get the value $\{F(\{S\langle i\rangle\})\}$ as near to zero as desired.

The formal solution for the adjustment vector is indicated in terms of a matrix inversion and a matrix product. Although this is correct, for computational purposes it is better to consider this as a problem in the solution of a system of linear equations. The direction solution of the system of equations has advantages in terms of reducing both computation time and round-off error. This matter is discussed also in Appendix A1.4.

If the process just described is to be done manually, the judgement as to when to quit undoubtably will be made by the person performing the calculation. If the calculation is to be made by a computer, then it is necessary to tell the machine when to quit. For most purposes, the process should be discontinued

1. When the residual has been reduced to a sufficiently small value as to be considered approximately zero;
2. When the adjustment is less than some minimum allowable step, indicating that further calculation will not significantly improve the result;
3. When some maximum allowable number of iterations has been exceeded, indicating that the process is not converging or is unusually slow to converge.

The residual is a vector function. Consequently, it is necessary to assign a meaning to the idea of reducing the residual to a suitably small quantity. The simplest measure of the residual, from a conceptual point of view, is as the largest of the absolute values of the vector components. Using this approach, the calculation is continued until the largest component absolute value is less than a chosen allowable error value, typically in the range 10^{-3} to 10^{-5} times a representative value. The representative value defines the scale of the problem, thus relating the allowable error to the dimensions of the system. Another approach is to measure the residual by the square of the Euclidean norm of the residual vector, denoted $|\{F\}|^2$, which is simply the sum of the squares of the components of the residual vector

$$|\{F\}|^2 = f_1{}^2 + f_2{}^2 + \cdot\cdot\cdot + f_{n2}{}^2$$

With this method, the computation is terminated when the square of the norm is less than an assigned value. Because of the squaring operation, different limits are appropriate, with 10^{-6} to 10^{-10} times the representative value being roughly equivalent to the previous range. This latter test is recommended primarily because of its ease of implementation and relative speed of execution.

In some cases, the process should be terminated because further calculation will not refine the estimate significantly. If an adjustment of 10^{-4} has been calculated, for example, then further calculation will not be expected to affect the first four digits to the right of the decimal point. This situation happens in cases where the slope of the functions is very large. Because the adjustment is a vector quantity, the previous comments regarding the measure of the residual vector apply here also. It is recommended that the computation cease when the square of the norm of the adjustment vector drops below an assigned level.

In the event that the calculation is terminated because it reaches the maximum allowable number of iterations, the results should be considered suspect; in most cases this indicates that the process did not converge. It is helpful to review the results of each iteration to determine if the process is, in fact, converging slowly or if it is actually diverging.

This process is summarized in the following diagram. The steps in this diagram are then implemented in the computer program that follows.

Newton–Raphson Calculation Sequence

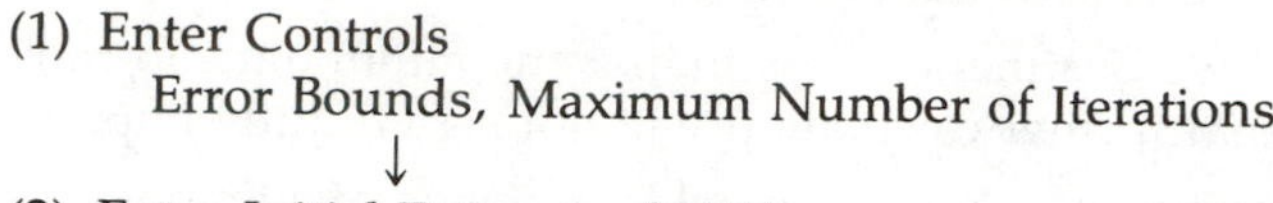

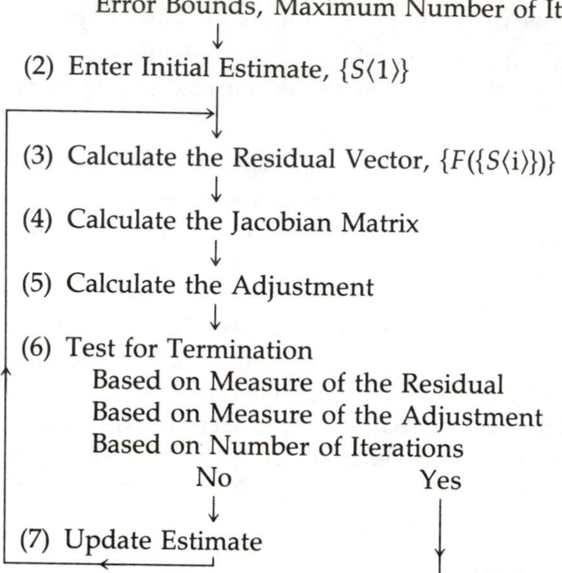

Newton–Raphson Example Program

This example program uses the Newton–Raphson method to generate a solution for the following equations:

$$C_1 \cos q_1 + B \cos A - C_2 \cos q_2 - C_3 = 0$$

$$C_1 \sin q_1 + B \sin A - C_2 \sin q_2 = 0$$

In these equations, q_1 and q_2 are considered as assigned values, and A and B are unknown. In the program, the unknowns make up the vector $\{S\} = \text{Col}(A, B)$. The equations are actually the position vector loop equations for a mechanism involving two input cranks and a sliding coupler link. For more details, see Section 3.1 for the discussion of the Sliding Four-Bar Mechanism.

```
1000 REM  NEWTON-RAPHSON EXAMPLE
1010 REM  EQUATIONS TO BE SOLVED AND
1020 REM  JACOBIAN EVALUATION ARE
1030 REM  PROGRAMMED AT LINE 2000
1040 REM
1050 DIM  F(2),S(2),DS(2),JM(2,2)
1060 REM  F  = VECTOR OF FUNCTION VALUES, RESIDUAL
1070 REM  S  = VECTOR OF SECONDARY COORDINATES
1080 REM  DS = ADJUSTMENT VECTOR, DELTA S
```

```
1090 REM  JM = JACOBIAN MATRIX
1100 REM  STEP 1 ENTER CONTROLS
1110 REM  E1 = MAX ALLOWABLE RESIDUAL
1120 REM  E2 = MIN ALLOWABLE ADJUSTMENT
1130 REM  I9 = MAX NUMBER OF ITERATIONS
1140 E1=0.0001
1150 E2=0.0001
1160 I9=10
1170 REM
1200 REM  STEP 2 ENTER KNOWN VALUES AND
1210 REM  INITIAL ESTIMATE FOR UNKNOWNS
1230 HOME
1240 PRINT"ENTER Q1"
1250 INPUT Q1
1260 PRINT"ENTER Q2"
1270 INPUT Q2
1280 PRINT"ENTER INITIAL ESTIMATE FOR S1=A"
1290 INPUT S(1)
1300 PRINT"ENTER INITIAL ESTIMATE FOR S2=B"
1310 INPUT S(2)
1320 REM
1400 REM BEGIN THE ITERATION
1410 FOR I=1 TO I9
1420 REM STEPS 3 & 4 EVALUATE THE
1430 REM RESIDUAL AND THE JACOBIAN
1440 GOSUB 2000
1500 REM STEP 5 EVALUATE THE ADJUSTMENT
1510 REM USING CLOSED FORM SOLUTION SECT A1.5
1520 D=JM(1,1)*JM(2,2)-JM(1,2)*JM(2,1)
1530 DS(1)=-(1/D)*(JM(2,2)*F(1)-JM(1,2)*F(2))
1540 DS(2)=-(1/D)*(-JM(2,1)*F(1)+JM(1,1)*F(2))
1550 REM
1600 REM  TEST FOR TERMINATION
1610 REM  FORM NORMS FOR RESIDUAL AND ADJUSTMENT
1620 N1=F(1)^2+F(2)^2
1630 N2=DS(1)^2+DS(2)^2
1640 IF N1>E1 THEN GOTO 1670
1650 V=1
1660 GOTO 1800
1670 IF N2>E2 THEN GOTO 1700
1680 V=2
1690 GOTO 1800
1700 REM  UPDATE SOLUTION ESTIMATES
1710 S(1)=S(1)+DS(1)
1720 S(2)=S(2)+DS(2)
1730 NEXT I
1740 REM
1750 PRINT"MAX NUMBER OF ITERATIONS"
1760 REM
1800 REM  OUTPUT
```

```
1810 PRINT"SOLUTION VALUES"
1820 PRINT
1830 PRINT"S(1) = A =";S(1)
1840 PRINT"S(2) = B =";S(2)
1850 PRINT
1860 PRINT"FUNCTION VALUES"
1870 PRINT"F(1)=";F(1)
1880 PRINT"F(2)=";F(2)
1890 PRINT"NUMBER OF ITERATIONS =";I
1900 IF V=1 THEN PRINT"TERMINATION BASED ON MAGNITUDE
     OF RESIDUAL"
1910 IF V=2 THEN PRINT"TERMINATION BASED ON MAGNITUDE
     OF ADJUSTMENT"
1920 END
1930 REM
2000 REM  PROBLEM GEOMETRY
2010 C1=3
2030 C2=4
2040 C3=6
2050 F(1)=C1*COS(Q1)+S(2)*COS(S(1))-C2*COS(Q2)-C3
2060 F(2)=C1*SIN(Q1)+S(2)*SIN(S(1))-C2*SIN(Q2)
2070 JM(1,1)=-S(2)*SIN(S(1))
2080 JM(1,2)=COS(S(1))
2090 JM(2,1)=S(2)*COS(S(1))
2100 JM(2,2)=SIN(S(1))
2110 RETURN
2120 END
```

This completes the Newton–Raphson example program.

REFERENCES

Dahlquist, G., and Bjorck, A., *Numerical Methods,* trans. by N. Anderson. Englewood Cliffs, N.J.: Prentice–Hall, 1974.

Hildebrand, F. B., *Introduction to Numerical Analysis.* New York: McGraw–Hill, 1956.

APPENDIX A3

Numerical Solution of Differential Equations

For most mechanisms, the differential equations of motion are nonlinear, coupled, ordinary differential equations with variable coefficients. Generally, closed-form solutions are not possible, but numerical solutions are readily available. The purpose of this appendix is to provide a numerical method of solution, the fourth-order Runge–Kutta method, without attempting to provide a comprehensive survey of available methods. The Runge–Kutta method is relatively reliable and is suitable for all of the problem types addressed in this text. There are many variations on the Runge–Kutta algorithm, and the particular forms presented here are those given by Abramowitz and Stegun (see References). For those who wish to pursue the topic further, there is a large body of literature available, often referenced under such headings as "Numerical Analysis of Differential Equations" or "Simulation."

A3.1 THE MARCHING SOLUTION

Before discussing the Runge–Kutta algorithm, it will be useful to consider the basic philosophy for the numerical solution of differential equations. This is presented most simply in terms of the Euler method applied to a single differential equation of the form

$$\dot{Y} = f(t, Y)$$

with the initial condition $Y(0) = Y_0$. The slope of the solution at $t = 0$ can be determined by using the initial condition to evaluate the slope as expressed by the differential equation; this information is indicated in Figure A3.1(a). To obtain an estimate of the solution at $t = h$, where h is a small

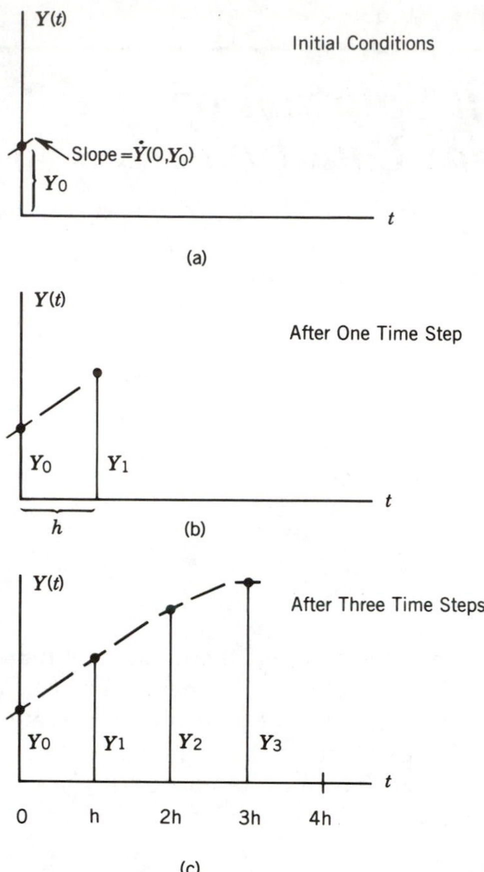

FIGURE A3.1 The Marching Solution
(a) Initial Conditions
(b) After One Time Step
(c) After Three Time Steps

time step, the tangent line at $t = 0$ can be projected to the right to intersect the line $t = h$. That intersection defines the new (approximate) solution point, Y_1, as shown in Figure A3.1(b).

The slope of the solution curve for $t = h$ can be determined from the differential equation, using $t = h$ and $Y = Y_1$. A tangent line can then be projected to intersect the line $t = 2h$, which defines a new solution point, Y_2. This process of projecting a tangent line repeatedly, called the Euler method, can be continued indefinitely and generates an approximate solution for the differential equation in the form of a table of values. From the description of the process, it is seen that the approximate solution is "marched out" from the initial condition under the guidance of the differential equation. The Euler method is not recommended for actual application, but it serves as a good basis for understanding the idea of a "marching solution." The Runge–Kutta algorithm, to be presented shortly, operates in a similar manner. The major difference is that a more involved and more accurate procedure is used to project each new solution point from the previous point.

A3.2 SINGLE SECOND-ORDER DIFFERENTIAL EQUATIONS

The differential equation developed for the motion of a single degree of freedom system is expressible as a single second-order differential equation of the form

$$\ddot{q} = f(t, q, \dot{q})$$

The Runge–Kutta algorithm provides a method to determine new points on the solution curve, each based on a previous point and four evaluations of the second derivative function, $f(t, q, \dot{q})$. The resulting approximation for the next solution value will agree with the result of a Taylor series expansion about the present solution point through the term involving the fourth degree in the time step, h. In terms of the previous values, the new values are

$$q_{n+1} = q_n + h[\dot{q}_n + (1/6)(m_1 + m_2 + m_3)]$$
$$\dot{q}_{n+1} = \dot{q}_n + (1/6)(m_1 + 2m_2 + 2m_3 + m_4)$$
$$t_{n+1} = t_n + h$$

The m-values are the result of derivative evaluations for four argument sets:

$$m_1 = h\, f(t_n, q_n, \dot{q}_n)$$
$$m_2 = h\, f(t_n + .5\, h, q_n + .5\, \dot{q}_n\, h + .125\, m_1\, h, \dot{q}_n + .5\, m_1)$$
$$m_3 = h\, f(t_n + .5\, h, q_n + .5\, \dot{q}_n\, h + .125\, m_2\, h, \dot{q}_n + .5\, m_2)$$
$$m_4 = h\, f(t_n + h, q_n + \dot{q}_n\, h + .5\, m_3\, h, \dot{q}_n + m_3)$$

Examples of this method applied to machinery dynamics problems are given in Section 7.6.

A3.3 SYSTEMS OF DIFFERENTIAL EQUATIONS

Multidegree of freedom systems give rise to systems of second-order differential equations. It is cumbersome notationally to present the application of the Runge–Kutta algorithm to a general system of N differential equations, so the process will be illustrated here in terms of two second-order differential equations. The extension to any larger number of equations is then evident.

Consider the following system of differential equations:

$$\ddot{q}_1 = f_1(t, q_1, q_2, \dot{q}_1, \dot{q}_2)$$
$$\ddot{q}_2 = f_2(t, q_1, q_2, \dot{q}_1, \dot{q}_2)$$

The first step is to replace this system with an equivalent system of first-order differential equations, based on the substitutions

$$x_1 = q_1 \qquad v_1 = \dot{q}_1$$
$$x_2 = q_2 \qquad v_2 = \dot{q}_2$$

so that the equivalent system is

$$\dot{x}_1 = v_1$$
$$\dot{x}_2 = v_2$$
$$\dot{v}_1 = f_1(t, x_1, x_2, v_1, v_2)$$
$$\dot{v}_2 = f_2(t, x_1, x_2, v_1, v_2)$$

Applying the Runge–Kutta algorithm to this system, the new solution values are given by

$$x_{1,n+1} = x_{1,n} + (1/6)(m_{11} + 2m_{12} + 2m_{13} + m_{14})$$
$$x_{2,n+1} = x_{2,n} + (1/6)(m_{21} + 2m_{22} + 2m_{23} + m_{24})$$
$$v_{1,n+1} = v_{1,n} + (1/6)(m_{31} + 2m_{32} + 2m_{33} + m_{34})$$
$$v_{2,n+1} = v_{2,n} + (1/6)(m_{41} + 2m_{42} + 2m_{43} + m_{44})$$
$$t_{n+1} = t_n + h$$

These new values are based on the following derivative evaluations:

$$m_{11} = h\, v_{1,n}$$
$$m_{21} = h\, v_{2,n}$$
$$m_{31} = h\, f_1(t_n, x_{1,n}, x_{2,n}, v_{1,n}, v_{2,n})$$
$$m_{41} = h\, f_2(t_n, x_{1,n}, x_{2,n}, v_{1,n}, v_{2,n})$$

$$m_{12} = h\,(v_{1,n} + .5\, m_{11})$$
$$m_{22} = h\,(v_{2,n} + .5\, m_{21})$$
$$m_{32} = h\, f_1(t_n + .5h, x_{1,n} + .5m_{11}, x_{2,n} + .5m_{21}, v_{1,n} + .5m_{31}, v_{2,n} + .5m_{41})$$
$$m_{42} = h\, f_2(t_n + .5h, x_{1,n} + .5m_{11}, x_{2,n} + .5m_{21}, v_{1,n} + .5m_{31}, v_{2,n} + .5m_{41})$$

$$m_{13} = h\,(v_{1,n} + .5\, m_{12})$$
$$m_{23} = h\,(v_{2,n} + .5\, m_{22})$$
$$m_{33} = h\, f_1(t_n + .5h, x_{1,n} + .5m_{12}, x_{2,n} + .5m_{22}, v_{1,n} + .5m_{32}, v_{2,n} + .5m_{42})$$
$$m_{43} = h\, f_2(t_n + .5h, x_{1,n} + .5m_{12}, x_{2,n} + .5m_{22}, v_{1,n} + .5m_{32}, v_{2,n} + .5m_{42})$$

$$m_{14} = h\,(v_{1,n} + m_{13})$$
$$m_{24} = h\,(v_{2,n} + m_{23})$$
$$m_{34} = h\,f_1(t_n + h,\, x_{1,n} + m_{13},\, x_{2,n} + m_{23},\, v_{1,n} + m_{33},\, v_{2,n} + m_{43})$$
$$m_{44} = h\,f_2(t_n + h,\, x_{1,n} + m_{13},\, x_{2,n} + m_{23},\, v_{1,n} + m_{33},\, v_{2,n} + m_{43})$$

The need for repeated evaluations of the second derivative expressions suggests that these evaluations should be programmed as a subroutine. This subroutine can then be called with the various argument combinations required. The actual application of these equations is not as difficult as might first appear. An example of this approach for a multidegree of freedom system is given in Section 8.5.

Several FORTRAN-based simulation languages have been developed to reduce the labor associated with the numerical solution of ordinary differential equations. These programs allow the user to specify the problem as a system of first-order differential equations with initial conditions. The processes of setting up the derivative evaluations and updating the solution as the problem time progresses are preprogrammed. These programs are applicable to differential equation systems of all sizes up to some maximum, usually quite large. Evaluating the kinematic solutions by the iterative Newton–Raphson technique within a simulation using one of the simulation languages may be difficult, as well as quite time consuming. The interpolation technique presented in the example of Section 7.6.2 may also be attractive for simulations using a simulation language. Simulation languages generally require a large computer and are relatively expensive to run. Some of the commercial simulation languages are MIMIC, CSMP (an IBM product), and ACSL (marketed by Mitchell & Gauthier Associates, Inc.). There is also a simulation language program available for microcomputers called TUTSIM (a product of Applied I).

REFERENCES

Abramowitz, M. and Stegun, L. A., eds., *Handbook of Mathematical Functions*. Washington, D.C.: National Bureau of Standards, Applied Mathematics Series 55, US Government Printing Office, 1964.

Hildebrand, F. B., *Introduction to Numerical Analysis*. New York: McGraw–Hill, 1956.

APPENDIX A4

Two-Argument Arctangent Function

Often it is necessary to determine angles in terms of inverse trigonometric functions. Most versions of BASIC allow only the inverse tangent function, ATN(X), which returns the principal value of the arctangent, $-\pi/2 < X < \pi/2$. Although this is adequate for some purposes, there are situations in which the angle must be placed in the correct quadrant. For that case, a two-argument arctangent function, ATN2(Y, X), is needed to compute $\arctan_2(Y, X)$ where both X and Y values are individually specified (such as is available in FORTRAN). The following short program demonstrates a subroutine for determining the correct angular position in any of the four quadrants. It is based on the principal value of the arctangent and tests on the sign of X and the sign of Y.

Two-argument arctangent demonstration

```
1000 REM  ATN2 DEMONSTRATION
1010 REM  ATN2 SUBROUTINE IS PROGRAMMED
1020 REM  IN LINES 2000 - 2130
1030 HOME
1040 PRINT"ENTER X"
1050 INPUT X
1060 PRINT"ENTER Y"
1070 INPUT Y
1080 GOSUB 2000
1090 PRINT
1100 PRINT"ANGLE =";A
1110 END
1120 REM
```

```
2000 REM  SUBROUTINE ATN2
2010 REM  X, Y ARE INPUTS
2020 IF X<=0 THEN GOTO 2050
2030 A=ATN(Y/X)
2040 GOTO 2120
2050 PI=3.1415926535
2060 S=+1
2070 IF Y<0 THEN S=-1
2080 IF X=0 THEN GOTO 2110
2090 A=ATN(Y/X)+S*PI
2100 GOTO 2120
2110 A=S*PI/2
2120 RETURN
2130 END
```

This completes the two-argument arctangent demonstration.

APPENDIX A5

Interpolation

In the kinematic study of a mechanism, it is often convenient to tabulate positions, velocity coefficients, and so forth for various values of the primary kinematic variable (the input, or generalized, coordinate). Linear interpolation, in which the function is approximated as a straight line between two adjacent tabular values, is the easiest method to understand. However, it is also relatively crude. Generally, better results are obtained using polynomial approximations that fit the tabulated data at several points. Four such interpolating polynomials are described in the following sections.

A5.1 QUADRATIC APPROXIMATION TO UNEVENLY SPACED DATA

Consider a tabulation of the function $F(X)$ consisting of the data pairs (X_1, F_1), (X_2, F_2), (X_3, F_3), . . . , (X_n, F_n), where $X_1, X_2, X_3, \ldots, X_n$ are not evenly spaced; a typical data set is plotted in Figure A5.1. If a value is to be estimated for $F(q)$ where q is not one of the tabulation points, the first step is to locate the interval containing q. This is done with a loop containing an IF test that compares the tabulation positions X_j to the value q. Suppose that the tabulation points adjacent to q are X_r and X_{r+1}. Three tabulation points are then required for quadratic interpolation, so either X_{r-1} or X_{r+2}, whichever is closer to q, should be used in addition to X_r and X_{r+1}. Let the three selected tabulation points (X_r, X_{r+1}, and X_{r+2}) or (X_{r-1}, X_r, and X_{r+1}) be denoted U_1, U_2, and U_3; the associated function values are denoted G_1, G_2, and G_3.

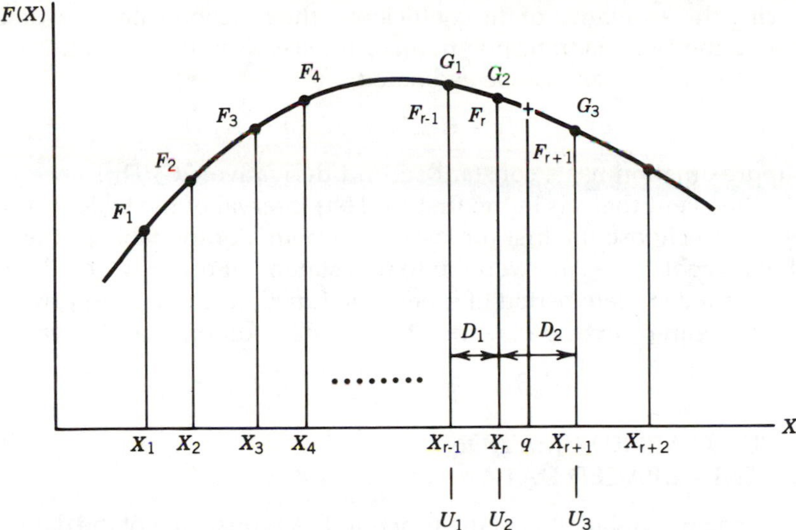

FIGURE A5.1 Unevenly Spaced Data Describing a Function of One Variable

The objective is to determine a quadratic polynomial that exactly passes through the three tabulated data points just chosen; this polynomial can then be evaluated at intermediate points to provide the required estimates. The quadratic polynomial will be of the form

$$f(q) = C_o + C_1\left(\frac{q - U_1}{D_1}\right) + C_2\left(\frac{q - U_1}{D_1}\right)^2$$

where $D_1 = U_2 - U_1$; the notation $D_2 = U_3 - U_2$ will also be used. The objective will be met if the coefficients of the polynomial are determined by

$$\begin{Bmatrix} C_0 \\ C_1 \\ C_2 \end{Bmatrix} = \frac{D_1^2}{D_2(D_1 + D_2)} \begin{bmatrix} S_{11} & 0 & 0 \\ S_{21} & S_{22} & S_{23} \\ S_{31} & S_{32} & S_{33} \end{bmatrix} \begin{Bmatrix} G_1 \\ G_2 \\ G_3 \end{Bmatrix}$$

where the matrix elements are

$$S_{11} = D_2(D_1 + D_2)/D_1^2$$

$$S_{21} = 1 - (D_1 + D_2)^2/D_1^2$$

$$S_{22} = (D_1 + D_2)^2/D_1^2$$

$$S_{23} = -1$$

$$S_{31} = D_2/D_1$$

$$S_{32} = -(D_1 + D_2)/D_1$$

$$S_{33} = 1$$

Following the evaluation of the coefficients, the function value is estimated by evaluating the quadratic polynomial. If necessary, the derivative of the tabulated function can also be estimated,

$$F'(q) \doteq f'(q) = C_1/D_1 + 2C_2(q - U_1)/D_1^2$$

The approximation has a constant second derivative, $2C_2/D_1^2$.

In the event that q is in the first (or last) interval of the table, it will be necessary to choose the first (or last) three points of the table as the basis for the interpolation. An exception to this statement occurs when the table represents a complete period of a periodic function. In that case, the ends of the table can be extended readily to provide additional tabular values as required.

A5.2 CUBIC APPROXIMATION TO EVENLY SPACED DATA

Consider again a tabulation of the function $F(X)$ consisting of the data pairs (X_1, F_1), (X_2, F_2), (X_3, F_3), . . . , (X_n, F_n), this time with equally spaced tabulation points X_1, X_2, X_3, . . . , X_n. A typical data set is plotted in Figure A5.2. As before, the function value $F(q)$ is to be estimated. For this case the quadratic polynomial just described can be used, but better results are generally expected if a cubic polynomial is used. For cubic polynomial interpolation, the interval location process proceeds as before, except that four tabulation points must be selected, denoted U_1, U_2, U_3, and U_4. Preferably, two are on either side of the required argument q,

$$U_1 < U_2 < q < U_3 < U_4$$

with the four associated function values G_1, G_2, G_3, and G_4 taken from the table. Let H denote the common difference:

$$H = U_2 - U_1 = U_3 - U_2 = U_4 - U_3$$

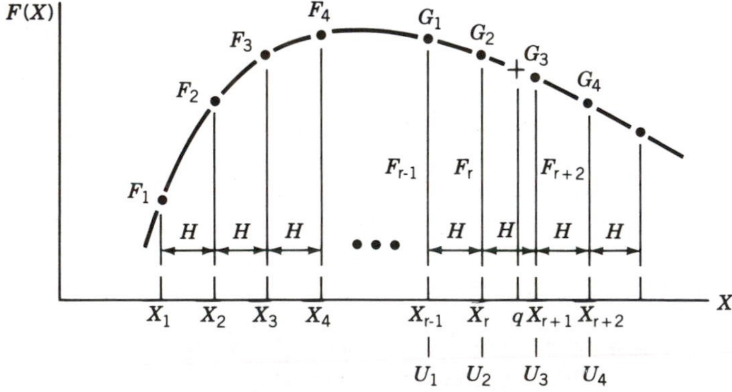

FIGURE A5.2 Evenly Spaced Data Describing a Function of One Variable

The polynomial to be fitted in this case has the form

$$f(q) = C_0 + C_1\left(\frac{q - U_1}{H}\right) + C_2\left(\frac{q - U_1}{H}\right)^2 + C_3\left(\frac{q - U_1}{H}\right)^3$$

This cubic polynomial will pass through the four points selected if the coefficients are determined by the following calculation:

$$\begin{Bmatrix} C_0 \\ C_1 \\ C_2 \\ C_3 \end{Bmatrix} = \begin{bmatrix} 1 & 0 & 0 & 0 \\ -11/6 & 3 & -3/2 & 1/3 \\ 1 & -5/2 & 2 & -1/2 \\ -1/6 & 1/2 & -1/2 & 1/6 \end{bmatrix} \begin{Bmatrix} G_1 \\ G_2 \\ G_3 \\ G_4 \end{Bmatrix}$$

After the coefficients are evaluated, the function value may be estimated by evaluating the polynomial $f(q)$. If derivatives of the tabulated function are required, they can be estimated by differentiating the polynomial approximation:

$$F'(q) \doteq f'(q) = \frac{C_1}{H} + \frac{2C_2}{H}\left(\frac{q - U_1}{H}\right) + \frac{3C_3}{H}\left(\frac{q - U_1}{H}\right)^2$$

$$F''(q) \doteq f''(q) = \frac{2C_2}{H^2} + \frac{6C_3}{H^2}\left(\frac{q - U_1}{H}\right)$$

The approximation has a constant third derivative, $6C_3/H^3$.

The comments made in the previous section regarding points near the ends of the table and the exception for periodic functions apply for the cubic interpolation as well, except that four points are required rather than three.

A5.3 INTERPOLATION FOR FUNCTIONS OF TWO VARIABLES

Unevenly Spaced Data. The ideas just presented for functions of a single variable can be extended to functions of two variables as well. Consider a function $F(X, Y)$ tabulated at X–values $X_1, X_2, X_3, \ldots, X_n$ and Y–values $Y_1, Y_2, Y_3, \ldots, Y_n$, where neither the X–values nor the Y–values are necessarily evenly spaced; a typical set of these uneven tabulation positions is shown in Figure A5.3. The irregular spacing suggests the use of quadratic approximation in each direction to estimate $F(q_1, q_2)$.

As before, the first step is to locate the relevant section of the data table by a search procedure. Suppose that q_1 lies among three X–values now denoted U_1, U_2, and U_3, while q_2 lies among three Y–values now denoted V_1, V_2, and V_3. The associated function values from the table are cast as a (3×3) matrix called $[G]$.

$$[G_{ij}] = [F(U_i, V_j)] \qquad i = 1, 2, 3 \qquad j = 1, 2, 3$$

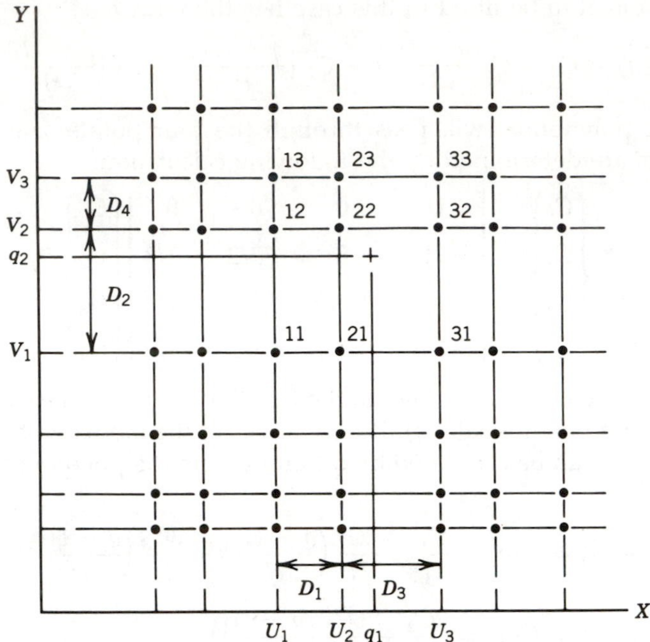

FIGURE A5.3 Unevenly Spaced Tabulation Locations for a Function of Two Variables

The intervals are denoted as follows:

$$D_1 = U_2 - U_1 \qquad D_2 = V_2 - V_1$$

$$D_3 = U_3 - U_2 \qquad D_4 = V_3 - V_2$$

The polynomial approximation to be fitted is of the form

$$f(q_1, q_2) = \{R_1(q_1)\}^t[C]\{R_2(q_2)\}$$

where $[C]$ is a square matrix of coefficients to be determined and $\{R_1(q_1)\}$ and $\{R_2(q_2)\}$ are vectors:

$$\{R_1(q_1)\} = \text{Col}(1, (q_1 - U_1)/D_1, (q_1 - U_1)^2/D_1^2)$$

$$\{R_2(q_2)\} = \text{Col}(1, (q_2 - V_1)/D_2, (q_2 - V_1)^2/D_2^2)$$

The polynomial describes a surface passing through the nine tabular values, provided that the coefficient matrix is determined by

$$[C] = B[S_h][G][S_v]$$

where

$$B = \frac{D_1^2 D_2^2}{D_3 D_4 (D_1 + D_3)(D_2 + D_4)}$$

The elements of $[S_h]$ and $[S_v]$ are as follows:

$$S_{h11} = D_3(D_1 + D_3)/D_1^2$$
$$S_{h12} = 0$$
$$S_{h13} = 0$$
$$S_{h21} = 1 - (D_1 + D_3)^2/D_1^2$$
$$S_{h22} = (D_1 + D_3)^2/D_1^2$$
$$S_{h23} = -1$$
$$S_{h31} = D_3/D_1$$
$$S_{h32} = -(D_1 + D_3)/D_1$$
$$S_{h33} = 1$$

$$S_{v11} = D_4(D_2 + D_4)/D_2^2$$
$$S_{v12} = 1 - (D_2 + D_4)^2/D_2^2$$
$$S_{v13} = D_4/D_2$$
$$S_{v21} = 0$$
$$S_{v22} = (D_2 + D_4)^2/D_2^2$$
$$S_{v23} = -(D_2 + D_4)/D_2$$
$$S_{v31} = 0$$
$$S_{v32} = -1$$
$$S_{v33} = 1$$

After the coefficients have been evaluated, the function value is estimated by evaluating the polynomial:

$$F(q_1, q_2) \doteq f(q_1, q_2) = \{R_1(q_1)\}^t[C]\{R_2(q_2)\}$$

Various partial derivatives can also be estimated by differentiating the polynomial approximation.

Evenly Spaced Data. In the event that the data is evenly spaced, cubic approximation in each direction is not too difficult. Locations for a typical set of evenly spaced data are shown in Figure A5.4. Suppose that the relevant section of the table has been located, and that

$$U_1 < U_2 < q_1 < U_3 < U_4$$

$$V_1 < V_2 < q_2 < V_3 < V_4$$

where U_1, U_2, U_3, and U_4 are X–values found in the table, whereas V_1, V_2, V_3, and V_4 are Y–values also found in the table. Let the 16 associated function values, which are taken from the table, be denoted

$$[G_{ij}] = [F(U_i, V_j)] \qquad \begin{matrix} i = 1, 2, 3, 4 \\ j = 1, 2, 3, 4 \end{matrix}$$

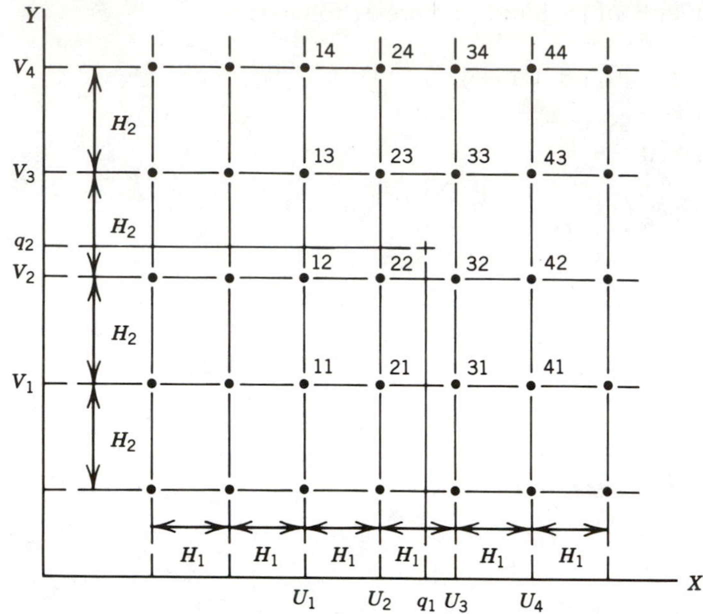

FIGURE A5.4 Evenly Spaced Tabulation Locations for a Function of Two Variables

The spacing in the X–direction is H_1, while in the Y-direction the spacing is H_2. The function to be fitted for this case is of the form

$$f(q_1, q_2) = \{R_1(q_1)\}^t[C]\{R_2(q_2)\}$$

where $\{R_1(q_1)\}$ and $\{R_2(q_2)\}$ are vectors

$$\{R_1(q_1)\} = \begin{Bmatrix} 1.0 \\ (q_1 - U_1)/D_1 \\ (q_1 - U_1)^2/D_1^2 \\ (q_1 - U_1)^3/D_1^3 \end{Bmatrix} \qquad \{R_2(q_2)\} = \begin{Bmatrix} 1.0 \\ (q_2 - V_1)/D_2 \\ (q_2 - V_1)^2/D_2^2 \\ (q_2 - V_1)^3/D_2^3 \end{Bmatrix}$$

The surface defined by this polynomial will pass through the 16 selected tabular values when the coefficient matrix is determined by

$$[C] = [S]^t[G][S]$$

where

$$[S] = \begin{bmatrix} 1 & -11/16 & 1 & -1/6 \\ 0 & 3 & -5/2 & 1/2 \\ 0 & -3/2 & 2 & -1/2 \\ 0 & 1/3 & -1/2 & 1/6 \end{bmatrix}$$

After the coefficients have been evaluated, the function value $F(q_1, q_2)$ is estimated by evaluating $f(q_1, q_2)$. As before, various partial derivatives can be estimated by differentiating the polynomial approximation.

A5.4 GENERAL COMMENTS ON INTERPOLATION

The interpolation schemes just discussed work well and, in themselves, can be executed quickly. The most time-consuming part of any interpolation is the interval location step that must be completed at the beginning. Any programming measures that can be taken to save time in this operation will improve the overall performance significantly. One such measure is starting the search near a previous value, rather than at the beginning of the tabulation. Another is to reduce the number of searches by repeating only the approximation evaluation (rather than also repeating the entire search and coefficient evaluation process) for a group of arguments that are relatively close together. These devices must be used with care and their validity tested for each application, but they will improve the overall computation time, particularly for relatively slow microcomputers.

APPENDIX A6

Kinetic Energy of a Rigid Body

Consider the rigid body shown in Figure A6.1, and in particular the differential element of mass, *dm*. This mass element is located by the position vector $\mathbf{R}$ relative to an inertial coordinate system. The position may also be expressed in terms of the position of the center of mass, $\mathbf{R}_{cm}$, and the relative position of the mass element with respect to the center of mass, $\mathbf{R}'$,

$$\mathbf{R} = \mathbf{R}_{cm} + \mathbf{R}'$$

Next consider the velocity of the mass element, *dm*. Because the body is rigid, the length of $\mathbf{R}'$ cannot change. The only change that can occur in $\mathbf{R}'$ is a change in direction. The velocity of the mass element is

$$\mathbf{V} = \mathbf{V}_{cm} + \boldsymbol{\omega} \times \mathbf{R}'$$

where $\mathbf{V}_{cm}$ is the velocity vector for the center of mass, and $\boldsymbol{\omega}$ is the angular velocity vector of the rigid body. With this information available, the kinetic energy of the mass element can be written as

$$\begin{aligned} dT &= 0.5\, \mathbf{V} \cdot \mathbf{V}\, dm \\ &= 0.5\, (\mathbf{V}_{cm} + \boldsymbol{\omega} \times \mathbf{R}') \cdot (\mathbf{V}_{cm} + \boldsymbol{\omega} \times \mathbf{R}')\, dm \\ &= 0.5\, [\mathbf{V}_{cm} \cdot \mathbf{V}_{cm} + 2\mathbf{V}_{cm} \cdot \boldsymbol{\omega} \times \mathbf{R}' + (\boldsymbol{\omega} \times \mathbf{R}') \cdot (\boldsymbol{\omega} \times \mathbf{R}')]\, dm \end{aligned}$$

To obtain the total kinetic energy of the rigid body, the contributions from every element are summed in an integration over the total mass.

$$\begin{aligned} T &= \int_M dT \\ &= 0.5\, \mathbf{V}_{cm} \cdot \mathbf{V}_{cm} \int_M dm + \mathbf{V}_{cm} \cdot \boldsymbol{\omega} \times \int_M \mathbf{R}'\, dm \\ &\quad + 0.5 \int_M (\boldsymbol{\omega} \times \mathbf{R}') \cdot (\boldsymbol{\omega} \times \mathbf{R}')\, dm \end{aligned}$$

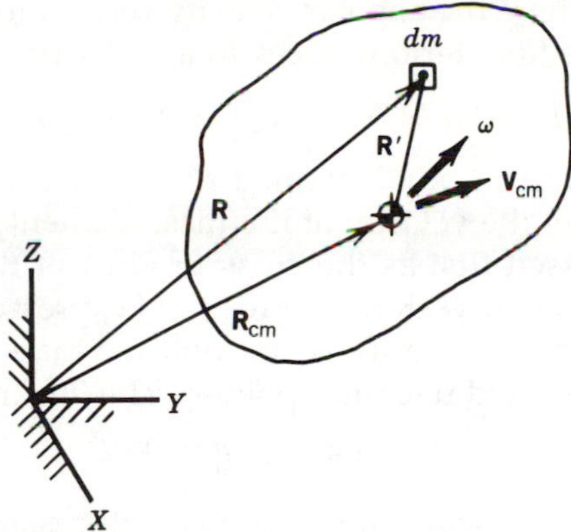

FIGURE A6.1 Rigid Body Moving in Three Dimensions

The first integral simply gives the mass of the body. The second integral vanishes. (Why?) The third integral requires rearrangement of the integrand using the vector identity:

$$\mathbf{A} \cdot \mathbf{B} \times \mathbf{C} = \mathbf{B} \cdot (\mathbf{C} \times \mathbf{A})$$

where the first factor in parentheses, $(\boldsymbol{\omega} \times \mathbf{R}')$, is taken for $\mathbf{A}$, $\boldsymbol{\omega}$ is taken for $\mathbf{B}$, and $\mathbf{R}'$ is taken for $\mathbf{C}$. The result makes $\boldsymbol{\omega}$ the leading factor that can be taken outside the integral; for the remaining integrand, the integral is the angular momentum with respect to the center of mass. To justify this statement, consider a motion consisting only of rotation about the center of mass, in which case the product $(\boldsymbol{\omega} \times \mathbf{R}')\,dm$ is the linear momentum of the mass element. When $\mathbf{R}'$ is crossed into the linear momentum, the result is the angular momentum of the mass element with respect to the center of mass. The integral then expresses the summation of the angular momentum contributions from each mass element. Including a non-zero velocity for the center of mass modifies the linear momentum for the mass element, but the additional term contributes nothing to the final angular momentum because the integral is zero.

$$\begin{aligned}
\int_M (\boldsymbol{\omega} \times \mathbf{R}') \cdot (\boldsymbol{\omega} \times \mathbf{R}')\,dm &= \boldsymbol{\omega} \cdot \int_M \mathbf{R}' \times (\boldsymbol{\omega} \times \mathbf{R}')\,dm \\
&= \boldsymbol{\omega} \cdot \mathbf{H}_{cm} \\
&= \{\omega\}^t[\mathrm{I}_{cm}]\{\omega\}
\end{aligned}$$

In the final line, the angular momentum with respect to the center of mass is written as the matrix product of the mass moment of inertia with respect to the center of mass multiplied with the angular velocity vector. The final result is now available:

$$\begin{aligned}
T &= 0.5\,M\,\mathbf{V}_{cm} \cdot \mathbf{V}_{cm} + 0.5\,\boldsymbol{\omega} \cdot \mathbf{H}_{cm} \\
&= 0.5\,M\,\mathbf{V}_{cm} \cdot \mathbf{V}_{cm} + 0.5\,\{\omega\}^t[\mathrm{I}_{cm}]\{\omega\}
\end{aligned}$$

For plane motion, where the angular velocity vector must be normal to the plane of motion, the final term reduces to a scalar product:

$$T = 0.5\, M\, \mathbf{V}_{cm} \cdot \mathbf{V}_{cm} + 0.5\, I_{cm}\, \omega^2$$

In the event that the body is pinned at some point, a similar analysis applies. For that case, the velocity of the mass element with respect to the fixed point is expressed first as the cross product of the angular velocity with the relative position vector. The result is expressed in a single term that involves the angular velocity vector and the mass moment of inertia matrix referred to the fixed reference point o. The final result in that case is

$$T = 0.5\, \{\omega\}^t [I_o] \{\omega\}$$

Again, for the case of plane motion, this reduces to the single scalar product

$$T = 0.5\, I_o\, \omega^2$$

These two cases are sufficient for all kinetic energy expressions that may be required.

REFERENCES

Beer, F. P., and Johnston, E. R., *Vector Mechanics for Engineers,* 4th ed. New York: McGraw–Hill, 1984.

Goldstein, H., *Classical Mechanics.* Reading, MA: Addison–Wesley, 1959.

Lass, H., *Vector and Tensor Analysis.* New York: McGraw–Hill, 1950.

APPENDIX A7

Geometric Calculations for Irregular Planar Areas

One of the common difficulties in dealing with the dynamics of actual machines is the need to determine the mass, mass moments of inertia, and center of mass location for actual machine part shapes. The purpose of this appendix is to present a numerical approach to this difficulty that is useful when the parts can be considered as irregular planar figures. For an irregular plane area, the program listed at the end of this appendix determines area, centroid location, and the area moments of inertia. Provided the mass distribution is uniform, it is then straight forward to obtain mass, center of mass location, and mass moments of inertia.

In the program, the irregular boundary is described in "dot-to-dot" form. For each point, coordinates must be specified in reference to any convenient rectangular cartesian coordinate system. In this manner, the actual area is approximated by an irregular polygon bounded by a sequence of straight line segments. For boundary segments that are straight, only the end points need be given; there is no benefit to giving intermediate points along a straight edge. For curves, the points should be located sufficiently close together so that a straight line satisfactorily approximates the actual boundary between the two points. The methods used in this program determine the various properties by approximating line integrals along the boundary. These particular equations were published by F. Wojciechowski (see References); the program was written by this author.

In preparing to use the program, boundary points should be numbered 0 through N, where both 0 and N represent the same point. The sequence of ascending point numbers should traverse the boundary with positive area on the right side of the curve. If the boundary is not entirely connected (that is, if there are holes in the interior) the boundary may be modified with a cut inward to the hole and a return to the outer boundary along the same cut line. This will not affect any of the computed values, and is illustrated by the following example problem.

Note that the computations are all geometric in the sense that all integrations are with respect to area, not to mass. To get mass and mass moments of inertia, the geometric values must be multiplied by the appropriate factor—the product of mass density and the thickness of the body, t. Thus,

$$Mass = \int_M dm = \rho\, t \int_A da = \rho\, t\, Area$$

$$I_{\underset{mass}{xx}} = \int_M y^2\, dm = \int_A y^2 \rho\, t\, da = \rho\, t\, I_{\underset{area}{xx}}$$

If the polar mass moment of inertia is required for an axis normal to the plane, it is the sum

$$I_{\underset{mass}{zz}} = I_{\underset{mass}{xx}} + I_{\underset{mass}{yy}}$$

Example. A machine part has the shape shown in Figure A7.1. To compute the area, centroid location, and moments of inertia for the shape, the boundary is approximated by the following sequence of coordinates. Notice that the points numbered 16 and 17 represent the cut from the outer boundary to the inner boundary, points 17 through 29 are on the inner boundary, and points 29 and 30 represent the return passage along the cut to return to the outer boundary. When the program is executed with the data in Table A7.1, first the data are printed out for verification followed by the results:

```
                    SECTION PROPERTIES

AREA=68.4891001
LOCATION OF CENTROID IN INPUT COORDINATES
```

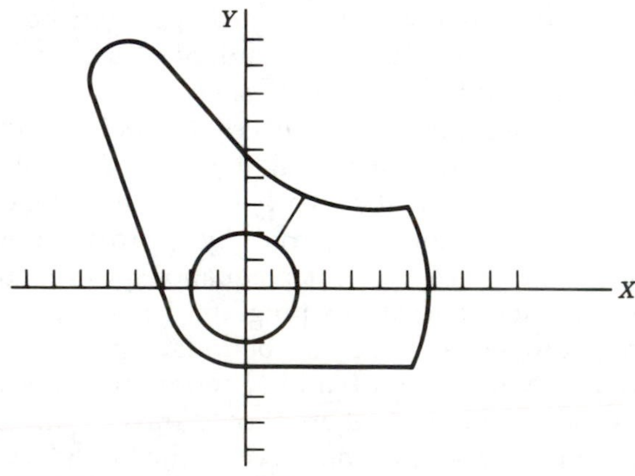

FIGURE A7.1 Irregular Plane Area

TABLE A7.1 Example Problem Data

i	X_i	Y_i	i	X_i	Y_i
0	6.15	2.93	17	1.06	1.74
1	6.65	1.42	18	0.0	2.0
2	6.80	0.0	19	−0.92	1.78
3	6.65	−1.42	20	−1.70	1.0
4	6.15	−2.93	21	−2.0	0.0
5	0.0	−2.93	22	−1.78	−0.97
6	−0.97	−2.79	23	−1.12	−1.70
7	−1.90	−2.21	24	0.0	−2.0
8	−2.65	−1.23	25	1.1	−1.74
9	−5.72	7.10	26	1.8	−0.95
10	−5.72	8.09	27	2.0	0.0
11	−5.14	8.76	28	1.84	0.88
12	−4.18	8.94	29	1.06	1.74
13	−3.34	8.47	30	1.65	3.48
14	−0.42	5.07	31	2.97	3.08
15	0.65	4.10	32	4.27	2.93
16	1.65	3.48	33	6.15	2.93

```
XC=.461007686
YC=2.06584708
CENTROIDAL MOMENTS OF INERTIA
AXES PARALLEL TO INPUT COORDINATES
ICX=626.898357
ICY=849.17048
ICXY=-489.396834
CENTROIDAL PRINCIPAL AXES
ROTATION FROM INPUT AXES (CCW+)
ANG=.897048547 RAD
   =51.3970959 DEG
ICPX=1239.89145
ICPY=236.177382
ICPXY=-1.76858157E-06
THIS LAST VALUE SHOULD BE ZERO
```

The program listing follows.

```
1000 REM    "SECPRP"
1010 REM PROPERTIES OF IRREGULAR CROSS SECTIONS
1020 REM
1030 REM SOURCE:
1040 REM "PROPERTIES OF PLANE CROSS SECTIONS"
1050 REM. F. WOJCIECHOWSKI
1060 REM MACHINE DESIGN, 22 JAN 1976
1070 DIM X(200),Y(200)
1080 PI=3.14159265
```

```
1090 REM
1100 REM *** INSTRUCTIONS ***
1110 HOME
1120 PRINT"PROPERTIES OF IRREGULAR CROSS SECTIONS"
1130 PRINT"COMPUTED BY TRAPEZOIDAL INTEGRATION"
1140 PRINT
1150 PRINT"ENTER DATA POINTS IN ORDER"
1160 PRINT"TRAVERSING THE BOUNDARY"
1170 PRINT"WITH POSITIVE AREA ON THE RIGHT"
1180 PRINT
1190 PRINT"ENTER 'E' FOR THE X VALUE"
1200 PRINT"TO TERMINATE DATA ENTRY"
1210 REM
1220 REM *** DATA ENTRY ***
1230 N=-1
1240 FOR I=0 TO 200
1250 PRINT"X(";I;"),Y(";I;")=?"
1260 INPUT X$,Y(I)
1270 IF X$="E" THEN GOTO 1380
1280 N=N+1
1290 X(I)=VAL(X$)
1300 IF I>0 THEN GOTO 1330
1310 X1=X(0)
1320 Y1=Y(0)
1330 IF X(I)<X1  THEN X1=X(I)
1350 IF Y(I)<Y1 THEN Y1=Y(I)
1360 NEXT I
1370 REM *** PRINT INPUT DATA ***
1380 FOR I=0 TO N
1390 PRINT I;"  X(";I;")=";X(I);"  Y(";I;")=";Y(I)
1400 NEXT I
1410 REM
1430 REM TRANSFORM TO FIRST QUADRANT
1440 FOR I=0 TO N
1450 X(I)=X(I)-X1
1460 Y(I)=Y(I)-Y1
1470 NEXT I
1480 REM
1490 REM AREA, 1ST & 2ND MOMENTS IN FIRST QUAD COORD
1500 A0=0
1510 M1=0
1520 M2=0
1530 I1=0
1540 I2=0
1550 I3=0
1560 FOR J=0 TO N-1
1570 J1=J+1
1580 X4=X(J1)-X(J)
1590 X5=X(J1)+X(J)
1600 Y4=J(J1)-Y(J)
```

```
1610 Y5=Y(J1)+Y(J)
1620 A0=A0-Y4*X5/2
1630 M1=M1-Y4/8*(X5^2+X4^2/3)
1640 M2=M2+X4/8*(Y5^2+Y4^2/3)
1650 F1=X4*Y5/24*(Y5^2+Y4^2)
1660 I1=I1+F1
1670 F2=Y4*X5/24*(X5^2+X4^2)
1680 I2=I2-F2
1690 IF X4=0 THEN GOTO 1750
1700 I6=Y4^2*X5*(X(J1)^2+X(J)^2)/8
1710 P=X(J1)*Y(J)-X(J)*Y(J1)
1720 I7=Y4*P*(X(J1)^2+X(J1)*X(J)+X(J)^2)/3
1730 I8=P^2*X5/4
1740 I3=I3+(I6+I7+I8)/X4
1750 NEXT J
1760 REM *** CENTROID IN 1ST QUAD COORD ***
1770 X0=M1/A0
1780 Y=M2/A0
1790 REM *** CENTROIDAL MOMENTS OF INERTIA ***
1800 I4=I1-A0*Y0^2
1810 I5=I2-A0*X0^2
1820 I6=I3-A0*X0*Y0
1860 GOSUB 2630
1870 GOSUB 2750
1880 REM *** OUTPUT RESULTS ***
1890 PRINT
1900 PRINT"     SECTION PROPERTIES"
1910 PRINT
1920 PRINT"AREA=";A0
1930 PRINT"LOCATION OF CENTROID IN INPUT COORDINATES
1940 PRINT"XC=";X0+X1
1950 PRINT"YC=";Y0+Y1
1960 PRINT"CENTROIDAL MOMENTS OF INERTIA"
1970 PRINT"AXES PARALLEL TO INPUT COORD AXES"
1980 PRINT"ICX=";I4
1990 PRINT"ICY=";I5
2000 PRINT"ICXY=";I6
2010 PRINT"CENTROIDAL PRINCIPAL AXES"
2020 PRINT"ROTATION FROM INPUT AXES (CCW+)"
2030 PRINT"ANG=";L;" RAD"
2040 PRINT"   =";L1;" DEG"
2050 PRINT"ICPX=";I7
2060 PRINT"ICPY=";I8
2070 PRINT"ICPXY=";I9
2080 PRINT"THIS LAST VALUE SHOULD BE ZERO"
2090 END
2630 REM *** ROTATION TO PRINCIPAL COORD ***
2640 M7=-2*I6
2650 M8=I4-I5
2660 IF M8<>0 THEN GOTO 2720
```

```
2670 IF M7=0 THEN GOTO 2700
2680 Q7=1
2682 IF M7<0 THEN Q7=-1
2684 L=Q7*PI/4
2690 GOTO 2730
2700 L=0
2710 GOTO 2730
2720 GOSUB 2800
2730 L1=180*L/PI
2740 RETURN
2750 REM *** MOMENTS OF INERTIA WRT ROTATED COORD ***
2760 I7=I4*COS(L)^2+I5*SIN(L)^2-I6*SIN(2*L)
2770 I8=I5*COS(L)^2+I4*SIN(L)^2+I6*SIN(2*L)
2780 I9=.5*(I4-I5)*SIN(2*L)+I6*COS(2*L)
2790 RETURN
2800 REM *** CALC L AS ATN2 FUNCT ***
2810 IF M8<0 THEN GOTO 2840
2820 L=0.5*ATN(M7/M8)
2830 RETURN
2840 S=+1
2850 If M7<0 THEN S=-1
2860 L=0.5*(ATN(M7/M8)+S*PI)
2870 RETURN
2880 END
```

REFERENCES

Kaplan, W., *Advanced Calculus.* Reading, MA: Addison–Wesley, 1952.

Wojciechowski, F., "Properties of Plane Cross Sections," *Machine Design,* 22 Jan., 1976.

APPENDIX A8

Computer Graphics

Suppose that a computer program generates data points representing a function $y(x)$, that is, a set of values for x denoted $X(i)$, and a corresponding set of values for $y(x)$ denoted $Y(i)$. Often it is useful to see this functional relationship plotted as a curve on a graph in addition to, or instead of, seeing a tabulation of the values. Many microcomputers can draw screen plots, and then can make permanent copies if connected to a suitable printer. A few microcomputers have powerful graphics commands in their BASIC language, whereas others require significant additional programming for plots. Both of these approaches to screen graphics are discussed in the following sections.

A8.1 BASIC LANGUAGE GRAPHICS COMMANDS

The following commands are typical of those found in engineering-oriented microcomputers. This list is representative only; the specific commands available should be determined for any particular machine.

GCLEAR clears the graphics screen

SCALE X1, X2, Y1, Y2 defines the graphics window to be from X1 to X2 in the horizontal direction, and from Y1 to Y2 in the vertical direction

XAXIS YI, XT specifies a horizontal axis to be drawn that intersects the vertical axis at YI, and which has tick marks at intervals XT

YAXIS XI, YT specifies a vertical axis to be drawn intersecting the horizontal axis at XI, and having tick marks at intervals YT

MOVE X,Y moves the plotting point from its current position to (X, Y) without drawing a line

PLOT X, Y moves the plotting point from its current position to (X, Y), drawing a straight line between them

For the data set described at the beginning of this discussion, assume that there are N values for each of $X(i)$ and $Y(i)$. Let X1 and Y1 denote the minimum values for $X(i)$ and $Y(i)$, and let X2 and Y2 be the maximum values of $X(i)$ and $Y(i)$. Provided that the commands just described are available, the following code would be sufficient to generate a plot of $Y(i)$ versus $X(i)$:

```
1720 GCLEAR
1730 SCALE X1,X2,Y1,Y2
1740 XAXIS 0,1
1750 YAXIS 0,10
1760 MOVE X(0),Y(0)
1770 FOR I = 0 TO N
1780 PLOT X(I),Y(I)
1790 NEXT I
```

The plot would have both X– and Y–axes drawn, with tick marks at each unit on the X–axis and at 10-unit intervals on the Y–axis. The loop that plots the curve would draw a sequence of straight-line segments from one plotting point to the next, a "dot-to-dot" plot.

Before the plotting code can be written, some information about the nature of the function to be plotted is necessary to scale appropriately. Although writing code for automatic scaling, in which the computer determines the values for X1, X2, Y1, Y2 and the tick mark intervals, is possible, it can lead to unusual scale factors and strange intervals for the tick marks. The need for such code can often be avoided simply by looking at a set of tabular results and choosing the parameters accordingly.

A8.2 PLOTTING WITHOUT GRAPHICS COMMANDS

Pixell Plotting. For computers that lack the BASIC graphics commands just described a user can create screen plots by using commands that light particular dots on the screen. This is called "pixell plotting" because each dot on the screen is a pixell. It is useful to think of the desired plot as already existing on a plane behind the computer screen. The computer screen then represents a "window" through which the plot plane is viewed. The window must be set to show the appropriate part of the plot plane, and the pixells corresponding to points on the curve must be lighted.

The location of a particular pixell is specified by its *screen coordinates*, a pair of non-negative integers that locate the pixell on a rectangular cartesian coordinate system fixed on the screen. The screen coordinate axes are denoted P and Q, with the P–axis horizontal across the top of the screen,

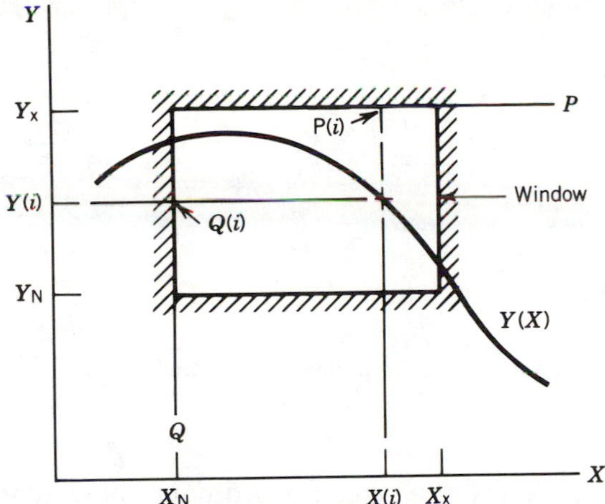

FIGURE A8.1 Relation Between X-Y Coordinates and the Screen Coordinates P-Q

and the Q–axis coming down the left edge of the screen. In Figure A8.1, these axes are shown as two edges of the window looking onto the plot plane. Consequently, the plotting command will take the form

```
PLOT P1,Q1 TO P2,Q2
```

where (P1, Q1) and (P2, Q2) are the screen coordinates of the two pixells corresponding to two plotting points. The computer then identifies and lights other intermediate pixells to approximate a straight line from the first point to the second.

Transformation to Screen Coordinates. The points to be plotted are known in terms of coordinate values ($X(i)$, $Y(i)$), and the screen coordinates for each plotting point are needed. This requires a transformation of coordinates from (X, Y) to (P, Q)—the transformation to screen coordinates.

The first step is to define the window, which is to identify what part of the X–Y plane is to be displayed on the screen. This is done by specifying, for each variable, the maximum and minimum values to be seen on the screen. Let XN and YN denote the minimum visible values of X and Y, while the maxima are XX and YX, respectively. Also let the maximum screen coordinate values be PX and QX, while the minimum for each of these is zero. The screen coordinates corresponding to the point ($X(i)$, $Y(i)$) are then

$$\mathrm{P(i)} = \mathrm{PX}\left(\frac{\mathrm{X}(i) - \mathrm{XN}}{\mathrm{XX} - \mathrm{XN}}\right)$$

$$\mathrm{Q(i)} = \mathrm{QX}\left(\frac{YX - Y(i)}{\mathrm{YX} - \mathrm{YN}}\right)$$

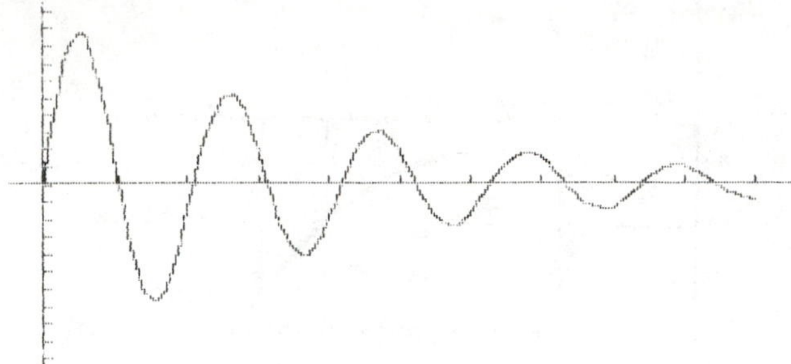

FIGURE A8.2 Results From Plotting Demonstration

It is useful to know the screen coordinate values corresponding to $X = 0$ and $Y = 0$, as they will be required to draw the coordinate axes. These values can be obtained from the transformation equations and are denoted simply Q and P (to be consistent with the computer program listing that follows later):

$$X = 0: \qquad P = -\frac{PX * XN}{XX - XN}$$

$$Y = 0: \qquad Q = \frac{QX * YX}{YX - YN}$$

One matter remains at this point—the screen coordinates computed for the plotting points and the values corresponding to the axis lines may not be integers. Consequently, they all must be rounded off so as to make the best possible plot. If they are not rounded, when the non-integer values are used in the PLOT statement, they will simply be truncated, possibly resulting in a one pixell placement error on the screen.

This pixell plotting process is demonstrated in the following short program. Figure A8.2 at the end of the listing is the result of executing the program. The plotting routine, beginning at line 6000, can be removed easily and adapted to other programs. With minor alterations, it can be used to plot multiple curves on a single set of axes. For this purpose, the first call should be GOSUB 6200, as this will calculate the required scale factors, draw the axes and tick marks, and plot the first curve. Later calls for the second and subsequent curves should use GOSUB 6500. Logic must be added to RETURN after 6610 until the last curve has been plotted.

```
1000 REM   PLOTTING DEMO
1010 REM   PLOTS THE FUNCTION Y = 10 E^(-.25*X) SIN(3X)
1020 REM   OVER THE INTERVAL 0 < X < 10
1030 REM   WITH TICK MARKS AT EACH UNIT BOTH AXES
1040 DIM   X(100), Y(100), XT(10), YT(21)
```

```
1100 REM  GENERATE THE DATA SET
1110 FOR I = 0 TO 100
1120 X (I) = I/10
1130 Y(I) = 10 * EXP(-.25 * X(I)) * SIN (3*X(I))
1140 NEXT I
1150 M = 100        :  REM TOTAL NUMBER OF DATA POINTS
1200 REM  SET THE TICK MARKS
1210 FOR I = 1 TO 10
1220 XT(I) = I
1230 NEXT I
1240 XT(0) = 10     :  REM TOTAL NUMBER OF X-TICKS
1250 FOR I = 1 TO 21
1260 YT(I) = I - 11
1270 NEXT I
1280 YT(0) = 21     :  REM TOTAL NUMBER OF Y TICKS
1300 REM  SET PLOT PARAMETERS
1310 XN = - 0.5
1320 XX = 10.5
1330 YN = -10.5
1340 YX = 10.5
1400 REM  CALL FOR PLOTTING ROUTINE
1410 GOSUB 6200
1420 END
6000 REM  *****  PLOTTING SUBROUTINE *****
6010 REM  XX        MAX X VALUE FOR WINDOW
6020 REM  XN        MIN X VALUE FOR WINDOW
6030 REM  YX        MAX Y VALUE FOR WINDOW
6040 REM  YN        MIN Y VALUE FOR WINDOW
6050 REM  PX        MAX NUMBER HORIZONTAL PIXCELS
6060 REM  QX        MAX NUMBER VERTICAL PIXCELS
6070 REM  X(I), Y(I)  ARE POINTS TO BE PLOTTED
6080 REM  0  <  I  <  M
6090 REM  XT(J), 1 < J < XT(0)  :  X-AXIS TICK MARKS
6100 REM  YT(J), 1 < J < YT(0)  :  Y-AXIS TICK MARKS
6200 REM  INITIAL SET-UP
6210 PX=279
6220 QX=159  :  REM THIS CAN BE SET TO 191
6230 HGR
6240 HCOLOR=3
6250 Q=YX*QX/(YX-YN)
6260 P=-XN*PX/(XX-XN)
6270 PP = PX / (XX-XN)
6280 QQ = QX / (YX-YN)
6290 REM  PLOT THE AXES
6300 IF Q>QX THEN GOTO 6320
6310 HPLOT 0,Q TO PX,Q
6320 IF P<0 THEN GOTO 6340
6330 HPLOT P,0 TO P,QX
6340 REM  PLOT AXIS TICK MARKS
```

```
6350 IF Q>QX THEN GOTO 6410
6360 NT = XT(0)
6370 FOR I = 1 TO NT
6380 XT(I) = INT ( PP * (XT(I)-XN) + 0.5)
6390 HPLOT XT(I),Q TO XT(I),Q-3
6400 NEXT I
6410 IF P < 0 THEN GOTO 6470
6420 NT = YT(0)
6430 FOR I = 1 TO NT
6440 YT(I) = INT( QQ * (YX-YT(I)) + 0.5)
6450 HPLOT P, YT(I) TO P+3, YT(I)
6460 NEXT I
6470 REM  SET-UP IS COMPLETE
6500 REM  SCALE THE DATA SET
6510 FOR I=0 TO M
6520 X(I) = INT(PP * (X(I)-XN) + 0.5)
6530 Y(I) = INT(QQ * (YX-Y(I) + 0.5)
6540 IF X(I)>PX THEN X(I)=PX
6550 IF X(I)<0 THEN X(I)=0
6560 IF Y(I)>QX THEN Y(I)=QX
6570 IF Y(I)<0 THEN Y(I)=0
6580 NEXT I
6590 FOR I=1 TO M
6600 HPLOT X(I-1),Y(I-1) TO X(I),Y(I)
6610 NEXT I
6620 PRINT : PRINT : PRINT : PRINT : PRINT
6630 PRINT : PRINT : PRINT : PRINT : PRINT
6640 PRINT : PRINT : PRINT : PRINT : PRINT
6650 PRINT : PRINT : PRINT : PRINT : PRINT
6660 PRINT : PRINT : PRINT : PRINT : PRINT
6670 PRINT : PRINT : PRINT : PRINT : PRINT
6680 PRINT "KEY RETURN TO GO BACK TO TEXT MODE"
6690 INPUT V$
6700 TEXT
6710 RETURN
6720 END
```

APPENDIX A9

Derivation of the Lagrange Equation

The derivation of the Lagrange form for the equation of motion by means of the Principle of Virtual Work is the subject of this appendix. The problem is formulated in terms of N particles, subject to external forces and, perhaps constraint forces. The constraints are limited to what are called *workless constraints,* that is, forces of constraint that do no work as the motion progresses. These are the common type of constraints, and this is not a severe limitation. Including constraint forces makes the result applicable to rigid bodies, because a rigid body may be considered as a collection of a large number of individual particles with a similarly large number of constraint forces.

Consider a collection of N particles, each with a position vector of the form

$$\mathbf{R}_i = \mathbf{R}_i(q_1, q_2, q_3, \ldots)$$

where $q_1, q_2, q_3, \ldots$ are generalized coordinates. The total force on the ith particle is composed of the externally applied force and the force of constraint (which may be a rigid body internal force):

$$\mathbf{F}_i = \mathbf{F}_{ei} + \mathbf{F}_{ci}$$

The motion of the particle is described by Newton's second law:

$$M_i\dot{\mathbf{V}}_i = \mathbf{F}_i = \mathbf{F}_{ei} + \mathbf{F}_{ci}$$

or

$$\mathbf{F}_{ei} + \mathbf{F}_{ci} - M_i\dot{\mathbf{V}}_i = 0$$

The virtual work of the terms in the preceding line is necessarily zero, and the sum of such terms over all of the particles remains zero:

$$\delta W = \sum_{i=1}^{N} (\mathbf{F}_{ei} + \mathbf{F}_{ci} - M_i \dot{\mathbf{V}}_i) \cdot \delta \mathbf{R}_i = 0$$

Now, because the constraints are assumed to be workless, the sum of the virtual work of the constraints is zero:

$$\sum_{i=1}^{N} \mathbf{F}_{ci} \cdot \delta \mathbf{R}_i = 0$$

In writing the remaining expression, the subscript *e* is dropped, as external forces are understood to be the only forces of interest:

$$\sum_{i=1}^{N} \mathbf{F}_i \cdot \delta \mathbf{R}_i - \sum_{i=1}^{N} M_i \dot{\mathbf{V}}_i \cdot \delta \mathbf{R}_i = 0$$

Now focus on the latter term, the virtual work of the momentum derivative. The virtual displacement of the ith particle is

$$\delta \mathbf{R}_i = \sum_j \frac{\partial \mathbf{R}_i}{\partial q_j} \delta q_j$$

where the sum extends over all of the generalized coordinates. The virtual work of the momentum derivative is then

$$\sum_{i=1}^{N} M_i \dot{\mathbf{V}}_i \cdot \delta \mathbf{R}_i = \sum_{i=1}^{N} M_i \dot{\mathbf{V}}_i \sum_j \frac{\partial \mathbf{R}_i}{\partial q_j} \delta q_j$$

$$= \sum_j \sum_i \left(M_i \dot{\mathbf{V}}_i \frac{\partial \mathbf{R}_i}{\partial q_j} \right) \delta q_j$$

Now look at the inner sum, and consider it as one term of the derivative of a product, so that

$$\sum_i M_i \dot{\mathbf{V}}_i \frac{\partial \mathbf{R}_i}{\partial q_j} = \frac{d}{dt} \sum_i M_i \mathbf{V}_i \frac{\partial \mathbf{R}_i}{\partial q_j} - \sum_i M_i \mathbf{V}_i \frac{d}{dt} \left(\frac{\partial \mathbf{R}_i}{\partial q_j} \right)$$

Two substitutions are needed for use in this expression for the virtual work of the momentum derivative. The position of the ith particle is

$$\mathbf{R}_i = \mathbf{R}_i(q_i, q_2, q_3, \ldots)$$

and the velocity is obtained by a time differentiation:

$$\mathbf{V}_i = \sum_j \frac{\partial \mathbf{R}_i}{\partial q_j} \dot{q}_j$$

If the partial derivative of $\mathbf{V}_i$ is then taken with respect to $\dot{q}_j$, all terms of the sum are zero except for one, by the independence of the generalized coordinates:

$$\frac{\partial \mathbf{V}_i}{\partial \dot{q}_j} = \frac{\partial \mathbf{R}_i}{\partial q_j}$$

In the expression for the virtual work of the momentum derivative, this will replace the first term on the right-hand side.

For the second substitution, consider once again the expression for the position of the ith particle and differentiate:

$$\mathbf{R}_i = \mathbf{R}_i(q_1, q_2, q_3 \ldots)$$

$$\frac{\partial \mathbf{R}_i}{\partial q_j} = \frac{\partial \mathbf{R}_i}{\partial q_j}$$

$$\frac{d}{dt}\frac{\partial \mathbf{R}_i}{\partial q_j} = \sum_k \frac{\partial^2 \mathbf{R}_i}{\partial q_k \partial q_j}\dot{q}_k$$

$$= \frac{\partial}{\partial q_j}\sum_k \frac{\partial \mathbf{R}_i}{\partial q_k}\dot{q}_k$$

$$= \frac{\partial \mathbf{V}_i}{\partial q_j}$$

This is used as a replacement in the final term of the virtual work of the momentum derivative. When these substitutions are made, the momentum derivative factor becomes

$$\sum_i M_i\dot{\mathbf{V}}_i\frac{\partial \mathbf{R}_i}{\partial q_j} = \frac{d}{dt}\sum_i M_i\mathbf{V}_i\cdot\frac{\partial \mathbf{V}_i}{\partial \dot{q}_j} - \sum_i M_i\mathbf{V}_i\cdot\frac{\partial \mathbf{V}_i}{\partial q_j}$$

$$= \frac{d}{dt}\frac{\partial}{\partial \dot{q}_j}\left(0.5\sum_i M_i\mathbf{V}_i\cdot\mathbf{V}_i\right) - \frac{\partial}{\partial q_j}\left(0.5\sum_i M_i\mathbf{V}_i\cdot\mathbf{V}_i\right)$$

$$= \frac{d}{dt}\frac{\partial T}{\partial \dot{q}_j} - \frac{\partial T}{\partial q_j}$$

The virtual work of the momentum derivative is then simply

$$\sum_i M_i\dot{\mathbf{V}}_i\cdot\delta\mathbf{R}_i = \sum_j\left(\frac{d}{dt}\frac{\partial T}{\partial \dot{q}_j} - \frac{\partial T}{\partial q_j}\right)\delta q_j$$

This is the most significant part of the derivation; it shows the relation of the momentum derivative (the mass · acceleration term of Newton's Second Law) to be a combination of derivatives of the kinetic energy.

The virtual work of the external forces is expressed in the usual manner, leading to the definition of the generalized forces Q_j (see Section 6.2):

$$\mathbf{R}_i = \mathbf{R}_i(q_1, q_2, q_3, \ldots)$$

$$\delta\mathbf{R}_i = \sum_j \frac{\partial \mathbf{R}_i}{\partial q_j}\delta q_j$$

$$\sum_i \mathbf{F}_i\cdot\delta\mathrm{R}_i = \sum_i \mathbf{F}_i\cdot\sum_j\frac{\partial \mathbf{R}_i}{\partial q_j}\delta q_j$$

$$= \sum_j \sum_i \mathbf{F}_i \cdot \frac{\partial \mathbf{R}_i}{\partial q_j} \delta q_j$$

$$= \sum_j Q_j \, \delta q_j$$

where

$$Q_j = \sum_i \mathbf{F}_i \cdot \frac{\partial \mathbf{R}_i}{\partial q_j}$$

After the virtual work of the workless constraints is dropped, the virtual work equation reads

$$\sum_i \mathbf{F}_i \cdot \delta \mathbf{R}_i - \sum_i M_i \dot{\mathbf{V}}_i \cdot \delta \mathbf{R}_i = 0$$

The several results developed earlier in this section can then be used in this equation to give

$$\sum_j Q_j \, \delta q_j - \sum_j \left(\frac{d}{dt} \frac{\partial T}{\partial \dot{q}_j} - \frac{\partial T}{\partial q_j} \right) \delta q_j = 0$$

or

$$\sum_j \left(Q_j - \frac{d}{dt} \frac{\partial T}{\partial \dot{q}_j} - \frac{\partial T}{\partial q_j} \right) \delta q_j = 0$$

The generalized coordinates are independent, so the coefficient of each δq_j must vanish to assure the vanishing of the sum. This gives the first form for the Lagrange equation of motion, which is written as follows:

$$\frac{d}{dt} \frac{\partial T}{\partial \dot{q}_j} - \frac{\partial T}{\partial q_j} = Q_j$$

In the preceding development, the generalized coordinates have been considered as independent, as they have throughout this book. This is true for *holonomic systems,* which are systems that involve only *holonomic constraints* (discussed in Section 2.4). The extension of the Lagrange equation to nonholonomic systems is accomplished using the Lagrange multiplier technique.[1] This technique involves introducing the nonholonomic constraint equations, each with an unknown multiplier. When the solution is obtained, the multiplier values are determined, and they can be interpreted as the forces required to enforce the constraints.

REFERENCES

Goldstein, H., *Classical Mechanics.* Reading, MA.: Addison–Wesley, 1959.

[1] A. F. D'Souza and V. K. Garg, *Advanced Dynamics,* pp. 139–141, Prentice–Hall, Englewood Cliffs, NJ, 1984.

APPENDIX A10

Example Programs in FORTRAN

The example programs incorporated in the text are given in BASIC, but some readers are more familiar with FORTRAN and would prefer to have examples in that language. The programs in this appendix are provided for that purpose. All of the examples given in the text are repeated here in FORTRAN with two exceptions. The first omission is the short BASIC program in Appendix A4 for the two-argument arctangent calculation. There is no need for such a program in FORTRAN because FORTRAN provides the entire operation in a single function, ATAN2(Y,X). The second omission is the plotting program presented in Appendix A8. Such programs are of necessity machine specific, and must be adapted for the particular machine to be employed. The program in Appendix A8 can serve as a model for many systems, but it may require significant adaptation in some cases.

The programs presented here in FORTRAN are identified in the following list along with a reference to the text section where they were previously presented in BASIC. These programs were developed and executed on a VAX 11/780 system.

Previous Reference	*Program Title*
Sect 2.2	Slider–Crank Kinematic Analysis
Sect 2.3	Four-Bar Linkage Analysis
Sect 2.5	Four-Bar/Toggle Linkage
Sect 5.10	Continued Fraction Approximation
Sect 5.10	Brocot's Table Calculation
Sect 5.10	Prime Factor Decomposition
Sect 7.6	Simulation of Rocker Response
Sect 7.6	Four-Bar Mechanism Simulation
Sect 8.5	Induction Motor Starting a Blower
App A1	Matrix Operations Package
App A1	Verification Program
App A2	Newton–Raphson Example
App A7	Properties of Irregular Crosssections

```
C  *****  SLIDER-CRANK KINEMATIC ANALYSIS  *****
C         USING CLOSED FORM KINEMATIC RELATIONS
C         AT ALL LEVELS
C
C  R        = CRANK RADIUS
C  FL       = CONNECTING ROD LENGTH
C  C        = SLIDER PIVOT OFFSET
C  A        = CONNECTING ROD ANGLE
C  Q        = CRANK ROTATION ANGLE
C             ASSOCIATED WITH THE SDOF
C  X        = SLIDER POSITION
C
C  **  SYSTEM  DATA  **
      R=3.5
      FL=10.3
      C=0.35
C  **  POSITION ANALYSIS  **
      WRITE(6,20)
   20 FORMAT(1X,'SLIDER-CRANK KINEMATIC ANALYSIS',/
     2  1X,'ENTER CRANK ANGLE IN RADIANS',/)
      READ *,Q
      FN=R*SIN(Q)-C
      D=SQRT(FL**2-FN**2)
      A=ATAN(FN/D)
      X=R*COS(Q)+FL*COS(A)
C  **  VELOCITY ANALYSIS  **
      FKA=R*COS(Q)/(FL*COS(A))
      FKX=-(C+X*TAN(A))
C  **  ACCELERATION ANALYSIS  **
      FLA=FKA**2*TAN(A)-R*SIN(Q)/(FL*COS(A))
```

```
      FLX=-R*COS(Q)-FKA**2*FL*COS(A)-FLA*FL*SIN(A)
C  **  OUTPUT  **
      WRITE(6,161) Q
      WRITE(6,162) A
      WRITE(6,163) X
      WRITE(6,164) FKA
      WRITE(6,165) FKX
      WRITE(6,166) FLA
      WRITE(6,167) FLX
  161 FORMAT(1X,'FOR Q=',F9.5,' RAD')
  162 FORMAT(1X,'      A=',F9.5)
  163 FORMAT(1X,'      X=',F9.5)
  164 FORMAT(1X,'     KA=',F9.5)
  165 FORMAT(1X,'     KX=',F9.5)
  166 FORMAT(1X,'     LA=',F9.5)
  167 FORMAT(1X,'     LX=',F9.5)
      STOP
      END
```

```
C  *****  FOUR BAR LINKAGE ANALYSIS  *****
C         USING NEWTON-RAPHSON SOLUTION
C         FOR POSITION ANALYSIS
C  F    = VECTOR OF FUNCTION VALUES
C  A    = VECTOR OF ANGLE VALUES
C  D    = VECTOR OF ANGLE CHANGES
C  FA   = SQUARE MATRIX OF PARTIAL
C         DERIVATIVES OF F WRT A
C  AA   = SCRATCH MATRIX
      COMMON C1,C2,C3,C4
      DIMENSION F(2),A(2),D(2),FA(2,2),AA(2,3)
C  **  FUNCITON ERROR LIMIT, EF
      EF=0.0001
C  **  MIN CORRECTION, EA
      EA=0.0001
C  **  MAX NUMBER OF ITERATIONS, I9
      I9=20
C  **  INITIALIZATION
      WRITE(6,42)
   42 FORMAT(//////,1X'FOUR BAR MECHANISM KINEMATIC ANALYSIS',//
     2   1X,'ENTER INPUT CRANK ANGLE, Q')
      READ *,Q
      WRITE(6,43)
   43 FORMAT(//,1X,'ENTER INITIAL ESTIMATES FOR A2 & A3')
      READ *,A2,A3
      A(1)=A2
      A(2)=A3
C  **  BEGIN POSITION SOLUTION
      DO  91  I8=1,I9
      CALL RANDJ(Q,A,F,FA)
```

```
      E=0.
      DO  56  I=1,2
      IF(E.LT.F(I)**2)  E=F(I)**2
   56 CONTINUE
      IF(E.LT.EF)  GOTO 98
C  **  NEWTON-RAPHSON DETAILS
      DO  58  I=1,2
      DO  57  J=1,2
      AA(I,J)=FA(I,J)
   57 CONTINUE
      AA(I,3)=-F(I)
   58 CONTINUE
      CALL MOS(-1,2,1,AA,DET)
      DO  59  I=1,4
      D(I)=AA(I,3)
   59 CONTINUE
C  **  SEARCH OUT & TEST LARGEST CHANGE
      E=0.
      DO  82  I=1,2
      IF(E.LT.D(I)**2)  E=D(I)**2
   82 CONTINUE
      IF(E.LT.EA)  GOTO 98
C  **  IF E<EA THEN THE MIN STEP CRITERION HAS BEEN SATISFIED
C  **  MAKE THE CHANGE
      DO  90  I=1,2
      A(I)=A(I)+D(I)
   90 CONTINUE
   91 CONTINUE
C  **  IF THE PROCESS GETS TO THIS POINT,
C  **  IT HAS FAILED TO CONVERGE
```

```
      WRITE(6,94)
   94 FORMAT(///,1X,'*****  POSITION SOLUTION FAILED  *****')
      GOTO 199
C  **  SATISFACTORY POSITION SOLUTIONS HAVE BEEN DETERMINED
   98 A2=A(1)
      A3=A(2)
C  **  VELOCITY SOLUTION
C  **  RE-EVALUATE THE JACOBIAN
      CALL RANDJ(Q,A,F,FA)
C  **  SET-UP RIGHT HAND SIDE ELEMENTS
      AA(1,3)=C1*SIN(Q)
      AA(2,3)=-C1*COS(Q)
      DO  109  I=1,2
      DO  108  J=1,2
      AA(I,J)=FA(I,J)
  108 CONTINUE
  109 CONTINUE
      CALL MOS(-1,2,1,AA,DET)
      FK2=AA(1,3)
      FK3=AA(2,3)
C  **  ACCELERATION SOLUTION
C  **  SET-UP COEF OF Q-DOT SQUARED
      AA(1,3)=C1*COS(Q)+C2*FK2**2*COS(A2)+C3*FK3**2*COS(A3)
      AA(2,3)=C1*SIN(Q)+C2*FK2**2*SIN(A2)+C3*FK3**2*SIN(A3)
      DO  162  I=1,2
      DO  161  J=1,2
      AA(I,J)=FA(I,J)
  161 CONTINUE
  162 CONTINUE
      CALL MOS(-1,2,1,AA,DET)
```

```
      FL2=AA(1,3)
      FL3=AA(2,3)
C  **  RECALL THAT COEF OF Q-DOT-DOT
C       ARE FK2, FK3, THE VEL COEF
C  **  OUTPUT
      WRITE(6,188) Q,A2,A3,FK2,FK3,FL2,FL3
  188 FORMAT(//,15X,'RESULTS',/,1X,'FOR Q=',F10.4,/,
     2    1X,'A2 =',F10.4,5X,'A3 =',F10.4,/,
     3    1X,'FK2=',F10.4,5X,'FK3=',F10.4,/,
     4    1X,'FL2=',F10.4,5X,'FL3=',F10.4,/)
  199 CONTINUE
      STOP
      END
      SUBROUTINE RANDJ(Q,A,F,FA)
      COMMON C1,C2,C3,C4
      DIMENSION A(2),F(2),FA(2,2)
C  **  SYSTEM GEOMETRY
      C1=5.
      C2=9.
      C3=7.
      C4=10.
C  **  EVALUATION OF RESIDUAL
      F(1)=C1*COS(Q)+C2*COS(A(1))+C3*COS(A(2))-C4
      F(2)=C1*SIN(Q)+C2*SIN(A(1))+C3*SIN(A(2))
C  **  EVALUATION OF JACOBIAN
      FA(1,1)=-C2*SIN(A(1))
      FA(1,2)=-C3*SIN(A(2))
      FA(2,1)=C2*COS(A(1))
      FA(2,2)=C3*COS(A(2))
      RETURN
      END
```

```
C   *****  FOUR BAR / TOGGLE LINKAGE  *****
C   GOVERNING EQUATIONS AND JACOBIAN MATRIX
C   ARE EVALUATED IN A SUBROUTINE
      COMMON R,C2,C3,C5,X1,Y1
      DIMENSION  F(4),S(4),DS(4),FJM(4,4),AA(4,8)
C   **  SET CONTROLS
C   E1    = MAX ALLOWABLE RESIDUAL
C   E2    = MIN ALLOWABLE ADJUSTMENT
C   I9    = MAX NUMBER OF ITERATIONS
      E1=0.0001
      E2=0.0001
      I9=10
C   **  ENTER CRANK POSITION AND INITIAL
C        ESTIMATE FOR UNKNOWNS
      WRITE(6,25)
   25 FORMAT(/////,15X,
     2  'FOUR BAR / TOGGLE LINKAGE KINEMATIC ANALYSIS',//
     3  1X,'ENTER Q')
      READ *,Q
      WRITE(6,26)
   26 FORMAT(1X,'ENTER  INITIAL  ESTIMATE  FOR  A2=S(1)')
      READ *,S(1)
      WRITE(6,27)
   27 FORMAT(1X,'ENTER  INITIAL  ESTIMATE  FOR  B3=S(2)')
      READ *,S(2)
      WRITE(6,31)
   31 FORMAT(1X,'ENTER  INITIAL  ESTIMATE  FOR  B5=S(3)')
      READ *,S(3)
      WRITE(6,33)
   33 FORMAT(1X,'ENTER  INITIAL  ESTIMATE  FOR  Y=S(4)')
```

```
      READ *,S(4)
C  **  BEGIN THE ITERATION
      DO  84  I=1,I9
      CALL RESJAC(Q,S,F,FJM)
C  **  EVALUATE THE ADJUSTMENT
      DO  56  K=1,4
      AA(K,5)=-F(K)
      DO  55  L=1,4
      AA(K,L)=FJM(K,L)
   55 CONTINUE
   56 CONTINUE
      CALL MOS(-1,4,1,AA,DET)
      DO  61  II=1,4
      DS(II)=AA(II,5)
   61 CONTINUE
C  **  TEST FOR TERMINATION
C      FORM NORMS FOR RESIDUAL AND ADJUSTMENT
      FN1=F(1)**2+F(2)**2+F(3)**2+F(4)**2
      FN2=DS(1)**2+DS(2)**2+DS(3)**2+DS(4)**2
      IF(FN2.GT.E1)  GOTO 77
      IV=1
      GOTO 90
   77 IF(FN2.GT.E2)  GOTO 80
      IV=2
      GOTO 90
C  **  UPDATE SOLUTION ESTIMATES
   80 DO  83  K=1,4
      S(K)=S(K)+DS(K)
   83 CONTINUE
   84 CONTINUE
```

```
      WRITE(6,86)
   86 FORMAT(1X,'*****  SOLUTION FAILED - MAX NUMBER OF ITER')
C  **  OUTPUT
   90 WRITE(6,91)
   91 FORMAT(//,1X,'SOLUTION VALUES',/)
      WRITE(6,96) S(1),S(2),S(3),S(4)
   96 FORMAT(1X,'S(1) = A2 = ',F10.6,/,
     2        1X,'S(2) = B3 = ',F10.6,/,
     3        1X,'S(3) = B5 = ',F10.6,/,
     4        1X,'S(4) = Y  = ',F10.6,/)
      WRITE(6,98)
   98 FORMAT(3X,'FUNCTION VALUES')
      WRITE(6,99)  F(1),F(2),F(3),F(4)
   99 FORMAT(1X,'F(1) = ',E12.6,/,
     2        1X,'F(2) = ',E12.6,/,
     3        1X,'F(3) = ',E12.6,/,
     4        1X,'F(4) = ',E12.6,/)
      WRITE(6,103) I
  103 FORMAT(1X,'NUMBER OF ITERATION = ',I2)
      IF(IV.EQ.1) WRITE(6,104)
  104 FORMAT(1X,'TERMINATION BASED ON MAGNITUDE OF RESIDUAL')
      IF(IV.EQ.2) WRITE(6,105)
  105 FORMAT(1X,'TERMINATION BASED ON MAGNITUDE OF ADJUSTMENT')
C  **  VELOCITY COEFFICIENT ANALYSIS  **
C  SET-UP RIGHTHAND SIDE VECTOR
      AA(1,5)=R*SIN(Q)
      AA(2,5)=-R*COS(Q)
      AA(3,5)=0.
      AA(4,5)=0.
C  RE-EVALUATE JACOBIAN
```

```
      CALL RESJAC(Q,S,F,FJM)
      DO  120  I=1,4
      DO  118  J=1,4
      AA(I,J)=FJM(I,J)
  118 CONTINUE
  120 CONTINUE
C  **  SOLVE FOR VELOCITY COEFFICIENTS
      CALL MOS(-1,4,1,AA,DET)
      WRITE(6,129) AA(1,5),AA(2,5),AA(3,5),AA(4,5)
  129 FORMAT(/,1X,'VELOCITY COEFFICIENTS',/,
     2    1X,'K2 = ',F10.6,/,1X,'K3 = ',F10.6,/,
     3    1X,'K5 = ',F10.6,/,1X,'KY = ',F10.6,/)
      STOP
      END
      SUBROUTINE RESJAC(Q,S,F,FJM)
      COMMON R,C2,C3,C5,X1,Y1
      DIMENSION  F(4),S(4),FJM(4,4)
      R=5.
      C2=14.
      C3=27.
      C5=44.
      X1=13.6
      Y1=18.
C **  FUNCTION EVALUATIONS
      F(1)=X1+R*COS(Q)-C2*COS(S(1))-C3*SIN(S(2))
      F(2)=Y1+R*SIN(Q)+C2*SIN(S(1))-C3*COS(S(2))
      F(3)=C3*SIN(S(2))-C5*SIN(S(3))
      F(4)=S(4)-C3*COS(S(2))-C5*COS(S(3))
C  **  JACOBIAN EVALUATION
      FJM(1,1)=C2*SIN(S(1))
```

```
FJM(1,2)=-C3*COS(S(2))
FJM(1,3)=0.
FJM(1,4)=0.
FJM(2,1)=C2*COS(S(1))
FJM(2,2)=C3*SIN(S(2))
FJM(2,3)=0.
FJM(2,4)=0.
FJM(3,1)=0.
FJM(3,2)=C3*COS(S(2))
FJM(3,3)=-C5*COS(S(3))
FJM(3,4)=0.
FJM(4,1)=0.
FJM(4,2)=C3*SIN(S(2))
FJM(4,3)=C5*SIN(S(3))
FJM(4,4)=1.
RETURN
END
```

```
C   *****  CONTINUED FRACTION APPROXIMATION  *****
      INTEGER  N(50),D(50),Q(50),X(50)
      WRITE(6,20)
   20 FORMAT(//,1X,'    *****  CONTINUED FRACTION APPROXIMATION  *****')
      WRITE(6,30)
   30 FORMAT(1X,'ENTER THE REQUIRED RATIO, 0 < RO < 1.0')
      READ *,RO
      E7=0
   60 V1=RO*10**E7
      V2=AINT(V1)
      IF(ABS(V2-V1).LT.0.1) GOTO 100
      E7=E7+1
      GOTO 60
C   SET FIRST TWO VALUES IN EACH COLUMN TO TABLE
  100 X(1)=INT(10**E7+0.5)
      X(2)=RO*X(1)
      Q(2)=INT(X(1)/X(2))
      X(3)=X(1)-Q(2)*X(2)
      N(2)=1
      D(2)=Q(2)
      J=2
      FND=REAL(N(J))/(D(J))
      WRITE(6,80) J-1,N(J),D(J),FND
   80 FORMAT(1X,I3,3X,I9,3X,I9,3X,F12.9)
      Q(3)=INT(X(2)/X(3))
      X(4)=X(2)-Q(3)*X(3)
      N(3)=Q(3)
      D(3)=Q(2)*Q(3)+1
      J=3
      FND=REAL(N(J))/(D(J))
```

```
      WRITE(6,80) J-1,N(J),D(J),FND
C  EXTEND THE TABLE
  126 J=J+1
      Q(J)=INT(X(J-1)/X(J))
      X(J+1)=X(J-1)-Q(J)*X(J)
      N(J)=Q(J)*N(J-1)+N(J-2)
      D(J)=Q(J)*D(J-1)+D(J-2)
      FND=REAL(N(J))/(D(J))
      WRITE(6,80) J-1,N(J),D(J),FND
      IF (X(J+1).EQ.0)  GOTO 134
      GOTO 126
  134 STOP
      END
```

```
C   *****  BROCOT'S TABLE CALCULATION  *****
      INTEGER N(100),D(100)
      WRITE(6,30)
   30 FORMAT(////,1X,'BROCOT TABLE CALCULATION',
     2  //,1X,'ENTER FIRST RATIONAL FRACTION',
     3  /,1X,'ENTER  A  FOR FORM A/B')
      READ *,A1
      WRITE(6,60)
   60 FORMAT(1X,'ENTER  B  FOR FORM A/B')
      READ *,B1
      WRITE(6,90)
   90 FORMAT(//,1X,'ENTER SECOND RATIONAL FRACTION',
     2  /,1X,'ENTER  C  FOR FORM C/D')
      READ *,C1
      WRITE(6,120)
  120 FORMAT(1X,'ENTER  D  FOR FORM C/D')
      READ *,D1
      WRITE(6,150)
  150 FORMAT(//,1X,'ENTER BASE, BO')
      READ *,BO
C   **  BEGIN THE BROCOT TABLE  **
      N(1)=A1
      N(2)=C1
      D(1)=B1
      D(2)=D1
      K=2
  230 F=0
      DO  380  I=1,K-1
      I1=I+1
      IF(D(I)+D(I1).GT.BO) GOTO 380
```

```
      F=F+1
C  **  OPEN A SPACE IN THE TABLE  **
      DO  330  J=K,I1,-1
      J1=J+1
      N(J1)=N(J)
      D(J1)=D(J)
  330 CONTINUE
      K=K+1
C  **  CALCULATE NEW TABLE ENTRIES
      N(I1)=N(I)+N(I+2)
      D(I1)=D(I)+D(I+2)
  380 CONTINUE
C  **  QUIT THE TABLE EXPANSION ONLY WHEN ALL DENOM
C      SUMS EXCEED THE BASE
C      THIS WILL BE INDICATED BY F=0 A THE END
C      OF THE LOOP ON I
      IF (F.GT.0) GOTO 230
C  **  PRINT THE BROCOT TABLE  **
      WRITE(6,460)
  460 FORMAT(///,1X,'          BROCOT TABLE',/
     2  1X,'        N           D       RATIO',/)
      DO  500  I=1,K
      RT=REAL(N(I))/D(I)
      WRITE(6,510) N(I),D(I),RT
  500 CONTINUE
  510 FORMAT(1X,I8,3X,I8,3X,F12.7)
      STOP
      END
```

```
C   *****  PRIME FACTOR DECOMPOSITION   *****
      INTEGER F,NO
      REAL N
      WRITE(6,20)
   20 FORMAT(///////,1X,'        PRIME FACTOR DECOMPOSITION')
      WRITE(6,40)
   40 FORMAT(//,1X,'ENTER NUMBER TO BE FACTORED')
      READ *,NO
      N=NO
      F=1
   80 F=F+1
      RN=SQRT(N)
      IF (F.GT.RN)  GOTO 116
      T=N/F
      IT=INT(T)
      IF (IT.NE.T)  GOTO 80
      WRITE(6,112)  F
  112 FORMAT(5X,I7)
      N=T
      F=1
      GOTO 80
  116 NN=INT(N+.0000005)
      IF (N.EQ.NO)  WRITE(6,117) NN
  117 FORMAT(5X,I7,' IS PRIME')
      IF (N.NE.NO)  WRITE(6,118) NN
  118 FORMAT(5X,I7)
      WRITE(6,119)
  119 FORMAT(1X,'    FACTORING IS COMPLETE')
      STOP
      END
```

```
C   *****  SIMULATION OF ROCKER RESPONSE  *****
C   *****  USING RUNGE-KUTTA INTEGRATION  *****
      COMMON  R,U1,V1,U2,V2,UC,VC,FIC,FM,SO,FK,B,FO,X3,Y3
C     X1,Y1  BASE COORD FOR PT1
C     X2,Y2  BASE COORD FOR PT2
C     X3,Y3  BASE COORD FOR PT3
C     R      RADIUS OF SUPPORT
C     F      APPLIED FORCE
C     FK     SPRING RATE
C     B      DASHPOT COEFF
C     S      LENGTH OF SPRING/DASHPOT
C     SO     FREE LENGTH OF SPRING
C   FKX,FKY  VEL COEF FOR CENTER OF MASS
C   FLX,FLY  VEL COEF DERIV FOR CM
C   FK1,FK2  K1X,K1Y VEL COEFF FOR PT1
C   FK3,FK4  K2X,K2Y VEL COEFF FOR PT2
C     KS     VEL COEF FOR LENGTH S
C     UC,VC  BODY COORD FOR CENTER OF MASS
C     U1,V1  BODY COORD FOR PT1
C     U2,V2  BODY COORD FOR PT2
C     FM     BODY MASS
C     FIC    CENTROIDAL MASS MOMENT OF INERTIA
C   **  PROBLEM DATA  **
      X3=3.
      Y3=-4.
      R=1.5
      FO=50.
      FK=55.
      B=3.5
      SO=2.5
```

```
      UC=1.27684237
      VC=1.58609453
      U1=0.6
      V1=5.4
      U2=1.3
      V2=-0.8
      FM=2.27904E-2
      FIC=8.46351E-2
C   **  END OF PROBLEM DATA  **
C   **  ENTER INITIAL VALUES  **
      WRITE(6,152)
  152 FORMAT(////////,1X,'ROCKER RESPONSE SIMULATION',//,
     2   1X,'ENTER INITIAL ANGLE, Q')
      READ *,Q0
      WRITE(6,154)
  154 FORMAT(//,1X,'ENTER INITIAL ANGULAR VEL, QDOT')
      READ*,QD
C  INITIAL TIME IS ZERO
      T0=0.
      WRITE(6,158)
  158 FORMAT(//,1X,'ENTER TIME STEP, H')
      READ *,H
C   **  PRINT HEADER & INITIAL VALUES  **
      WRITE(6,161)  H
  161 FORMAT(///,1X,'ROCKER RESPONSE SIMULATION',/,
     2    1X,'USING RUNGE-KUTTA INTEGRATION',/,
     3    1X,'      INTEGRATION TIME STEP = ',F10.4,/)
  165 FORMAT(1X,'T0=',F10.4,5X,'Q=',F10.4,5X,'QD=',F10.4)
      WRITE(6,165) T0,Q0,QD
      DO  208  II=1,100
```

```
C  FIRST EVAL OF DERIV
      CALL SECDER(T0,Q0,QD,Q2)
      FM1=H*Q2
C  SECOND EVAL OF DERIV
      CALL SECDER(T0+H/2.,Q0+H*QD/2.+FM1/8.,QD+FM1/2.,Q2)
      FM2=H*Q2
C  THIRD EVAL OF DERIV
      CALL SECDER(T0+H/2.,Q0+H*QD/2.+FM2/8.,QD+FM2/8.,Q2)
      FM3=H*Q2
C  FOURTH EVAL OF DERIV
      CALL SECDER(T0+H,Q0+H*QD+H*FM3/2.,QD+FM3,Q2)
      FM4=H*Q2
C  UPDATE ALL VALUES
      Q0=Q0+H*(QD+(FM1+FM2+FM3)/6.)
      QD=QD+(FM1+2.*FM2+2.*FM3+FM4)/6.
      T0=T0+H
C  PRINT UPDATED RESULTS
      WRITE(6,165)  T0,Q0,QD
C  TEST FOR TERMINATION
      IF(Q0.GT.0.)  GOTO 208
      IF(QD.LT.0.)  GOTO 210
  208 CONTINUE
  210 CONTINUE
      STOP
      END
      SUBROUTINE SECDER(T,Q,Q1,Q2)
C  EVALUATION OF SECOND DERIVATIVE
C  INPUTS ARE TIME, GEN'L COORD & GEN'L VEL, (T,Q,Q1)
C  OUTPUT IS GEN'L ACCEL, Q2
      COMMON  R,U1,V1,U2,V2,UC,VC,FIC,FM,S0,FK,B,F0,X3,Y3
```

```
      X2=(R+U2)*COS(Q)+(R*Q-V2)*SIN(Q)
      Y2=(R+U2)*SIN(Q)+(V2-R*Q)*COS(Q)
      S=SQRT((X2-X3)**2+(Y2-Y3)**2)
      FK1=(R*Q-V1)*COS(Q)-U1*SIN(Q)
      FK2=(R*Q-V1)*SIN(Q)+U1*COS(Q)
      FK3=(R*Q-V2)*COS(Q)-U2*SIN(Q)
      FK4=(R*Q-V2)*SIN(Q)+U2*COS(Q)
      FKX=(R*Q-VC)*COS(Q)-UC*SIN(Q)
      FKY=(R*Q-VC)*SIN(Q)+UC*COS(Q)
      FLX=(R-UC)*COS(Q)+(V2-R*Q)*SIN(Q)
      FLY=(R-U2)*SIN(Q)+(R*Q-V2)*COS(Q)
      FKS=((X2-X3)*FK3+(Y2-Y3)*FK4)/S
      FIN=FM*(FKX**2+FKY**2)+FIC
      C=FM*(FKX*FLX+FKY*FLY)
      F=0.
      IF(T.LT.0.01)  F=F0
      QN=-FK1*F-B*Q1*FKS**2
      DV=FK*FKS*(S-S0)
      Q2=(QN-C*Q1**2-DV)/FIN
C  THIS IS THE SECOND DERIVATIVE VALUE
      RETURN
      END
```

```
C   *****  FOUR BAR MECHANISM SIMULATION  *****
C   *****  USING INTERPOLATED KINEMATIC SOLUTIONS  *****
C   *****  WITH CUBIC INTERPOLATION FOR EVENLY SPACED DATA  *****
        REAL J1,J2,J3
        COMMON DD(11),PC(4,7)
        COMMON /PARAM/  C1,C2,C3,C4,J1,J2,J3,FMM,UC,VC,BB,XB,YB
C   **  PROBLEM  DATA  **
C   LINK LENGTHS: C1, C2, C3, C4
       C1=5.
       C2=9.
       C3=7.
       C4=10.
C   LINK MASS MOI: J1,J2,J3
       J1=0.075
       J2=0.213444
       J3=0.15
C   MASS OF LINK 2
       FMM=1.81416E-2
C   BODY COORD FOR CM
       UC=5.11111112
       VC=1.33333333
C   DASHPOT COEF AND ANCHOR COORD
       BB=5.5
       XB=13.
       YB=1.2
C   **  END OF PROBLEM DATA  **
C
C   **  INITIAL VALUES  **
       Q0=0.2
       QD=0.0
```

```
      TO=0.0
      H=0.005
C  PRINT HEADER AND BEGIN SOLUTION TABULATION
      WRITE(6,191)  H
  191 FORMAT(//////,1X,'FOUR BAR MECHANISM SIMULATION',/,
     2     1X,'USING INTERPOLATED KINEMATIC SOLUTIONS',/,
     3     1X,'TIME STEP = ',F10.4,//,
     4     1X,'TIME',12X,'Q',10X,'QDOT',/)
  192 FORMAT(1X,F6.4,5X,F8.4,5X,F8.4)
      WRITE(6,192)  TO,Q0,QD
C  **  RUNGE-KUTTA INTEGRATION OF EQUATION OF MOTION  **
C  FIRST EVAL OF SECOND DERIV
  202 CALL SECDER(1,TO,Q0,QD,Q2)
      FM1=H*Q2
C  SECOND EVAL OF SECOND DERIV
      CALL SECDER(2,TO+H/2.,Q0+H*QD/2.+H*FM1/8.,QD+FM1/2.,Q2)
      FM2=H*Q2
C  THIRD EVAL OF SECOND DERIV
      CALL SECDER(3,TO+H/2.,Q0+H*QD/2.+H*FM2/8.,QD+FM2/2.,Q2)
      FM3=H*Q2
C  FOURTH EVAL OF SECOND DERIV
      CALL SECDER(4,TO+H,Q0+H*QD+H*FM3/2,QD+FM3,Q2)
      FM4=H*Q2
C  UPDATE THE SOLUTION VALUES
      Q0=Q0+H*(QD+(FM1+FM2+FM3)/6.)
      QD=QD+(FM1+2.*FM2+2.*FM3+FM4)/6.
      TO=TO+H
C  PRINT UPDATED VALUES
      WRITE(6,192)  TO,Q0,QD
C  TEST FOR TERMINATION
```

```
      IF(QD.LT.0.05)  GOTO 290
C  OTHERWISE, RETURN TO BEGINNING OF R-K CYCLE
      GOTO 202
  290 CONTINUE
      STOP
      END
      SUBROUTINE SECDER(IIC,T,Q,Q1,Q2)
      REAL J1,J2,J3
      COMMON DD(11),PC(4,7)
      COMMON /PARAM/ C1,C2,C3,C4,J1,J2,J3,FMM,UC,VC,BB,XB,YB
C  EVALUATION OF THE SECOND DERIVATIVE
C  T       TIME, INPUT
C  Q       GEN'L COORD, INPUT
C  Q1      GEN'L VEL, QDOT, INPUT
C  Q2      GEN'L ACCEL, QDOTDOT, OUTPUT
C  FIN     GEN'L INERTIA
C  CC      CENTRIPETAL COEF
C  FK2     KA2, VEL COEF
C  FK3     KA3, VEL COEF
C  FKX     KCM2X, CM VEL COEF
C  FKY     KCM2Y, CM VEL COEF
C  FKL     KL, DASHPOT LENGTH VEL COEF
C  FL2     LA2, VEL COEF DERIV
C  FL3     LA3, VEL COEF DERIV
C  FLX     LCM2X, CM VEL COEF DERIV
C  FLY     LCM2Y, CM VEL COEF DERIV
C  UC      BODY COORD FOR CM
C  VC      BODY COORD FOR CM
C  IIC     CALL NUMBER COUNTER
      IF(IIC.GT.1)  GOTO 320
```

```
C  FIRST CALL ENTRY POINT
      CALL INTERVAL(Q,I1)
      DO  319  JJ=2,7
      CALL COEF(JJ,I1,G0,G1,G2,G3)
      PC(1,JJ)=G0
      PC(2,JJ)=G1
      PC(3,JJ)=G2
      PC(4,JJ)=G3
  319 CONTINUE
C  ENTRY POINT FOR SECOND AND LATER CALLS
  320 SS=0.174533
      D=(Q-DD(I1))/SS
      A2=PC(1,2)+D*(PC(2,2)+D*(PC(3,2)+D*PC(4,2)))
      A3=PC(1,3)+D*(PC(2,3)+D*(PC(3,3)+D*PC(4,3)))
      FK2=PC(1,4)+D*(PC(2,4)+D*(PC(3,4)+D*PC(4,4)))
      FK3=PC(1,5)+D*(PC(2,5)+D*(PC(3,5)+D*PC(4,5)))
      FL2=PC(1,6)+D*(PC(2,6)+D*(PC(3,6)+D*PC(4,6)))
      FL3=PC(1,7)+D*(PC(2,7)+D*(PC(3,7)+D*PC(4,7)))
      FKX=-(C1*SIN(Q)+FK2*(UC*SIN(A2)+VC*COS(A2)))
      FKY=C1*COS(Q)+FK2*(UC*COS(A2)-VC*SIN(A2))
      FLX=-C1*COS(Q)-FL2*(UC*SIN(A2)+VC*COS(A2))
     2       -FK2**2*(UC*COS(A2)-VC*SIN(A2))
      FLY=-C1*SIN(Q)+FL2*(UC*COS(A2)-VC*SIN(A2))
     2       -FK2**2*(UC*SIN(A2)+VC*COS(A2))
      X2=C1*COS(Q)+C2*COS(A2)
      Y2=C1*SIN(Q)+C2*SIN(A2)
      FLL=SQRT((XB-X2)**2+(YB-Y2)**2)
      FKL=(C1*SIN(Q)+C2*FK2*SIN(A2))*(XB-X2)/FLL
     2       -(C1*COS(Q)+C2*FK2*COS(A2))*(YB-Y2)/FLL
C  EVALUATION OF GENL INERTIA & CENTRIP COEF
```

```
      FIN=J1+J2*FK2**2+FMM*(FKX**2+FKY**2)+J3*FK3**2
      CC=J2*FK2*FL2+FMM*(FKX*FLX+FKY*FLY)+J3*FK3*FL3
C  EVALUATION OF GENL FORCE
      TA=0.
      IF(T.LT.0.075)  TA=25.
      QQ=TA-BB*Q1*FKL**2
C  EVALUATION OF SECOND DERIVATIVE
      Q2=(QQ-CC*Q1**2)/FIN
      RETURN
      END
      SUBROUTINE INTERVAL(X,I1)
      COMMON DD(11),PC(3,7)
C  POSITIONS FOR THE TABULATED KINEMATIC SOLUTIONS
      DATA(DD(I),I=1,11)/0.,0.174533,0.349066,0.523599,0.698132,
     2  0.872665,1.047198,1.221730,1.396263,1.570796,1.745329/
C  X IS THE INDEPENDENT VARIABLE VALUE
C  I1  IS THE INDEX FOR THE FIRST OF THE FOUR REQ'D VALUES
C  LOCATE THE INTERVAL
      DO  387  II=1,11
      IF(X.LT.DD(II))  GOTO 388
  387 CONTINUE
  388 I1=II-2
      RETURN
      END
      SUBROUTINE COEF(JJ,I1,G0,G1,G2,G3)
      DIMENSION DD(11,7)
C  KINEMATIC SOLUTION DATA FOLLOWS:  A2,A3,KA2,KA3,LA2,LA3
      DATA(DD(1,J),J=2,7)/.884943,4.612222,-1.,-1.,
     2   .201008,1.636776/
      DATA(DD(2,J),J=2,7)/.718139,4.466885,-.889107,-.647550,
```

```
     2    .963387,2.249333/
      DATA(DD(3,J),J=2,7)/.578997,4.387930,-.701629,-.263037,
     2    1.097732,2.058619/
      DATA(DD(4,J),J=2,7)/.472529,4.370945,-.524802,.053105,
     2    .904271,1.551687/
      DATA(DD(5,J),J=2,7)/.393422,4.401303,-.388285,.281256,
     2    .664561,1.081449/
      DATA(DD(6,J),J=2,7)/.334725,4.464951,-.289718,.438151,
     2    .475192,.737204/
      DATA(DD(7,J),J=2,7)/.290661,4.551333,-.218995,.544950,
     2    .343894,.501803/
      DATA(DD(8,J),J=2,7)/.257188,4.653189,-.167084,.617648,
     2    .257113,.341125/
      DATA(DD(9,J),J=2,7)/.231627,4.765557,-.127434,.666764,
     2    .201519,.227781/
      DATA(DD(10,J),J=2,7)/.212260,4.884936,-.095460,.698806,
     2    .167988,.142919/
      DATA(DD(11,J),J=2,7)/.198053,5.008710,-.067820,.717550,
     2    .151259,.073582/
C   COEF ARE DETERMINED FOR CUBIC POLYNOMIAL INTERPOLATION
C   JJ  DENOTES THE DATA COLUMN INDEX
C   I1  DENOTES FIRST OF FOUR VALUES FOR INTERPOLATION
C   G0,G1,G2,G3  ARE THE COEFFICIENTS TO BE RETURNED
C   EXTRACT DATA VALUES
      I2=I1+1
      I3=I1+2
      I4=I1+3
      V1=DD(I1,JJ)
      V2=DD(I2,JJ)
      V3=DD(I3,JJ)
```

```
      V4=DD(I4,JJ)
C  EVALUATE THE COEF
      G0=V1
      G1=-11.*V1/6.+3.*V2-3.*V3/2.+V4/3.
      G2=V1-5.*V2/2.+2.*V3-V4/2.
      G3=-V1/6.+V2/2.-V3/2.+V4/6.
      RETURN
      END
```

```
C   *****      INDUCTION MOTOR STARTING A BLOWER      *****
C   *****  COUPLED THROUGH A CLUTCH AND A GEAR BOX   *****
      REAL J1,J2
      INTEGER F7
      COMMON C0,C1,FM5,T5,C5,F7,J1,J2
      DIMENSION A5(100),A6(100),S6(100),B6(100)
      PI=3.1415926535
C   **  SYSTEM PARAMETERS  **
C       ***  MOTOR  ***
C   MOTOR SHAFT INERTIA, CLUTCH OUTPUT SHAFT INERTIA
      J1=1.341656
      J2=1.61399958
C   MOTOR NO LOAD SPEED, RATED SPEED, RATED POWER, RATED TORQUE
      U0=1790.*2.*PI/60.
      U1=1700.*2.*PI/60.
      P1=50.*550.*12.
      FM1=P1/U1
C   **  LINEAR APPROX TO MOTOR TORQUE - SPEED RELATION
C        MOTOR TORQUE = C0 - C1 * V1
C        WHERE V1 IS MOTOR SHAFT SPEED
      C0=FM1*U0/(U0-U1)
      C1=FM1/(U0-U1)
C       ***  CLUTCH PARAMETERS  ***
C   SLIP TORQUE (IN-LB), RISE TIME (SEC)
      FM5=2050.
      T5=0.45
C   **  CTC = CLUTCH TORQUE CAPACITY
C             = 0.5 * FM5 * (1 - COS(PI*T/T5))    0 < T < T5
C             = FM5                               T5 < T
C        ***  GEAR BOX PARAMETERS  ***
```

```
C  SINGLE STAGE GEAR RATIO
      FN = 29./114.
C        ***  BLOWER PARAMETERS  ***
C  RATED POWER, RATED SPEED
      P4=43.*550.*12.
      U4=110.*2.*PI/60.
C  **  C4 = BLOWER COEF, C5 = EFF BLOWER COEF
      C4=P4/U4**3
      C5=C4*FN**6
C  INITIAL OUTPUT
      WRITE(6,140)
  140 FORMAT(//////,10X,'SIMULATION OF AN INDUCTION MOTOR',/,
     2    10X,'STARTING A BLOWER COUPLED THROUGH',/,
     3    10X,'A CLUTCH AND A GEAR BOX',//)
      WRITE(6,149)  C0,C1
  149 FORMAT(10X,'CALCULATED SYSTEM PARAMETERS',/,
     2   10X,'MOTOR COEFFICIENTS',/,
     3   15X,'C0 = ',F10.4,/,15X,'C1 = ',F10.4)
      WRITE(6,153) C4
  153 FORMAT(10X,'BLOWER COEFFICIENT',/,15X,'C4 = ',F10.4)
C  **  INTEGRATION TIME STEP
      H=0.01
C  **  F7 IS THE FLAG FOR ONE VS TWO DEGREES OF FREEDOM
C       F7=1  ---> ONE DOF
C       F7=2  ---> TWO DOF
C  **  INITIAL CONDITIONS: TIME, FLAG, MOTOR SHAFT SPEED,
C       CLUTCH OUTPUT SPEED, CLUTCH TORQUE
      T0=0.
      F7=2
      A1=U0
```

```
      A2=0.
      S0=0.
      WRITE(6,170)
  170 FORMAT(//,10X,'ENGAGEMENT HISTORY',/,
     2 1X,'DOF      TO         A1           A2           S0',/,
     3 1X,'         SEC        R/S          R/S          IN-LB',/)
      WRITE(6,223)  F7,T0,A1,A2,S0
C  **  SAVE INITIAL VALUES  **
      A5(1)=A1
      A6(1)=A2
      S6(1)=S0
      B6(1)=0.
C  B6 IS BLOWER LOAD TORQUE
C
C  **  RUNGE - KUTTA INTEGRATION  **
      I=0
  179 I=I+1
C  FIRST DERIVATIVE EVALUATION
      CALL ACCEL(T0,A1,A2,V3,V4,S0)
      FK1=H*V3
      FK2=H*V4
C  SECOND DERIVATIVE EVALUATION
      CALL ACCEL(T0+H/2.,A1+FK1/2.,A2+FK2/2.,V3,V4,S0)
      FK3=H*V3
      FK4=H*V4
C  THIRD DERIVATIVE EVALUATION
      CALL ACCEL(T0+H/2.,A1+FK3/2.,A2+FK4/2.,V3,V4,S0)
      FK5=H*V3
      FK6=H*V4
C  FOURTH DERIVATIVE EVALUATION
```

```
      CALL ACCEL(T0+H,A1+FK5,A2+FK6,V3,V4,S0)
      FK7=H*V3
      FK8=H*V4
C  UPDATE THE SOLUTION
      T0=T0+H
      A1=A1+(FK1+2.*(FK3+FK5)+FK7)/6.
      A2=A2+(FK2+2.*(FK2+FK6)+FK8)/6.
C  TEST TO CONTINUE AS 2DOF OR NOT
      IF(A2.LT.A1) GOTO 217
      F7=1
      A2=A1
  217 CONTINUE
C  STORE SOLUTION VALUES
      A5(I)=A1
      A6(I)=A2
      S6(I)=S0
      B6(I)=S3
C  WRITE RESULTS
      WRITE(6,223)  F7,T0,A1,A2,S0
C  223 FORMAT(1X,I2,3X,F5.2,2X,F9.4,2X,F9.4,2X,F6.0)
  223 FORMAT(1X,I2,3X,F5.2,5X,F6.2,5X,F6.2,5X,F6.0)
C  TEST FOR TERMINATION
      IF (V4.LT.0.01)  GOTO 230
      GOTO 179
  230 CONTINUE
C  NOTE FINAL VALUE OF INTEGRATION STEP INDEX, I9=MAX(I)
      I9=I
C  CALCULATE THE CLUTCH ENERGY LOSS, G6
      G6=0
      DO  237  I=2,I9
```

```
      G1=S6(I-1)*(A5(I-1)-A6(I-1))
      G2=S6(I)*(A5(I)-A6(I))
      G6=G6+(G1+G2)/2.
  237 CONTINUE
      G6=H*G6
      WRITE(6,240) G6
  240 FORMAT(///,15X,'CLUTCH ENERGY LOSS = ',F12.6)
      STOP
      END
      SUBROUTINE ACCEL(T,V1,V2,V3,V4,S0)
C  **  EVALUATION OF THE SHAFT ACCELERATIONS
C  **  INPUTS TO THIS ROUTINE ARE T, V1, V2
C  **  OUTPUTS FROM THIS ROUTINE ARE V3, V4, S0
C             V3 = V1 DOT
C             V4 = V2 DOT
C      S0 = TORQUE THROUGH THE CLUTCH
C      S1 = MOTOR TORQUE VALUE
C      S2 = CLUTCH TORQUE CAPACITY
C      S3 = EFFECTIVE BLOWER LOAD TORQUE
      REAL J1,J2
      INTEGER F7
      COMMON C0,C1,FM5,T5,C5,F7,J1,J2
      PI=3.13415926535
      S1=C0-C1*V1
      S2=FM5
      IF (T.GT.T5)  GOTO 314
      S2=0.5*FM5*(1-COS(PI*T/T5))
  314 S3=C5*V2**2
      IF (F7.EQ.2) GOTO 321
C  F7=1  --->  SDOF
```

```
      V3=(S1-S3)/(J1+J2)
      V4=V3
      S0=J2*V4+S3
      RETURN
C  F7=2  --->  2DOF
  321 V3=(S1-S2)/J1
      V4=(S2-S3)/J2
      S0=S2
      RETURN
      END
```

```
C   *****  MATRIX OPERATIONS PACKAGE  *****
C
C
C   *****  GENERAL MATRIX PRODUCT  *****
C           MAT 04   = MAT 02  *  MAT 03
C         (IZ1,IZ3) (IZ1,IZ2)  (IZ2,IZ3)
      SUBROUTINE GMP(02,03,04,IZ1,IZ2,IZ3)
      DIMENSION 02(1),03(1),04(1)
      DO  18  I5=1,IZ1
      DO  18  I6=1,IZ3
      Z9 = 0.
      I56=I5+IZ1*(I6-1)
      DO  15  I7=1,IZ2
      I57=I5+IZ1*(I7-1)
      I76=I7+IZ2*(I6-1)
      Z9=Z9+02(I57)*03(I76)
   15 CONTINUE
      04(I56)=Z9
   18 CONTINUE
      RETURN
      END
C
C   *****  PRODUCT OF A MATRIX WITH A VECTOR  *****
C           MAT 04   =  MAT 02   *  VECTOR 01
C         (IZ1,1)     (IZ1,IZ2)     (IZ2,1)
      SUBROUTINE MVP(01,02,04,IZ1,IZ2)
      DIMENSION 01(1),02(1),04(1)
      DO  16  I5=1,IZ1
      Z9 = 0.
      DO  14  I6=1,IZ2
```

```
      I56=I5+IZ1*(I6-1)
      Z9=Z9+O2(I56)*O1(I6)
   14 CONTINUE
      O4(I5)=Z9
   16 CONTINUE
      RETURN
      END
C
C  ***** MATRIX INVERSION &  LINEAR SYSTEM SOLUTION   *****
C  FOR THE LINEAR SYSTEM      C         X      =    B
C                          (N X N)(N X NSYS) (N X NSYS)
C      INPUT:
C              A = LOCATION OF THE AUGMENTED MATRIX C:B
C              N = ORDER OF  A  MATRIX
C          IFLAG = 0  FOR DET(C) ONLY,  C = (NXN) ARRAY
C                > 0  FOR INV(C) AND DET(C),  A = (NX2N) ARRAY
C                < 0  FOR THE SOLUTIONS OF THE SYSTEM  C X = B
C                     WHERE  A = (N X (N+NSYS) ) ARRAY
C           NSYS = NUMBER OF SYSTEMS TO BE SOLVED
C      OUTPUT:
C            DET = DETERMINANT OF COEF MATRIX
C          IFLAG = 0, THERE IS NO OTHER OUTPUT
C          IFLAG > 0, INV(C) IS IN RIGHTMOST N COLS OF A
C          IFLAG < 0, SOLUTIONS ARE IN RIGHTMOST NSYS COLS OF A
      SUBROUTINE MOP(IFLAG,N,NSYS,A,DET)
      DIMENSION A(1),X(10)
      SIGN=1
      MARK=0
      NMI=N-1
      NN=2*N
```

```
      NPY=N+NSYS
      IF (IFLAG.LE.0)  GOTO 10
C  **  INITIALIZE  B  MATRIX FOR INVERSE CALCULATION  **
      DO  8  I=1,N
      DO  6  J=N+1,NN
      IJ=I+N*(J-1)
      A(IJ)=0.0
    6 CONTINUE
      INI=I+N*(N+I-1)
      A(INI)=1.0
    8 CONTINUE
      NPY=NN
C  **  LOCATE PIVOT ELEMENT  **
   10 DO  200  I=1,NMI
      MAX=I
      II=I+N*(I-1)
      AMAX=ABS(A(II))
   20 DO  50  K=I+1,N
      KI=K+N*(I-1)
      IF(ABS(A(KI)).LE.AMAX)  GOTO 50
      MAX=K
      AMAX=ABS(A(KI))
   50 CONTINUE
C  **  ROW INTERCHANGES  **
      IF(MAX.EQ.I)  GOTO 70
      DO  60  L=I,NPY
      IL=I+N*(L-1)
      MAXL=MAX+N*(L-1)
      TEMP= A(IL)
      A(IL)=A(MAXL)
```

```
      A(MAXL)=TEMP
   60 CONTINUE
      SIGN=-SIGN
   70 DO  190  J=I+1,N
      JI=J+N*(I-1)
      II=I+N*(I-1)
      IF (A(JI).EQ.0)  GOTO 190
      CONST=-A(JI)/A(II)
      DO  180  L=I,NPY
      JL=J+N*(L-1)
      IL=I+N*(L-1)
      A(JL)=A(JL)+CONST*A(IL)
  180 CONTINUE
  190 CONTINUE
  200 CONTINUE
C  **  COMPUTE DET(A)  **
      TEMP=1
      DO  220  I=1,N
      II=I+N*(I-1)
      IF (A(II).EQ.0) GOTO  230
      TEMP=A(II)*TEMP
  220 CONTINUE
      DET=SIGN*TEMP
      GOTO 240
  230 MARK=1
  240 IF (IFLAG.EQ.0)  GOTO 999
      IF (IFLAG.NE.1)  GOTO 250
      IF (MARK.NE.1)  GOTO 250
      GOTO 999
C  **  BACK SUBSTITUTION  **
```

```
250 DO   360   I=N+1,NPY
    DO   330   K=N,1,-1
    KI=K+N*(I-1)
    KK=K+N*(K-1)
280 X(K)=A(KI)
    IF (K.EQ.N)   GOTO 325
    DO   320   J=K+1,N
    KJ=K+N*(J-1)
    X(K)=X(K)-X(J)*A(KJ)
320 CONTINUE
325 X(K)=X(K)/A(KK)
330 CONTINUE
    DO   340   L=1,N
    LI=L+N*(I-1)
    A(LI)=X(L)
340 CONTINUE
360 CONTINUE
999 CONTINUE
    IF(MARK.NE.0)   WRITE(6,1001)
1001 FORMAT(1X,'********    MATRIX IS SINGULAR    ********')
    IF(MARK.NE.0)   WRITE(6,1002)DET, MARK
1002 FORMAT(1X,'              DET = ',E10.4,'   MARK = ',I4)
    RETURN
    END
```

```
C  *****  VERIFICATION PROGRAM   *****
C  TEST OF THE MATRIX OPERATIONS PACKAGE
      DIMENSION A(3,3),AO(3,6),B(3,2),C(3,3),D(3),E(3),F(3,2)
C  **  DATA FOR MAT  A  **
      DATA ((A(I,J), J=1,3),I=1,3)/-1.,2.,3.,2.,3.,4.,3.,4.,5./
C  **  DATA FOR MAT  B  **
      DATA ((B(I,J), J=1,2),I=1,3)/1.,2.,2.,5.,4.,3./
C  **  DATA FOR VECTOR  D  **
      DATA (D(I),I=1,3)/1.,2.,3./
C
C  **  TEST GENERAL MATRIX PRODUCT  **
      WRITE(6,32)
   32 FORMAT(/////,10X,'GENERAL MATRIX PRODUCT TEST',/)
      WRITE(6,34)
   34 FORMAT(10X,'CORRECT RESULT IS',//,
     2  1X,'-1 2 3     1 2     15  17',/,
     3  1X,' 2 3 4  X  2 5  =  24  31',/,
     4  1X,' 3 4 5     4 3     31  41',//,
     5  10X,'COMPUTED RESULT',/)
      CALL GMP(A,B,F,3,3,2)
      DO  56  I=1,3
      WRITE(6,57)  (F(I,J), J=1,2)
   56 CONTINUE
   57 FORMAT(5X,2(F10.4,5X))
      WRITE(6,59)
   59 FORMAT(//,15X,'KEY ANY NUMBER, FOLLOWED BY RETURN TO CONTINUE',/)
      READ *,V
C  **  TEST PRODUCT OF RECTANGULAR MATRIX WITH VECTOR  **
      WRITE(6,63)
   63 FORMAT(/////,10X,'MATRIX - VECTOR PRODUCT TEST',/)
```

```
      WRITE(6,67)
   67 FORMAT(10X,'CORRECT RESULT IS',//,
     2    1X,'-1  2  3      1      12',/,
     3    1X,' 2  3  4   X  2  =   20',/,
     4    1X,' 3  4  5      3      26',//,
     5    10X,'COMPUTED RESULT',/)
      CALL MVP(D,A,E,3,3)
      DO  84  I=1,3
      WRITE(6,85) E(I)
   84 CONTINUE
   85 FORMAT(7X,F10.4)
      WRITE(6,59)
      READ *,V
C  **  MATRIX INVERSION TEST  **
      WRITE(6,91)
   91 FORMAT(/////,10X,'MATRIX INVERSION TEST',/)
      WRITE(6,92)
   92 FORMAT(12X,'CORRECT RESULT IS',//,
     2    1X,'      -1  2  3      -0.5  1.0 -0.5',/,
     3    1X,'INV    2  3  4  =    1.0 -7.0  5.0',/,
     4    1X,'       3  4  5      -0.5  5.0 -3.5',//,
     5    15X,'COMPUTED RESULT',/)
      DO  99  I=1,3
      DO  98  J=1,3
      A0(I,J)=A(I,J)
   98 CONTINUE
   99 CONTINUE
      CALL MOP(1,3,1,A0,DET)
      DO  107  I=1,3
      WRITE(6,108) (A0(I,J), J=4,6)
```

```
  107 CONTINUE
  108 FORMAT(5X,3(F10.4,5X))
      WRITE(6,59)
      READ *,V
C  **  TEST LINEAR SYSTEM SOLUTION  **
      WRITE(6,114)
  114 FORMAT(/////,10X,'LINEAR SYSTEM SOLUTION',/)
      WRITE(6,115)
  115 FORMAT(10X,'CORRECT RESULT IS',//,
     2   1X,'-1  2  3      X1      12',/,
     3   1X,' 2  3  4      X2  =   20',/,
     4   1X,' 3  4  5      X3      26',//,
     5   5X,'GIVES  X1,X2,X3 = 1, 2, 3',//,
     6   10X,'COMPUTED RESULT',/)
      DO  121  I=1,3
      DO  120  J=1,3
      AO(I,J)=A(I,J)
  120 CONTINUE
  121 CONTINUE
      AO(1,4)=12.
      AO(2,4)=20.
      AO(3,4)=26.
      CALL MOP(-1,3,1,AO,DET)
      DO  135  I=1,3
      WRITE(6,136) AO(I,4)
  135 CONTINUE
  136 FORMAT(6X,F10.4)
      STOP
      END
```

```
C   *****  NEWTON RAPHSON EXAMPLE   *****
C   EQUATIONS TO BE SOLVED AND JACOBIAN
C   EVALUATION ARE PROGRAMMED AS A SUBROUTINE
       DIMENSION  F(2),S(2),DS(2),FJM(2,2)
C   F     = VECTOR OF FUNCTION VALUES
C   S     = VECTOR OF SECONDARY COORDINATES
C   DS    = ADJUSTMENT VECTOR, DELTA S
C   FJM   = JACOBIAN MATRIX
C   **  STEP 1  ENTER CONTROLS   **
C   E1    = MAX ALLOWABLE RESIDUAL
C   E2    = MIN ALLOWABLE ADJUSTMENT
C   I9    = MAX NUMBER OF ITERATIONS
       E1=0.0001
       E2=0.0001
       I9=10
C   **  STEP 2  ENTER KNOWN VALUES AND
C         INITIAL ESTIMATE FOR UNKNOWNS   **
       WRITE(6,24)
    24 FORMAT(//////,1X,'ENTER Q1')
       READ *,Q1
       WRITE(6,26)
    26 FORMAT(/,1X,'ENTER Q2')
       READ *,Q2
       WRITE(6,28)
    28 FORMAT(/,1X,'ENTER INITIAL ESTIMATE FOR S1=A')
       READ *,S(1)
       WRITE(6,30)
    30 FORMAT(/,1X,'ENTER INITIAL ESTIMATE FOR S2=B')
       READ *,S(2)
C   **  BEGIN THE ITERATION   **
```

```
      DO  73  I=1,I9
C  **  STEPS 3 & 4  EVALUATE THE
C       RESIDUAL AND THE JACOBIAN  **
      CALL RANDJ(Q1,Q2,S,F,FJM)
C  **  STEP 5  EVALUATE THE ADJUSTMENT
C       USING CLOSED FORM SOLUTION  **
      D=FJM(1,1)*FJM(2,2)-FJM(1,2)*FJM(2,1)
      DS(1)=-(1/D)*(FJM(2,2)*F(1)-FJM(1,2)*F(2))
      DS(2)=-(1/D)*(-FJM(2,1)*F(1)+FJM(1,1)*F(2))
C  **  TEST FOR TERMINATION  **
C       FORM NORMS FOR RESIDUAL AND ADJUSTMENT
      FN1=F(1)**2+F(2)**2
      FN2=DS(1)**2+DS(2)**2
      IF(FN1.GT.E1) GOTO 67
      IV=1
      GOTO 80
   67 IF(FN2.GT.E2) GOTO 70
      IV=2
      GOTO 80
C  **  UPDATE SOLUTION ESTIMATES  **
   70 S(1)=S(1)+DS(1)
      S(2)=S(2)+DS(2)
   73 CONTINUE
C  **  SOLUTION FAILURE OUTPUT  **
      WRITE(6,75)
   75 FORMAT(//,10X,'*****  MAX NUMBER OF ITERATIONS  *****')
C  **  SOLUTION OUTPUT  **
   80 WRITE(6,83) S(1),S(2)
   83 FORMAT(////,10X,'SOLUTION VALUES',/,
     2   1X,'S(1) = A = ',F10.4,/,
```

```
     3    1X,'S(2) = B = ',F10.4)
      WRITE(6,86) F(1),F(2)
   86 FORMAT(/,10X,'FUNCTION VALUES',/,
     2    1X,'F(1) = ',F10.4,/,
     3    1X,'F(2) = ',F10.4)
      WRITE(6,89)  I
   89 FORMAT(1X,'NUMBER OF ITERATIONS = ',I2)
      IF(IV.EQ.1) WRITE(6,90)
   90 FORMAT(1X,'TERMINATION BASED ON MAGNITUDE OF RESIDUAL')
      IF(IV.EQ.2) WRITE(6,91)
   91 FORMAT(1X,'TERMINATION BASED ON MAGNITUDE OF ADJUSTMENT')
      STOP
      END
      SUBROUTINE RANDJ(Q1,Q2,S,F,FJM)
      DIMENSION F(2),FJM(2,2),S(2)
C  **  PROBLEM GEOMETRY  **
      C1=3.
      C2=4.
      C3=6.
      F(1)=C1*COS(Q1)+S(2)*COS(S(1))-C2*COS(Q2)-C3
      F(2)=C1*SIN(Q1)+S(2)*SIN(S(1))-C2*SIN(Q2)
      FJM(1,1)=-S(2)*SIN(S(1))
      FJM(1,2)=COS(S(1))
      FJM(2,1)=S(2)*COS(S(1))
      FJM(2,2)=SIN(S(1))
      RETURN
      END
```

```
C  *****  PROPERTIES OF IRREGULAR CROSS SECTIONS   *****
C
C  SOURCE:  PROPERTIES OF PLANE CROSS SECTIONS
C           BY  F. WOJCIECHOWSKI
C           MACHINE DESIGN, 22 JAN 1976
      DIMENSION X(200),Y(200)
      PI =  3.1415926535
C  **  INSTRUCTIONS  **
      WRITE(6,112)
  112 FORMAT(///////,1X,'PROPERTIES OF IRREGULAR CROSS SECTIONS',/,
     2    1X,'COMPUTED BY TRAPEZOIDAL INTEGRATION',/)
      WRITE(6,115)
  115 FORMAT(1X,'ENTER DATA POINTS IN ORDER',/,
     2    1X,'TRAVERSING THE BOUNDARY',/,
     3    1X,'WITH POSITIVE AREA ON THE RIGHT',//)
      WRITE(6,119)
  119 FORMAT(1X,'ENTER THE NUMBER OF DATA POINTS, N')
      READ *,N
C  **  DATA ENTRY  **
  125 FORMAT(//,1X,'ENTER COORDINATES FOR BOUNDARY DESCRIPTION')
  126 FORMAT(/,1X,'X(',I2,') = ?,  Y(',I2,') =?')
      WRITE(6,125)
      WRITE(6,126)  1,1
      READ *,X(1),Y(1)
      X1=X(1)
      Y1=Y(1)
      DO  136  I=2, N
      WRITE(6,126) I,I
      READ *,X(I),Y(I)
      IF (X(I).LT.X1)  X1=X(I)
```

```
      IF (Y(I).LT.Y1)  Y1=Y(I)
  136 CONTINUE
      NP1=N+1
      X(NP1)=X(1)
      Y(NP1)=Y(1)
C  **  PRINT INPUT DATA  **
      WRITE(6,138)
  138 FORMAT(///,1X,'INPUT DATA RECAP',/)
      DO  140  I=1,N
      WRITE(6,141)  I,X(I),Y(I)
  140 CONTINUE
  141 FORMAT(1X,'  I = ',I2,5X,'X(I) = ',F10.4,5X,'Y(I) = ',F10.4)
C  **  TRANSFORM TO FIRST QUADRANT  **
      DO  147  I=1,NP1
      X(I)=X(I)-X1
      Y(I)=Y(I)-Y1
  147 CONTINUE
C  **  AREA, 1ST & 2ND MOMENTS IN FIRST QUAD COORD
      A0=0.
      FM1=0.
      FM2=0.
      FI1=0.
      FI2=0.
      FI3=0.
      DO  175  J=1,N
      J1=J+1
      X4=X(J1)-X(J)
      X5=X(J1)+X(J)
      Y4=Y(J1)-Y(J)
      Y5=Y(J1)+Y(J)
```

```
      A0=A0-Y4*X5/2
      FM1=FM1-Y4/8*(X5**2+X4**2/3)
      FM2=FM2+X4/8*(Y5**2+Y4**2/3)
      F1=X4*Y5/24*(Y5**2+Y4**2)
      FI1=FI1+F1
      F2=Y4*X5/24*(X5**2+X4**2)
      FI2=FI2-F2
      IF (X4.EQ.0)  GOTO 175
      FI6=Y4**2*X5*(X(J1)**2+X(J)**2)/8
      P=X(J1)*Y(J)-X(J)*Y(J1)
      FI7=Y4*P*(X(J1)**2+X(J1)*X(J)+X(J)**2)/3
      FI8=P**2*X5/4
      FI3=FI3+(FI6+FI7+FI8)/X4
  175 CONTINUE
C  **  CENTROID IN 1ST QUAD COORD  **
      X0=FM1/A0
      Y0=FM2/A0
C  **  CENTROIDAL MOMENTS OF INERTIA  **
      FI4=FI1-A0*Y0**2
      FI5=FI2-A0*X0**2
      FI6=FI3-A0*X0*Y0
C  **  PERFORM ROTATION TO PRINCIPAL COORD  **
      CALL  ROTR(FI4,FI5,FI6,AR,AD)
C  **  EVAL MOI WITH RESPECT TO ROTATED COORD  **
      CALL  MOI(AR,FI4,FI5,FI6,FI7,FI8,FI9)
C  **  PRINT RESULTS  **
      WRITE(6,190) A0
  190 FORMAT(//,20X,'SECTION PROPERTIES',/,
     2  1X,'AREA = ',F10.4)
      XC=X0+X1
```

```
      YC=Y0+Y1
      WRITE(6,194) XC,YC
  194 FORMAT(1X,'LOCATION OF CENTROID IN INPUT COORDINATES',/
     2    1X,'XC = ',F12.6,/,1X,'YC = ',F12.6)
      WRITE(6,199)  FI4,FI5,FI6
  199 FORMAT(1X,'CENTROIDAL MOMENTS OF INERTIA',/,
     2    1X,'AXES PARALLEL TO INPUT AXES',/,
     3    1X,'ICX = ',F12.6,/,1X,'ICY = ',F12.6,/,
     4    1X,'ICXY = ',F12.6)
      WRITE(6,201) AR,AD
  201 FORMAT(1X,'CENTROIDAL PRINCIPAL AXES',/,
     2  1X,'ROTATION FROM INPUT AXES (CCW+)',/,
     3  1X,'ANG = ',F12.6,' RADIANS',/,
     4  1X,'     = ',F12.6,' DEGREES')
      WRITE(6,205) FI7,FI8,FI9
  205 FORMAT(1X,'ICPX = ',F12.6,/,
     2  1X,'ICPY = ',F12.6,/,
     3  1X,'ICPXY = ',F12.6,/,
     4  1X,'THIS LAST VALUE SHOULD BE ZERO')
      STOP
      END
      SUBROUTINE ROTR(FI4,FI5,FI6,AR,AD)
C  **  ROTATION TO PRINCIPAL COORD  **
      PI = 3.1415926535
      FM7=-2*FI6
      FM8=FI4-FI5
      IF(FM8.EQ.0)  GOTO 50
      AR=.5*ATAN2(FM7,FM8)
      AD=180*AR/PI
      RETURN
```

```
50 IF (FM7.EQ.0)  GOTO 60
   Q7=1
   IF(FM7.LT.0)  Q7=-1
   AR=Q7*PI/4
   AD=180*AR/PI
   RETURN
60 AR=0
   AD=0
   RETURN
   END
   SUBROUTINE MOI(A,FI4,FI5,FI6,FI7,FI8,FI9)
   FI7=FI4*COS(A)**2+FI5*SIN(A)**2-FI6*SIN(2*A)
   FI8=FI5*COS(A)**2+FI4*SIN(A)**2+FI6*SIN(2*A)
   FI9=.5*(FI4-FI5)*SIN(2*A)+FI6*COS(2*A)
   RETURN
   END
```

Index